The Biology and Culture of Tilapias

This volume is dedicated to

Dr. Ethelwynn Trewavas

in recognition of her outstanding contributions to fisheries science and to mark the occasion of her eightieth birthday, 5 November 1980

The Biology and Culture of Tilapias

Proceedings of the International Conference on the
Biology and Culture of Tilapias, 2-5 September 1980
at the Study and Conference Center of the
Rockefeller Foundation, Bellagio, Italy, sponsored by the
International Center for Living Aquatic Resources Management
Manila

Edited by

R.S.V. Pullin and R.H. Lowe-McConnell

1982

INTERNATIONAL CENTER FOR LIVING AQUATIC RESOURCES MANAGEMENT

The Biology and Culture of Tilapias

Proceedings of the International Conference on the Biology and Culture of Tilapias, 2-5 September 1980 at the Study and Conference Center of the Rockefeller Foundation, Bellagio, Italy sponsored by the International Center for Living Aquatic Resources Management, Manila

EDITED BY

R.S.V. PULLIN
R.H. LOWE-MCCONNELL

1982

Reprinted 1983, 1984

Printed in Manila, Philippines

Published by ICLARM, MCC P.O. Box 1501, Makati, Metro Manila, Philippines

Pullin, R.S.V. and R.H. Lowe-McConnell, Editors. 1982. The biology and culture of tilapias. ICLARM Conference Proceedings 7, 432 p. International Center for Living Aquatic Resources Management, Manila, Philippines.

ISSN 0115-4389

ISBN 971-04-0003-7 cloth
ISBN 971-04-0004-5 paper

Cover: Stylized tilapia artifact from Indonesia.

Preface

Tilapias are a major protein source in many of the developing countries. Although endemic to Africa, their distribution has been widened by artificial introductions, mainly since the 1950s, to include much of the tropics and subtropics. Tilapias have many attributes that recommend them for culture. They show excellent growth rates on low protein diets, whether cropping natural aquatic production or receiving supplementary food. They tolerate wide ranges of environmental conditions, show little susceptibility to disease and are amenable to handling and captivity. They have a short generation time and breed in captivity. Most important of all, they enjoy wide acceptance as food fish because of their high palatability and history of use from inland fisheries. With all these advantages, tilapias could become prime domesticated species for fish culture.

The culture of tilapias however, is still beset with problems of rearing and general husbandry. These result from an inadequate research base on their biology, particularly behavior and physiology, and insufficient cooperation between fish biologists and culturists. The former have often pursued limited, essentially academic studies within their specialist fields and the latter have paid inadequate attention to published information. In particular, there is a great deal of information gathered by field biologists which is relevant to the behavior, growth and reproduction of tilapias in culture systems.

This conference was convened to bring biologists and culturists together to present reviews of existing information in their specialist fields; to discuss current research areas and culture methods; to define future research requirements and to comment on any other measures which would help the future development of tilapia culture.

It was seen as a unique opportunity to address some of the unresolved problems for researchers and culturists alike ranging from standardization of nomenclature and techniques for the development of known genetic strains and hybrids to optimization of growth and feeding and control of reproduction.

When editing this volume we decided to accept the division of the tilapias into the genera *Sarotherodon* (mouthbrooders) and *Tilapia* (substrate spawners). However, as explained in the addendum to Dr. Trewavas' paper, the subgenus *Oreochromis* has been raised to generic status. The maternal mouthbrooding tilapias so far used in fish culture should now be referred to as *Oreochromis* (e.g., *Oreochromis mossambicus*, *O. niloticus*, *O. aureus*). The generic name *Sarotherodon* is now restricted to a West African group (which includes *S. melanotheron*) and to *S. galilaeus*.

ICLARM wishes to thank the Rockefeller Foundation, not only for providing the superb facilities of the Bellagio Study and Conference Center, but also for their generous sponsorship of the conference participants. The administrator and staff of the Bellagio Center also deserve special thanks for their hard work and help in ensuring the success of this conference. Finally, we wish to thank Dr. Roy Harden Jones for preparing the excellent indexes and Linda Temprosa for assembling the reference list.

R. S. V. PULLIN
R. H. LOWE-MCCONNELL

Contents

Introduction

This conference brought together 19 tilapia biologists and experimental culturists from 10 countries. It was designed from the outset as a technical conference on the basic biology of the tilapias and applications in culture systems. The conference did not consider the commercial aspects of tilapia culture. A further conference will be held in Israel, 8-13 May 1983.*

The conference consisted of fifteen major review papers grouped under three main session headings:

Biology — taxonomy and speciation, ecology and distribution, tilapias in fish communities, and life histories.

Physiology — environmental, reproductive, feeding and digestion, metabolism and growth.

Culture — ponds, cages, diseases, hybridization, genetic markers, control of reproduction and mass fry production systems.

The proceedings follow this arrangement with each session preceded by a chairman's overview. The culture session was split into two parts: culture systems and culture-related topics. About half the conference was devoted to discussion sessions and a summary of these follows the papers presented. The conference produced a consensus statement and a list of research requirements, which may stimulate the further development of tilapia culture.

The references cited have been combined into a single bibliography to provide a useful information source on both cultured and wild tilapias. The indexing is designed for easy referencing on systematics, biology and culture.

*Details from the Organizing Committee, International Symposium on Tilapia in Aquaculture, P.O. Box 3054, Tel Aviv, 61030, Israel.

SESSION I: BIOLOGY

Chairman's Overview

R. H. LOWE-McCONNELL

At the start of this first session of the Conference it is very important to stress the aims of the meeting, as set out in Roger Pullin's letter of invitation, namely, "the urgent need to bring tilapia biologists and culturists together to exchange views and information and to define approaches and priorities for future work". We need to keep this firmly in mind and aim to produce the most constructive guidelines for future research.

The two groups of people mentioned, the tilapia biologists who have been working mainly under field conditions or in laboratories (or both) and those with practical experience of aquaculture, have much to give one another of very different experiences that can help to solve one another's problems. This is a two-way process. As a field worker myself, I know only too well the complexity of working conditions in natural waters in the tropics, where so many species interact and where, for example, it is often so difficult to determine fish growth rates. We have to look to those who grow tilapias in ponds to solve many problems posed in the field, in particular, those concerning the switch from growth to reproduction (nanism, dwarfing) which is such a special feature of tilapia biology. I regard this as possibly the key problem for our attention. If we could really understand the mechanism which controls the switch from growth to reproduction, this would be one of the biggest steps towards improving tilapia culture.

This session is concerned mainly with field studies, and should provide a good background to our subsequent consideration of the physiology of tilapias from both field and laboratory work. For those who grow tilapia in ponds, we must stress the vast scale of the theatre for field studies. The equatorial Lake Victoria in East Africa is over 6.5 million ha and the Kafue floodplain, another site for much tilapia research, around 121,000 ha. Sampling such huge areas presents special problems, and errors arise in extrapolation of results. Catch statistics are of very variable reliability compared with emptying a pond and weighing all the fish. Furthermore, what is loosely referred to as 'fish production' from these vast areas is really the fish catch and only a very small percentage of the total biological production (unlike ponds from which most of the total production may be channelled into spawners).

In this session we have four papers.

Dr. Ethelwynn Trewavas first introduces the taxonomy and speciation of tilapias, and says why it is so important to know which species we are dealing with when making studies of any kind. We all owe a great debt of gratitude to Dr. Trewavas who has toiled away for very many years putting together a definitive monograph on the tilapias. I see her herculean efforts at first hand—I share her habitat among the tilapia bottles at the Natural History Museum in London—and I have seen (and have benefited from) the ever-ready help she has given so unstintingly to numerous biologists from all over the world.

Second, we have the contribution on ecology and distribution prepared by Dr. J-Cl. Philippart with Professor J-Cl. Ruwet, who is unable to attend this meeting. They have selected certain areas of ecology, believing that aspects such as growth and production would be treated in other papers. They have considered the physico-chemical factors involved in tilapia distribution; feeding behavior, e.g., trophic spectra, variations in feeding regimes and the periodicity of feeding activity (subjects which lead onto our later discussions on physiological aspects of feeding); reproductive behavior, comments on practical implications and the choice of microhabitats and spatial segregation. They have then assessed the ecological consequences of introductions and transfers of tilapias in Africa and elsewhere.

Dr. D.L.G. Noakes then presents a paper that he and Dr. E. Balon, who is also unable to be here, have prepared on life histories of tilapias. This seems to be a rather theoretical paper but it leads up to a practical suggestion of culture methods which might help to produce the large fish with late onset of breeding and fast growth rates that we all desire.

Lastly, the paper which I was asked to prepare on tilapias in fish communities was originally aimed to look at the partitioning of resources among sympatric tilapias and at some of the factors controlling tilapia numbers in nature (with special attention to the switch from growth to reproduction in *Sarotherodon niloticus*). It became clear, however, that much of what I had written had considerable overlap with other manuscripts, so I have attempted, at very short notice, to fill in some gaps by collating information on tilapia growth rates, production and yields under natural conditions.

Tilapias: Taxonomy and Speciation

E. TREWAVAS

Fish Section
British Museum (Natural History)
Cromwell Road, South Kensington
London, S.W. 7, England

TREWAVAS, E. 1982. Tilapias: taxonomy and speciation, p. 3-13. *In* R.S.V. Pullin and R.H. Lowe-McConnell (eds.) The biology and culture of tilapias. ICLARM Conference Proceedings 7, 432 p. International Center for Living Aquatic Resources Management, Manila, Philippines.

The tilapias are here grouped into two genera, *Tilapia* and *Sarotherodon*, characterized by both their feeding and their reproductive habits and structures. There is some overlap in feeding habits, some species of both genera feeding on detritus and periphyton, but substrate-spawning and mouthbrooding respectively characterize the two genera.

The mouthbrooding of the species of *Sarotherodon* is of two main types, biparental or parental in subgenus *Sarotherodon* (in the restricted sense) and maternal brooding with mating of a lek pattern in the rest. The maternal mouthbrooders have been divided into several (5 to 7) subgenera: *Oreochromis, Danakilia, Alcolapia, Nyasalapia* and *Neotilapia* are recognized here. Speciation, both allopatric and sympatric, of subgenus *Sarotherodon* has taken place in west Africa, and of the other subgenera mostly in east and central Africa. There is no doubt that the division between subgenus *Sarotherodon* and other subgenera is deeper than the divisions within the latter. For this reason some authors (Thys, Peters) prefer a generic grouping that goes no lower than *Tilapia*, but the mouthbrooding subgenera have so much in common that they must have descended from one or a few closely related substrate-spawning species, and grouping them all at the generic level seems to have both theoretical and practical justification.

An example is given of the importance of recognizing specific distinctions.

Biology and Taxonomy

All the tilapias, in the broad sense, have in common a mainly herbivorous diet, in distinction to the majority of fishes which feed predominantly on small invertebrates or on young or small-sized fishes. They are therefore only one step from the primary producers (plant life) and as they grow to a good size they are a valuable food source for man, the omnivore.

Structural adaptations to this diet are the long, coiled intestine, which may be up to fourteen times the body length, the bicuspid and tricuspid teeth of the jaws and the small, sharp pharyngeal teeth used to prepare the food by shredding the coarser materials and breaking some of the cell walls before passing it on to the stomach. Since the preferred diet of the different species varies from coarse vegetation (grasses, young shoots and leaves of water weeds) to unicellular algae and even bacteria, the teeth also vary in the degree of coarseness and movability.

Tilapias all exhibit a high degree of parental care and in this function they are sharply divided into substrate-spawners and guarders of the brood on the one hand and mouthbrooders on the other (Lowe-McConnell 1959).

A division based on diet roughly coincides with one based on reproductive habits. I say 'roughly' partly because tilapias are flexible and opportunistic in their feeding habits, but also because of two species that fail to conform.

Tilapia mariae is a substrate-spawner, but its usual diet consists of epiphytic algal growths and its jaw teeth are fine, slender-shafted and movable. Its pharyngeal teeth are fine, on a bone whose anterior blade may be as long as the toothed area and its standard gill-raker count is 12-16: high for a *Tilapia*.

Sarotherodon niloticus niloticus is a maternal mouthbrooder, but its jaw teeth are nearly as stout as those of *Tilapia zillii* and its pharyngeal teeth are firm, on a triangular area whose length is often greater than the length of the blade. But the gill-rakers are more numerous than in *T. mariae*, the standard count being 18-26.

But this overlap is merely an inconvenience when we try to define the two groups and no one would doubt to which group each species is related.

The substrate-spawners constitute the genus *Tilapia* Smith (1840), the mouthbrooders *Sarotherodon* Rüppell (1852).

Tilapia was an effort by A. Smith, its author, to spell the Bushman word for 'fish', which began with a click, rendered 'Til'. *Sarotherodon* means 'brush-toothed'.

Examples of the two genera familiar to fish culturists are: *Tilapia sparrmanii* A. Smith, *T. rendalli* Boulenger, *T. zillii* (Gervais), *T. cabrae* Boulenger, *Sarotherodon melanotheron* Rüppell, *S. galilaeus* (Linnaeus), *S. niloticus* (Linnaeus), *S. aureus* (Steindachner), *S. mossambicus* (Peters), *S. niger* (Günther) and *S. macrochir* (Boulenger).

The four examples of *Tilapia* were named after persons who collected the originals or *(T. sparrmanii)* had explored their region of origin. The word *melanotheron* means 'black-hinged' or 'black-chinned', *galilaeus* and *niloticus* are self-explanatory, '*aureus*' refers to the golden-yellow color of the preserved fish (not the living) and '*macrochir*', meaning 'big hand', refers to the long pectoral fin, the homolog of the forelimb (arm and hand) of man.

T. sparrmanii feeds on the coarser algae and grasses and also on the small animal life among them. It is sometimes described as omnivorous. It is valued for pond use in South Africa because its wide temperature tolerance enables it to withstand the winters of the High Veldt.

T. rendalli and *T. zillii* have strong cusped teeth in jaws and pharynx and are specialized feeders on vascular plants. Ruwet (1963a) has given a vivid description of the voracious attack by *T. rendalli* on the newly flooded grasses of the Mwadingusha dam on River Lufira, Zaïre.

These two species, which are not very easily distinguished at sight, have largely complementary areas of distribution. *T. zillii* is a Soudanian form, extending from west Africa through the Chad basin to the Nile, Lake Albert and Lake Turkana into Israel and the Jordan Valley. *T. rendalli*, one of the forms formerly included in '*T. melanopleura*', has a wide distribution in the Upper Zaire and Kasai, Lakes Tanganyika and Malawi, the Zambezi basin and the Limpopo. At Yangambi, near Kisangani, where the River Zaïre turns westwards, the natural distributions of the two species overlap. In west Africa *T. rendalli* is replaced by *T. dageti*, which appears to be less abundant than *T. zillii*.

At Yangambi *T. zillii* is in nature less abundant than *T. rendalli*. Gosse (1963), who administered the fishponds at Yangambi, did not at first realize that his *T. rendalli* ponds included an admixture of *T. zillii*, but after a while the more robust *T. zillii* dominated the pond populations. The same thing evidently happened in the ponds at Yaounde, Cameroon, to which Gosse had inadvertently sent a mixture of the two species.

In Lake Victoria *T. zillii* was introduced from Lake Albert and *T. rendalli* either from Shaba or Zambia. At first they were believed to have interbred, but later, when introductions were made into many farm dams in Uganda and Kenya, the *Tilapia* were named as *T. zillii*.

Gosse was emphatic that *T. rendalli* was by far superior as a food fish. It is therefore important for pisciculturists to know one from the other. The differences are:

Color. *T. zillii* usually has two horizontal dark stripes, one mid-lateral, the other nearer the dorsal outline. These are crossed by vertical bars and the strongest marks are blotches at the intersections. The caudal fin is usually covered by a grey network with pale interstices. The eggs are green.

T. rendalli usually shows dark vertical bars only and the scales have each a dark basal crescent. The caudal fin has a spotted upper half and a red or yellow lower half. This red or yellow pigment may also color the anal fin. The eggs are yellow.

Both species frequently have a carmine flush on the lower flanks, behind the pectoral fin, and this is not confined to one sex or to mature fishes.

General shape. *T. rendalli* is typically more deep-bodied than *T. zillii* but this is an average difference and not suitable for individual identification.

Meristic characters (counts). The number of gill-rakers is the same in both species: 8-10 below the joint of the first arch. This distinguishes them at once from *Sarotherodon*. The mean numbers of scales, vertebrae and dorsal fin-rays provide good distinctions and the modal numbers of dorsal spines are sufficiently well-marked to allow a quick field test on a number of specimens. *T. zillii* usually has 28 vertebrae, *T. rendalli* 29. In *T. zillii* the commonest number of dorsal spines is XV, in *T. rendalli* XVI (Table 1). For the whole dorsal fin, among 130 *T. rendalli* I found 26% with XVI 12, 66.9% with XVI 13; other combinations were found in only one or two fishes. Among 40 *T. zillii* 18 (45%) had XV 12 and 10 (25%) XV 13.

There are about 29 species of *Tilapia* and most of this speciation has taken place in the western rivers of Africa. The euryhaline *T. guineensis* is found in lagoons and estuaries along the coast from Angola to Senegal. Other species are little known beyond their native rivers, although *T. tholloni* has been much used in laboratory studies.

Table 1. Frequency distribution of vertebrae and dorsal spines in *Tilapia zillii* and *T. rendalli*.

		T. zillii	*T. rendalli*
Vertebrae	27	5	
	28	33	1
	29	2	31
	30		1
Dorsal spines	XIII	2	
	XIV	60	
	XV	467	4
	XVI	63	124
	XVII		2

In the eastern rivers north of the Zambezi *Tilapia* was never reported before the widespread introductions of the last three decades. But the fish fauna of Mozambique and Tanzania was little known and we cannot now be sure that the presence of *T. rendalli* in the rivers of Tanzania is not natural. We do know, however, that Kenyan waters were devoid of this genus until the introduction of *T. zillii.*

The tilapias most cultivated are species of *Sarotherodon*, the mouthbrooders.

It is now well known that the details of the mouthbrooding habit are not the same in all mouthbrooding tilapias (Heinrich 1967; Peters and Berns 1978). The type-species of *Sarotherodon*, *S. melanotheron*, is, as it happens, the most atypical. In nature it lives in brackish water and it is a paternal mouthbrooder. Many studies of its behavior show that in comparison with *S. mossambicus* for example, it has retained vestiges of some of the reproductive habits and even structures of the substrate-spawners. So has its relative in fresh waters, *S. galilaeus.* When we subdivide the mouthbrooders, it is these two species *(S. melanotheron* and *S. galilaeus)* that typify *Sarotherodon* in the strict sense, that is, subgenus *Sarotherodon. S. mossambicus, S. niloticus, S. aureus* and other east and central African species have a reproductive schema that excludes the males from the care of the brood. They are in breeding time sexually dimorphic and the female takes the eggs as soon as they are fertilized to special nursery areas where she holds them in her mouth until the yolk is sufficiently reduced for them to swim freely. Even then, for several days she takes them back into her mouth at night or when danger threatens. The east African maternal brooders have received a second subgeneric name, the earlier available being *Oreochromis* (now raised to genus; see Addendum).

The meaning of *Oreochromis* is 'mountain cichlid'. (*Chromis* was an early name given to both cichlids and a marine genus and is now restricted to the latter). It was first given to *Oreochromis hunteri* Günther (1889), an inhabitant of a small rocky crater-lake on the slopes of Mount Kilimanjaro. Little is known about the habits of this species, but from its structure and geographic situation it is believed to be a maternal mouthbrooder like its neighbors *S. pangani* and *S. mossambicus.*

It is commonly believed that systematic conclusions are based on anatomical considerations involving dissection and the use of the microscope. Such evidence is indeed most important, but there is no aspect of an organism's biology that a systematist can afford to ignore. In defining the subgenera *Sarotherodon* and *Oreochromis* the main evidence is behavioral, backed by geographical. The geographical evidence is well set out by Thys (1963). Many observers have contributed to the behavioral evidence briefly summarized above.

I do not know of any structural character that trenchantly separates the two subgenera. Some members of *Oreochromis*, including *S.(O.) hunteri* have four anal spines, either in all or in some individuals, but others have only three; all species of subgen. *Sarotherodon* have three. In no species of subgen. *Sarotherodon* are the jaws enlarged in sexually mature males. They are enlarged in several species of subgen. *Oreochromis* but not in all. The vertebral numbers range from 26 to 32 in subgen. *Sarotherodon*, 27 to 34 in *Oreochromis.* Parallel adaptations to various feeding regimes have developed in species of the two subgenera. If we do not know the breeding habits of a species we can place it in a subgenus only if its overall characters add up to a general resemblance to a geographically-close member of one or other o the subgenera.

The development of a tassel on the male genital papilla is a conspicuous feature that characterizes a number of related species *(S. macrochir*, the Malawi flock, etc.) and for these we may use a subgeneric name *Nyasalapia* Thys (1968).

In the case of *S. tanganicae*, of whose breeding habits little is known, the fact that its ventral scales are very small suggests relationship to *Oreochromis.* But it has very specialized teeth and as it has been given a subgeneric name of its own, I propose to use it and to call it *Sarotherodon (Neotilapia) tanganicae.*

Another peculiar species in the same kind of limbo is the little *S. franchettii* of Lake Afrera (= Giulietti) in Dancalia, eastern Ethiopia. Its outer teeth as well as the inner are tricuspid and it has only 27 vertebrae and 10-12 lower gill-rakers. It has been given its own subgeneric name by Thys (1968) and so I propose to leave it as *Sarotherodon (Danakilia) franchettii.* (*Danakilia* is now raised to genus; see Addendum).

For zoologists who are not systematists and for fish culturists there is no need to use subgeneric names, and for these all the mouthbrooding tilapias may be referred to by the generic name *Sarotherodon* followed by the specific name.

Peters (1979) and Thys (1968, 1971a, 1971b) would prefer to leave the tilapias undivided for the general and applied zoologist and to lump substrate-spawners and mouthbrooders alike in a broad genus *Tilapia.* If I interpret them rightly their reason for this is:

> *S. melanotheron* and *S. galilaeus*, the two best known species of subgen. *Sarotherodon*, have breeding habits distinct not only from those of *Oreochromis* but also in some ways from each other. Specifically, *S. melanotheron* has large yellow eggs brooded normally by the father alone,

whereas *S. galilaeus* has small green eggs brooded by both parents. What further differences may exist among the species apparently related to these we do not know. Perhaps they even evolved from different species of *Tilapia* or from the same species at different times. Therefore we may eventually recognize different lines of evolution in these species and wish to call them by different generic names. Meanwhile let us be conservative and call them all *Tilapia*.

This is the position of Thys and Peters. My view (Trewavas 1973a, 1973b, 1980 and in press) is, however, that the mouthbrooding species showing affinity to *Tilapia* are sufficiently alike in their feeding adaptations and even in their breeding habits to show close interrelationship and that this justifies separating them from the substrate-spawning *Tilapia* under the one generic name. The substrate-spawners have themselves branched out into a number (about 29) of species and I find it difficult to lump all the mouthbrooders together with them as *Tilapia*. In the last few years I notice that fish culturists and others have followed my proposal to call the mouthbrooders *Sarotherodon*.

Speciation

The pattern of speciation in *Sarotherodon* and *Tilapia* leads to the conclusion that it has occurred in two ways: 1) by geographical isolation; allopatric and 2) by ecological diversification in one area; sympatric (see Figures 1 and 2).

1. Examples of allopatric speciation are:

a. *Tilapia zillii* and *T. rendalli*, described above. Their coming together at Yangambi is probably secondary. A number of allopatric and geographically more restricted species related to these are present in west Africa.

b. *Sarotherodon melanotheron* and its subspecies in areas along the coast, separated by steeper shores. (These have been ranked as species by Thys (1971b), a matter of policy).

c. The tasselled tilapias surrounding the Rift Valley: *S. variabilis*, *S. upembae* and its related form in the Malagarasi, *S. rukwaensis*, *S. macrochir* and the Malawi flock considered as a unit.

d. *S. mossambicus* and related forms in the eastern rivers.

2. Examples of sympatric speciation are:

a. The species of the Malawi flock, *S. squamipinnis*, *S. saka*, *S. lidole* and *S. karongae*, whose ecological differences are described by Lowe (1952, 1953).

b. The two species of the Malagarasi swamps, *S. karomo* and the other species related to *S. upembae*. The highly specialized jaws and teeth of *S. karomo* must indicate a specialization in feeding on the epiphytic growths.

c. The species of the small lake Barombi Mbo in Cameroon, especially evident in *S. linnellii* and *S. caroli* (Trewavas et al. 1972).

When two species occur together but each has its nearest relatives elsewhere, this is regarded as allopatric speciation in two lines that have come together geographically, and not, of course, as sympatric speciation. Such are *S. esculentus* and *S. variabilis* in Lake Victoria related respectively to *S. niloticus* and the tasselled tilapias. Other examples are *S. shiranus* (related to *S. mossambicus*) and the *S. squamipinnis* group considered as a unit (related to the tasselled group) all inhabiting Lake Malawi; *S. andersonii* and *S. macrochir* in the upper Zambezi, Kafue and Okovango basins related respectively to *S. mossambicus* and the tasselled group.

That the sarotherodons have not formed such huge species flocks as have the haplochromine and other cichlids of the Great Lakes I put down to the fact that as colonizers they were already highly specialized for feeding on minute epiphytic and planktonic algae and had few or no rivals in that field. That they diversified at all within these limits is a matter for wonder.

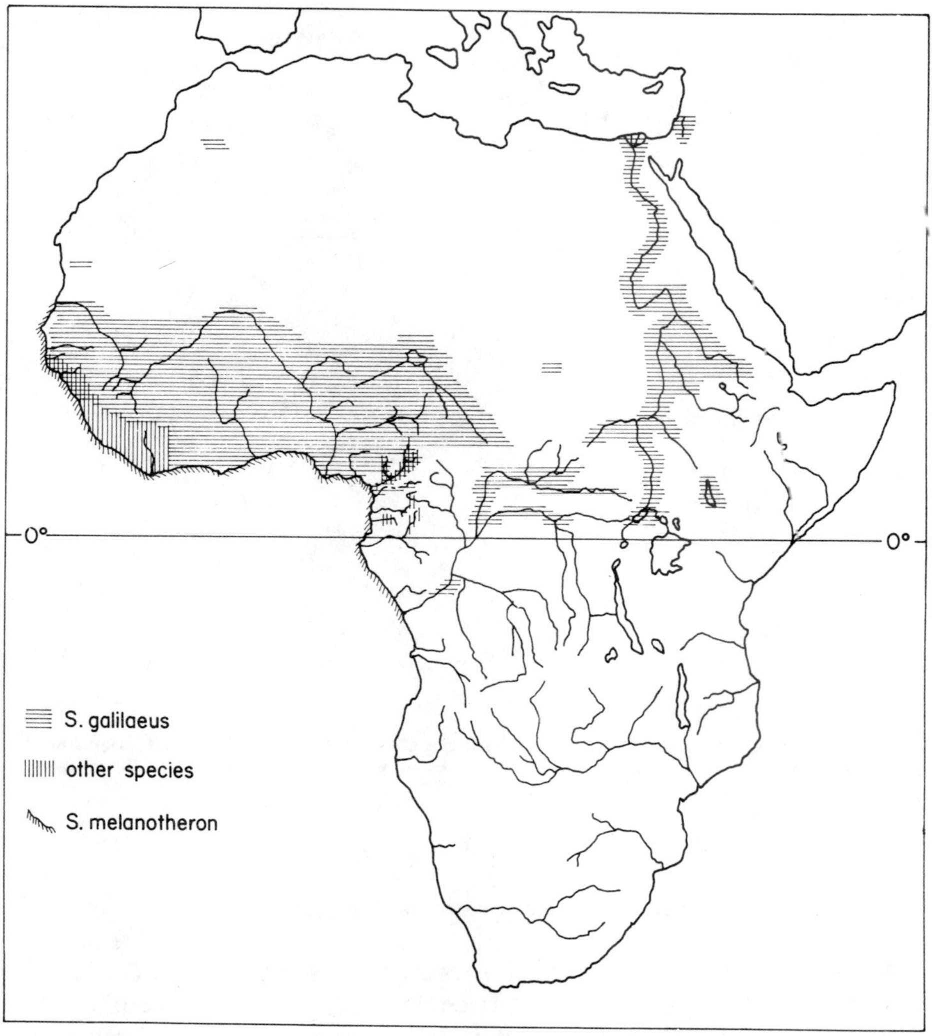

Figure 1. Natural distribution of the subgenus *Sarotherodon* of genus *Sarotherodon*. ≡ *S. galilaeus*, |||| other species except the brackish water *S. melanotheron*. (Trewavas, in press)

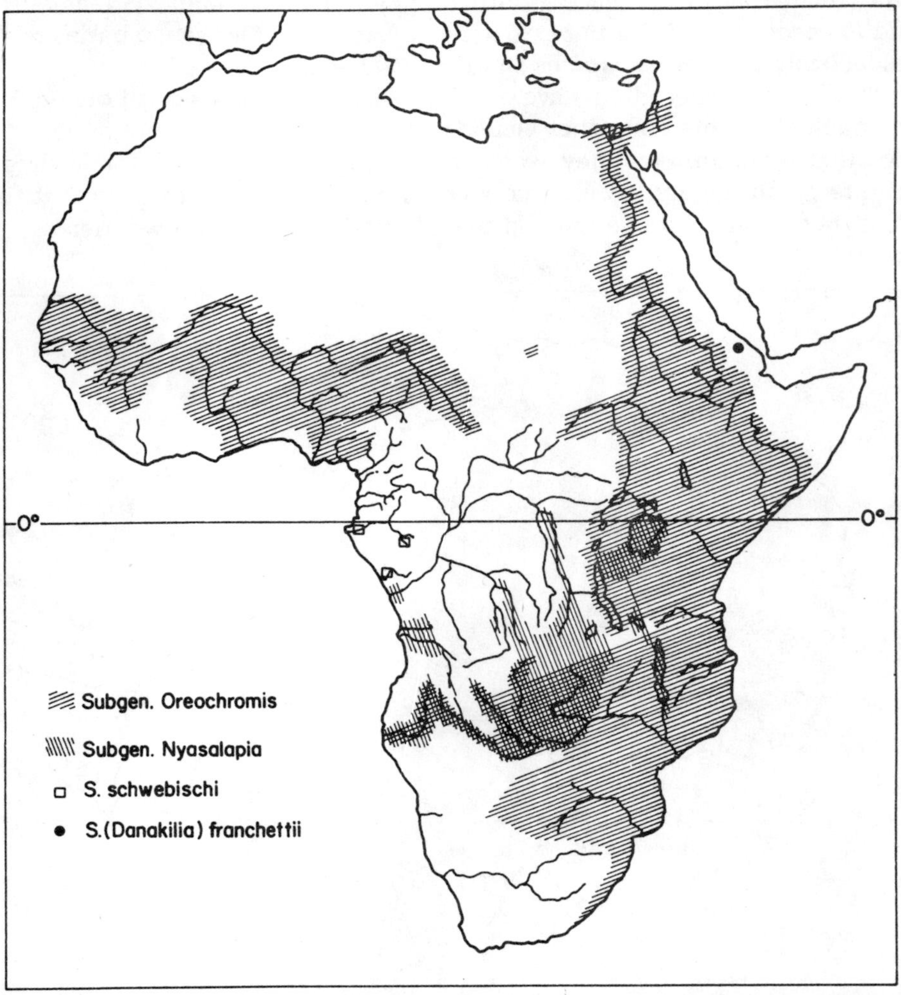

Figure 2. Natural distribution of the subgenera *Oreochromis* (////) and *Nyasalapia* (\\\\). □ *S. schwebischi* (Sauvage), of uncertain subgeneric affinity and ● *S. (Danakilia) franchettii* (Vinciguerra). (Trewavas, in press).

Sarotherodon Species and Fish Culture

S. galilaeus is the only species of subgenus *Sarotherodon* that has been used for fish culture. It has not been favored for ponds, mainly, I think, because of its slower growth, but it is (or was) so used in Cameroon, the stock being mainly from the Chad basin, but also from R. Sanaga, the latter perhaps a distinct subspecies. Moreover *S. galilaeus* has done well in man-made lakes and reservoirs in open water.

The other well-known cultured species are members of subgen. *Oreochromis* and the tasselled species, *Nyasalapia.* The familiar names are *S. niloticus* with seven subspecies, of which three have been cultured experimentally by Pruginin et al. (1975); *S. aureus*, which accompanies *S.n. niloticus* through much of its area of distribution; *S. mossambicus* and *S. hornorum* well known for their hybridization and its effect on sex ratio (Chen 1969; Hickling 1960); *S.s. spilurus* and *S.s. niger*, which were studied in Kenya by Van Someren (1962) and Van Someren and Whitehead (1959a, 1959b, 1960a, 1960b, 1961a, 1961b, 1961c); *S. macrochir*, which was studied in Zambia and Shaba and transported to Ivory Coast where it was experimentally interbred with *S. niloticus* (Jalabert et al. 1971; Lessent 1968); and *S. andersonii* studied in Zambian ponds by Mortimer (1960) and in the Kafue floodplains by Dudley (1979) and Kapetsky (1974).

The taxonomy of the species of *Sarotherodon* is dealt with fully in a new work (Trewavas, in press).

Although the practical fish farmer cannot be expected to go into the niceties of specific distinctions, I suggest that key suppliers of stock should know what species they are using and whether these are pure. It is only necessary to recall that the sex-ratio may be affected by mixing the species, many of which interbreed easily. There is also the possibility that one species, and not necessarily the most desired, may oust the other as *T. zillii* did *T. rendalli* at Yangambi (Gosse 1963).

Further I suggest that in Africa itself fisheries officers should look carefully at the native species before importing stock from other regions. If on due consideration they decide to import stock they should keep a careful record of it.

Addendum: Alternative and Preferred Classification

Since writing the above, I have had discussions with fish culturists in Kenya. They are using a brood tank whose design is based on the breeding habits of members of subgenera *Oreochromis* and *Nyasalapia*, namely the 'lek' system with migration of the brooding mother to a 'nursery' area. They refer to this as the system characteristic of *Sarotherodon*. It is not, however, the system practiced by species of *Sarotherodon sensu stricto*, exemplified by *S. galilaeus*, and I have advised them to refer to it as the Oreochromis system of breeding. This and other considerations now lead me to believe that we should place more emphasis on the gap between *Sarotherodon* and the other subgenera by placing the latter in a separate genus, the earliest available name for which is *Oreochromis*. *Danakilia* is also generically distinct.

The mouthbrooding tilapias used for fish culture accordingly would be grouped in two genera, *Sarotherodon* and *Oreochromis* the latter comprising four subgenera as shown in the accompanying table.

Genus	Subgenus	Type species	Other examples
Tilapia A. Smith	(three to six)	*T. sparrmanii* A. Smith	*T. rendalli* Boulenger *T. zillii* Gervais
Sarotherodon Rüppell		*S. melanotheron* Rüppell	*S. galilaeus* (Linn.) *S. linnellii* (Lönnberg)
Oreochromis Günther		*O. hunteri* Günther	
	Oreochromis Günther	*O. hunteri* Günther	*O. (O.) niloticus* (Linn.) *O. (O.) mossambicus* (Peters) *O. (O.) aureus* (Steindachner) *O. (O.) spilurus* (Günther)
	Nyasalapia Thys	*O. (N.) squamipinnis* (Günther)	*O. (N.) macrochir* (Boulenger) *O. (N.) variabilis* (Boulenger) *O. (N.) angolensis* (Trewavas)
	Alcolapia Thys	*O. (A.) grahami* (Boulenger) = *O. alcalicus grahami*	*O. (A.) alcalicus alcalicus* (Hilgendorf)
	Neotilapia Regan	*O. (N.) tanganicae* (Boulenger)	None
Danakilia Thys		*O. (D.) franchetti* (Vinciguerra)	None

Discussion

GUERRERO: Is it difficult to inter-breed *Tilapia* species or *Sarotherodon* species with *Tilapia* species such as *T. zillii?*

TREWAVAS: Have you ever tried to interbreed them?

GUERRERO: No, but the literature suggests incompatibility between *Sarotherodon* and *Tilapia.* One is a mouthbrooder and the other a bottom spawner. I was wondering if it is difficult to interbreed them. Do you think this is possible or that *Tilapia* species will hybridize?

TREWAVAS: *T. zillii* and *T. rendalli* do hybridize or they are believed to hybridize in Lake Victoria where they have both been introduced.

GUERRERO: We have not tried it, but I want to find out if it is possible because of the differences between the two groups. *Tilapia* species seem more difficult to interbreed as opposed to *Sarotherodon* species.

TREWAVAS: The two genera have never been known to interbreed in nature, nor freely if they are placed together in one tank, but hybrids have been obtained by artificial fertilization at Tübingen. Some attempted crosses were unsuccessful, but in one case hybrids structurally intermediate have been reared to adult size. They were all females.

AVTALION: We have been testing immunological techniques in my laboratory for the different species of *Sarotherodon.* All of them are crossreacting antigenically. We could not find any antigenic differences with these species. In mammalians we know there is no antigenic crossreactivity between different species. Now, to what extent the species should be called subspecies, I do not know, but several tilapias cannot be termed different species *per se* if we refer to immunological criteria. In this respect, we found some differences between *T. zillii* and *Sarotherodon* species, but no differences between *S. galilaeus*, *S. aureus*, and *S. niloticus*. However, this matter needs to be studied in more detail.

LOWE-McCONNELL: So the particular question is, do you think that the latter are truly separate species as you have been unable to find any immunological differences?

ROBERTS: I think we will have to distinguish between antigenic characteristics (which means tissue matching) and serum and muscle enzymes, because there are certainly differences between all of the *Sarotherodon* species in terms of the numbers of enzyme loci.

HEPHER: But you can also find such differences between sexes.

ROBERTS: Yes, this has been reported, but not to anything like the same degree as the differences between species.

HEPHER: But there are differences.

ROBERTS: We did not find differences between sexes. I think the important point is that, while someone like Dr. Trewavas has devoted so much time to distinguishing morphologically between these different species, Dr. Avtalion, from his antigenic and immunological studies, finds that there is no difference. I think we have to be very careful with our definition of a species and recognize a problem here.

LOWE-McCONNELL: This is a particular subject for people who are interested in speciation.

PULLIN: Dr. Avtalion, you said that you found a large difference between *T. zillii* and *Sarotherodon* species which tends to support the generic split.

AVTALION: Yes, there is quite a significant difference.

TREWAVAS: It is interesting that you bracket both kinds of *Sarotherodon*, that is, *S. (Sarotherodon) galilaeus* together with *S. (Oreochromis) niloticus* and *aureus.*

HEPHER: We have noticed that it is very easy to hybridize *S. niloticus* and *S. aureus*, but it is very difficult to cross *S. galilaeus* with *S. aureus* or *S. niloticus.* Maybe this has some significance.

LOWE-McCONNELL: I think the important thing here is that there are "tilapias," there are "sarotherodons" and there is a group in west Africa that are more difficult to define. We do not know much about these.

TREWAVAS: Yes, I think that *S. galilaeus* is the only species that fish culturists have had anything to do with from that group.

Ecology and Distribution of Tilapias

J-CL. PHILIPPART AND J-CL. RUWET

Service d'ethologie animale
Institut Ed. van Beneden
Université de Liège
Quai van Beneden, 22
B-4020 Liège
Belgium

PHILIPPART, J-CL. AND J-CL. RUWET. 1982. Ecology and distribution of tilapias, p. 15-59. *In* R.S.V. Pullin and R.H. Lowe-McConnell (eds.) The biology and culture of tilapias. ICLARM Conference Proceedings 7,432 p. International Center for Living Aquatic Resources Management, Manila, Philippines.

This paper reviews three aspects of the ecology of tilapias: i) their geographical distribution, ii) the physical (current velocity, depth, temperature) and chemical (salinity, alkalinity, pH, oxygen and other dissolved gases) factors which influence this and iii) aspects of their behavioral ecology namely feeding behavior, reproductive behavior, selection of microhabitat, schooling behavior and movement. Maps are given showing the natural distribution of the principal substrate-spawning genus *Tilapia* and the mouthbrooding genus *Sarotherodon* and the history of their transfers in Africa and introductions to other continents are tabulated.

The natural distribution of the tilapia species depends upon: i) historico-geological factors which led to geographical isolation and speciation, ii) ecological factors which demonstrate the requirements and preferences of the species mainly with regard to temperature and salinity and iii) behavioral mechanisms relating to feeding and reproductive strategies. Data on the natural distribution, ecological amplitude and eco-ethological characteristics could help in the selection of species for culture and introductions for fisheries and in the prevention of unfortunate consequences, such as unwanted hybridization, competition for food or spawning grounds, and contamination of pure wild strains or failure of the introduced stocks (e.g., due to unsuitable temperature conditions, overpopulation and nanism, etc.). The dangers of introductions made without sufficient knowledge of the biology of the species and recipient ecosystems and the necessity for better control of fish transfers and future research on the basic and applied ecology of tilapias are all stressed.

Introduction

In their recent review Balarin and Hatton (1979) concentrated on the biology of tilapias in relation to their uses in aquaculture. However, apart from their considerable economic importance for aquaculture and fisheries, tilapias are important elements of biocenoses, involved in the workings of tropical aquatic ecosystems. As such they raise a multitude of problems in basic ecology, the study of which could have, if it had been done in advance, prevented the large number of unsuccessful attempts to culture or to introduce tilapias.

This paper is written from a broad ecological perspective, although practical applications are shown whenever these arise. Five aspects of tilapia ecology are discussed:

1. Geographical distributions, with special emphasis on assessing the success of introductions and transfers.
2. Analysis of physical and chemical factors which influence the natural distribution of tilapias and characterize their ecological range.
3. Feeding behavior.
4. Reproductive behavior.
5. Selection of microhabitats and dynamics of spatial separation.

Zoogeography and Distribution

Original Distributions of Tilapias

The family Cichlidae is widely distributed in Africa (including Madagascar) and Palestine, in South and Central America, in southern India and in Sri Lanka (Lagler et al. 1977). The tilapias, however, originated exclusively from the African continent (excluding Madagascar) and from Palestine (Jordan Valley and coastal rivers). In Africa they are absent only from the Northern Atlas and from south west Africa. This is an old pan-African group with a slow or average rate of speciation, primarily riverine, secondarily lacustrine. About seventy species are nowadays distinguished as two genera: *Tilapia* (macrophagous, substrate-spawners) and *Sarotherodon* (microphagous mouthbrooders). Their original distributions are given in Figures 1a, b and 2a, b, c and in Table 1. The data used for this review stem from Thys (1963, 1964, 1968, 1971a, 1971b), Trewavas (1966), Fryer and Iles (1972), Ruwet and Voss (1974), Ruwet et al. (1975) and Balarin and Hatton (1979), which relate specifically to tilapias, and from other works on African ichthyology, Günther (1889), Pellegrin (1921), Poll (1957, 1973), Ruwet (1961, 1962), Greenwood (1965, 1966, 1974, 1976), Jubb and Farquaharson (1965), Bell-Cross (1968, 1972, 1973), Daget (1968), Gee (1968b), Petr (1968, 1969, 1975), Lowe-McConnell (1969, 1975, 1979), Jubb and Gaigher (1971), Gaigher (1973), Green et al. (1973, 1978), Trewavas (1973a, 1973b), Johnson (1974), Eccles (1975), Ottobo (1976), Landau (1979), and other studies cited below.

The natural distributions of the species reflect two types of factors:

1. those historico-geological factors which led to geographical isolation and to speciation (endemic species in lakes or stretches of rivers): a knowledge of the past geology and hydrography of Africa is necessary to understand the species distributions, while, conversely, a knowledge of the species distributions helps in confirming hypotheses concerning the evolution of hydrographic basins and their former connections.
2. those ecological factors which demonstrate the requirements and preferences of the species with regard to temperature, salinity, the chemical composition of the water, and the behavioral mechanisms which reflect feeding and reproductive strategies.

Dealing first with the historical factors: for those affecting distribution, our ideas are based essentially on the synthesis by Thys (1963) on the tilapias of Zaïre and adjacent regions, and on Beadle (1974) for the paleogeography. The maps (Figures 1 and 2) show that:

a. The genus *Tilapia* has a very wide distribution except for the eastern slope of the eastern Rift valley and river basins facing the Indian Ocean. They are abundant in west and central Africa. Speciation is slow; they are 'stable', river-dwelling, close to the original stock. The species are separated not by geographical or hydrographic barriers, but rather by ecological or behavioral barriers.
b. The genus *Sarotherodon* is rare in west Africa, absent from the Zaïre basin, but abundant, diversified and specialized in small local populations in the Rift lakes. The subgenus *Oreochromis (S. mossambicus, S. mortimeri, S. andersonii* and *S. spilurus niger)* occurs in the basins facing the Indian Ocean. The various *Sarotherodon* species are segregated by geographical and hydrographical barriers and generally have a small range of distribution, except for *S. galilaeus, S. niloticus* and *S. mossambicus.*

The examples discussed below show the relationship between past geological-hydrological events and the natural distribution of tilapias.

DISTRIBUTION OF *T. ZILLII, S. GALILAEUS* AND *S. NILOTICUS*

These three 'soudanian' species have a very wide common range (Senegal, Niger, Chad, Nile, Jordan, Lake Turkana (Rudolf), Lake Albert) resulting from former interconnections of the Chad and Nile basins.

S. niloticus, originating from the upper Nile in Uganda evidently moved southwards, colonizing all the western Rift lakes down to Lake Tanganyika. It also colonized central and western Africa, via the Chad and Niger basins. Its expansion is still taking place; it has not yet reached some of the tributaries of the upper Niger and it is rare in the coastal rivers of western Africa. Conversely, *T. zillii* and *S. galilaeus* have spread eastwards towards the Nile and the first Rift lakes.

It seems likely that the Chad-Nile connection came into existence later than the southward push of *S. niloticus* along the Rift lakes, and later than the separation of Lake Albert from Lake Edward, which could explain why *S. niloticus* occurs in the latter lake though *T. zillii* and *S. galilaeus* did not reach it. *T. zillii* and *S. galilaeus* were present in the Ubangi and Uele Rivers, now tributaries of the Zaïre river, before these two rivers were captured by the Zaïre. This capture must have taken place before the Chad-Nile connection since *S. niloticus* is absent from the Ubangi-Uele system. The Ituri River (Zaïre, central Africa) is a particular case: *T. zillii* evidently colonized it in an eastward direction, but not *S. galilaeus*, which found itself competing against *S. niloticus* already established. Before its integration with the Zaïre basin, the Ituri was linked with Lake Albert, which was populated with *S. niloticus*. In the great Zaïre basin, the expansion of *T. zillii* was prevented by the presence of *T. tholloni*.

The *S. galilaeus* coming from the north did not manage to invade the Zaïre basin, which seems to be an environment hostile to microphagous cichlids, though *S. galilaeus* reached, and was able to establish itself in the Malebo (Stanley) Pool upriver from Kinshasa.

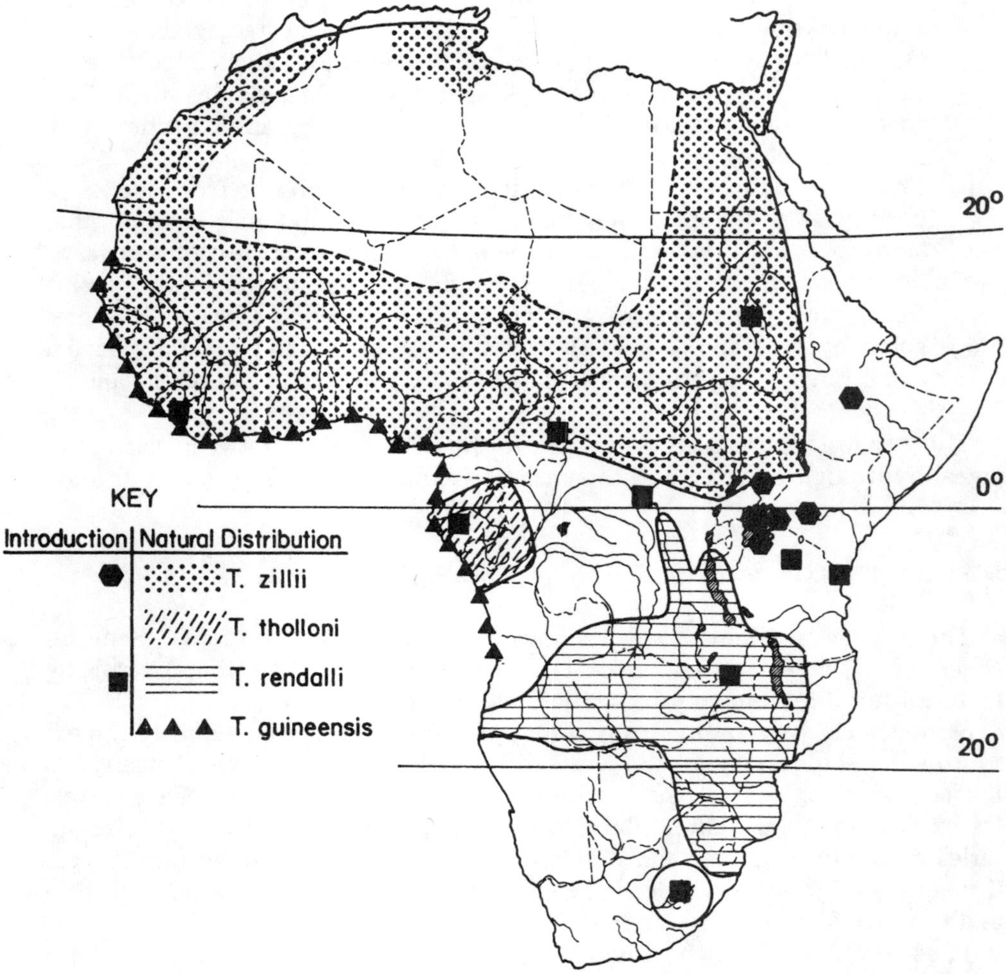

Figure 1a. Substrate-breeding tilapias (genus *Tilapia*): natural distribution and introductions of *T. zillii*, *T. tholloni*, *T. rendalli* and *T. guineensis*. The many transfers of *T. rendalli* in South Africa are not recorded precisely but only suggested.

DISTRIBUTION OF *S. MACROCHIR* AND *S. ANDERSONII*

These two species are sympatric in part of their range. *S. macrochir* lives in the rivers and lakes of the Upper Zaïre: Bangweulu-Luapula-Mweru, Luvua, Lufira and Upper Lualaba, which were part of the Zambezi basin prior to their capture by the Zaïre. It also occurs in the upper Zambezi and in the Kafue. The absence of *S. andersonii* from the system now belonging to the Zaïre river north of the present Zaïre/Zambezi divide is evidence that its arrival in the upper Zambezi occurred after this divide was established. Below the Victoria Falls on the Zambezi, below the Kafue Gorges and the Luangwa Falls, *S. andersonii* is replaced by *S. mortimeri*. But since the construction of the Kariba and Cabora Bassa dams, *S. andersonii* has been recorded in the Middle Zambezi (Balon 1974; Jackson and Rogers 1976).

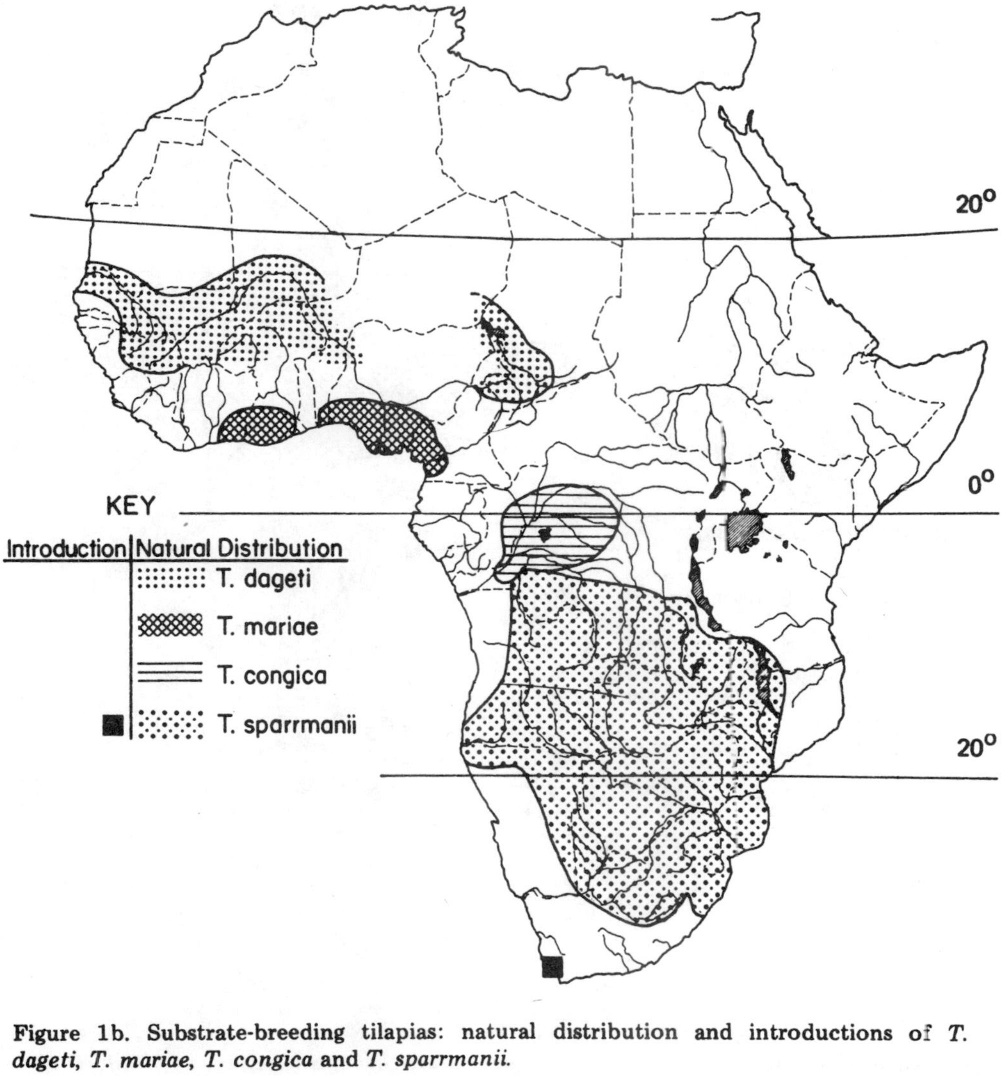

Figure 1b. Substrate-breeding tilapias: natural distribution and introductions of *T. dageti, T. mariae, T. congica* and *T. sparrmanii.*

DISTRIBUTION OF *T. RENDALLI* AND *T. SPARRMANII*

The distribution of these species is quite similar and quite 'Zambezian'. The presence of *T. rendalli* and *T. sparrmanii* in the Zaïre basin can be explained by the capture of a number of rivers in the Zambezi system by the Zaïre river and its tributaries in Shaba and Kasai. But while *T. sparrmanii* could not colonize the Lualaba below the Kongolo (probably because of the high temperatures), *T. rendalli* has colonized the Lualaba to the limits of its basin. Also it has colonized the Tanganyika basin and its tributaries, via the Likuga. *T. rendalli* possibly came into Lake Tanganyika after the penetration by *S. niloticus*, which came from Lake Kivu after the latter was captured by the Ruzizi.

Distributions of *T. sparrmanii* and *T. rendalli* overlap markedly, but the former is more southern, while the latter reaches north to the edges of the equatorial basin.

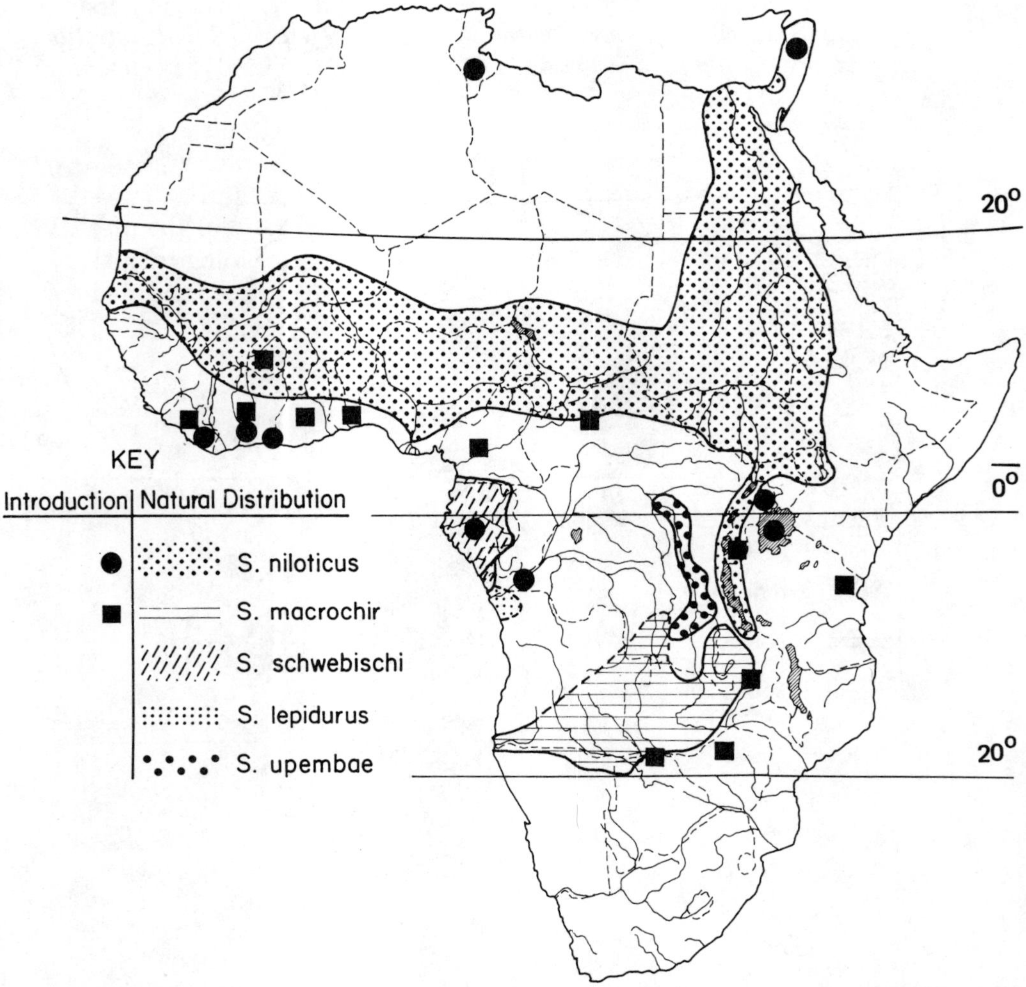

Figure 2a. Mouthbrooding tilapias: natural distribution and introductions of *S. niloticus* (genus *Sarotherodon*), *S. macrochir*, *S. schwebischi*, *S. lepidurus* and *S. upembae*.

DISTRIBUTION OF *T. GUINEENSIS*, *S. MELANOTHERON* AND *S. MOSSAMBICUS*

These very euryhaline species have a wide distribution, the first two on the western (e.g., Baron 1974), the third on the eastern coast. They can move from one estuary to the next, especially during floods.

It is particularly noteworthy that the macrophagous substrate-spawning tilapias of the *Coptodon* subgenus (*T. zillii, T. rendalli, T. tholloni, T. congica, T. guineensis*) all have a wide distribution and exclude each other geographically.

Species of the *Sarotherodon* group—with the exception of *S. galilaeus, S. niloticus* and *S. mossambicus*—are specialized and more localized; many species overlap (*S. esculentus* and *S. variabilis* in Lake Victoria; *S. leucostictus* and *S. niloticus* in Lakes Edward and George; *S. macrochir* and *S. andersonii* in the Zambezi; *S. galilaeus* and *S. niloticus* in the Nile; *S. niloticus, S. aureus* and *S. galilaeus* from Senegal to Chad; several species in Lake Malawi). This overlapping suggests particular preferences and requirements which guarantee ecological isolation. From this, it follows that a knowledge of the natural geographical distribution is necessary to understand the origin of the present situation by explaining it in relation to the geological-hydrobiological history of the continent, select intelligently the species to be cultured, and deduce certain ecological requirements and preferences.

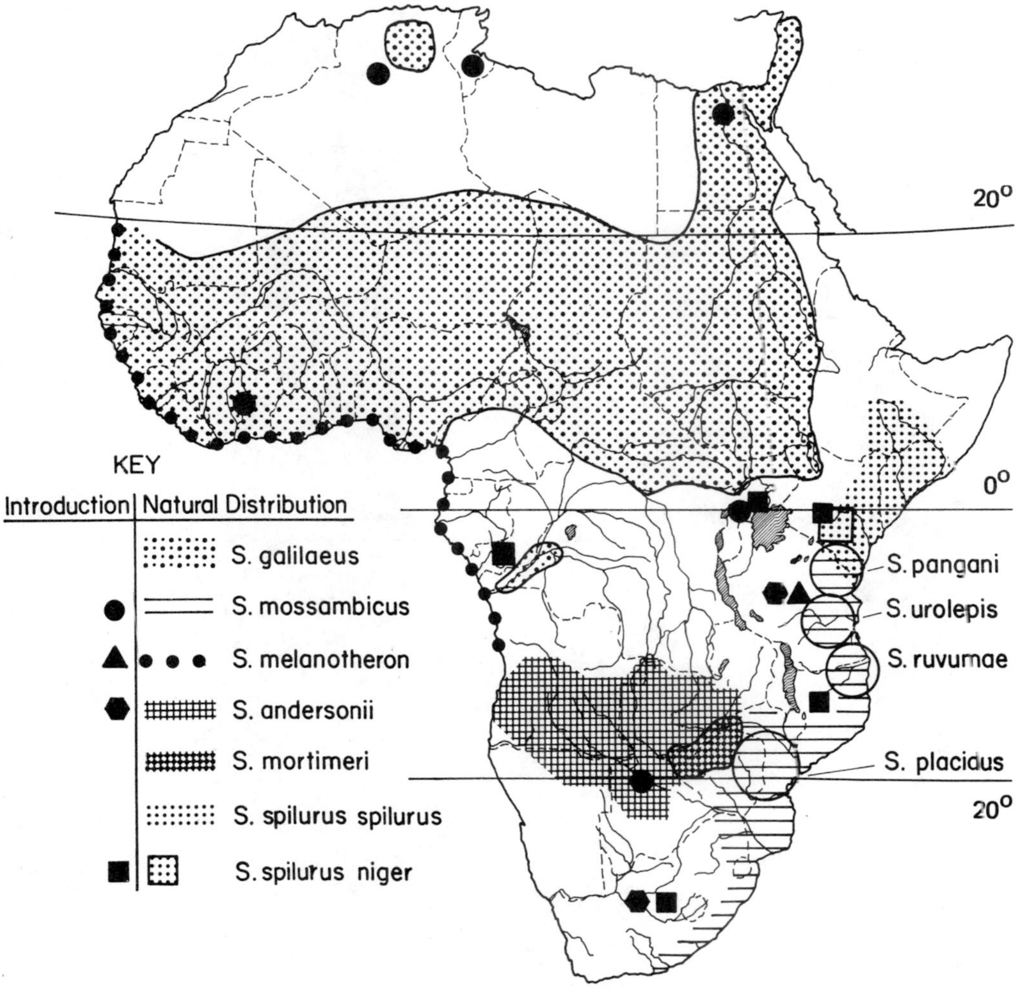

Figure 2b. Natural distribution and introductions of *S. galilaeus, S. mossambicus, S. melanotheron, S. andersonii, S. mortimeri, S. spilurus spilurus, S. spilurus niger, S. pangani, S. urolepis, S. ruvumae, S. placidus.*

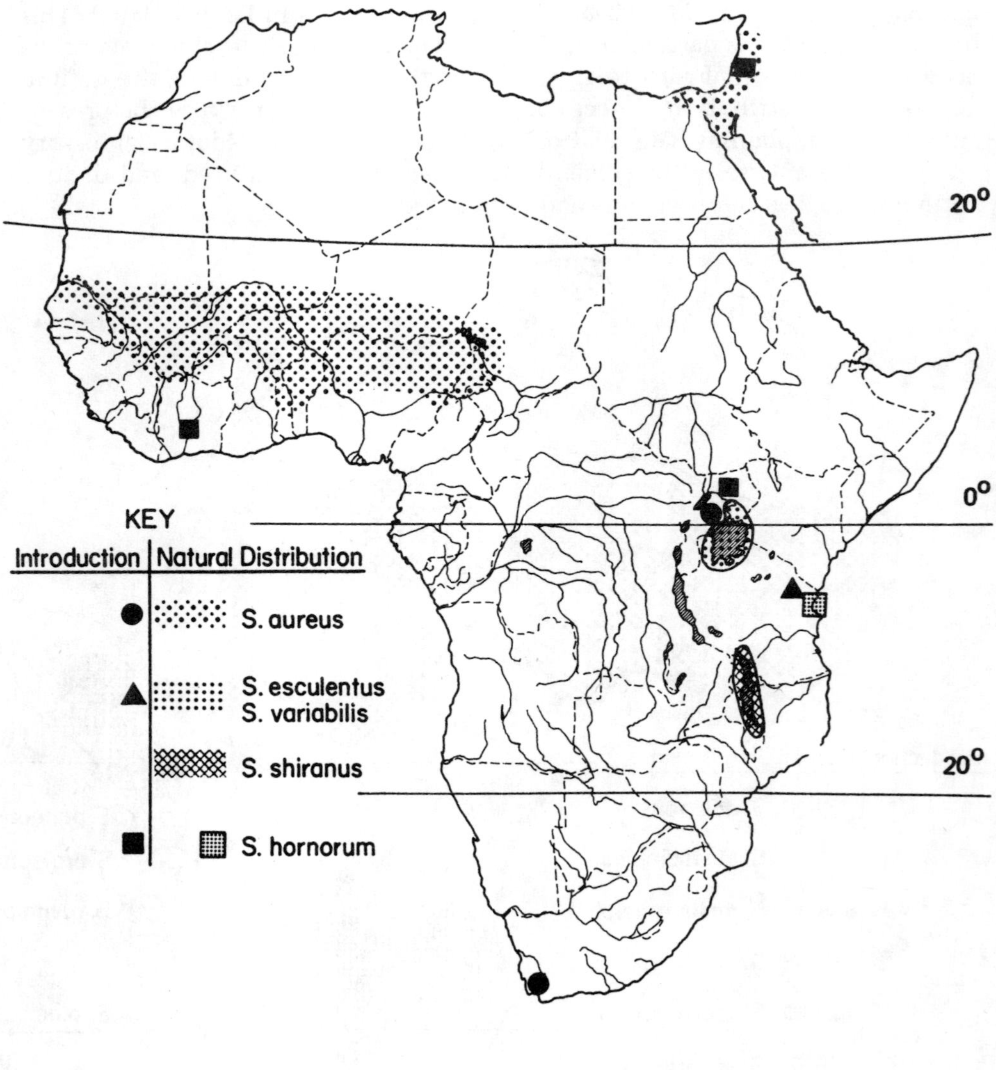

Figure 2c. Natural distribution and introductions of *S. aureus, S. esculentus, S. variabilis, S. shiranus* and *S. hornorum.*

Table 1. Tilapia species with a relatively restricted or endemic geographical distribution (species not included in the maps of Figures 1 and 2).

Species	Distribution
A. Lakes of eastern Africa	
S. leucostictus	L. Albert and Edward/George (Zaïre, Uganda)
S. esculentus, S. variabilis	L. Victoria, Kyoga, Nabugabo (Uganda, Kenya, Tanzania)
S. alcalicus alcalicus	L. Natron (Tanzania)
S. alcalicus grahami	L. Magadi (Kenya)
S. amphimelas	L. Manyara, L. Kitangiri (Tanzania)
S. karomo	Malagarasi swamps (Tanzania)
S. lidole, S. saka, S. squamipinnis, S. karongae	L. Malawi (Malawi)
S. shiranus shiranus	L. Malawi, upper R. Shire (Malawi)
S. shiranus chilwae	L. Chilwa and L. Chiuta (Malawi/Moz.)
S. rukwaensis	L. Rukwa (Tanzania)
S. tanganicae	L. Tanganika
S. hunteri	L. Chala (Kenya)
S. jipe, S. girigan	L. Jipe (Kenya)
B. Other lakes	
T. guinasana	L. Guinas (S.W. Africa)
T. gefuensis	L. Gefu (Angola)
T. multiradiata (= *T. zillii*)	L. Chad (Chad)
T. bemini	L. Bemin (Cameroon)
T. dekerti	L. Ejagham (Cameroon)
T. kottae	L. Barombi ba Kotto (Cameroon)
T. discolor, T. bosumana, S. galilaeus multifasciatus	L. Bosumtwi (Ghana)
S. linnellii, S. lohbergeri, S. caroli, S. steinbachi	L. Barombi Mbo (Cameroon)
C. Rivers of the eastern coast	
S. urolepis	R. Rufigi, Kingani, Mbemkuru (Tanzania)
S. ruvumae	R. Ruvuma (Mozambique-Tanzania)
S. pangani, S. korogwe	R. Pangani (Tanzania)
D. Rivers of the Guinea-Ghana coast	
T. brevimanus, T. walteri, T. louka, T. coffea, T. joka, T. rheophila, S. caudomarginatus, S. tournieri	
E. Rivers of the Cameroon-Congo-Angola coast	
T. camerunensis	R. Meme (Cameroon)
T. ogowensis	R. Ogooué (Cameroon)
S. mvogoi, T. margaritacea	R. Nyong (Cameroon)
T. cabrae	Rs. Ogooué, Chiloango, Kwilu, Bengo, Quanza (lower)
S. angolensis	R. Quanza, Bengo (Angola)
S. sanagaensis	R. Sanaga (Cameroon)
T. cameronensis	R. Sanaga (Cameroon)

The original distribution of the African tilapias has been modified by the deliberate or unplanned introduction of several species outside their distribution area (within the same basin, or from one basin to the next). Figures 1 and 2 summarize the main data from the literature (Bardach et al. 1972; Fryer and Iles 1972; Beadle 1974; Lowe-McConnell 1975; George 1975; Vincke 1979; Coche 1978; Balarin and Hatton 1979; Moreau 1979a, 1979b), while Table 2 presents a synthesis of the best known examples of introductions into natural habitats and into reservoirs.

These transfers had different objectives:

i. Stocking natural lakes in which no tilapias occurred (*S. alcalicus grahami* into Lake Nakuru, *S. spilurus niger* and *T. zillii* into Lake Naivasha, *S. spilurus niger*, *S. esculentus* and *S. niloticus* into many lakes in Uganda and Ruwanda; *S. macrochir* and *T. rendalli* into Lake Lusiwashi).

ii. Introduction into a natural habitat to fill an ecological niche not occupied by any of the tilapias present (*T. zillii* and *S. niloticus* into Lake Victoria and Lake Kyoga) in the hope of increasing fishery yields.

iii. Introduction into artificial water bodies to develop new fisheries based on the introduced tilapias (*S. mossambicus* and *S. niloticus* into the reservoir of a southern Tunisian oasis), or to utilize the plankton production of reservoirs (*S. macrochir* into Lakes Kariba and McIlwaine; *S. placidus* into Lake Kyle; *S. mossambicus* and *S. mortimeri* into the lakes of the Zimbabwe eastern highlands).

iv. Biological control of aquatic vegetation (*T. rendalli* into Sudanese irrigation channels and some artificial lakes of Shaba), or for the control of mosquitos (*S. niloticus*, see George 1975).

v. Fish culture in rice fields (Vincke 1979; Khoo and Tan 1980), in floating cages, and especially in ponds (Bardach et al. 1972; Coche 1978). Although at first confined to these aquaculture systems, the tilapias have inevitably escaped into natural waters (during draining of ponds, break-up of dykes, floods). For example, *T. rendalli* escaped into the Pangani River in Tanzania (Trewavas 1966) and *S. mossambicus* into the middle Zambezi basin (Jubb 1974).

vi. Involuntary introductions occurring during the deliberate introduction of another species (*S. leucostictus* into Lake Naivasha, *T. rendalli* into Lakes Victoria and Kyoga), or because of the confusion between sympatric species which have hardly begun to differentiate (*S. niloticus* and *S. aureus; S. mossambicus* and *S. placidus; S. mossambicus* and *S. mortimeri*, and the *Sarotherodon* species of the east coast rivers in general).

The reasons for the success, or lack of it, in these introductions, their effects on the various ecosystems and the aquacultural aspects will be discussed in a later section in the light of information on the ecology of the various species. On the strictly biogeographical level, the older, unrecorded introductions, or those made in a doubtful manner (lacking present-day taxonomic and hydrographic knowledge) all represent small catastrophes. In several cases it is indeed impossible to know if the presence of a given species is a natural occurrence (which could have far-reaching consequences in biogeographical and paleographical terms) or simply the results of human activities.

The following examples may be cited:

i. Natural occurrence or old introduction: *S. hornorum* in Zanzibar; *S. hunteri* in Lake Chala, Kenya (Trewavas 1966).

ii. Endemicity of *S. girigan* and *S. jipe* in Lake Jipe (these two species were introduced into the Pangani River and now occur in the whole basin, Trewavas 1966).

iii. The biogeographical relevance of the presence of *S. mossambicus* in the Thalamakane River, Botswana (Jubb and Gaigher 1971), in small water bodies in the Namib desert (Dixon and Blom 1974), and in two tributaries of the middle Zambezi (Jubb 1974) where this species was considered to be absent. Another similar case is the presence of *S. macrochir* in the middle course of the Buzi River, Mozambique (Bell-Cross 1973), completely outside the rest of its range.

To add to the confusion, during the fifties many aquaculturists called any macrophagous substrate-spawning tilapia 'melanopleura', whatever its geographical origin. Thus *T. zillii*, *T. rendalli*, *T. tholloni* and *T. guineensis* have often been spread around in Africa confused under the name "*melanopleura*".

INTRODUCTIONS OUTSIDE AFRICA

The first, probably accidental, introduction of a tilapia outside Africa was that of *S. mossambicus* prior to 1939 in Java where a few individuals were caught in the Serang River (Atz 1954; Riedel 1965). From then on, *S. mossambicus* was introduced into various countries (see Riedel 1965). Then the practice spread to other species, and now tilapias occur in natural waters throughout the tropics, even in Australia (Queensland). Figure 3, drawn from data in Balarin and Hatton (1979), Welcomme (1979a, 1979b), Rosenthal (1976), Courtenay and Robins (1973), Bardach et al. (1972), FAO (1977c) and some other authors, shows the worldwide pattern of introductions of the species of major importance: *S. mossambicus*, *S. niloticus*, *S. aureus*, *S. hornorum*, *T. rendalli* and *T. zillii*. Other species introduced outside Africa are dealt with in Table 3.

The objectives of these introductions were, as in the case of Africa, pond culture (see Bardach et al. 1972; Coche 1978; Vincke 1979), the stocking of natural habitats, of man-made lakes and water storage reservoirs, of mining pits, and the control of aquatic vegetation. For twenty years there have also been introductions to stock cooling waters and geothermally heated waters in temperate regions, in Japan, the USA and in Europe (e.g. TVA 1978; Kuroda 1977; Mélard and Philippart 1980).

Table 2. Tilapia introductions in natural waters of Africa.

Lake Victoria (Lake Kyoga) (Uganda-Kenya-Tanzania)

- two indigenous endemic species: *S. esculentus* and *S. variabilis*
- four introduced species: *S. leucostictus, S. niloticus, T. zillii* and *T. rendalli* (accidentally with *zillii*) between 1951 and 1954.
- good acclimatization of introduced species favored by the existence of empty ecological niches, notably the niche for a macrophage herbivorous tilapia: *T. zillii* underwent a large development and became almost as abundant as *S. variabilis* which was one of the main exploited species (Welcomme 1967a).
- but negative effects (see Fryer and Iles 1972; Fryer 1961a; Welcomme 1964, 1967a):
  - i. in the first stages of its growth, *T. zillii* shared the same niche as the alevins of *S. variabilis*, then competed with them and supplanted them due to more rapid growth and greater aggression.
  - ii. *T. zillii* competed with and also supplanted *S. variabilis* for the occupation of breeding grounds.
  - iii. *S. niloticus* competed with *S. esculentus* and hybridized with *S. variabilis*.
  - iv. *T. zillii* hybridized with *T. rendalli*, naturally allopatric.

Koki lakes (Uganda)

- no indigenous tilapias
- complete failure with the introduction of *S. spilurus niger* and *S. esculentus* but satisfactory success with *S. niloticus* (see Lowe (McConnell) 1958)

Lake Bunyoni-Lake Nkugute (Uganda)

- no indigenous tilapias
- complete failure of *S. spilurus niger* introduction (Lake Bunyoni) and *S. esculentus* (Lake Nkugute) (Lowe (McConnell) 1958) and success of *S. niloticus* in the two lakes
- but hybridization *S. niloticus* x *S. spilurus niger* and *S. niloticus* x *S. esculentus*; excessive development of *S. niloticus* populations with, in consequence, retarded growth, nanism, infestation by parasites and finally a very poor result from the fishery point of view (Beadle 1974, p. 83).

Lake Naivasha (Kenya)

- no indigenous tilapias
- introduction of *S. spilurus niger* in 1925 and of *T. zillii* and *S. leucostictus* (accidental) in 1956
- good initial acclimation in *S. spilurus niger* which formed the mainstay of the fishery during the years 1950-60; in 1961 appearance of hybrids of *S. spilurus niger* x *S. leucostictus*; increase in their number (57% of catches in 1962) followed by their regression particularly marked after 1971, date of the total disappearance of *S. spilurus niger*; after this date, great development of *S. leucostictus* which became the dominant species in the lake and of *T. zillii* (Elder et al. 1971; Siddiqui 1979b)

Lake Lusiwashi (Zambia)

- no indigenous tilapias
- introduction of *S. macrochir* (1949) and *T. rendalli* (1959) successful (catches: 80 t, Fryer and Iles 1972) because ecologically complementary species (microphagous and macrophagous) transplanted close to their original area of distribution in a milieu offering vacant niches.

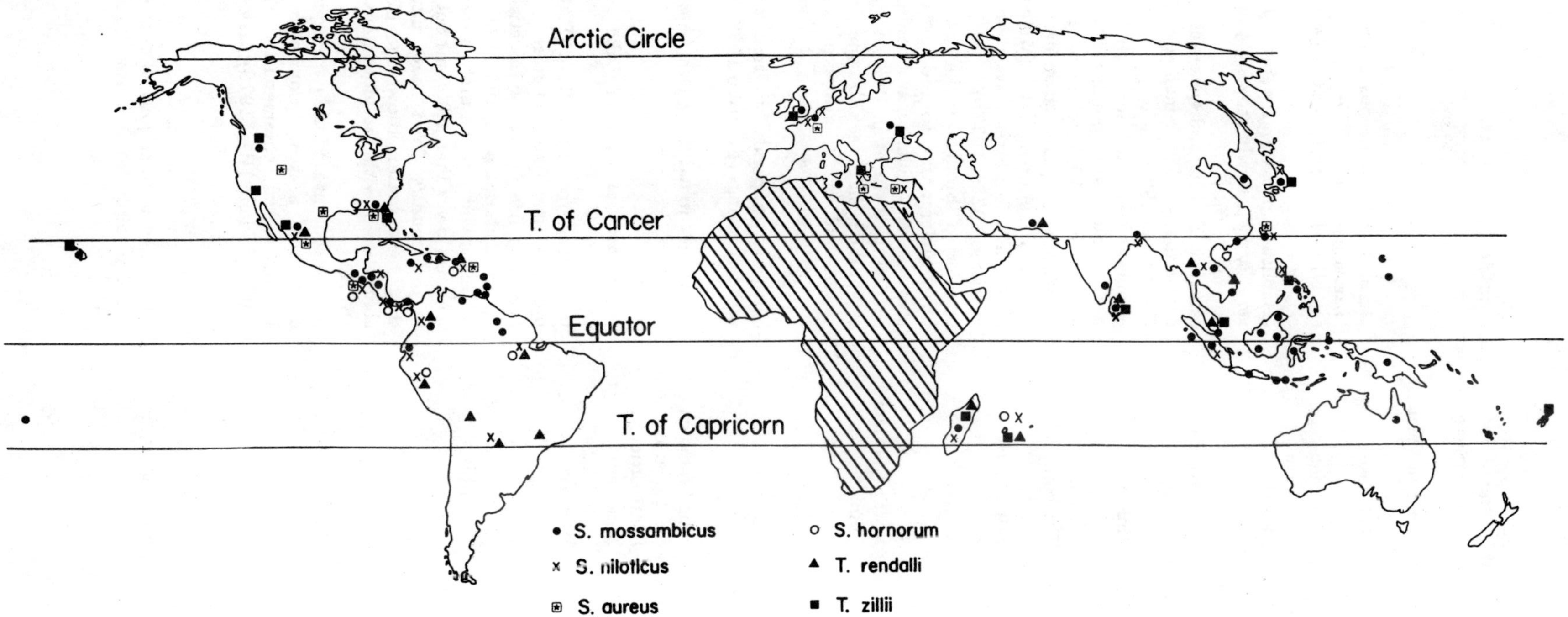

Figure 3. Introductions of tilapias (six principal species) outside Africa (see also Table 3).

Table 3. Results of introductions of tilapias outside the African continent.

	S. mossambicus	
Africa	Madagascar	established in numerous waters of the country (Lamarque et al. 1975)
SE Asia	Java, Malaysia, Sri Lanka	established in natural waters (Atz 1954)
	Indonesia	established in natural water but often forming very dense, slowly-growing populations (Welcomme 1979b)
	New Guinea	established in natural waters at same time as 21 exotic species (Glucksman et al. 1976; West and Glucksman 1976)
	Philippines	established in brackish water but considered a pest in the *Chanos* ponds (Pillai 1972; Rabanal and Hosillos 1957, cited by Rosenthal 1976)
	Thailand, Bangladesh	initially well-established in natural waters then regressed spontaneously (Welcomme 1979b)
	Hong Kong	established in Plover Cove Reservoir (Man and Hodgkiss 1977a, 1977b)
	Taiwan	established throughout the country (rice fields) but problems due to too low winter temperatures, overpopulation and nanism (Chen 1976)
	North Vietnam	acclimatized (Le et al. 1961)
	China	established in brackish and marine coastal waters and in the rivers of the central and southern provinces, e.g., Fukien (Borgstrom 1978)
	Japan	cultivated in ponds but transferred in winter into hot springs (Fukusho 1968)
America	Haiti, Jamaica, St. Lucia, Grenada, Puerto Rico	established in natural waters (Atz 1954)
	Nicaragua	established, at least in Lake Moyua (Riedel 1965)
	Mexico, El Salvador	introduced for fish culture (FAO 1977c) and probably established in nature
	Ecuador	reared in fish culture with satisfactory results up to altitudes of 2,400 m (!) (FAO 1977c)
	Colombia	establishment in natural waters of certain regions (lower than 1,000 m) but dissemination in the country from there interrupted until they have better knowledge of the risks of competition with the indigenous species (FAO 1977c)
	U.S.A.	introduced into Florida by aquariophiles and considered a pest (Courtenay and Robins 1973); reared commercially in geothermal water in Idaho (Ray 1978); Hawaii (Neil 1966)
	S. niloticus	
Africa	Madagascar	established in numerous lakes (e.g., Lake Itasy, 1,200 m altitude) and rivers (Lamarque et al. 1975)

Table 3 (cont'd)

SE Asia	Indonesia	established in natural waters where found to be economically very interesting (Welcomme 1979b)
	Bangladesh, Thailand	established in natural waters but the populations declined some years after the introduction (Welcomme 1979b)
	Japan	principal species of tilapia raised in industrial warm water (Kuroda 1977)
America	Mexico	cultivated in numerous fish stations and established in numerous artificial lakes in the center and south of the country (FAO 1977c)
Europe	Cyprus	established (?) in a reservoir (Welcomme 1979b)
	Belgium	cultivated intensively in warm water of a nuclear power plant with possibility of surviving in the cooling stream in summer (Mélard and Philippart 1980)
	S. aureus	
SE Asia	Philippines, Taiwan	introduced for hybridization
America	El Salvador	introduced for rearing in ponds and in cages with risks of escapes
	Puerto Rico	established in quarry lakes used for rearing in cages (Pagan-Font 1977)
	U.S.A.	introduced with a view to biological control of the vegetation in lakes of central Florida in 1961 and actually established in at least 20 counties where it is the object of commercial exploitation (Langford et al. 1978) but catches are small and it is responsible for the destruction of the vegetation and native species of fish in the waters colonized (Courtenay and Robins 1973); probably established in certain lakes in south Texas but maintained artificially in cooling water of electric power stations (Stickney and Hesby 1978); surviving in a natural lake in Alabama only when the winter is exceptionally warm (Habel 1975); reared in geothermal water in Colorado (Lauenstein 1978)
Europe	Cyprus	established (?) in a reservoir (Welcomme 1979b)
	S. macrochir	
Africa	Madagascar	established all over the country in certain mangrove swamps (e.g., Nemakia) (Lamarque et al. 1975)
	S. melanotheron	
America	U.S.A.	reproducing in natural waters in Florida but does not seem destructive (Courtenay and Robins 1973; Welcomme 1979b)
	S. spilurus niger	
Africa	Madagascar	failed introduction because temperature too low at altitude (in Balarin and Hatton 1979)

Table 3 (cont'd)

	S. shiranus chilwae	
Africa	Madagascar	only cultured experimentally (George 1976)
	S. hornorum	
America	Puerto Rico, Brazil, U.S.A	introduced for hybridization in numerous experimental fishery and fish culture stations
	S. galilaeus	
Europe	Germany	experiments for rearing in industrial thermal effluents (in Balarin and Hatton 1979)
	T. rendalli	
Africa	Madagascar	established in numerous lakes and rivers but has seriously perturbed the ecology of Lake Kinkony (Lamarque et al. 1975)
	Mauritius	introduced for fish culture in 1956, it escaped into rivers and reservoirs where it occasionally has serious effects on the indigenous flora and fauna (George 1976)
Asia	Sri Lanka	well-established in natural waters where a vacant ecological niche existed (Welcomme 1979b)
	Thailand	at first well-established in natural waters then progressively regressing and disappearing, probably due to competition from local species (Welcomme 1979b)
America	Brazil	reared intensively in the northeast and the central-south of the country and established in natural waters, notably Lake Pinheiro in Brazilia and in numerous hydroelectric reservoirs in Sao Paulo State; but many reports of overpopulation and destruction of the vegetation (Nomura 1976, 1977)
	Colombia	cultivated in ponds and established in natural waters in the Valle del Cauca between 1,000 m and 1,400 m; but the dissemination into other regions is forbidden until more is known of the effects on the ecosystems and the native species (Norena 1977)
	Peru	cultivated in ponds and established in certain lakes (FAO 1977c)
	Bolivia, Paraguay	only cultivated in ponds (FAO 1977c)
	Mexico	cultivated in ponds up to an altitude of 1,500 m in the center and south of the country (FAO 1977c)
	T. zillii	
Africa	Madagascar	established in numerous ponds, lakes and rivers (Lamarque et al. 1975)
SE Asia	Malacca, Fiji	introduced but established (?)
	Taiwan	little appreciated by fish culturists because of its aggressiveness (Chen 1976)
America	U.S.A.	introduced to Hawaii (?); established in natural waters in Florida; reared commercially in

Table 3 (cont'd)

		geothermal water in Idaho (Ray 1978; TVA 1978); experiments on biological control of vegetation in California but progressively abandoned because of mortalities and slow feeding activity due to low temperature (Hauser 1977; Platt and Hauser 1978)
Europe	Great Britain	introduced accidentally and considered acclimatized in a canal receiving thermal effluents from an electricity station (Wheeler and Maitland 1973)
	T. mariae	
America	U.S.A.	established in natural waters in Florida (Hogg 1976)
	T. guineensis	
Europe	Belgium	experiments (abandoned) for rearing in industrial thermal effluents (Mélard and Philippart 1980)
	T. sparrmanii	
Asia	Japan	experiments for rearing in industrial effluents

Physical and Chemical Factors Affecting Tilapia Distribution

HABITAT DIVERSITY

Within their original areas of distribution, the tilapias have colonized widely different habitats: permanent and temporary rivers, rivers with rapids, large equatorial rivers (Zaïre), tropical and subtropical rivers (Senegal, Niger, Nile, Zambezi, Limpopo), deep lakes (Albert, Kivu, Tanganyika, Malawi), swampy lakes (Bangweulu, Mweru, Victoria, Kyoga, Rukwa, Chad), highly alkaline and saline lakes (Magadi, Natron, Manyara, Mweru Wantipa, Chilwa, Chiuta, Turkana, Tana), other saline lakes (Lake Quarun), hot springs (for example in Lake Magadi), volcanic crater lakes (Lakes Chala, Barombi Mbo, Barombi ba Kotto) or meteoritic crater lakes (Lakes Bosumtwi), lakes with low mineral content (Lake Bangweulu, Lake Nabugabo), sometimes very acidic waters (Lake Tumba), permanent water bodies in the Sahara (Pellegrin 1921 in Beadle 1974) and in the Namib desert (Dixon and Blom 1974), open or closed estuaries, lagoons and coastal brackish lakes that often become hypersaline, marine habitats with normal salinity of Atlantic water and hypersaline in the Red Sea (Gulf of Suez).

All these different habitats represent (both in terms of absolute amplitude and in terms of the speed at which fluctuations take place) an extraordinary varied range of physical parameters (depth, current velocity, turbidity), of temperature and of chemical composition, especially salinity, pH, dissolved oxygen (DO) and other gases.

Balarin and Hatton (1979) have collated the extensive literature concerning tolerance limits and the preferences of tilapias for the physico-chemical properties of their habitats. Here we shall recall some of these data and complement them with more recent results (post-1976), dealing mainly with data pertaining to ecological conditions actually encountered in the natural habitats and some semi-natural habitats (fish ponds). Data obtained under laboratory conditions are discussed only when this is necessary to define better the ecological amplitude of certain species.

CURRENT VELOCITY AND DEPTH

Several tilapias are perfectly adapted to fast-flowing rivers with rapids, for example, *T. rheophila* endemic to the Konkoure river, Liberia (Thys 1969) and *T. busumana* of Lake Bosumtwi, Ghana and the rivers flowing into it, especially the Ebo river where *T. busumana* occurs in places where the slope ranges from 13 to 60‰ (Lelek 1968). *S. andersonii* and *T. sparrmanii* may be found in the upper and middle reaches of the Kalomo river (a tributary of the Zambezi) where the slope has an average of 2.21‰ (min. 0.9‰, max. 10‰) and 5.73‰ (min. 3‰, max. 19‰) respectively, but not in the lower reaches where the slope is very steep (average 6.63‰) (Balon 1974, p. 459). Even for such species well adapted to river life, rapids and falls are hostile zones which often represent ecological barriers (mechanical obstacles, excessive current velocities, oxygen and nitrogen supersaturations) which prevent the mixing of neighboring but different ichthyofaunas.

Quite a number of tilapias, especially those with a wide area of distribution, may be encountered in both rivers and lakes where they tend to remain in shallow inshore waters (for reproduction and feeding) and in the pelagic epilimnion (for their nutrition in the case of plankton-feeders). This depth limitation of tilapia distribution (see Table 4) can be found even in the majority of species that are endemic in the African Great Lakes.

According to Caulton and Hill (1973) for *S. mossambicus*, and Caulton (1975a, 1975b) for *T. rendalli* and *T. sparrmanii*, tilapias should be physiologically unable to adapt to the increased pressure that goes with increasing depth. However, the depth distribution of lake-dwelling tilapias is also influenced by temperature and oxygen gradients, as well as by concentration of dissolved toxic gases such as CO_2, and especially H_2S and NH_3. The temporal dynamics of distributions with regard to habitats are examined below because it is obvious that these are the result of several factors, some of which are interacting, for example, the influence of temperature on the DO and other dissolved gases, on the toxicity of NH_3 and H_2S, on the innate abilities to compensate and on the speed of adaptation to depth/pressure; in *S. mossambicus* the maximum adaptation depth is 20 m at 30°C but only 7 m at 15°C (Caulton and Hill 1975).

Whether a tilapia is rheophilic (current loving), limnophilic, or indifferent to current velocity, may be readily appreciated from the results of introductions (deliberate or not) of typically riverine species into lakes, of typically lacustrine species into rivers, and from the evolution of tilapia populations after the transformation of habitats with a fast flow to ones with a slow flow (e.g. man-made lakes). This transformation, however, affects not only current velocity but also the depth *per se*, the thermal regime, various physical and chemical factors and the structure of the whole ecosystem (see Balon 1974, for Lake Kariba on the Zambezi).

Table 4. Data on depth distribution of tilapias.

S. variabilis and *S. esculentus*	maximum 35 to 40 m in Lake Victoria from bottom trawling (Bergstrand and Cordone 1970, in Fryer and Iles 1972).
S. variabilis	present to 13 m in Lake Victoria (Gee 1968, in Fryer and Iles 1972).
S. esculentus	present to 30 m but most abundant at less than 13 m in Lake Victoria (Gee 1968, in Fryer and Iles 1972).
S. tanganicae	common in Lake Tanganyika but rarely captured at less than 10 m (Poll 1956).
S. mossambicus	adults absent in less than 12 m but juveniles (max. 15 m) and alevins capable of descending to a greater depth in Lake Sibaya, S. Africa (Bruton and Boltt 1975).
S. macrochir	observed by diving to 12-14 m in Lake McIlwaine, Zimbabwe and other lakes in that area (Caulton, pers. comm.).
S. niloticus, S. galilaeus and *T. zillii*	captured in gillnets to 7 m in Lake Kainji, Nigeria (Ita 1978).
S. multifasciatus, T. discolor and *T. busumana*	maximum 30 m, 10 m and 7 m respectively in Lake Bosumtwi, Ghana (Whyte 1975).
T. sparrmanii	observed in diving to 30 m in Lake Sibaya, S. Africa (Bruton and Boltt 1975) (but maximum depth compensation of 15 m at 22°C as experimentally determined by Caulton (1975b).
T. rendalli	maximum 7.5 m in Lake Kariba (in Caulton 1975a, 1975b)

TEMPERATURE

Tilapias are thermophilic fishes, and their geographical distributions are closely determined by temperature, particularly by low temperatures. Thus a natural population will be able to maintain itself if:

a. there is, during part of the year, a temperature high enough to allow for reproduction and for sufficient growth;
b. the temperature at no time drops below values that are lethal for all individuals.

Figure 4 summarizes the main published data on the temperature ranges to which tilapias are exposed in nature and their thermal tolerance limits. It should be noted, however, that the latter are influenced by acclimation (thermal history) and were not determined under uniform conditions (see also Chervinski, this volume).

The most northerly natural occurrence of tilapias is in Lake Huleh, Israel, in which the *T. zillii* population has to withstand temperatures of 6 to 7°C during the coldest nights (Kirk 1972). Observations reported by Hauser (1977) in California suggest that *T. zillii* can survive for two weeks at 13°C, but some began to die at 11.2°C. Their lower tolerance limit was 6.5°C, which corresponds well to the naturally occurring minimum in Lake Huleh. Hauser (1977) also reports a minimum temperature for reproduction of 20°C in *T. zillii. S. aureus, S. niloticus* (introduced) and *S. galilaeus*, which occur with *T. zillii* in Lake Huleh and Lake Kinneret, Israel, are almost as resistant to low temperatures as this latter species. In ponds *S. niloticus* and *S. galilaeus* tolerate winter temperatures of 8°C for several hours each night (Yashouv 1958), while *S. aureus* died at 8-9°C during experiments (Sarig 1969). It may be mentioned, however, that Chervinski and Lahav (1976) reported an improved low temperature resistance of the local Israeli *S. aureus* over that of *S. niloticus* imported from Africa.

The southernmost natural occurrences of tilapia include *S. mossambicus* in coastal waters near Port Elizabeth (33°42′S). This has been attributed by Jubb (1967, in Beadle 1974) to the fact that the southeastern coast of Africa is swept by a warm current (the Agulhas) which maintains subtropical conditions at a latitude at which tilapias would not normally occur (cf. their absence at the same latitude on the west coast). In Africa *T. sparrmanii* is the species which occurs at the lowest latitude naturally, tolerating winter temperatures of 7°C and needing a minimum of 16°C for reproduction (Chimits 1957). As a result of introductions, *T. sparrmanii* and *S. aureus* have been established in the Cape region (Caulton pers. comm.).

Table 5 lists the tilapias which occur naturally, or after having been introduced, in lakes of Africa and Madagascar at altitudes higher than 1,000 m. In natural environments 2,000 m seems to represent the limiting altitude. The species which tolerate best the climates occurring at 1,500 to 2,000 m are *S. niloticus, S. leucostictus* and *T. zillii.* In ponds *T. zillii* has been reported from an altitude of 2,500 m in Uganda (Chimits 1957), *T. rendalli* from up to 1,400 to 1,500 m in Mexico and Colombia (FAO 1977c) and *S. mossambicus* from up to 2,400 m in Ecuador (FAO 1977c). For similar altitudes rivers tend to have minimum temperatures during the cold season that are lower than those in lakes, which explains why few tilapias occur in rivers above an altitude of 1,300 m. Studies by Gaigher (1973) have shown that *T. sparrmanii* occurs in rivers above 1,220 m in the upper reaches of the rivers of the Limpopo basin, but that *T. rendalli* and *S. mossambicus* are limited to warmer rivers, below 1,220 m.

In addition to their temperature tolerance, the distributions of tilapias can be related to their maximum swimming performance, which occurs at 24°C in *T. sparrmanii*, 28°C in *T. zillii* and *S. macrochir*, 28 to 32°C in *S. niloticus* and 32°C in *S. mossambicus* and *S. galilaeus* (Fukusho 1968).

A few tilapias have colonized hot springs. *S. spilurus percivali* occurs in the hot springs of the northern Uaso Nyiro system where the temperature reaches 42.4°C (Trewavas 1966). *S. alcalicus grahami* is endemic in Lake Magadi where it can live close to the hot springs (35 to 40°C), tolerating short periods at 40 to 44°C (Coe 1966, 1967; Reite et al. 1974), and *S. shiranus chilwae*, endemic to Lake Chilwa (Malawi) can tolerate 40 to 42°C for short periods (Morgan 1972). It may be mentioned for comparison that North American species such as *Notropis lutrensis* and *Barbus callensis* occur in hot springs of 40°C (Castenholz and Wickstrom 1975) and that species of desert pupfish (Cyprinodontidae) may be found at temperatures of 42°C (Brown 1971).

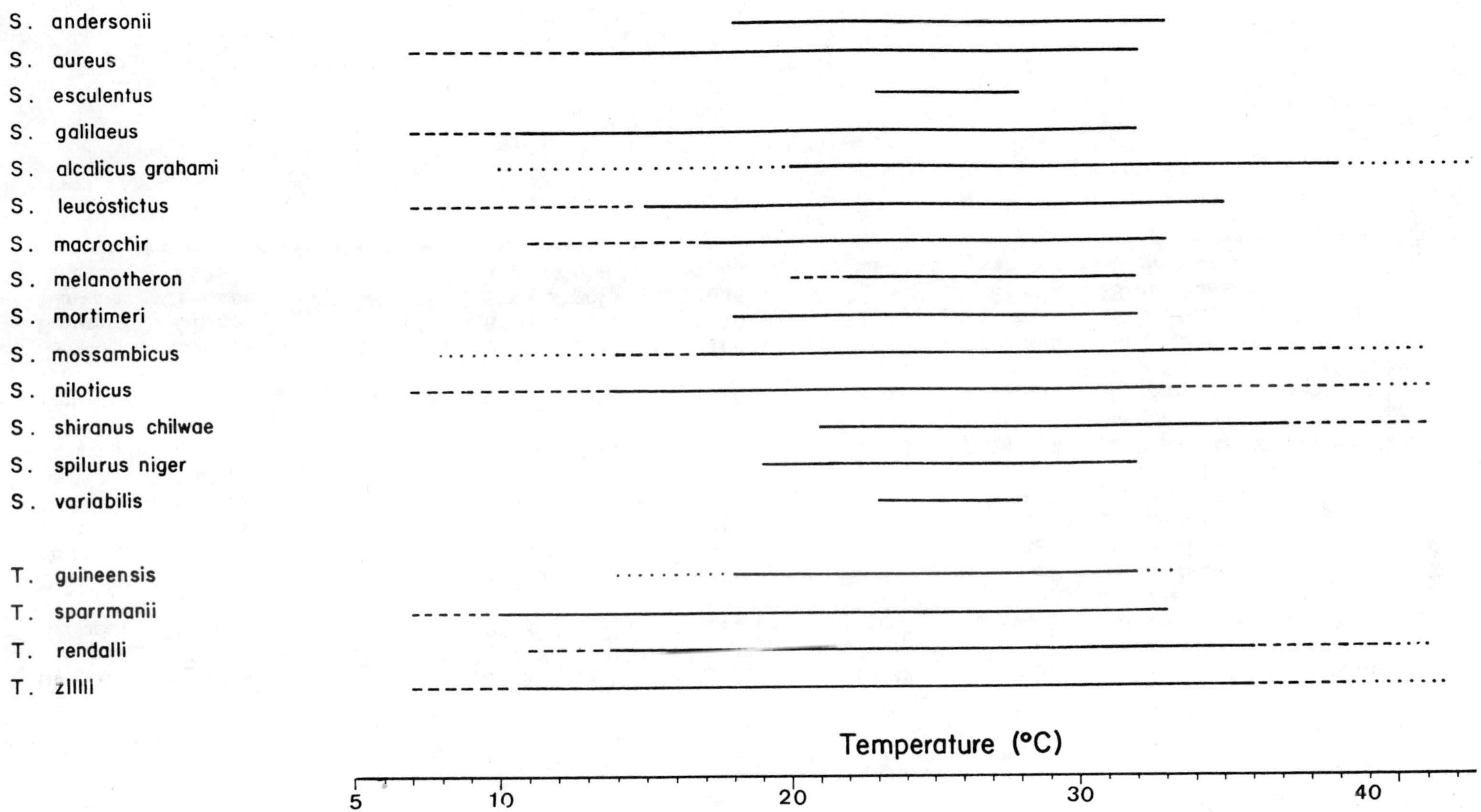

Figure 4. Thermal tolerance ranges of tilapias (*Sarotherodon* and *Tilapia*). —— normal range of temperature variation in natural habitats (after Welcomme 1972); – – – extreme temperatures tolerated in some natural habitats and in culture ponds (after Balarin and Hatton 1979); · · · · · physiological limits of tolerance as shown by laboratory experiments (after Balarin and Hatton 1979).

Table 5. Tilapias indigenous or introduced with success in high altitude lakes in Africa (C_{20} = conductivity at 20°C in μS/cm; * signifies an introduction which failed).

Lake	Country	Latitude	Altitude (m)	Temperature (°C)	C_{20}	
Bunyoni	Uganda	1°16′ S	1973	21-22	99-262	*S. niloticus* (S. spilurus niger*, S. esculentus*)*
Naivasha	Kenya	0°46′ S	1890	—	318-400	*S. leucostictus*, T. zillii* (S. spilurus niger*)*
Tana	Ethiopia	12°00′ N	1829	—	—	*S. niloticus*
Mutanda	Uganda	1°12′ S	1790	—	200-230	*S. niloticus**
Kivu	Zaïre-Rwanda	1°30′-2°30′ S	1463	24-25	1240-4000	*S. niloticus*
Mohasi	Rwanda		1450	—	—	*S. niloticus**
Itasy	Madagascar		1200	18-29	—	*T. rendalli*, S. niloticus**
Bangweulu	Zambia	10°15′-12°304 N	1160	18-26	14-152	*S. macrochir, T. rendalli, T. sparrmanii*
Victoria	Kenya, Uganda, Tanzania	0°20′N-3°0′ S	1136	23-28	91-98	*S. esculentus, S. variabilis, S. niloticus*, S. leucostictus*, T. zillii**
Kyoga	Uganda	0°36′-2°0′ N	1100	28	245-365	idem Lake Victoria
Bosumtwi	Ghana	6°30′ N	1070	27-28	—	*S. multifasciata, T. discolor*
Manyara	Tanzania	3°25′-3°90′ S	1045	—	94	*S. amphimelas*
Upemba	Zaïre	8°25′-9°0′ S	1000	23-33	145-255	*S. niloticus, T. rendalli, T. sparrmanii*

Experimental studies, aquacultural and field observations (Balarin and Hatton 1979) demonstrate that many tilapias which are not especially adapted to warm waters can tolerate temporary temperatures of 35 to 42°C. For *T. zillii*, mortality begins when the temperature exceeds 39.5°C and the upper lethal temperature is 42.5°C (Hauser 1977). For *T. rendalli*, there is a sublethal zone of tolerance between 37 and 40°C (Caulton 1977) and an upper tolerance limit of 40.6 to 41.9°C after adaptation at 24°C (Morgan 1972). For *S. niloticus*, there is an upper tolerance limit of 42°C (Denzer 1968 cited by Aston and Brown 1978). *S. niloticus* also tolerates 36°C for 6 hours (with peaks of 40°C during brief periods) in industrial cooling waters in Belgium (Mélard and Philippart 1980). Lowe (McConnell) (1958) reports the capture of an *S. niloticus* individual from a hot spring of the Turkwel River, Kenya (40°C).

On the basis of their thermal preferences and their tolerances (both determined experimentally and from their geographical distributions), three categories of tilapias emerge:

1. Eurythermal species which tolerate a wide range of temperatures—*T. zillii* (6.5 to 42.5°C), *S. mossambicus* and *S. niloticus* (8 to 42°C), and also probably *S. aureus* and *S. galilaeus*.
2. Species that are eurythermal but seem less tolerant than 1. to high temperatures—*T. sparrmanii* (7°C to ?)—or to low temperatures, *S. spilurus niger* (8 to 10°C to ?), *T. rendalli* (11 to 41°C), *S. macrochir* (11°C to ?) and *S. alcalicus grahami* (10 to 41°C).
3. More stenothermal species such as *T. guineensis* (14 to 33°C, Mélard, pers. comm.) and *S. melanotheron* which lives in a temperature range of 18 to 33°C in West African lagoons (Fagade 1974; Pauly 1975, 1976) and is reported to die at 20°C in aquaria (Sterba 1967, in Balarin and Hatton 1979).

Within a species, the temperature preferences and tolerances depend on size, the young fish generally being more tolerant to higher, and often to lower temperatures than the adults. Thus Bruton and Boltt (1975) reported 16.5 to 39°C as the tolerated range, and 19 to 35°C as preferred range for fingerlings (less than 15 cm) compared to 19 to 32 and 22 to 30°C respectively, for adult *S. mossambicus* in Lake Sibaya, South Africa. These intraspecific variations in thermal tolerance have a physiological basis (see Whitfield and Blaber 1976 for *T. rendalli*) and play an important role in the spatial and dynamic aspects of the fish's distribution.

Because of their adaptations to specific environmental conditions, different populations of the same species may display different characteristics with regard to temperature which can be used when selecting strains for aquaculture or introductions.

Finally, the relationships between the thermal resistance of tilapias and other ecological factors, especially salinity, must be discussed. Thus *S. mossambicus* (Allanson et al. 1971) and *S. aureus* (Chervinski and Lahav 1976) tolerate low temperatures in saline water (5% seawater) better than in freshwater. This phenomenon helps to explain how the limit of the range of *S. mossambicus* can be so far south (to the estuary of the Kongie River, Port Alfred, South Africa) where, in spite of the influence of the sea, the winter temperature drops to 12°C (Allanson et al. 1971).

SALINITY

Several tilapias are euryhaline, able to live and reproduce at salinities higher than 30‰. These occur naturally in estuaries and coastal lagoons along the coast of west Africa *(T. guineensis, S. melanotheron)* or along the east coast of Africa (*S. mossambicus* and the related *S. hornorum* and *S. placidus*). Tilapias also occur in hypersaline habitats such as the Bay of Suez (42 to 43‰) in the Red Sea where *T. zillii* is now established (Bayoumi 1969, in Balarin and Hatton 1979), and in the Bardawil Lagoon in Israel (41 to 45‰) where *T. zillii* and *S. aureus* occur (Chervinski and Yashouv 1971; Chervinski and Hering 1973). The high salinity of certain lakes is also tolerated: Lake Magadi and Lake Natron (salinity 30 to 40‰) have endemic species *S. alcalicus grahami* and *S.a. alcalicus* (Coe 1966, 1967), and Lake Manyara, Tanzania (58‰), the endemic species *S. amphimelas*.

Several species, though less euryhaline, are capable of maintaining populations in habitats where the salinity reaches 30‰: *S. niloticus* and *S. galilaeus* in the Bitter Lakes of Egypt (13 to 29‰) (Kirk 1972), *S. niloticus, S. aureus* and *T. zillii* in Lake Qarun, Egypt (11 to 29‰) (Fryer and Iles 1972) and *S. shiranus chilwae* in Lake Chilwa (12 to 29‰). The last species, however, leaves the lake when the salinity becomes too high: conductivity at 20°C (C_{20}) = 5,000 μS/cm (Morgan and Kalk 1970, in Beadle 1974).

Other species are less euryhaline and do not tolerate salinities above 20‰, such as *T. sparrmanii, S. andersonii, S. macrochir* and *T. rendalli.* These species occur essentially in freshwater, although some populations prosper in saline habitats, notably *S. macrochir* which (as the result of an introduction) occurs in the Nemakia mangrove swamp in Madagascar, in places where the conductivity reaches 350 to 10,000 μS/cm (Lamarque et al. 1975), as well as in Lake Mweru Wantipa, Zambia (a natural population) where the salinity sometimes reaches 7‰ and the electrical conductivity 60,000 μS/cm (Fryer and Iles 1972). However, *S. macrochir* occurs in the lake during only part of the year, when the salinity is lower than the extreme value given above. During high salinity periods, the populations take refuge in the rivers flowing into the lakes (Fryer and Iles 1972, in Beadle 1974).

A similar phenomenon may be observed in Lake Chilwa, Malawi, where salinity fluctuations occur (0.3 to 16.7‰ for the 1966 to 1970 period according to Morgan and Kalk 1970, cited by Beadle 1974). *S. shiranus chilwae* leaves the lake and takes refuge in the rivers and coastal pools when the salinity increases above 5‰ (C_{20} = 5,000 μS/cm), at salinities which were unfortunately not recorded precisely.

Also, it is well established that *T. rendalli* has colonized certain brackish habitats, Lake Poelela, Mozambique among others, where the salinity reaches 8‰ (Whitfield and Blaber 1976). These authors have demonstrated experimentally that *T. rendalli* (isosmotic at 10‰), can tolerate a maximum of 19‰, the salinity tolerance being maximal at 20 to 28°C. Temperature and salinity are considered to be the determining factors for the distribution of this species. It is interesting, in this context, that *T. rendalli* has been introduced successfully into rivers and reservoirs south of the Pongolo River, South Africa ($>$ 29°S), which marks the southern limit of the natural distribution of this species. Whitfield and Blaber (1976) suggested that it is because of its inability to tolerate salinities above 19‰ that *T. rendalli* never succeeded in colonizing, by way of estuaries, coastal lagoons and the sea, the more southern rivers where the temperature regime is still sufficient for its survival and reproduction. In this *T. rendalli* differs from *S. mossambicus* (Jubb and Jubb 1967, in Beadle 1974).

Other extreme situations with regard to salinity are represented by the rivers and lakes containing very low mineral contents in which tilapias occur: *S. macrochir*, *T. sparrmanii* and *T. rendalli* in Lake Mweru (salinity, 0.023‰; C_{20} = 14 to 35 μS/cm; Beadle 1974), *S. variabilis*, *S. esculentus*, *T. zillii* and *S. niloticus* (the latter two species introduced) in Lake Nabugabo near Lake Victoria, Uganda (salinity, 0.015‰; C_{20} = 25 μS/cm) and *T. congica* in Lake Tumba, Zaire (salinity = 0.016‰; C_{20}= 24 to 32 μS/cm) Dubois (1959).

Several species can adapt to a wide range of salinities: *T. zillii* (0.16 to 44‰), *S. mossambicus* (0 to 120‰) (Whitfield and Blaber 1979), and *S. amphimelas*, endemic in the hypersaline Lake Manyara (salinity 58‰; C_{20} = 94,000 μS/cm) and in Lake Kitangiri, Tanzania where a normal C_{20} = 185 μS/cm (Fryer and Iles 1972; Beadle 1974).

The different ranges of tolerance with regard to salinity of different tilapias often show when natural or man-made changes occur in certain habitats. Thus the construction of a dam across coastal rivers has isolated freshwater populations from species which normally prefer brackishwaters: *T. guineensis* in Lake Ayeme on the Bia River (Ghana/Ivory Coast border), and in Lake Mount Coffee on the St. Paul River (Liberia) (Thys 1971a).

The gradual increase of salinity in Lake Qarun (near Cairo, Egypt) (11‰ in 1920 and 22‰ in 1932) had led to the gradual disappearance of species that were abundant *(S. niloticus, S. aureus)* and their replacement by *T. zillii* which is much more euryhaline (Fryer and Iles 1972).

In the closed lagoons, lakes and coastal lagoons along the southeast coast of Africa, the irregular connections with the sea lead to a sequence of high and low salinities (Whitfield and Blaber 1979). The populations of *S. mossambicus* occurring in these habitats have thus to tolerate very wide ranges of salinity (0 to 120‰ in St Lucia Lake), the gradual character of which, however, allows the populations to adapt. This adaptation may be accompanied by a reduction of the biomass (*S. mossambicus* formed 12.3% of the catch when the salinity was 10‰ but only 1.2% when the salinity was 80%) or by a migration towards the upper parts of the estuaries where the salinity remains lower (Whitfield and Blaber 1979).

Studies by these authors demonstrate the absence of *S. mossambicus* from most of the estuaries that are permanently open to the sea and which are characterized by rapid variations in salinity because of the tides. *S. mossambicus* thus seems incapable of tolerating rapid changes of salinity, but tolerates seawater and/or slow changes in salinity very well. However, the absence of *S. mossambicus* from the estuarine systems studied by Whitfield and Blaber (1979) may not be caused by rapidly changing salinities, but could be linked to other negative factors also operating in these systems: rapid currents, bad conditions for nest construction, competition with marine fishes, predation by piscivorous fishes, etc.

pH AND ALKALINITY

S. alcalicus grahami in Lake Magadi tolerates a pH of 10.5 and an alkalinity of 80 g $CaCO_3$/l (Coe 1969; Reite et al. 1973). *S.a. alcalicus* of Lake Natron tolerates alkalinities that are even higher: up to 161 g $CaCO_3$/l (Morgan 1972). Reite et al. (1974) have shown experimentally that *S.a. grahami* can withstand a pH range of 5 to 11 for at least 24 h, but dies after 2 to 6 hours at pH's < 3.5 and > 12. Morgan (1972) reported mortalities at alkalinities of 1.1 to 6.2 g $CaCO_3$/l (at a pH of 9.0 to 9.9 and 22.5°C) in *S. shiranus chilwae*, and at 3.5 to 4.9 g $CaCO_3$/l in *T. rendalli*. In the Sudan, *S. niloticus* tolerates pH's of 8 to 11 in ponds (George 1975). In Lake Tumba (Zaïre), where *T. congica* lives, the pH is as low as 4.5 to 5.0 (Dubois 1959).

OXYGEN AND OTHER DISSOLVED GASES

The available observations in nature and in ponds, cages and tanks suggest that tilapias are very resistant to low levels of dissolved oxygen (DO) (see Balarin and Hatton 1979). DO's as low as 0.1 ppm are tolerated by *S. mossambicus* (Maruyama 1958) and *S. niloticus* (Magid and Babiker 1975; Mélard and Philippart 1980). This enables some tilapias to live and reproduce in swamps and shallow lakes where strong deoxygenations occur from time to time which are fatal to other fishes. The physiological basis for this tolerance to low oxygen may be one of several types: (1) using haemoglobins which bind oxygen at very low tensions, 0.1 ppm in *S. macrochir* (Dussart 1963) and 0.17 ppm at 24°C in *S. esculentus* (Fish 1956); (2) breathing just below the water surface, flushing the gills with oxygen-rich water (Dussart 1963) and (3) withstanding anaerobiosis (Kutty 1972; Magid and Babiker 1975).

Tilapias can withstand not only low DO's, but also very high levels of CO_2, with maximum tolerance levels of 72.6 ppm in *S. macrochir* (Dussart 1963), and 50 ppm in *S. esculentus* (Fish 1956). Other gases (NH_3, H_2S) which originate from the decomposition of organic matter are also well tolerated. A recent experimental study by Redner and Stickney (1979) demonstrated that *S. aureus* can tolerate 2.4 ppm of un-ionized NH_3 (LD_{50}, 48 hr).

Massive fish kills involving tilapias do occur, however, in deep lakes in cases where the seasonal turnover of water brings the deoxygenated deep water to the surface, and in shallow lakes where violent storms mix the well-oxygenated surface waters with deoxygenated bottom waters, and stir up anoxic and toxic bottom mud (containing H_2S, NH_3 and CO_2). Such phenomena have affected the *T. rendalli* populations in Lake Chilwa (Morgan 1972) in the Nampongue River, Zambia (Tait 1965), and have been reported from Lake George, Uganda (with mortalities of 1.3 million fish, predominantly *S. niloticus*, in a few hours in 1957, Beadle 1974) and also from Tanganyika and Lake Victoria, near Entebbe in July, 1963 (Fryer and Iles 1972).

Finally it should be mentioned that life in lakes characterized by high phytoplankton production presumes a marked tolerance of conditions of oxygen supersaturation (up to 400% in Lake Chilwa, Morgan 1972). Supersaturation of dissolved gases (oxygen and also nitrogen) often poses problems in the intensive rearing of tilapias in industrial heated effluents (Mélard and Philippart 1980).

OTHER FACTORS

Tilapias tolerate high turbidities and are rather resistant to pollution by toxic substances, whether organic or inorganic, natural or artificial (e.g., pesticides, see Balarin and Hatton 1979).

Feeding Behavior

FOOD COMPONENTS

Tilapias of the genus *Tilapia*, especially *T. rendalli*, *T. zillii*, *T. sparrmanii* and *T. tholloni* are macrophyte-feeders in which the adults feed preferentially on filamentous algae, aquatic macrophytes and vegetable matter of terrestrial origin (leaves, plants, etc.); but this specialization does not exclude certain stages of development (alevins) at certain times of year (winter), and in certain waters poor in aquatic vegetation, taking animal food (here including the alevins) as shown by Spataru (1978) for *T. zillii* in Lake Kinneret (Israel). In this lake, *T. zillii* also eats blue-green algae.

In the genus *Sarotherodon*, certain species (often endemic lacustrine species) seem very specialized feeders, notably *S. variabilis* of Lake Victoria (fine benthic sediments, Fryer and Iles 1972), *S. alcalicus grahami* of Lake Magadi (epilithic blue-green algae), *S. esculentus* (phytoplankton) and *S. macrochir* (phytoplankton and epilithic algae). But many species have a much more diversified feeding regime with a dominant vegetable component (epilithic, epiphytic and filamentous algae, phytoplankton, vegetable debris and fine sediments rich in diatoms and bacteria) and an animal component (zooplankton and benthic organisms such as insect larvae, crustaceans, molluscs). *S. aureus* (Spataru and Zorn 1978), *S. galilaeus* (Lauzanne and Iltis 1975) and *S. niloticus* (Moriarty 1973) eat blue-green algae. Moriarty and Moriarty (1973a) have demonstrated that *S. niloticus* can assimilate 70 to 80% of the carbon ingested in this form *(Microcystis, Anabaena* and *Nitzchia)*. *S. aureus* (Spataru and Zorn 1978), *S. shiranus chilwae* (Bourn 1974) and *S. mossambicus* (Bruton and Boltt 1975; Man and Hodgkiss 1977b) are equally able to eat vegetable debris and macrophytes. In certain waters *S. mossambicus* even catches terrestrial insects floating on the water surface and fish (Lake Sibaya, Bruton and Boltt 1975).

S. andersonii, *S. aureus*, *S. mossambicus* and *S. niloticus* appear to be omnivorous compared to *S. galilaeus* and *S. macrochir* which have a much more limited food spectrum. Studies made in Lake Kinneret, Israel (Spataru 1976; Spataru and Zorn 1978) show that feeding is much more selective in *S. galilaeus* (Pyrrophytes, *Peridinium* sp.) than in *S. aureus* (zooplankton, while this is abundant; phytoplankton and vegetable debris, when the zooplankton is less abundant).

In general a comparison of feeding in the same species of tilapia in a large range of water bodies reveals a very great variability of feeding regime. This is an element of the remarkable plasticity and ecological adaptability of tilapias. The general qualitative characteristics of the feeding regime depend on the following: 1) the type of organisms present, which depends on the limnological, physicochemical characteristics of the water body; 2) the accessibility of the food organisms according to their localization (for example, in certain lakes or reservoirs, some abundant types of food are situated too deep and are inaccessible to tilapias which are not able to descend to these depths (see Caulton and Hill 1973 for *S. mossambicus*) and 3) tne presence of competing species (tilapias or others) which forces each species to restrict its food spectrum and to exploit its specializations, e.g., in many tropical waters where competition for food is intense, *T. zillii* is strictly a macrophyte-feeder, whereas it eats plankton and benthos in Lake Kinneret, Israel (Sparatu 1978) and in Lake Qarun, Egypt (Alkholy and Abdel Malek 1972) where the competition is less.

VARIATIONS IN FEEDING REGIME WITHIN POPULATIONS

Within a given water body, the feeding regime of a species is extremely variable, depending on size and age, the microhabitats occupied by the fish and the time of year.

The alevins (both of species which are strictly phytoplankton-feeders or macrophyte-feeders when adult) generally have a diversified feeding regime extracting small organic particles from the sediments, phytoplankton, diatoms, periphyton, zooplankton and benthic organisms (Bruton and Boltt 1975 for *S. mossambicus;* Gophen 1980 for *S. galilaeus* and Whyte 1975 for the Lake Bosumtwi tilapias).

In Lake Sibaya, S. Africa (Bruton and Boltt 1975), the *S. mossambicus* adults captured in the marginal vegetation zones feed on diatoms, vegetable debris and mud, but those fish captured in open water (limnetic zone) feed on aerial insects (Coleoptera and Hemiptera). In Lake Kinneret, Israel (Spataru 1978), *T. zillii* captures prey from the surface in open water and rarely from the bottom except when breeding and guarding the young (April-May-June) when they feed equally on benthic organisms (chironomid larvae, ostracods, nematodes and sponge gemmules).

The qualitative seasonal variations of feeding regime depend partly on the annual cycle of production and availability of prey, and partly on the degree of feeding selectivity of the species and their distribution in different habitats at different times of year.

Seasonal changes in feeding have been studied in *T. zillii*, *S. galilaeus* and *S. aureus* in Lake Kinneret (Spataru 1976, 1978; Spataru and Zorn 1978), and in *S. mossambicus* in Plover Cove Reservoir, Hong Kong (Man and Hodgkiss 1977b).

In spring, *S. aureus* in Lake Kinneret feeds intensively and more or less selectively on zooplankton which is very abundant. Beginning midsummer, the zooplankton production slows down and its availability diminishes due to competition from other species. *S. aureus* then feeds more on benthic debris and phytoplankton, which seems on the whole to be ingested passively rather than sought actively. *S. galilaeus* feeds much more selectively (phytoplankton: Pyrrophytes, *Peridinium* sp.) and its feeding regime has a less great seasonal variation, in accordance with the availability of its food. *T. zillii* feeds mainly on Coleoptera and chironomid pupae plus green, brown and above all red algae (*Peridinium* sp.) in winter and spring and on zooplankton (Cladocera) in summer and autumn (Spataru 1978). While breeding, *T. zillii* modifies its feeding habits and captures benthic prey normally not eaten at other times of year.

In Plover Cove Reservoir, Hong Kong, *S. mossambicus* feeds almost exclusively on mud and vegetable debris (more than 80% by volume) throughout the year, without appearing to have any seasonal cycle.

FEEDING PERIODICITY

According to Man and Hodgkiss (1977b), in Plover Cove Reservoir, Hong Kong, *S. mossambicus* feeds during the day. The feeding intensity (measured by an index of stomach fullness) is maximal between 12:00 P.M. and 3:00 P.M. and then slows progressively so that stomachs are completely empty between 12:00 A.M. and 3:00 A.M. Diurnal feeding has also been observed in *S. shiranus chilwae* (Bourn 1974), *S. alcalicus grahami* (Coe 1966, 1967), *S. niloticus* in Lake George (Moriarty 1973) and *T. busumana* in Lake Bosumtwi (Whyte 1975), but several other species are nocturnal feeders, notably *T. discolor* and *S. multifasciatus* of Lake Bosumtwi (Whyte 1975).

Tilapia feeding activity varies seasonally according to various factors: temperature, reproduction, interspecific competition. In Plover Cove Reservoir, Hong Kong, the seasonal cycle of feeding activity follows the temperature cycle: minimum activity in January-February (16 to 17°C) and maximum in July-September (27 to 30°C). In Lake Kinneret, Israel (Spataru and Zorn 1978), the feeding activity of *S. aureus* (measured by index of stomach fullness) is maximal in summer and autumn (temperature maximal, 30°C) whereas that of *T. zillii* is maximal in spring and relatively constant at other seasons (Spataru 1978).

Feeding activity is not interrupted while guarding the young in substrate-spawners (for example *T. zillii*, Spataru 1978) but normally stops in mouthbrooding females, though not always, because food has been found in the stomach of a mouthbrooding female of *S. alcalicus grahami* (Coe 1966, in Fryer and Iles 1972, p. 124).

Reproductive Behavior

PUBLISHED INFORMATION

Numerous authors have studied tilapia breeding behavior in the laboratory, in culture systems and in the field. The following notable references are not an exhaustive list:

S. esculentus (Cridland 1961)

S. galilaeus (Fishelson and Heinrich 1963; Iles and Holden 1969)
S. karomo (Lowe (McConnell) 1956a)
S. leucostictus (Lowe (McConnell) 1957; Welcomme 1967a; Hyder 1970a, 1970b; Siddiqui 1977a, 1977b)
S. macrochir (De Bont 1950; Ruwet 1962, 1963b; Carey 1965; Marshall 1979a, 1979b; Voss and Ruwet 1966; Ruwet and Voss 1966)
S. melanotheron (Aronson 1949; Oppenheimer and Barlow 1968; Pauly 1976)
S. mossambicus (Baerends and Baerends-van Roon 1950; Neil 1966; Bruton and Boltt 1975; Russock and Schein 1977; Baerends and Blokzijul 1963)
S. niloticus (Lowe (McConnell) 1958; El-Zarka et al. 1970a; Babiker and Ibrahim 1979)
S. variabilis (Lowe (McConnell) 1956a; Fryer 1961a)
T. guineensis (Voss and Ruwet 1966; Voss 1969)
T. rendalli (De Bont 1950; Ruwet 1962, 1963a; Donnelly 1969; Kenmuir 1973; Monfort and Ruwet 1968)
T. sparrmanii (Voss 1972a, 1972b)
T. tholloni (Monfort-Braham and Voss 1969)
T. zillii (Daget 1952; El Zarka 1956; Voss 1969; Loiselle 1977; Siddiqui 1979a)

Papers concerning tilapias endemic to the African Great Lakes are synthesized in Fryer and Iles (1972) and more recently Balarin and Hatton (1979) have collated information in the literature up to 1976.

Periodicity of Reproduction

In order to breed, most tilapias need a water temperature of at least 20°C (Huet 1970; Bardach et al. 1972; Balarin and Hatton 1979), but certain species are able to reproduce at a much lower temperature (for example *T. sparrmanii* at 16°C, Chimits 1957). Other factors also seem to play a releasing role, notably photoperiodicity and light intensity (see Balarin and Hatton 1979, p. 29) as well as the rainy season, via water temperature (temperature linked with rains), water level (accessibility of certain spawning grounds) or other mechanisms. Table 6 presents some data on the periodicity of tilapia reproduction at different latitudes and altitudes.

In equatorial and tropical waters where the temperature is high throughout the year, one observes numerous cases of continuous reproduction (e.g., *S. leucostictus* and *T. zillii* in Lake Naivasha) but the intensity often varies seasonally (e.g., *S. variabilis* in Lake Victoria). The same occurs in waters relatively constant from the point of view of photoperiodicity and temperature, where certain species have a well-defined breeding season generally associated with the rainy season (e.g., *S. esculentus* of Lake Victoria and *S. squamipinnis* of Lake Malawi) or the hot season which precedes it (*S. saka* of Lake Malawi). In this case, the length of the breeding season and the number of spawnings per female are very variable. *S. lidole*, *S. saka* and *S. squamipinnis* of Lake Malawi represent an extreme case as they produce only one spawning a year (but see also Berns et al. 1978).

In certain tropical and subtropical regions, low temperature inhibits reproduction during part of the year, the length of time depending on the latitude and altitude of the place: the duration of the breeding season then conditions the number of spawnings a year, and consequently the overall fecundity of the populations. The breeding seasons coincide with the hottest times of year (*T. zillii* of Lake Qarun, Egypt, *S. mossambicus* of Lake Sibaya) sometimes just before the rainy season (*S. mortimeri* in the Zambezi River and *S. macrochir* and *T. rendalli* in Shaba).

Table 6. Data on the periodicity of reproduction of some tilapia species in natural waters at different latitudes and altitudes.

Coastal Lake Sibaya (S. Africa) 27°25′S

The breeding season of *S. mossambicus* starts in September (20-26°C) and is prolonged until March (duration: 7 months) but the maximum activity is concentrated in September-December; a complete cycle of reproduction lasts about 7 weeks (20 to 22 days for incubation of embryos and guarding the young and about 14 days for the maturation of ovules); the same female should therefore theoretically reproduce 3 to 4 times per year (Bruton and Boltt 1975).

The man-made Lake McIlwaine (Zimbabwe) 18°S

The breeding season of *S. macrochir* (introduced) lasts principally from September to December, that is to say during the hottest months of the year, just before the rainy season; but a rise in lake level later in the year (January-March) is able to stimulate a second phase of reproduction (Marshall 1979b).

Middle Zambezi 15-18°S

The short breeding season (one or more spawnings) of *S. mortimeri* coincides with the rainy season (Fryer and Iles 1972).

The man-made Lake Lufira (Shaba, Zaïre) 11°S, 1,100 m

S. macrochir and *T. rendalli* reproduce throughout the year but there is a very slight slowing down of reproduction during the cold dry season (May-July) and an intense activity in the rainy season (Ruwet 1962); observations made in fish culture in the same region (De Bont 1950), showed that the duration of the rainy season determines the number of spawnings (3 to 4 per year at intervals of 7 weeks in *T. rendalli*).

Lake Malawi 9°30′-14°40′S, 471 m

S. saka, *S. squamipinnis* and *S. lidole* produce only one spawning a year; *S. saka* breeds during the hot season (August-November) before the rainy season, *S. squamipinnis* during the rainy season (December-April) and *S. lidole* in October-November, just before the rainy season (Lowe 1952; Fryer and Iles 1972). Apart from the rains, the temperature conditions and photoperiodicity vary relatively little through the year.

Lake Naivasha (Kenya) 0°46′S, 1,890 m

The photoperiod is practically constant (12 h) and the temperature favorable throughout the year. Reproduction of *S. leucostictus* (introduced) is continuous (nonseasonal) and one cannot observe any correlation between its intensity and the seasonal precipitations (Siddiqui 1977b). Previously another author (Hyder 1970a) had concluded that a seasonal reproduction was correlated with certain conditions of temperature and light intensity. The observations of Siddiqui (1979a) indicated that *T. zillii* (introduced) reproduced equally throughout the year, with little seasonal variation in the intensity of reproduction.

Table 6 (cont'd)

Lake Victoria 03°00′S-0°20′N, 1,136 m

Fryer (1961a) indicated that the reproduction of *S. variabilis* is continuous but there are seasonal variations in intensity which are not clearly correlated with climatic conditions (rains and temperature); there are at least three spawnings in a period of eight months. In *S. esculentus*, there exists a correlation between the rainy season and the periodicity of reproduction (Garrod 1959, cited by Fryer and Iles 1972); Welcomme (1967b) reported two reproductions or more a year.

Lake Bosumtwi (Ghana) 06°30′N, 1,070 m

The endemic tilapias of this lake breed throughout the year but the reproductive activity is maximal at a determined period of year: October-February *(T. busumana)*, November-April *(S. multifasciatus)*, December-May *(T. discolor)* (Whyte 1975).

Lake Moyua (Nicaragua) 12°35′N, 420 m

S. mossambicus (introduced) reproduces during 5-6 months corresponding to the rainy season (Reidel 1965).

Plover Cove Reservoir (Hong Kong)

S. mossambicus (introduced) is here at the extreme northern limit of its range for reason of temperature conditions (less than 20°C and practically without growth during 3 to 4 winter months); the breeding season extends from May to October and the same female lays at least twice per year (Hodgkiss and Man 1978).

Lake Qarun (Egypt) 29°29′N, 45 m

The breeding season of *T. zillii* coincides with the hottest months, May-November (El-Zarka 1956, cited by Siddiqui 1979a).

Choice of Spawning Grounds, Territoriality, Spawning and Guarding the Young

In the substrate-spawners the mature males first billet themselves for nesting along the shores where each delimits and defends a territory on the bottom; the females join them and after a fairly long courtship of several hours to some days, the couples are formed. The two partners participate in preparing a nest by digging holes of varying depth some of which may shelter an adult (De Bont 1950; Daget 1952; Ruwet 1962). In the Lufira reservoir (Ruwet 1962) *T. rendalli* established its territories at a depth of 20 to 80 cm (especially 50 cm) preferentially on ground with a marked slope and near to the marginal fringe of vegetation; the territories, which attain one meter diameter, may be adjoining, giving the impression of true colonies. In Lake Kinneret, Israel, *T. zillii* prefers bottoms with pebbles or sand with abundant vegetation *(Phragmites)*; the territories are smaller (2.0 to 2.5 m^2) in habitats with dense vegetation than in open water (5 to 6 m^2) (obstructing vision curbs aggression by restricting the sight of neighbors) (Fishelson 1966a, 1966b).

Oviposition and fertilization are long operations: males and females alternately apply the genital papilla to the bottom, at the side of a hole, the one depositing a batch of eggs, the other covering them with milt, and the application forms a plaque of spawn composed of several thousand eggs stuck to the bottom.

Males and females then guard the eggs, and ventilate them with the aid of their fins. After hatching, they move the alevins frequently from one hole to another. As soon as they know how to swim, the alevins organize themselves into a cloud which stays in the immediate neighborhood of the nest and retreats into a hole in time of danger. They are constantly regrouped by the jolting movements, accentuated by contrasting colors, of one parent while the other mounts guard at the boundary of the territory (Voss and Ruwet 1966). After 2 to 3 weeks, the alevins leave the territory and the parents are able to recommence spawning.

In mouthbrooders, the males remain in the nesting zones. Each carrying a marked, often brilliant, color, delimits and defends a territory and makes a nest where he tries to attract and retain a female (Lowe-McConnell 1959). The organization of the breeding arenas varies. In Lake Victoria and the Malagarasi swamps the arenas of *S. variabilis* and *S. karomo* observed by Lowe-McConnell (1956a, 1959) were situated in open water, on a sandy bottom, at a depth of 0.15 to 1.0 m *(S. variabilis)* and 0.5 to 1.5 m *(S. karomo)*. In the Lufira reservoir (Ruwet 1962) *S. macrochir* nest at a depth of 0.8 m to 1.5 m (mode, 1.2 to 1.3 m) on a flat bottom, devoid of vegetable debris from the beaches, and on high alluvial banks situated towards the open water. In Lake McIlwaine, the areas of *S. macrochir* are established in shallow water but certain nests are made exceptionally up to 5 m deep (Marshall 1979b). Bruton and Boltt (1975) have studied in detail (mainly by diving) the distribution and the characteristics of *S. mossambicus* nests in Lake Sibaya, according to type of habitat, depth and vegetation. Most of the nests are established at the edge of the littoral terrace, just before the steep slope. The nests are situated at a depth of 0.4 to 8.5 m but with a maximum concentration between 0.5 and 5.0 m; the size of the nest increases with depth; they are associated with scattered vegetation of medium density, composed mainly of *Scirpus* sp. and *Potamogeton* sp.

The females make only brief visits onto the arenas. Passing from one territory to another, they are courted by several successive males, until the moment when, stopping in front of a nest basin, a female deposits a batch of eggs, the male fertilizes them immediately and the female takes them in her mouth to brood them. The operation is very brief (50 to 60 seconds) and may be repeated with the same male, or with another in a neighboring territory (successive polygyny and polyandry, Ruwet 1963b). Finally, the female leaves the arena, where the males stay billeted, and she carries several hundred eggs in her mouth (see Balarin and Hatton 1979) which she broods, staying in a zone sheltered with vegetation (brooding area) described by Bruton and Boltt (1975) for *S. mossambicus*. As soon as the yolk sac is resorbed (see Table 7) the female lets the alevins escape from her mouth. They form a cloud which orientates itself constantly in contact with the mother, following her slow movements and taking refuge in her mouth at the least sign of danger, warned by her movements (see Voss and Ruwet 1966). When the alevins reach a certain size (9 to 10 mm in *S. mossambicus*, Bruton and Boltt, 1975) the females leave the brooding areas and sometimes gather in schools of several hundred individuals, notably in *S. variabilis* (Fryer and Iles (1972), *S. lidole* (Lowe (McConnell) 1957) and *S. mossambicus* (Bruton and Boltt 1975). Finally, after a variable time according to the species (22 days in *S. mossambicus*, Bruton and Boltt 1975; Russock and Schein 1977), the young, totally independent, are liberated by the female in shallow water where they live in schools and continue their growth.

In almost all mouthbrooders, guarding the young is undertaken by the female (uniparental maternal family) though the male may participate exceptionally, notably in *S. mossambicus* (Bruton and Boltt 1975). But in *S. melanotheron*, it is the male alone which incubates (uniparental paternal family) (Oppenheimer and Barlow 1968). In *S. galilaeus* (Iles and Holden 1969; Fishelson and Heinrich 1963) and *S. multifasciatus* (Fryer and Iles 1972) the two sexes form a stable couple and both practice mouthbrooding. This seems, therefore, to be an intermediate stage between the biparental family of substrate-spawners and the uniparental family of mouthbrooders (Ruwet 1968).

Table 7. Principal characteristics of development of eggs and alevins in eight species of tilapia (after Hanon 1975). (Temperature: 26 to 27°C in aquarium).

Species	Length of eggs (mm)	Hatching age (days)	Hatching length (mm)	First day of swimming age (days)	First day of swimming length (mm)
Tilapia					
T. guineensis	2.7	2	5.0-5.5	7	7.5
T. rendalli	1.8	2	3.9	6	6.0
T. sparrmanii	1.6	2½	3.5	5-6	5.0
T. tholloni	1.6	2	3.6	6½	6.1
Sarotherodon					
S. galilaeus	2.5	4	—	11	—
S. melanotheron	3.5	6	6.5-6.8	12-13	11.0
S. niloticus	2.8	4-5	4.5	11	8.0
S. saka	2.7	5	4.7	11	9.2

Guarding the young increases considerably the survival of the alevins and reproductive success, but mouthbrooding represents an even more remarkable specialization which confers on *Sarotherodon* great independence from the milieu. Mouthbrooders have a better capacity for adaptation (in cases of introduction) and resistance in water bodies subjected to rapid changes in water level (for example, the advantage of *S. macrochir* over *T. rendalli* in the man-made Lake Mwadingusha, Shaba Zaïre: Ruwet 1961-1962, 1963b, 1968, and the success of the introduction of *S. macrochir* in the artificial Lake McIlwaine, Zimbabwe, Marshall 1979b).

The territoriality of substrate-spawners is a regulatory eco-ethological mechanism which allows a certain number of spawners—generally the largest, the best-motivated and the most aggressive—to occupy the breeding space offering optimal conditions for survival of the young (protection against predators and exploitation of the best physical sites for reproduction). The smaller and less motivated individuals occupy marginal zones of the spawning grounds, where habitat conditions are suboptimal and the exposure to predators is greater.

The social structure of the mouthbrooders, and their arenas comprise a system whereby those males with the strongest reproductive drive occupy the territories at a given time and attract the females. Less motivated and younger individuals are pushed to the edge of the spawning ground or outside it (marginal individuals).

Both in the substrate-spawners and the mouthbrooders, territorial aggression is shown mainly by displays and symbolic conflicts and rarely by true fights leading to wounds. However, the latter are known in nature, for example in *S. mossambicus* where, of a total of 323 territorial individuals, Bruton and Boltt (1975) recorded 10% carrying wounds. Individual aggression depends mainly on size of fish, the size of the territories (which varies with size of male, density of the population and the nature of the substratum) and the degree of visual isolation from the neighbors (influenced by the nature and the variety of the bottom). One can also see specific differences; certain species are considered aggressive *(S. niloticus)*, others intermediate *(S. macrochir)* and others feebly aggressive *(S. galilaeus)*. The degree of aggression is an important factor to consider in the intensive culture of tilapias in cages or in ponds (Mélard and Philippart 1980).

PRACTICAL IMPLICATIONS

The methods of reproduction and the social organization of tilapias have practical implications for culture, which we will review in passing. Substrate-spawners form stable territorial pairs and one can only stock a spawning pond with the number of pairs corresponding to the space available. Any surplus will not participate in breeding, and will disturb the spawners. For mouthbrooders, however, with breeding arenas, there is interest in planning a spawning pond comprising three zones with distinct uses: 1) a flat bottom, muddy or sandy, of medium depth (ca. 1 m) which one can stock with a number of males higher than the space available for nest making, in such a way that always the most motivated occupy the bottom territories and ensure continuous reproduction; 2) a vegetated zone to shelter the females separately from the males during the period of buccal incubation and 3) a shallow zone (10 to 20 cm) to harvest the alevins before sexing them and transferring them to growing ponds or floating cages (see Ruwet and Voss 1974).

A precise knowledge of the breeding habitat (depth, nature of substrate, slope, vegetation) would allow the design for each species of a model grow-out pond limiting breeding as far as possible and, in consequence reducing overpopulation and dwarfing. One can also envisage the possibility of harvesting the alevins by attracting them into traps by means of the sound emissions (Lanzing 1974) possibly used by the parents for reassembling the clouds of alevins, both in substrate-spawners and in mouthbrooders.

In taking into account the eco-ethology of reproduction, it is often possible to explain subsequently, or to foresee, the differential dynamics of substrate-spawning and mouthbrooding species in natural water bodies subject to fluctuations (for example changes in water level, development of vegetation) in man-made lakes or artificial reservoirs, or since the introductions of species. The rapid falls in level in certain man-made lakes can be catastrophic for populations of substrate-spawning species (of which the breeding males and females as well as the eggs and alevins are strongly attached to the nest and, moreover, often situated in sites close to the shores) but have little or no effect on mouthbrooding species in which only the males (which stay billeted on the arenas and always at greater depths than substrate-spawners) are eventually affected, for the females are capable of moving the alevins towards deeper water. In Lake Mwadingusha, ex-Katanga, Zaïre (Ruwet 1962, 1968) the variations in level affected *T. rendalli* much more than *S. macrochir*, a species which also coped successfully with water level fluctuations in Lake McIlwaine, Zimbabwe (Marshall 1979b).

In Lake Naivasha, Kenya, the development of the (introduced) population of *T. zillii* is associated with an evolution of the environment (rise in level and appearance of a six-year flood cycle) which has permitted the development of a fringe of riverine and marginal vegetation favorable for the reproduction of this species; this evolution has also favored *S. leucostictus* to the detriment of *S. spilurus niger*, a fluviatile species unable to adapt to the disappearance of bottoms suitable for its reproduction (Siddiqui 1979a, 1979b).

In Lake Kainji, Nigeria created in 1969, the tilapias are represented by *S. galilaeus*, *S. niloticus* and *T. zillii*, in the proportion 16:5:1. The predominance of *S. galilaeus* is partly explained by its less specialized requirements for breeding than those of *S. niloticus* and *T. zillii*, which need special bottoms and vegetation (Ita 1978). One can see, however, evolution from the original riverine situation in tilapia populations in several other man-made lakes (Lake Nasser, Latif 1976) characterized by a progressive substitution of *S. galilaeus* for *S. niloticus*, initially the more abundant.

Certain deceptive results in fish culture are explained largely by a modification of the frequency of spawning in species transplanted into an environment climatically different from their original environment. Thus, the good results obtained with *T. rendalli* in fish culture in Shaba where the existence of a cold dry season limits the numbers of spawnings (and in consequence population of the ponds and dwarfing) are not also found in the Zaïre basin at Yangambi, where reproduction is continuous (Gosse 1963). A comparable phenomenon is probably responsible for the failure of the transfer of *S. andersonii* from Zambia (a single reproduction per year and maturity at 12 to 15 months old) to Tanzania (maturity at 3 to 5 months old and excessive breeding) (Lema and Ibrahim 1975).

The relative aggression of the species at intraspecific (intensive monoculture) and interspecific (introductions, polyculture) levels and its variations with sex, size, age and environmental factors such as temperature, must be taken into consideration by fish culturists. In Taiwan, *T. zillii* is considered of little interest because it is too aggressive (Chen 1976); in ponds in Zaïre *T. zillii* competes with *T. rendalli* (Gosse 1963) and in Lake Victoria it competed with, and supplanted, *S. variabilis* (Table 2).

Microhabitats and the Dynamics of Spatial Separation

SELECTION OF MICROHABITAT

Information on spatial separation (horizontal, vertical) of numerous tilapias and their preferences for a habitat has been obtained either by analyzing statistics of commercial and artisanal fisheries, or by employing different sampling methods: trawling, gill nets, traps or rotenone (Ita 1978; Bruton and Boltt 1975; Balon 1974), echo-sounding (Bruton and Boltt 1975; Capart 1955), direct underwater observations by diving (Bruton and Boltt 1975) and also by electric fishing in lakes and rivers (Lamarque et al. 1975; Lamarque and Closset 1975; Pienaar 1968).

The general characteristic of the spatial distribution of tilapias in lakes is their restriction to the shallow littoral zones (see Table 4). Thus in Lake Kainji, Nigeria (maximum depth ca. 50 m), *S. galilaeus*, *S. niloticus* and *T. zillii* are caught in the littoral zone, at a depth of 0 to 7 m. The biomass there attains 107 kg/ha (Ita 1978).

The littoral distribution of lacustrine tilapias originates fundamentally from their physiological incapacity to descend to great depths (Caulton and Hill 1973, 1975), a characteristic which supports the riverine origin of tilapias. But ecological factors also affect this:

- temperature, through on one hand the existence of tolerance limits and specific temperature preferences, and on the other hand a direct influence on the physiological capacity of tilapias to adapt to depth (Caulton and Hill 1975), or to withstand certain unfavorable conditions of DO, toxic gases, etc.;
- DO and the presence of dissolved toxic gases (H_2S, NH_3, CO_2) which restrict the depth penetration to varying degrees according to the species and their tolerance and preferences;
- demands and specific preferences for a habitat or determined biotope in accordance with feeding (feeding grounds), nesting (spawning grounds), guarding the young in mouthbrooders (brooding areas) and protection against predators (refuges and shelters: escape range, see Gerking 1959).

On the whole, substrate-spawners with macrophyte feeding habits are more dependent on shoreline habitats (macrophytic vegetation has a maximum depth of 7 to 8 m) than are the mouthbrooders with their microphagous feeding and more pelagic habits, especially in waters rich in plankton (Lowe-McConnell 1959). It is interesting to note that in certain shallow lakes very productive of plankton, for example, Lake George, Uganda (maximum depth 2.5 m), the tilapias *(S. niloticus* and *S. leucostictus)* preferentially occupy a littoral band of 50 m even though the plankton food is abundant throughout the lake (Burgis et al. 1973; Gwahaba 1975). Caulton (1978a, 1978b, 1978c) has put forward a bioenergetic interpretation of this phenomenon.

The microseparation of tilapias in rivers is, as in most species of fish, influenced by current speed (in accordance with the swimming capacity) and by ecological factors which are closely associated with this: the depth, the nature of the substrate and the cover of vegetation (Hynes 1970; Brown 1975; Huet 1959). Gaigher (1973) showed that *S. mossambicus* avoided the rapids and fast-flowing water and preferred still water and pools. In the same way, electric fishing in the Shire River, Malawi (Lamarque and Closset 1975) showed the absence of tilapias (mainly *S. mossambicus*) from the central part of the river where the current is rapid, and their concentration along the sheltered banks provided with vegetation.

The habitat of a species is especially more difficult to characterize as it differs with size, age and stage of sexual maturity and it varies for each intraspecific group with the seasons and even the time of day (variations which lead to daily movements and seasonal cyclical migrations).

a. Influence of size, age and stage of sexual maturity. The size of fish limits the physical accessibility of certain habitats (depth) in such a way that the alevins, juveniles and adults distribute themselves in depth zones from the shore towards open water; this spatial separation favors complementary exploitation of the feeding niches of different habitats, avoiding specific competition. The alevins and the juveniles generally have a much larger temperature tolerance range than the adults (16.5 to 39° C, with a preference for 19 to 35° C in the alevins and juveniles of *S. mossambicus*, compared with a tolerance range of 19 to 32° C and a preference for 22 to 30° C in the adults, Bruton and Boltt 1975), which explains why they are able to live throughout the year in the littoral zone (warmer than open water in summer but colder in winter) while the adults only stay there during the summer breeding period.

b. Seasonal changes of habitat. Bruton and Boltt (1975) studied the seasonal changes in the distribution of *S. mossambicus* of Lake Sibaya. During the cold season (May to July, 20° C), the juveniles and adults live in the pelagic zone at a depth of generally less than 12 to 13 m. When the water warms up (August to September), first the juveniles, then the adults migrate towards the shallower littoral zone which forms the feeding and breeding grounds until the end of the hot season. In January, there is a return migration towards deeper water. Seasonal migrations of the same type, between littoral and pelagic zones have been observed in *S. lidole* of Lake Malawi (Lowe 1953), *S. variabilis* of Lake Victoria (Fryer (1961a) and *S. macrochir* of Lake Mweru (Carey 1965).

Changes (seasonal or more irregular) in the spatial distribution of tilapias are introduced by the variations of salinity in certain saline lakes (*T. rendalli* and *S. shiranus chilwae* of Lake Chilwa; *S. macrochir* of Lake Mweru Wantipa) and by variations in the turbidity of certain river systems (Pienaar 1968).

Whitefield and Blaber (1979) showed equally important displacements of *S. mossambicus* populations in estuaries which take into account their open and closed phases. *S. mossambicus* are abundant in the lower reaches during the closed phase (salinity stabilized and current weak) but retreat into the upper reaches when the estuary is open (strong daily variations in salinity and faster current).

Finally, it is well known (Lowe-McConnell 1975) that important seasonal changes in habitat are produced in all rivers subject to alternating floods (colonization of the flood plain, partly for reproduction) and dry periods (withdrawn into the river bed and into the permanent pools of the intermittent streams). Whyte (1975) has studied the seasonal changes in the spatial distribution of *T. busumana*, *T. discolor* and *S. multifasciatus* in Lake Bosumtwi and in the permanent and intermittent rivers flowing into it.

c. Daily changes of habitat. In Lake Magadi, Kenya, *S. alcalicus grahami* feeds mainly during the day in lake zones where the temperature is 25 to 28°C and where food is abundant; it then moves toward hotter zones (35 to 40°C) which it occupies at night (Coe 1966, 1967 in Caulton 1978b).

In tilapias living in lakes or lagoons (*S. variabilis* in Lake Victoria, Welcomme 1964; *S. mossambicus* in Lake Sibaya, Bruton and Boltt 1975; *T. rendalli* in Lake Kariba, Donelly 1969; *T. rendalli*, Caulton 1978b), the juveniles, but above all the alevins, make daily cyclical movements between the shallower littoral zones, occupied during the day, and the deeper zones towards the open water, occupied by night. The juveniles of *S. mossambicus* of Lake Sibaya move thus from 0.15-1.0 m to a depth of more than 1.0 m (alevins from less than 0.15 m to 0.15-1.50 m). In the course of these displacements the fishes undergo a large temperature difference (up to 10°C) between the shallow water heated up by day (sometimes to more than 35°C) and the deeper water which is colder during the night. The change of habitat occurs at nightfall when the littoral waters become cooler than the deeper water (Caulton 1978b for *T. rendalli*; Bruton and Boltt 1975 for *S. mossambicus* in Lake Sibaya). These daily movements are made possible because the little fish, unlike the adults, are capable of a very rapid adaptation to variations in pressure and depth (Caulton and Hill 1973), a phenomenon further facilitated by the high temperatures (Caulton and Hill 1975).

Concerning the ecological significance and the survival value of these daily changes in habitat, several interpretations have been proposed:

1. They are tactics for avoiding predators, present in deep water during the day (Fryer 1961a, 1965b; Donnelly 1969) or in shallower water during the night (case of *Clarias gariepinus*, preying on *S. mossambicus* alevins and juveniles of less than 15 cm in Lake Sibaya, Bruton and Boltt 1975) although such movements also exist in the absence of aquatic predators (Welcomme 1964; Caulton 1975a, 1975b), or in spite of very heavy predation by day in shallow water by birds (Bruton and Boltt 1975).
2. They reduce feeding competition by the successive exploitation in the course of the day of abundant food in several habitats (Welcomme 1964).
3. They are required by physiological adaptations to give the best growth of the juvenile population (Caulton 1978a).

Caulton *(op. cit.)* has supported the last hypothesis by bioenergetic study of *T. rendalli* and has stressed its interest for the distribution (density and biomass) of the cichlid populations in lakes (i.e., their concentration in the littoral zones where temperature fluctuations are more marked, and this despite more abundant food in open water, for example, in Lake George) as well as certain natural examples of dwarfing apparently connected with a homothermal regime and due, for example, to water supply from a spring. On the practical level, it should be useful to find out if rearing tilapias would give better results if the temperatures were varied strongly during the course of the day (influence of pond depth).

SCHOOLING BEHAVIOR

After their liberation by the female, the alevins of many mouthbrooding tilapias remain in schools until a stage of development more or less advanced according to the species (see Fryer and Iles 1972). In *S. mossambicus* (Bruton and Boltt 1975) the juveniles (less than 8 cm) form schools of some hundreds of individuals (several thousands when they group together after an alarm) which mix closely and disperse during the night. Observations made on *S. variabilis* of Lake Victoria (Fryer 1961a) suggest that the schools of alevins and juveniles keep a certain coherence for up to 6 to 8 months. It seems, however, that the schools of juveniles finish by breaking up, except in some pelagic plankton-feeding species of Lake Malawi (*S. saka*, *S. squamipinnis* and above all *S. lidole*) where they are maintained throughout life (Fryer and Iles 1972; Berns et al. 1978).

In the breeding season, many types of extremely structured schools are to be found: migrating schools (with possible presence of a leader) in *S. macrochir* of Lake Mweru (Carey 1965), schools of males (*S. mossambicus*, Bruton and Boltt 1975) or of females (*S. lidole*, Lowe 1952) near the spawning grounds, schools of mouthbrooding females (*S. variabilis*, Fryer 1961a and *S. mossambicus*, Bruton and Boltt 1975).

It is also known that several *Sarotherodon* species (*S. galilaeus* of Lake Kinneret, Israel, Fryer and Iles 1972, and *S. niloticus* of Lake Edward) form very large schools of 100 m^2 or more at the lake surface. These are probably for feeding. Echo-soundings made in Lake Kivu (Capart 1955) showed that schools of *S. niloticus* which had dispersed during the night reformed at dawn.

Again, the existence is reported in certain species (e.g., *S. esculentus* of Lake Victoria) of schools of small individuals (immature juveniles) living at the surface and of schools of large individuals (adults) living on the bottom (Fryer and Iles 1972). Finally, it is interesting that *T. rendalli* of Lake Malawi, a macrophyte-feeding littoral species, does not form schools, even though this is the rule among the plankton-feeding pelagic species like *S. saka*, *S. squamipinnis* and *S. lidole*.

The ecological significance of schooling behavior in tilapias and cichlids in general has been discussed by Fryer and Iles (1972). These authors see it as a mechanism with multiple functions: protection against predators, facilitation of feeding and population regulation, i.e., information on the level of population density at the time of reproduction (see Wynne-Edwards 1962).

MOVEMENTS, HOME RANGE AND HOMING BEHAVIOR

There are relatively few studies on these aspects of tilapia ecology, even in lakes. We know with certainty that the reproductive migrations of *S. macrochir* in Lake Mweru (Carey 1965) and of *S. lidole* in Lake Malawi (Lowe 1952, cited by Fryer and Iles 1972) involve displacements of several kilometers. After the liberation of their young, the female schools of *S. variabilis* in Lake Victoria (Fryer 1961a) generally disperse a little way (less than eight km) from the place of capture and tagging on the brooding ground, but certain individuals make much longer journeys (maximum: 100 km in 30 days). Tags carried by *S. mossambicus* juveniles in Lake Sibaya (Bruton and Boltt 1975) showed that most of the returns found within two days were in a radius of 400 m from the tagging point; one individual was recaught 400 m away from the tagging point after two hours and another at 3,100 m after 14 days. Lowe (McConnell) (1956b) recorded important movements in *S. esculentus* in Lake Victoria (56 km in 40 days; 10 km in 3 days) but most of the individuals were recaptured at the marking place after 228 to 700 days, which suggested well-circumscribed populations, fixed to a home range. Marking experiments made by Holden (1963, cited by Lowe-McConnell 1975) showed great spatial stability of *S. niloticus* and *S. galilaeus* in the pools of the Sokoto River, Nigeria.

The first case of homing in the Cichlidae was reported by Fryer (1961a) in Lake Victoria, Jinja region. Eight *S. variabilis*, captured and marked on the brooding grounds, were recovered after 45 to 805 days on the same brooding ground or in its immediate proximity. In Lake Sibaya (Bruton and Boltt 1975), three *S. mossambicus* marked during a breeding season were captured during the following season: two individual males exactly on the breeding site of the previous year (homing) and one individual on another spawning ground three kilometers away.

Ecological Effects of Introductions and Transfers

AFRICA INCLUDING MADAGASCAR AND MAURITIUS (see Tables 2 and 3)

Introductions into natural waters not containing any tilapias or offering a vacant ecological niche for tilapias are often crowned with success when one uses a species for which the natural distribution area is near: e.g., *S. niloticus* in Uganda (Koki Lakes) and in Madagascar; *S. spilurus niger* in Kenya; *S. macrochir* and *T. rendalli* in Zambia (e.g., Lake Lusiwashi); *S. macrochir* in Lake McIlwaine; *T. zillii* in Lakes Naivasha, Kyoga and Victoria; *S. alcalicus grahami* in Lake Nakuru and *T. rendalli* in certain South African lakes.

On the other hand, many long-distance introductions, to a different climatic zone or to a different altitude, have been considered complete failures: *S. macrochir*, *S. andersonii*, *S. spilurus niger* and *T. rendalli* in certain regions too cold in South Africa; *S. spilurus niger* in Madagascar; *S. mossambicus* in the north of Egypt and in Malta and *S. spilurus niger* and *S. esculentus* in lakes at altitude in Uganda. This is explicable for the most part by the unsuitable temperature conditions. A certain number of other results disappointing for the control of vegetation (*T. rendalli* in the Sudan), for improving a fishery (*S. niloticus* in certain lakes at altitude in Uganda) or for fish culture are explained by 1) poor survival (e.g., fish kills of *S. mortimeri* and *S. mossambicus* in man-made lakes of the Zimbabwe plateau, Jubb 1974) and slow growth (for example in *T. rendalli* of Lake Itasy, 1,200 m altitude, Madagascar, Moreau 1975) observed in the species transferred into regions characterized by a well-marked cold season and 2) excessive reproduction (leading to overpopulation of ponds and dwarfing) in the species transferred from highlands where the reproduction is seasonal to lower and warmer regions where reproduction becomes continuous (e.g., *T. rendalli* from Shaba to the Zaïre basin and *S. andersonii* from Zambia to Tanzania).

The introduction of tilapias into waters already containing indigenous tilapias have had rapid and unfortunate consequences on aquatic ecology (e.g., hybridization, competition for food and breeding sites) and on the fisheries: for example, in Lake Victoria, hybridization of *S. niloticus* with the indigenous *S. esculentus*, and of *T. zillii* with *T. rendalli* and competition of *S. niloticus* (but above all of *T. zillii*) with the indigenous *S. variabilis* (Fryer 1961a; Welcomme 1967b; Fryer and Iles 1972).

Another interesting problem is posed by the failure of the introduction of *S. macrochir* into the man-made Lake Kariba, where *S. mortimeri* is indigenous. Based on the existence of a flourishing fishery for *S. macrochir* in Lake Mweru, the new lake was stocked with this species, but the source of supply was not Lake Mweru in the Zaïre basin but the Kafue River, a Zambezi affluent. *S. macrochir* never prospered in Lake Kariba, perhaps because the supply source was wrong, perhaps because the conditions of life in the lake were very different from the water of origin, or perhaps again because it was subject to the concurrence of *S. mortimeri*, which, a little against all expectation, is very well adapted to withstand the transformation of the fluviatile environment to a lacustrine one (Balon 1974). However, the *S. macrochir* from the Kafue were transplanted successfully into Lake McIlwaine situated on the Hunyani River, an affluent of the middle Zambezi.

Variations in the success of interspecific introductions are apparent from introductions made simultaneously or successively of tilapias capable of competing or presenting differential capacities of population resilience and eco-physiological adaptations to new environmental conditions (degree of eurycity). Thus *S. spilurus niger*, originally from Kenya rivers, was not able to acclimatize in the Koki Lakes in Uganda and in many other lakes at altitude in this region at a time when, on the contrary, the introduction of *S. niloticus*, a very euryoecious species, was crowned with success. *S. spilurus niger*, a fluviatile species, was at first very well adapted in Lake Naivasha until a change of environmental conditions (rise in water level plus cycles of littoral flooding plus the creation of lacustrine conditions) led to its disappearance and progressive replacement by *S. leucostictus*, a lacustrine species which had been accidentally introduced (Siddiqui 1979b). This accidental introduction also permitted the hybridization of the two species and the production of apparently all-male offspring (Fryer and Iles 1972, p. 168). Lamarque et al. (1975) showed an identical phenomenon in Lake Itasy, Madagascar, where *S. macrochir* (introduced in 1958) prospered for several years before disappearing and being replaced by *S. niloticus* (introduced in 1961-62); the hybridization of the two species produced slow-growing and sometimes deformed individuals.

The introduction and proliferation of macrophyte-feeding tilapias and especially of *T. rendalli* has seriously perturbed the ecology of certain water bodies, for example, Lake Kyle, Zimbabwe (in Jubb and Skelton

1974) and rivers of the island of Mauritius (George 1976). Also in Lake Kinkony in Madagascar, *T. rendalli*, accidentally introduced, in three years devastated nearly 3,000 ha of *Ceratophyllum* and *Nymphaea* beds with in consequence the almost total disappearance of a valuable indigenous fish, *Paretropus petiti* (Lamarque et al. 1975).

OTHER REGIONS (see Table 3)

In environments characterized by suitable temperature conditions and offering a vacant ecological niche, tilapias are able to establish themselves and form stable populations, contributing to the augmentation of the local fishery resources (e.g., *S. niloticus* in Indonesia, Bangladesh and Mexico; *T. rendalli* in Sri Lanka and *S. mossambicus* in Hong Kong).

In most cases, however, the establishment of tilapias in natural waters has engendered, as in Africa, unfavorable consequences: 1) initial growth of the population followed by spontaneous regression with some of these failures (e.g., *S. mossambicus* in Thailand and Bangladesh and *S. niloticus* and *T. rendalli* in Thailand) resulting from competition from local species or the influence of unfavorable environmental conditions, which, even if they do not occur every year suffice to hamper the long-term survival of the populations; 2) disappointing practical results due to continuous reproduction (overpopulation and dwarfing) in equatorial and tropical environments (e.g., *S. mossambicus* in Indonesia and *T. rendalli* in certain Brazilian lakes) and to slow growth and winter mortalities in regions situated at the limit of the natural distribution of tilapias (e.g., *S. mossambicus* in the rice fields of Taiwan and *T. zillii* for the biological control of aquatic vegetation in California and 3) excessive destruction of aquatic vegetation by macrophyte-feeding tilapias (*T. rendalli* in Brazil), damage to the indigenous fauna (*S. mossambicus* in Florida and in Colombia), competition for food and predation affecting other valuable species (*S. mossambicus* in the brackish *Chanos chanos* (milkfish) ponds in the Philippines).

Introductions sometimes prove to be failures in an economic scale as the species introduced was not accepted by the local population because of its unfavored taste (*S. mossambicus* in Thailand, Welcomme 1979b), small size, or because it was not the most judicious choice to fulfill its assigned role. Considered as a 'miracle fish' at the time of its first introduction to Southeast Asia, *S. mossambicus* has given poor results which explains why this species is now being progressively abandoned for use in culture: *S. niloticus*, *S. aureus* (Bowman 1977) and various faster growing hybrids with a higher proportion of males are preferred. *S. mossambicus* is in Asia what *S. macrochir* is in Africa.

Finally, the introduction of tilapias (and of exotic species in general) may assist the dispersal of diseases and undesirable parasites; thus the construction of tilapia ponds in Puerto Rico has favored the habitat of the molluscan vectors of bilharzia and this malady may reach Florida (Courtenay and Robins 1973).

Conclusions

Tilapias constitute a diversified and plastic group of great interest to ecologists. But until now most studies have been qualitative and limited to one particular aspect of their biology. Future ecological research should above all consist of quantitative overall studies of populations (cf. Bruton and Allanson 1974 and Bruton and Boltt 1975 for *S. mossambicus*) and studies of population dynamics: analyses of the causes of temporal fluctuations of the populations; demographic comparisons of several species in the same water body; demographic comparisons on the same species in contrasting milieu; evolution of populations in water bodies subject to artificial modifications (cf. Dudley 1979), and interactions between indigenous and introduced populations (cf. Noble et al. 1976).

Another important objective is to show the precise geographical distribution of each species and the introductions and transfers with their effects in Africa and in the other regions (i.e., build up a data bank); this would provide a reference point for the easier evaluation of the evolution of future populations.

It is equally necessary to develop quantitative research on the environment in order to understand better the ecological factors which influence the distribution of different species and their absolute abundance, as well as their preferred habitats and microhabitats. This type of study should lead to preparation of an inventory of populations (ecophenotypes) adapted for particular conditions (temperature, salinity, etc.) and interesting to exploit for culture. But parallel to this research on the environment, it appears indispensable to undertake, in a more systematic manner, experiments on environmental tolerance limits and preferences. For this it is necessary to first define a standard methodology in order to eliminate the extreme variability of the results which appear in the literature. A systematic study of preferences and tolerance limits in regard to temperature (McCauley 1980) is especially important.

Finally, concerning introductions, the errors made and the past failures (notably with *S. mossambicus*) by stocking without sufficient knowledge of the biology of species and ecosystems, encourage very great prudence concerning introductions of new species *(S. aureus, S. hornorum, S. niloticus)* in all parts of the world. It is good to report that many countries, without doubt because they are subject to the unpleasant consequences of ancient and uncontrolled introductions, are starting to take control measures on the spreading of tilapias (e.g., Colombia, U.S.A. and Papua New Guinea). The burden of decision in this matter ought to rest on an evaluation of the risks of perturbation of the native biocenoses and ecosystems. However, often in such ecological studies it is difficult to generalize one way or the other (total prohibition or total liberty). Numerous examples show that the same species introduced into different waters often reacts in an unpredictable and unequal manner depending on the milieu and the action of physicochemical factors and biocenotic characteristics. Only intensive research on the particular case will permit the understanding of the mechanisms involved and the implications for the management or conservation of the aquatic milieu.

Acknowledgments

Dr. D. Pauly of ICLARM and Dr. R. Lowe-McConnell translated our French manuscript and revised the final version of the paper; we are indebted to them for their kind cooperation. We are also grateful to Dr. E. Trewavas for comments on the distribution maps and Drs. M.S. Caulton, B. Hepher, L. Lovshin and J. Moreau for supplying additional information on tilapia distribution and introductions. Thanks are due to colleagues of the University of Liège, Dr. J. Voss, Mrs. C. Guyot-Hassé and Mr. Ch. Mélard who helped to gather the literature. The first author, Dr. J-Cl. Philippart prepared this review while a Research Associate of the Belgian National Foundation for Scientific Research (Fonds National de la Recherche Scientifique, FNRS).

Life Histories of Tilapias: An Evolutionary Perspective

D. L. G. NOAKES AND E. K. BALON

Department of Zoology
College of Biological Science
University of Guelph
Guelph, Ontario
Canada N1G 2W1

NOAKES, D.L.G. AND E.K. BALON. 1982. Life histories of tilapias: an evolutionary perspective, p. 61-82. *In* R.S.V. Pullin and R.H. Lowe-McConnell (eds.) The biology and culture of tilapias. ICLARM Conference Proceedings 7, 432 p. International Center for Living Aquatic Resources Management, Manila, Philippines.

The early ontogenies of tilapias are shown to be of major importance in the evolution of life history styles. By applying hierarchical systems for classification of reproductive guilds, and a uniform system of categorization for developmental intervals, we show that the life history styles of tilapias fall into two clearly defined categories, altricial and precocial, and that a significant evolutionary mechanism exists for changes between these. The major evolutionary trend has been from a more altricial style, the nest-spawning guarders (equivalent to substrate-spawners), to the more precocial style, the mouthbrooding external bearing species (equivalent to mouthbrooders). The major distinction between these is largely a consequence of heterochronous shifts in ontogeny, particularly the lack of a larval period in the ontogeny of the bearers. A survey of published information shows the predicted differences between the two types of species, with some overlap. The overlap exists because individual species retain the ability to adapt somewhat more towards an altricial or a precocial life history style by ontogenetic shifts within their own reproductive guild (i.e., guarder or bearer). This predicts the adaptive responses expected of fish under varying ecological conditions, including aquacultural practices. Stunting, perhaps the most serious problem in aquaculture, is not only interpreted in our theoretical framework, it is predicted by us. We make some tentative suggestions as to how stocks of tilapias might be manipulated to better purpose for aquaculture production.

I ask the reader to remember that what is most obvious may be most worthy of analysis. Fertile vistas may open out when commonplace facts are examined from a fresh point of view. L. L. Whyte (in Koestler 1967)

Introduction

This review of the life histories of tilapiine cichlid species of the breeding morphs called *Tilapia* and *Sarotherodon* will focus in particular on early life history and developmental features. We believe this approach brings profitable insights to the understanding of the biology of these fishes. We also believe that mouthbrooding likely evolved several times among the tilapias as an adaptation to stable and competitive conditions (Balon 1980, 1981a). If there has been this polyphyletic origin, it must be questioned as to whether to separate the generic names on this basis. An outline of our theories and some essential terms formulated by Balon (1979a, 1979b, 1980, 1981a) and of the theoretical framework we propose to apply will be given first. We will then consider the tilapias with some comparisons to related taxa and conclude with some practical considerations.

Although most terms we will use are by now established, confusion still exists as to precise meanings of some, or how some may be applied in particular circumstances. A number of recent reviews or textbooks, in addition to those cited above, have more than adequately discussed the basic ecological and epigenetical framework used later, so we refer interested readers to those sources (e.g., MacArthur and Wilson 1967; Emlen 1973; Ricklefs 1973; Løvtrup 1974; May 1976; Gould 1977; Stearns 1976, 1977; Krebs 1978; Pianka 1978).

Review of Terminology

The life history of any species (more properly, of individuals) is the sum of a number of events and processes with ecological and evolutionary consequences. Recent usage has introduced the term "strategy" in such discussions. While there is some confusion over both the etymology and application of this term, it can be a useful shorthand reference to evolutionary considerations. We do not propose to defend its uncritical acceptance or application, but find it a useful heuristic tool, at least in the early stages of discussing the evolution of life history patterns. For it is surely the evolutionary questions that will be of greatest interest at the level of the comparative approach we use. We employ the term and assume the phrase "as if", i.e., animals would be expected to respond to natural selection "as if" they were attempting to maximize fecundity, or minimize predation, or minimize energy expenditure, and so on. In all cases, the more formal argument could be made (Dawkins 1976) in a more orthodox fashion, to lead to the same conclusion.

A conventional consideration of life histories would likely discuss the biology of these species in terms of fecundity, growth and development, age at maturity, parental care, and longevity (Emlen 1973; Ricklefs 1973). The assumption is that natural selection acts so as to maximize the fitness of individual animals (but see Dawkins 1978) by adjusting some or all of these parameters in accordance with the ecological conditions faced, to produce an integrated whole, the life history strategy (= style).

The Pitfalls of "r" and "K"

A concept often applied to life histories, frequently uncritically or even inappropriately, is that of the so-called "r- and K-continuum" (MacArthur and Wilson 1967; Wilson and Bossert 1971; Emlen 1973; May 1976; Pianka

1978). These terms refer to the parameters of the widely-used logistic growth curve for populations, where "r" is the slope (growth rate of the population), and "K" the upper asymptote (sometimes referred to as the carrying capacity of the environment). We prefer the terms altricial and precocial life history style to "r-selected" and "K-selected", respectively, for reasons we explained earlier (Balon 1979b) and will elaborate below. First, some further implications of this general dichotomy in life histories are provided.

Species are said to be "r-selected" if they have a life history pattern that includes a short growth interval and early maturation, high fecundity, reduced parental care and short life (generation) span. These features characterize animals in environments where rapid colonization is favored, where catastrophic mortality is liable to act in a density-independent manner, or where environmental disruptions are liable to be serious and unpredictable. Species are said to have a "K-selected" strategy if they have a prolonged growth interval and deferred maturation, reduced fecundity, increased parental care and an extended life (generation) span. They are favored in environments where conditions are more stable, where competition is liable to be a more serious factor, where mortality is likely to be density-dependent, and where conditions are likely to be stable over long periods of time (May 1976; Horn 1978; Krebs 1978; Green 1980). In a relatively stable (numerically) population, with limiting resources and severe intraspecific competition, variation in juvenile survival is likely to be greater than variation in adult survival. In a population going through a series of colonizing episodes, the reverse is likely to be the case (Stearns 1977).

This is a simplification of what is clearly an idealization. But it is useful because it brings our attention to these aspects of life histories, and couches our considerations in evolutionary terms. It brings the important questions to our attention, even if it does not immediately provide the answers. And the consideration of "r- and K-selection" has its obvious limitations (Stearns 1976, 1977; Horn 1978; Green 1980). Certainly, the life history of any species, when considered in these terms may well be a compromise (i.e., a mixed rather than a pure strategy), and it may not even be correct to consider "r" and "K" as being on the same continuum (Horn 1978; Green 1980). As this concept has obvious limitations, we will refer to it largely because it has been so widely used, and carries at least some general connotations (Fryer and Iles 1972; Lowe-McConnell 1975, this volume).

Guilds and Ontogeny: Hierarchical Systems

Our consideration of life histories will focus on the ontogeny of individuals, especially early ontogeny. The reasons for believing that the early life history will be of critical importance and have the primary determining influence on the entire life history have been given elsewhere (Balon 1964, 1971, 1978, 1979a, 1979b, 1980, 1981a; Noakes 1978a, 1978b, 1981), and so need not be elaborated here. In particular, we will stress eco-ethological guilds (reproductive guilds), and an approach based upon an hierarchical model of ontogeny with a uniform system of nomenclature for development intervals (Balon 1975a, 1975b, 1981b).

The proposal (Balon 1975a) to define the ontogeny of fishes by uniform, objectively-recognized intervals in an hierarchical system, has lead to increasingly widespread reassessments (Balon 1978, 1980; Bond 1979; Noakes 1981), although the proposal itself has been considered by some as little more than a semantic exercise (Balon 1976; Richards 1976).

The life history (ontogeny) of any fish is composed of a series of intervals. By now most would accept this concept, but disagreements continue as to the definition and nomenclature for these intervals. The proposal of a uniform terminology for these intervals, based on a hierarchical system from the smallest recognizable stages in development to the broadest periods in the life history has significant and far-reaching implications (the fundamental implications of hierarchical systems in biology are well-established, Koestler 1967). These intervals are arranged, in sequence from broadest to smallest, as periods, phases, steps and stages. Each interval is composed of one or more at the next lower level in the hierarchy.

These terms overlap somewhat with previous usage, especially as they have been applied by some developmental biologists and ecologists, but they impose an order on otherwise ambiguous usage (e.g. Keenleyside 1979) and align with current theories of ontogeny.

The life history of a fish can consist of a maximum of five periods; embryonic, larval, juvenile, adult and senescent. These encompass the entire life of an individual, from beginning to the end of its life. The embryonic period begins with activation (insemination) of the egg, and terminates with the transition to exogenous feeding. The larval period begins with the transition to exogenous feeding and terminates with metamorphosis, when most of the temporary embryonic structures are replaced by permanent "adult" organs (typically including differentiation of the median fin fold, and ossification of vertebral centra). The juvenile period extends from this point until the first maturation of the gonads. The first gonadal maturation marks the beginning of the adult period, which continues until the last reproduction. The senescent period includes that time after the fish has ceased reproduction (or gametes are very few and of inferior quality), and ends with death.

At least some of these intervals may be greatly extended (e.g., larval period of several years in anguillid eels), or be entirely eliminated (as is the larval period in mouthbrooding cichlids (Fig. 1), the coelacanth, *Latimeria chalumnae*, and many cyprinodontids). Recognition of the alterations of life histories by the relative timing and duration of these intervals (usually referred to as heterochrony; Jones 1972; Gould 1979) leads to productive insights (Balon 1980). We must again emphasize that hatching (or parturition) plays no role in defining these intervals. This is perhaps the most common and persistent misunderstanding in considerations of ontogeny and life history features (Balon 1980, 1981a; Noakes 1981). We tend to date most things from birth (parturition) or hatching, and assume that the event has general significance in ontogeny, likely as a result of peculiarities of our own development (Noakes 1981). It does not.

Our terminology provides a uniform scheme to apply to any organism for the sake of comparisons. Typically, authors attempting such comparisons have assumed (usually implicitly) that hatching (or birth) should be the "zero point" at which to adjust the scales of measurement for the species in question, instead of activation, insemination, or fertilization. Our system allows us to recognize the comparable intervals for any species, and thus to make meaningful (biological) comparisons of life history styles across virtually all organisms (Balon 1975a, 1981a; Noakes 1981).

Reproductive Guilds

The second basic point for our discussion is the earlier (Balon 1975b, 1975c, 1978; Balon et al. 1977) proposal to define fishes according to reproductive guilds. Again, this has been described in detail elsewhere, and so will be outlined here only in summary fashion. Basically, the assumption is that the reproductive style, or guild, of each species will be determined by factors relating to oxygen supply for the developing young, and predation upon them (Kryzhanovsky 1956; Soin 1968). The full set of proposed guilds will not be discussed here, only the major categories (again, this system is hierarchical). The most inclusive category, the section, contains a number of sub-sections, each of which contains the individual reproductive guilds. There are three sections; nonguarders, guarders and bearers. Within these are, respectively: open spawning nonguarders and brood-hiding nonguarders, substrate-choosing guarders and nest-spawning guarders, and external and internal bearers (Table 1).

Table 1. Reproductive guilds of fishes (after Balon 1975b), with most guilds omitted for the sake of brevity.

Section	Subsection	Guild
A. Nonguarders	A.1 Open spawners	A.1.1 Pelagophils A.1.6 Psammophils
	A.2 Brood-hiders	A.2.1 Lithophils A.2.5 Xerophils
B. Guarders	B.1 Substrate-choosers	B.1.1 Lithophils B.1.4 Pelagophils
	B.2 Nest-spawners	B.2.1 Lithophils* B.2.2 Phytophils* B.2.3 Psammophils* B.2.5 Speleophils* B.2.8 Actiniariophils*
C. Bearers	C.1 External	C.1.1 Transfer brooders C.1.3 Mouthbrooders* C.1.6 Pouch brooders
	C.2 Internal	C.2.1 Ovi-ovoviviparous C.2.3 Viviparous

*Indicates guild known to contain cichlid species.

Overview of Tilapia Life Histories

The question of reproductive guilds is an obvious one to apply to tilapias. Several authors have proposed categorizations of these fishes on this basis (e.g., Baerends and Baerends-van Roon 1950; Wickler 1966a; Barlow 1974; Keenleyside 1979). The usual distinction has categorized the cichlids, including especially tilapias, as either substrate-spawning or mouthbrooding species (Lowe-McConnell 1975). The distinctions are obvious and clear, with few if any ambiguities assigning species to either category. The only problematic species are those few which behave initially as substrate-spawners (i.e., they do not take the fertilized eggs into the buccal cavity but direct custodial care towards embryos outside the body), but subsequently (up to as long as a few days later) take the embryos into the mouth and complete the cycle in a fashion essentially similar to that of a typical mouthbrooding species (Timms and Keenleyside 1975; Keenleyside 1979).

Further elaborations of these schemes include considerations as to whether young are taken back into the parent's mouth after they are first released (in mouthbrooding species), and the precise location of spawning. These generally assume (usually implicitly) that release from the mouth or hatching from the egg envelope(s) is a significant event for interspecific comparisons in ontogeny. However, as we have shown, this is misleading, as these events are not involved in defining any of the developmental intervals.

The distinction between mouthbrooding and substrate-spawning habits is also significant in taxonomic considerations of tilapiine fishes (Trewavas 1973a, 1978, this volume), consistent with the generic divisions of *Sarotherodon* and *Tilapia*, respectively. However, if mouthbrooding has evolved independently a number of times from nest-spawning guarders, it may not represent an evolutionary trend.

In our scheme of reproductive guilds, the substrate-spawning species are categorized as nest-spawning guarders (guarders, for convenience). The mouthbrooders are mouthbrooding external bearers (bearers, for convenience). There is little significant difference between the (usually) accepted classification of reproductive styles of these fish (Barlow 1974) and our guilds. The (apparently) intermediate species have been omitted at present, for lack of detailed information to decide on their guild. The utility of the guild concept extends beyond tilapias, however, so our adherence to it is more than just an attachment to our particular terminoloy.

The life histories of the tilapias can be profitably reviewed from the combined model of ontogeny and reproductive guilds. The close association of these, and their apparent relationship to other ecological aspects of life histories have been described for a *Labeotropheus* species (Balon 1977) and charrs (*Salvelinus* species; Balon 1980). There is good reason, both empirical and theoretical, for believing that heterochronous shifts in early development have been of major importance in the evolution of not only life history styles, but also consequently of independent species (Balon 1980, 1981a). This could be a plausible mechanism for sympatric speciation, perhaps to account for some of the endemism in African cichlids (Fryer and Iles 1969, 1972). We will return to this suggestion later.

We have surveyed published life history data for tilapias, especially as related to growth and development, age at maturity, fecundity, parental care and longevity (Lowe-McConnell 1959; Berns et al. 1978; Gwahaba 1978; Hodgkiss and Man 1978; Babiker and Ibrahim 1979; Dudley 1979; Marshall 1979a, 1979b; Vareschi 1979; and the more general works mentioned previously). At times we have had to paint with a broad brush to complete our picture, as necessary data were not always available, but the fit to our hypothesis has been gratifyingly close, as will be seen.

We wanted to see if life histories fell into categories in our consideration, and whether they fit an evolutionary progression as proposed from our model, i.e., an evolutionary sequence, or back and forth alteration from altricial to precocial in life history styles. We propose that selection should act through heterochronous shifts of character anlagen in early ontogeny, to favor increasingly precocial forms (i.e., species adapt to conditions which tend to become more uniform and stable over time). However, the mechanism exists for this trend to be reversed, so that species will not necessarily become "trapped" in an evolutionary "dead end" of a highly specialized (and therefore unadaptable) precocial form (Balon 1980, 1981a).

We use the terms altricial (generalist) and precocial (specialist) in the general ecological sense, i.e., altricial young being those that are relatively small and incompletely developed, whereas precocial young are relatively larger and more completely developed, at a particular time in ontogeny (the term precocial is not to be confused with *'precocious'* breeding at a small size in tilapias which can occur in both bearers and guarders, i.e., in association with either 'precocial' or 'altricial' young).

So, for example, if we compare young tilapias at the onset of exogenous feeding (Figure 1), young of mouthbrooding species are clearly more advanced and better developed, i.e., precocial. At the corresponding time in development (onset of exogenous feeding), young of guarding species are smaller and less well-developed, i.e., altricial. For this reason, we refer to the guarding species as having a more altricial life style, and the bearers a more precocial style. The same kind of comparison can be made between species within a guild, or between forms within a species, to determine which is more altricial or precocial (since the terms are relative).

The relative position of a species (in terms of its life history style) will be apparent in terms of the relative timing of the developmental intervals in ontogeny. Precocial forms will include a truncated larval period and delayed maturation, and a senescent period of some length. Altricial forms, on the other hand, will have the early periods (embryonic and/or larval) prolonged, an early maturation and a reduced senescent period.

The only study of ontogeny in a species similar enough to tilapia to serve directly as a model (Balon 1977 on *Labeotropheus* species) will form the basis for much of what we will say regarding bearers. There have been several studies of guarding tilapias (e.g., Fishelson 1966a, 1966b; Peters 1963, 1965, 1973) but most have focused on descriptions or compilations of stages of development (not corresponding to our saltatory steps of ontogeny, Balon 1979a), and descriptions of certain important features (e.g., adhesive organs, Ilg 1952) of interest in these species. Data on fecundity are available for a number of tilapias (although still surprisingly few species) and are by no means complete. Data on growth and development (including age at first reproduction) are available for even fewer species, and information on longevity is sketchy at best. Ironically, behavioral data (i.e., reproductive and parental behavior) are perhaps most nearly complete, partly as a result of the popularity of these fish with aquarists, and for interest in them as representative of some of the most specialized parental and social behavior among fishes (Balon 1975b; Maynard Smith 1977; Keenleyside 1979). Consequently, we summarize data for generalized guarders and bearers, as representative of the two alternative life history styles, altricial and precocial, while acknowledging that these data may at times be drawn from different species in a variety of circumstances, including non-tilapiine species in some cases (Balon 1959a, 1959b; Noakes and Barlow 1973a, 1973b; Noakes 1978a, 1978b, 1979; Lowe-McConnell, this volume).

Parental Roles and the Evolution of Parental Care Patterns in Tilapiine Fishes

There is convincing evidence from different sources (e.g., Fishelson 1966a, 1966b; Fryer and Iles 1972; Trewavas 1973a; Barlow 1974; Balon 1975b) that mouthbrooding tilapias (bearers) evolved from substrate-spawning (guarder) species (see Oppenheimer 1970 for a general discussion of mouthbrooding in fishes). As Keenleyside (1979) has noted, tilapias are remarkable, not so much for their well-developed parental care (which is in fact universal among the Cichlidae), but for the fact that uniparental custodial care (of whatever type) is almost invariably carried out by the female. This, paradoxically, resembles the pattern commonly encountered in birds and mammals, but is opposite that found in most other fishes which show custodial (parental) care. The typical pattern among teleosts is that the male is the responsible parent (Barlow 1974; Blumer 1979). In fact, this has given rise to a number of efforts (Dawkins and Carlisle 1976; Maynard Smith 1977; Baylis 1978, 1981; Barlow 1978; Ridley 1978; Perrone and Zaret 1979; Werren et al. 1980) to suggest why the male should so often be the responsible teleost parent.

The tilapias clearly differ from this pattern, and so require either a different, or an extended explanation. All guarding tilapias are biparental, the pattern typical for cichlids in this guild (Barlow 1974, Keenleyside 1979). The bearers are remarkable. All bearing (mouthbrooding) species have been assigned to the genus *Sarotherodon* (Trewavas 1973a, 1978, this volume) a designation we might question, as we have already mentioned, although the generic status is not critical to our discussion, as will become apparent. A few bearer species (e.g., *S. galilaeus)* are biparental (i.e., both male and female parent carry developing young in their buccal cavities) (Iles and Holden 1969; Fryer and Iles 1972). In a few others (e.g., *S. melanotheron)*, the male is the sole custodial parent (Aronson 1949; Barlow and Green 1970; Ridley 1978). All other *Sarotherodon* forms are maternal bearers (i.e., only the female carries the developing young).

Various proposals (e.g., Trewavas 1978) have been suggested as possible alternative evolutionary pathways for the different bearer strategies from the guarding ancestral condition. We do not propose to resolve that controversy, nor suggest a possible explanation for the overwhelming incidence of maternal bearer species beyond the following comments, as this issue is not central to our discussion.

All female-uniparental *Sarotherodon* forms are lek species in courtship and breeding (Fryer and Iles 1972; Lowe-McConnell 1975; Loiselle and Barlow 1978). Field observations on reproductive behavior of male-uniparental *Sarotherodon* forms are insufficient to draw any firm conclusions from (Lowe-McConnell, Trewavas, pers. comm.). This by itself is not a remarkable correlation, since a number of other teleosts also have lek breeding systems, but with uniparental male custodial care (e.g. Centrarchidae, Belontiidae, Gasterosteidae; Loiselle and Barlow 1978; Keenleyside 1979). Such communal mating displays (leks) have been noted (Southwood 1976) as typical of "K-strategists" but irrespective of whether this is a valid generalization, we shall see that it does not necessarily restrict these *Sarotherodon* forms to a strictly K-selected strategy.

Nor is it mouthbrooding (bearing) which is responsible for the parental roles in *Sarotherodon*. The general pattern of parental roles in mouthbrooding (bearing) teleosts does not appear significantly different from that in parental teleosts in general (nonbearers) (Oppenheimer 1970), although the data are very sketchy (Breder and Rosen 1966). There must be some aspect of the ecology and/or life history of the *Sarotherodon* forms which predisposes them to, or necessitates, the predominant role for the female in parental care (see also Maynard Smith 1977). Neither their position as cichlids, nor as lek breeding species nor as bearers can account for the female parental role. The proposal by Barlow (1974) of an evolutionary progression from biparental care, to polygyny and uniparental female care certainly agrees with the major trend in tilapias, assuming selection favors parental care.

If we assume that these species were likely adapted for riverine existence in their ancestral form(s) (Fryer and Iles 1972; Liem 1973; Greenwood 1974), they would be better adapted (over an evolutionary time scale) as altricial rather than precocial forms. The evidence of such fluvial species is that they tend to be adapted to breed seasonally, often in flooded or inundated habitats with strong seasonal and/or yearly fluctuations (Lowe-McConnell 1975; Welcomme 1979a). But if conditions did not permit guarding (e.g., unsuitable substrate for a nest, with danger of exposure, etc.) and hence bearing were favored, maternal bearing (mouthbrooding) would most likely be the end result of an extended evolutionary progression (Barlow 1974; Loiselle and Barlow 1978; Balon 1981b; but see Timms and Keenleyside 1975). Perrone and Zaret (1979) hypothesized that lake-dwelling cichlids, because of the rather constant environmental conditions, with extended or continuous breeding seasons, should be maternal mouthbrooders (bearers). They suggested that riverine cichlids should have a greater incidence of male parental care, including biparental guarding, because of stronger seasonal changes and more restricted breeding seasons. The correlation may be a general one, but may also be difficult to apply rigorously to all tilapias (e.g., Lowe-McConnell, this volume).

Bearing and its attendant adaptations are clearly an advanced, precocial life style, seemingly ill-suited to the kind of habitat we have described for a fluvial species invading a recently-inundated area (but see Philippart and Ruwet, this volume). The key point is the saltatory nature of development we have described. In a very real sense, these fish can have the best of two worlds and both life styles. Through heterochronous shifts in development during ontogeny they can quickly adapt towards a more altricial mode to take advantage of new habitats (e.g., seasonally flooded lagoons), or revert to a more precocial mode under other circumstances (e.g., more stable, long term conditions in large lakes, high competitive diversity of fish taxa) while remaining in the same reproductive guild.

Tilapiine Life Histories and Ontogenetic Inferences

The relevant life history data for guarding and bearing tilapias under natural conditions are summarized in Table 2. There is clearly a correspondence between the reproductive guild (and, hence the genus; Trewavas 1973a, 1973b, this volume) and other aspects of the life history style. Guarders are generally more altricial, bearers more precocial. There can be little doubt that the bearers are more specialized, and almost certainly derived from ancestral guarding species (Fishelson 1966a; Peters 1965; Lowe-McConnell 1959; Trewavas 1973a, this volume). The guarders are more primitive, also, as proposed in the system of reproductive guilds (Balon 1975b, 1981b). As already mentioned, bearing may have evolved several times independently among the tilapias (Trewavas, this volume), and if so it cannot likely be the basis for taxonomic distinctions. In any event, the multiple evolution of bearers is consistent with our hypothesis of altricial and precocial "twin forms".

Comparisons between tilapias in the two different guilds are informative, with at least some confirmation of our predictions. There are some apparent exceptions however. These seem to be the result of adaptive shifts in life history features within each guild, causing the ranges of certain values to overlap between guilds. This will become more apparent when we discuss

Table 2. Comparison of tilapiine species in natural populations*, according to reproductive guilds (modified from Fishelson 1966a; Fryer and Iles 1972; Peters 1973; Lowe-McConnell 1975; Balon 1975b, 1977; Balarin and Hatton 1979).

Characteristic	Guarders	Bearers
Fecundity (fertility)	up to 7,000 eggs	up to 1,000 eggs
Egg diameter (mm)	1-1.5	up to 5
Yolk—% lipid	less than 25	up to 45
Yolk color	clear, pale yellow	opaque orange
Size at onset of feeding (mm)	5-6	9-10
Age at first reproduction (years)	0.5-2	1-4
Total length at first reproduction (cm)	up to 20	up to 40
Courtship	prolonged, monogamous	brief, polygamous
Juvenile mortality	relatively higher	relatively lower
Longevity (years)	up to 7	up to 9

*Extreme values associated with fish stocked in non-native conditions, held in intensive rearing facilities, or other exceptional circumstances have been excluded.

such ecological (rather than evolutionary) shifts in life history features.

The bearing species have a relatively lower fecundity, larger ova (with higher yolk density), a longer growth interval, later age at maturity (with overlap between guilds), and possibly (the data are few) greater longevity than the guarding species (see also Lowe-McConnell, this volume). The bearers lack the larval period in their ontogeny (Balon 1977), an adaptation for increased protection of the young. The young are released in an advanced state of development (as juveniles), at a relatively large size (up to 15 per cent adult body size) (Figure 1). The guarding species have significantly higher fecundity, with correspondingly smaller, less energy-rich eggs (Fishelson 1966a; Fryer and Iles 1972; Hanon 1975; Balon 1977). Their development includes an abbreviated but distinct larval (alevin) period. Maturation (onset of the adult period) is relatively earlier, the growth interval shorter, and life span appears to be shorter (again, these data are not strong) (see also Shine 1978).

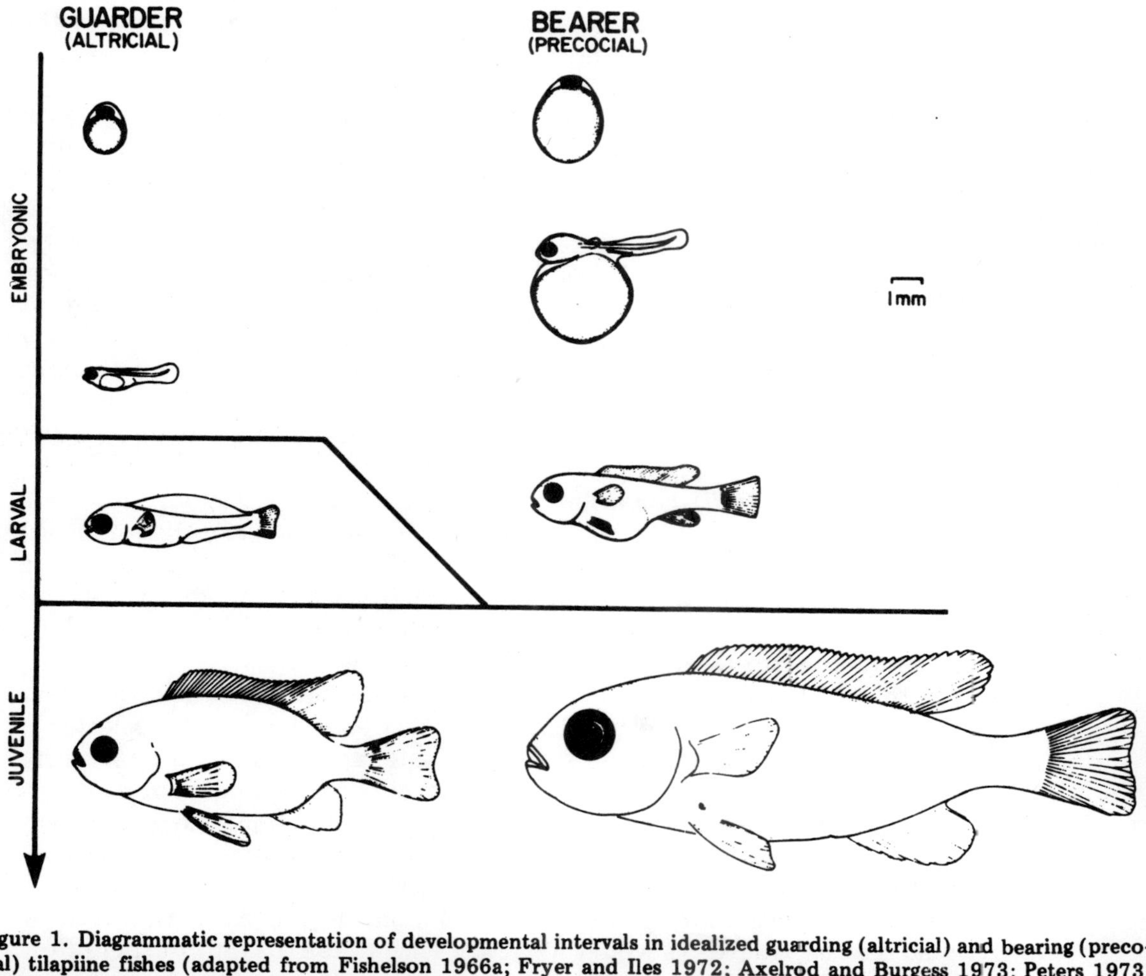

Figure 1. Diagrammatic representation of developmental intervals in idealized guarding (altricial) and bearing (precocial) tilapiine fishes (adapted from Fishelson 1966a; Fryer and Iles 1972; Axelrod and Burgess 1973; Peters 1973; Balon 1977, 1981a). Durations of intervals as indicated by vertical extent in the diagram, are not to scale.

This correspondence in life history features of the two groups of tilapias is not surprising, and at least some of these features have been noted by almost all previous investigators who have studied more than a few species (e.g., Baerends and Baerends-van Roon 1950; Fryer and Iles 1972; Lowe-McConnell 1975, this volume). We emphasize it since it agrees with our basic hypothesis, and leads to some interesting predictions. These are particularly relevant as they may apply to management of these fishes, and their responses to environmental perturbations over an ecological time frame (i.e., during the ontogeny of individual fish).

Saltation of Ontogeny, and Altricial to Precocial Tilapias

The hypothesis of saltatory development predicts that species should respond to environmental situations, both on the time scale of ontogeny and of phylogeny, by heterochronous rates in development. These rates will provide a more appropriate "match" to existing environmental conditions by moving the individual more towards the altricial or towards the precocial trend of the life history style. Since these conditions correspond, in the general way we have already described, to "r- and K-selection", respectively, they are of some considerable significance.

Heterochrony is the "shifting of gears" in development (Jones 1972; Gould 1977, 1979; Balon 1980), or change in ontogeny as a result of relative timing of developmental events and intervals. Since we have a uniform set of terminology to apply to ontogeny, we can ask whether heterochrony has occurred, and if so, at what point(s). The answers appear to be directly related to considerations of selective pressures acting on life history styles.

Heterochrony provides the ability to adapt to a fluctuating environment by shifting the timing of appearance of structures and functions. It likely varies among species, or even forms within a species, depending on previous evolutionary history. The trend in evolution will be from altricial to precocial (Balon 1980; Noakes 1981). This is based on the assumption that, other things being equal, communities will tend towards a more mature, competitive ("K-selected") condition through ecological succession, favoring a more precocial life style in the fishes. But juvenilization (paedomorphosis) remains as the process available for "retracing" phylogeny, and prolonging early ontogeny, with its concomitant flexibility and adaptability. By this mechanism, fish will retain some flexibility in ontogeny, and be able to shift back more towards an altricial life style.

In theory, at least, we can derive an altricial form (more or less equivalent to "r-selected"), or a precocial (more or less equivalent to "K-selected") form for a particular tilapiine species by altering the timing of ontogeny (heterochrony). A shortened (or absent) larval period would be related to a longer growth interval, to a later reproductive maturation, and consequent shifts in life history (and vice versa). Since size (a consequence of growth) and fecundity are typically positively correlated in fishes (bigger fish have bigger gonads with more gametes; Lowe-McConnell, this volume), fecundity would also be altered, as would longevity (Balon 1981b). But the net effect of such shifts in ontogeny would depend on the details of each case, as an earlier age of first reproduction will often dramatically increase "r" (Lewontin 1965; Green 1980). Thus earlier age of first reproduction will likely be more significant than changes in fecundity. Ecologists have provided extensive discussions of these considerations, particularly the concept of reproductive value, i.e., the allocation of resources to growth or reproduction at a particular time (Southwood 1976; Horn 1978; Krebs 1978; Hirschfield 1980). More altricial forms would either occupy habitats with persistent, unpredictable fluctuations, or invade "new" habitats before they stabilize (e.g., Courtenay and Hensley 1979). More precocial forms would be better suited for stable conditions and resultant competition, or specialization, or both (Baker and Stebbins 1965; Southwood 1976; Horn 1978; Balon 1981a). In this regard, it is worth noting the generalizations made by Philippart and Ruwet (this volume). They conclude that the macrophagous, substrate-spawning (guarders) *Tilapia* species of the *Coptodon* group all have wide distributions, exclude each other geographically, have been slow to speciate and are closer to the original stock. The microphagous, mouthbrooding (bearers) *Sarotherodon*, on the other hand, are diversified and specialized into small local populations with restricted distributions in the Rift Lakes (with a few exceptions). The distinctions they draw correspond well to the separation we have suggested between the altricial life style of the guarders and precocial life style of the bearers, with attendant adaptations as "generalist" and "competitive" species, respectively.

Also, the report by Lowe-McConnell (this volume) of the results of simultaneous stocking of *T. zillii* (a guarder) and *S. leucostictus* (a bearer) in the Teso dams of Uganda fits the same pattern. *T. zillii* predominated at first, but was later overtaken in numbers by *S. leucostictus*, even though the fecundity of the latter was lower. As we would predict, the more altricial guarder would be better suited (ecologically) for the initial invasion, but with time (and presumably increasing ecological maturity of the system) the more precocial life style of the bearer would be increasingly favored. However, not all such cases of apparent competition fit this pattern: see discussion of the advantage of *S. macrochir* over *T. rendalli* in initial colonization of the man-made Lake Mwadingushu, Shaba (Philippart and Ruwet, this volume) and ousting of *S. variabilis* by *T. zillii* in L. Victoria (Lowe-McConnell, this volume).

But the intrinsic appeal of our formulation is that it does not require dramatic or drastic alteration in a species to produce substantial evolutionary changes, i.e., a species would not necessarily require major new genetic "mutations" to shift its life history style. Such shifts could be the result of relatively minor, genetically-determined changes in the timing of ontogenetic processes. In fact, a more precise formulation provides a mechanism for evolutionary change in either direction between altricial and precocial forms. The sympatric occurrence of two forms of a species, altricial and precocial, has been discussed at length elsewhere (Balon 1980, 1981a, 1981b).

Such a heterochronous shift could make possible a kind of sympatric speciation (through allochronic speciation, perhaps), or only lead to adaptive changes in life histories. As an aside, however, as we mentioned previously, it is worth noting the possible significance of this as a mechanism for the proposed dichotomous splitting of taxa (Løvtrup 1974; Balon 1981b). The possibility that such a mechanism may have been responsible for at least some of the adaptive radiation in African cichlids, as seen in the Great Lakes, is obvious, and clearly warrants attention in the future.

The responses of tilapias during ontogeny to either natural or man-made changes in ecological conditions are perhaps the best evidence to support our hypothesis of heterochrony and saltatory development. Our hypothesis predicts, for example, that as environmental conditions change (or are changed by human intervention) towards either of the extremes described previously, the fish should alter their ontogenetic development accordingly. Sudden and marked alterations in environmental conditions (e.g., flooding of river banks or lake shore lagoons, increases in adult mortality and/or decreases in juvenile mortality) should all favor a more altricial life style. Consequently, we would expect that any or a combination of these environmental features (or changes) should lead to one or more of: an earlier maturation, increased fecundity (clutch size), smaller eggs, faster growth and shorter life span in the fish. Conversely any change(s) in the opposite direction in those environmental conditions would favor a more precocial ontogeny and life style. We would predict the fish should respond by showing some or all of: delayed maturation, decreased fecundity, with slower, larger eggs, and increased longevity.

The data available, both from field and aquaculture observations, appear to support these predictions (see, in particular, extensive reports summarized in Fryer and Iles 1972; Lowe-McConnell, this volume). The comparisons can be made in a number of ways. Fish living (native) in habitats differing in the above ways can be compared. Fish artificially stocked from one habitat to another can also provide data to test these predictions, especially since the introductions could take place either into a habitat the same as, or different from the original one. The responses of tilapias to naturally-occurring environmental changes can also be used here. The responses of these fish in aquaculture facilities, particularly to different types of management procedures, can also be interpreted to test the predictions (Table 3). A brief summary of some examples will illustrate our case.

Table 3. Size at sexual maturity (beginning of adult period), maximum size and longevity in different stocks of tilapiine species. A smaller size at first maturation (within a species, and to a lesser extent within a guild) indicates a more altricial life history style (see text for full discussion) (adapted from Worthington and Ricardo 1936; Ricardo 1939; Lowe (McConnell) 1955b, 1957, 1958; Lowe-McConnell 1975; Garrod 1959; Cridland 1961, 1962; P.J.P. Whitehead 1962; R.A. Whitehead 1962; Iles and Holden 1969; Hyder 1970a, 1970b; Iles 1971; Fryer and Iles 1972; Balon and Coche 1974; Ben-Tuvia 1978; Hodgkiss and Man 1978; Silverman 1978a, 1978b; Bruton 1979; Siddiqui 1979a; De Silva and Chandrasoma 1980). (All *Tilapia* species are guarders, all *Sarotherodon* species are bearers).

Species	Locality	Typical (T) or Dwarfed (D) Stock	Maturation Size (mm)	Maximum Size (mm)	Longevity (years)
T. mariae	Nigeria, Osse River	T	165	300	—
T. mariae	Jamieson River	D	100		
T. rendalli	Lake Kariba	T	—	450	5
T. zillii	L. Kinneret	T	135	270	7
T. zillii	Egypt, ponds	T	130	250	—
T. zillii	L. Naivasha	D	90		
T. zillii	aquaria	D	70		

Table 3 (cont.)

S. lidole	L. Malawi	T	285	390	5
S. mossambicus	Egypt	T	—	300	7
S. mossambicus	Hong Kong	T	165	310	4
S. mossambicus	Sri Lanka	T	150	340	—
S. mossambicus	L. Sibaya	D	100		
S. mossambicus	South Africa	T	—	390	11
S. mossambicus	Buffalo Springs	D	100		
S. mossambicus	aquaria	D	45		
S. m. mortimeri	L. Kariba	T	300	390	8
S. m. mortimeri	lower Malolo River	T	180		
S. m. mortimeri	upper Malolo River	D	90		
S. niloticus	Egypt	T	200	330	9
S. niloticus	L. George	T	—	400	—
S. niloticus	L. Rudolf	T	390	640	—
S. niloticus	L. Rudolf, Crater Lake A	D	250		
S. niloticus	L. Edward	T	170		
S. niloticus	L. Baringo	T	260	360	
S. alcalicus grahami	L. Magadi	D	25	100	
S. aureus	L. Kinneret	T	190	315	5
S. esculentus	L. Victoria	T	230	375	10
S. esculentus	Kavirondo Gulf	T	230	330	7
S. esculentus	Jinja	T	225	340	7
S. esculentus	Mwanza	T	240	325	7
S. esculentus	aquaria	D	105		
S. esculentus	pond	D	164		
S. galilaeus	L. Kinneret	T	220	345	7
S. galilaeus	Sokoto River	D	110		
S. grahami	L. Magadi	D	25	100	—
S. leucostictus	L. Naivasha	T	180	250	—
S. leucostictus	L. Edward	T	210	240	—
S. leucostictus	L. George	T	140	280	—
S. leucostictus	L. Albert	T	260	280	—
S. leucostictus	L. Albert, lagoon	D	100		
S. leucostictus	Uganda, pond	D	120		
S. leucostictus	Kenya, pond	D	70		
S. niloticus	L. Albert	T	280	500	—
S. niloticus	L. Albert, lagoon	D	100		
S. rukwaensis	L. Rukwa, tributary	D	280		
S. saka	L. Malawi	T	275	340	5
S. shiranus	L. Malawi	T	220	290	5
S. s. chilwae	L. Chilwa	T	200	290	5
S. squamipinnis	L. Malawi	T	240	330	5
S. variabilis	L. Victoria	T	220	300	7

(Only the maturation size is given for Dwarfed stocks, since the evidence strongly suggests that these stocks are only phenotypically dwarfed, and so the maximum size remains that for the Typical stocks, in theory.)

The situation of *S. mossambicus* in Lake Sibaya is a well-documented example of a shift in the life history of a species as an adaptive response to (native) environmental conditions (Bruton 1979; Bowen, this volume). It clearly fits into the category of "stunted" or "dwarfed" natural populations of tilapia, known from a number of regions (Fryer and Iles 1972; Bruton 1979).

Both juveniles and adults are dependent upon feeding on benthic detritus. Juveniles feed in shallower, more inshore water, adults in deeper offshore locations. There is a strong correlation between protein content of benthic detritus and depth (and therefore, distance offshore), and consequently, between depth and quality of diet. Thus, juveniles obtain a diet adequate in protein for growth, but as they become larger and feed in deeper waters, they progressively encounter protein deficiency, and as a result adults in the lake are severely emaciated (Bowen 1979, this volume).

The smallest juveniles are released by female *S. mossambicus* in the shallowest, inshore waters. Older (larger) fish are increasingly excluded from the richer (shallow) feeding areas, and so growth declines progressively with age, particularly in comparison with more typical *S. mossambicus* populations in other, nearby areas.

With water levels at or below normal, only the larger adult males are reproductively active. With higher than normal water levels, successively smaller (younger) males are recruited into the breeding stock, indicative of a shift towards a more altricial mode (i.e., earlier age of first reproduction). These smaller males attempt to breed in the most marginal waters around the edge of the lake, including particularly any recently inundated by flooding.

This shift in life history style, together with the typical food-restricted conditions in Lake Sibaya, is obviously the type of adaptive response we would predict in such circumstances. The generally harsh (i.e., food-restricted) environment, together with fluctuating lake levels, produces earlier maturation (onset of the adult period) in Lake Sibaya. *S. mossambicus* in this lake typically breed at about 0.43 (length at first breeding/maximum length), compared to the more typical value of about 0.70 for African cichlids (Iles 1971; Table 3). Others (most notably Iles 1971; Fryer and Iles 1972; Lowe-McConnell 1975) have also mentioned similar cases of dwarfed tilapia stocks, and have correctly interpreted them as representing adaptive responses to particular environmental conditions. Our proposed mechanism, saltatory development and heterochronous shifts in development (ontogeny), appears to be the mechanism involved in these cases.

An earlier age of first reproduction (thus a shorter generation time) tends to have the greatest effect on net reproductive effort, and to greatly accelerate population growth (Krebs 1978) (i.e., it increases "r" very markedly). If fish were to delay onset of sexual maturation, more of their available resources would be channelled into somatic growth. Thus, they would be larger (in size) at the time of their (later) maturation, and likely have higher fecundity (more or less equivalent to fertility, but see Welcomme 1967b, 1970). However, increased fecundity and longevity generally have much less significant effects on "r" than does an earlier age of first reproduction. Since food (and therefore potential for future survival, growth and maturation) is severely and progressively limited for *S. mossambicus* in Lake Sibaya, it is clearly adaptive for these fish to begin breeding when resources are still available to allow diversion into reproduction (and before adult mortality rate begins to increase because of protein deficiency) (Horn 1978).

Lowe-McConnell (1975, this volume) summarized a number of studies of growth and maturation in tilapia species throughout their range. She reports that the same species delay their maturation when living in large, deep lakes, whereas in smaller water bodies (floodplain pools, fish ponds, etc.) the fish breed at smaller sizes and younger ages. The growth rates of fish in the latter cases do not suggest physical "stunting", i.e. they are not small because they are slow-growing. For example, they may breed for the first time at an age of less than one year (compared to three or four years in a big lake); they produce smaller eggs in these situations, and they may spawn repeatedly at relatively short intervals. Inshore species in a lake tend to breed earlier (in age) than do offshore species (e.g., in Lake Malawi), and the same species living in smaller (and therefore more "harsh") lakes breeds at an earlier age (and smaller size) than when living in larger lakes. She also draws attention to an example we shall return to in some detail, that of tilapias in fish culture ponds. Under these conditions the fish (e.g., *S. mossambicus* and *S. niloticus)* can breed when only a few months old, at about one-half the typical size at first reproduction in the field, and produce relatively more, but smaller eggs than fish of comparable size would if they bred at that size in the field.

Fryer and Iles (1972) cite a number of examples, too numerous to repeat here in their entirety, but worth mentioning in a summary fashion. Again, they clearly drew attention to the adaptive nature of such responses in tilapias, for which we are proposing the mechanism. They reported the response of fish in "confined" conditions (i.e., dams, pools, ponds) to breed at smaller sizes (earlier ages), to breed more frequently, and to produce smaller eggs than normal (the extremes of these sometimes being reported from aquarium conditions). Under natural conditions, bigger species grow faster (but these data need to be extended), mature at a later age, and live longer (observations remarkably in accord with our description of the precocial life style) (see also Lowe-McConnell, this volume). From the evidence available, they concluded that these responses, especially the "stunting" of body size, were due to a phenotypic (rather than genetic) change in the fish (i.e., reversible). They also concluded, quite correctly, that the phenomenon is not one of "stunting", but of relatively earlier breeding in the fish (they described the situation as "neoteny") (but see Balon 1981b for a discussion of this term). They defined neoteny as the ability to breed successfully while still in the juvenile interval of development, and elaborated on the potential survival value of this property for tilapias. They suggested that this property is at least in part responsible for the widespread distribution, plasticity and success as colonizers of tilapias, particularly in shallow lakes subject to fluctuations in water levels. Others have noted such "generalist", "colonizing" abilities of these fishes, and have also drawn attention to the apparent relationship between this and the relative lack of adaptive radiation in these species (only about 100 species of tilapia *sensu lato)*, compared to the explosive speciation in genera adapted for stable, lake conditions (e.g., *Haplochromis, sensu lato*).

Some Comments on Culture

In a very real sense, the interests of fish culturists are working at odds with the biology of these fishes. The culturist wishes to raise the fish in a monoculture (or simple polyculture) system, with a high density of fish, high growth rate, and a harvest (yield) of large individuals in as short a time as possible.

As we have outlined, the biology of these fishes is such that under these conditions of rapid turnover, simple ecological community, and frequent environmental perturbations (e.g., draining ponds for stocking, harvest, etc.) they are forced strongly into a more altricial life style. This tends to result in fish showing earlier sexual maturation, and directing much of their resources into reproduction, with consequent lack of somatic growth (especially in females, in the bearers). In fact, these circumstances might well favor one of the guarding species over any of the bearers, since the former are, as a guild, more altricial.

The task in this case would be to select a guarder species that would not mature (sexually) so rapidly as to cancel the inherent benefits of its life style (high fecundity, general ability as an "invader" species). Furthermore, there might be disadvantages associated with the guarding guild (substrate requirements for breeding, decreased likelihood of producing hybrids between species because of prolonged, monogamous courtship and pair formation) that would mitigate against them. However, some evidence (Caulton, this volume, pers. comm.) suggests that under appropriate circumstances, a guarder species (in his case *T. rendalli)* may have considerable potential for aquaculture.

"Stunting": A Misunderstood Phenomenon

Perhaps the most striking aspect of the literature on the culture of tilapias is the so-called "stunting" problem (Balarin and Hatton 1979; Avtalion, this volume; Hepher, this volume; Jalabert and Zohar, this volume; Lowe-McConnell, this volume) encountered in many attempts at intensive aquaculture. In fact, as we and some others have indicated (e.g., Fryer and Iles 1972), the problem is not one of inhibited somatic growth but one of accelerated ontogeny (sexual maturation).

Not only has the problem apparently been widely misinterpreted, and therefore misunderstood, it is also perhaps the clearest possible verification of the predictions of our hypothesis of altricial and precocial trends in ontogeny. Such early maturation is the classic response of an altricial form. There is little doubt that an earlier age of first reproduction will have the greatest effect on the rate of population increase (Lewontin 1965; Wilson and Bossert 1971; Emlen 1973; Southwood 1976; Horn 1978; Krebs 1978; Pianka 1978). It is clearly a heterochronous shift towards a more altricial form, as we would predict.

Tilapiine fishes are almost certainly derived from fluvial ancestors and still retain (at least in some species) the ability to successfully invade newly-created marginal aquatic habitats (Welcomme 1970; 1979a; Lowe-McConnell 1975, this volume). Stocking these fish in a pond or other aquaculture facility under the usual conditions (Balarin and Hatton 1979; Lovshin, this volume) is as close to a replication of those natural (ecological) phenomena as one could imagine. A relatively small number of adults or juveniles is introduced into a previously unpopulated pond, creating a low density, non-competitive situation. Often supplemental food is also provided for the fish (Hepher, this volume; Lovshin, this volume). A natural invasion into a newly-created river floodplain pool or marginal lagoon (Welcomme 1970, 1979a) would be the ecological equivalent of this man-made event. The response of the fish is identical in both situations. There is a shift towards a more altricial life style, with a shorter interval of somatic growth (inferred from smaller body size), an earlier onset of reproductive maturity (again inferred from smaller body size) (Table 3), as well as production of more numerous, smaller eggs (Lowe (McConnell) 1955b; Fryer and Iles 1972).

The apparent misinterpretation of this response in aquaculture conditions as one of "stunting" has serious practical consequences (beyond any theoretical implications). The fish clearly are stunted, in that they are physically smaller than (large) adults known for the same species. But the real phenomenon is of course as we have described it. The fish are not "small for their age", they are "old for their size". The difference is not only semantic, and the consequences of the correct understanding are profound.

This also requires an understanding of the distinction between ultimate and proximate factors regulating these responses of the fish. Our discussion has centered on ultimate (i.e., evolutionary) factors. These must not be confused with proximate (i.e., immediate) factors influencing the animals. The latter might include food supply, water temperature, chemical signals from conspecifics and so on (Guerrero, this volume; Mires, this volume). For example, one result of the misunderstanding that the fish in ponds are "stunted" has been a misguided effort to provide ever more food for the animals. On the assumption that the fish are truly (somatically) stunted, from lack of sufficient food, the solution would appear to lie in providing more and more food. Ironically, this may worsen the problem. Not only may such attempts lead to serious problems of oxygen depletion (Balarin and Hatton 1979; Caulton, this volume; Hepher, this volume), but they may exacerbate the basic problem. If we assume that the fish are responding to what they perceive (by whatever means) as their invasion into a "vacant", low-competition habitat, then providing ever more food may simply add to that perception. Perhaps fish culturists should adopt the opposite strategy. Some evidence for the success of this exists in the literature, but it does not appear to have been fully appreciated or investigated (Hauser 1975; Balarin and Hatton 1979).

We would predict that if one wanted to produce large fish, with late onset of breeding (in ontogeny), and with a long interval of somatic growth, one should do things to push the fish more towards a precocial life style. This would mean that conditions should favor high intraspecific competition, with relatively higher mortality (rate) for younger fish and a lower rate for older individuals (see our previous discussion of the environmental conditions associated with precocial life history styles). We predict, for example, conditions such as restricted food, stable conditions over relatively long time periods, and high population density should favor the precocial life style.

Some evidence from aquacultural practices supports this. For example, in some cases, intensive stocking in aquaculture facilities has practically eliminated reproduction and led to production of desirable (i.e., large body size) fish in a relatively short time (Balarin and Hatton 1979). In fact, Allison et al. (1979) found an inverse relationship between stocking density and reproduction in *S. aureus* (but note the small tanks used by these authors from which it is probably invalid to extrapolate to large culture ponds—Editors). The so-called "paradoxical stunting" reported at lower or intermediate stocking densities, ascribed to the action of "living space on growth" seems at best a *post hoc* attempt to explain the situation. Our hypothesis not only predicts the supposedly anomalous responses at different densities, but makes specific predictions as to the manipulation of tilapias for management purposes.

We feel our approach may be more rational than the great variety of attempts to produce monosex cultures of tilapias for pond culture, to produce all-male hybrid broods, to chemically or surgically castrate fish to inhibit reproduction, or the search for growth-inhibiting factors that operate differentially at different densities. Since it is based on the evolutionary (and ecological) features of the life histories of these fish, and manipulates their biology with a minimum of intervention requiring high technology or direct human action, it should also be the desired course of action for those with a solely practical interest in these fish, particularly in developing countries.

It is rewarding to find that the activities of basic science, particularly taken from first principles and theory, may in the end yield something of value comparable to that normally expected only from so-called mission-oriented research. If our efforts have such an effect in this case, we would view them as a success.

Acknowledgments

The support for our research which has allowed us to develop our ideas in this area has come from the National Science and Engineering Research Council of Canada and the Ontario Ministry of Natural Resources. Final preparation of the manuscript was carried out by DLGN while in the Animal Behavior Research Group, Department of Zoology, University of Oxford, with the financial support of a Faculty Research Leave Award from the University of Guelph. Thanks to Pat and Jeff for their understanding, and to Dr. D.J. McFarland for providing space and facilities.

We thank collectively a number of colleagues who have suffered through and at times contributed to the development of our ideas. One of us (DLGN) profited from time spent as a Research Fellow at the Zentrum fur Interdisziplinäre Forschung, Universitat Bielefeld, B.R.D. in an interdisciplinary project on early development in animals and man. The other (EKB) benefitted from a research leave spent at the Senckenberg Institut and Museum, Frankfurt, B.R.D. and in Southeast Asia. We thank Joan Cunningham for her efforts in preparing the figure, and Marilyn White for her assistance with photographic material. In preparing this review we have drawn extensively, and depended heavily upon the published material of a number of well-known authorities on tilapias. Drs. Fishelson, Fryer, Greenwood, Iles, Lowe-McConnell and Trewavas deserve our special acknowledgement in this regard. All our extrapolations and flights of fancy from their data remain entirely our responsibility.

Discussion

LOVSHIN: I found the paper very interesting and stimulating as a culturist. I have thought along the same lines for quite a while, and I think the points you raise have importance. For example, why does a fish dwarf in a pond? What are the environmental, behavioral or density-related factors involved? However, I have a very strong resistance to your basic conclusion. You said that restrictive food and stable conditions would mean that the fish would not reproduce at such an early age, whereas Dr. Lowe-McConnell has said that tilapia in poor condition will switch to reproduction at a smaller size than those in better condition. I have also found that to be true in culture ponds. You also said there is some evidence that at high densities, reproduction is restricted. There is probably no doubt about that, but you have to understand that, at these high densities, if you do not feed fish they are not going to grow. You have to feed fish to get them to a large size particularly at high densities. This conflicts with your idea of restricting food. Your basic theory is very interesting, but I think we have to determine whether we can manipulate the environment, or manipulate something in some way so that we can eliminate the reproduction of tilapias in their first year. If so, we would in fact have solved many of our culture problems. This would be a tremendous advancement to tilapia culture, especially in developing countries. The high stocking densities to which you refer will reduce reproduction but these are only achievable with high cost, intensive, aerated circulation systems and they require much feeding.

NOAKES: I don't disagree with anything you have said. Obviously, fish are not going to grow unless you feed them. I will try to clarify that position. The question is not so much whether you feed, but given that you feed a certain amount, what leads the fish to channel that food into reproduction or into somatic growth? The problem is that we don't really know to what signal the animal is responding when making such a physiological shift.

LOVSHIN: But you make a very strong point in your paper—"We feel that our approach is not only more rational than a great variety of attempts to produce monosex culture of tilapias fishes for pond culture . . . to chemically or surgically castrate fish to inhibit reproduction . . . prospect for ultimate success." The way it is stated in this paper, it will not work at those high densities; not for culturing large fish.

HEPHER: I would like to support Len Lovshin. You said that you think that at a sufficiently high density, reproduction will stop. I would like to quote an experiment done at Auburn University by Allison who stocked 50,000, 100,000 and 200,000, densities you would suggest will make reproduction stop. Reproduction did decrease, but even at 200,000/ha, he got 38,380 fry/ha produced, which means there was still reproduction enough to affect the whole system and to cause stunting in the population. I would also like to quote the work of Miranova (1969). She held *S. mossambicus* in aquaria and the food was very limited. Fish of 6 g spawned in her aquarium. Of course, they didn't grow much because of lack of food, but they did spawn.

NOAKES: The particular density value we gave is simply a value we took from the literature. I am sure you are well aware that it is difficult to find data to test our ideas. I do not disagree with what you are saying. What I would like to say is that our hypothesis predicts that there should be a trend: if you decrease stocking density, it should tend to increase reproduction. The important thing is not so much to what extent stocking density is a factor, but that we find out what the fish are responding to—whether they spawn or whether they grow. It may be density; it may be any of a number of things.

BOWEN: I think we can divide the factors that are likely to be important into two categories: physicochemical types of cues, and cues related to food. When I need fry for experiments, I can always put juvenile *Sarotherodon mossambicus* of about 4 cm standard length in aquaria and raise the temperatures to 25°C and they will begin to reproduce, even in the presence of abundant food, at this very small size.

Some groups of fishes, such as cyprinids, are known to secrete pheromones when in densely packed communities. These have a negative feedback effect on growth and tend to trigger spawning. The food obviously has a role to play both in terms of quantity and quality. To my knowledge there has been very little attention paid to food quality with regard to this switching mechanism.

NOAKES: In temperate species, in highly seasonal conditions, food itself may not be so critical. The animal will be committed to enter a reproductive cycle. In the tropics, however, food may be much more important.

CAULTON: I would like to make one comment. About three years ago, we tried some high density rearing of *T. rendalli*. We found that they bred very easily, but we did not find many offspring in the pond because they are very cannibalistic. We were really recycling the gonadal products back into the food chain, and we got very good size fish from this system. As the fish has a short reproductive cycle, not too much energy is devoted to breeding and this is a useful way to combat it.

NOAKES: I would like to suggest that anything which alters the juvenile survivorship, or which decreases their mortality, will induce the fish to put more effort into growth and less into reproduction.

TREWAVAS: I would like to suggest that we avoid using the terms precocial and altricial in connection with tilapias. I do not think they help much, and they cause confusion as we already use the word 'precocious' for tilapias which breed at a small size, whether they are substrate-spawners or mouthbrooders—this dwarfing occurs in both groups. In Dr. Noakes' paper these terms appear to be used in two ways: (i) equated with substrate-spawners (altricial) and mouthbrooders (precocial), and (ii) with early and late development, a phenomenon which occurs in both groups. Thus, for example, the dwarf populations of *S. niloticus* breeding precociously in lagoons have, according to Dr. Noakes' terminology, become more altricial (hence *less* precocial) than the larger-growing lake populations.

Then there is heterochrony. The contact reaction of juveniles to parents is activated at an early stage. The reaction can then go on for some weeks. In *S. melanotheron*, which has a large egg, the young are said to exhibit no contact reaction after release. But Bauer's (1968) experiments with extra-orally reared embryos show that the reaction, though a weak one, is initiated at the usual time, but because of the large yolk, the embryos are normally then still in the parental mouth. This can hardly be called heterochrony.

It has been shown that certain structural and behavioral features develop at the same time in both substrate-spawners and mouthbrooders. These are incapable of expression in terms of function in the mouthbrooders, because the embryo is still attached to a heavy egg and is passive in the parental mouth. Hence, the temporal factor is constant, the key difference being the size of the egg, as Fishelson (1966b) showed.

Tilapias in Fish Communities

R. H. LOWE-MCCONNELL

Formerly Overseas Research Service
(Present address: c/o Fish Section, British Museum (Natural History)
Cromwell Rd., South Kensington, London, S.W. 7, England)

LOWE-MCCONNELL, R.H. 1982. Tilapias in fish communities, p. 83-113. *In* R.S.V. Pullin and R.H. Lowe-McConnell (eds.) The biology and culture of tilapias. ICLARM Conference Proceedings 7, 432 p. International Center for Living Aquatic Resources Management, Manila, Philippines.

This paper reviews the information available on the ecology and behavior of tilapias in natural fish communities, the species interactions and factors controlling tilapia numbers, particularly conditions under which tilapias switch from growth to reproduction. Maturation and maximum sizes tend to be smaller (i.e., the fish 'dwarf') in small bodies of water than in larger ones, and populations of fish with low weight for length switch to reproduction at a smaller size than whose in which the fish are in better condition. Males tend to grow larger than females in 'dwarf' populations, even in species in which the two sexes grow to comparable sizes in large lakes.

Comparison of tilapia growth rates from different lakes and rivers shows that within a water body the various species grow at different rates, the faster-growing species reaching a larger size. But the same species will grow at different rates in different water bodies, suggesting that environmental differences are more potent than genetic ones in determining maturation and maximum sizes.

Studies of biomass, production, yield and turnover rates from the Kafue floodplain and Lake Kariba show that only a small part of the total production is cropped as yield in natural waters; the turnover rate (production/biomass ratio) is higher in the substrate-spawning *Tilapia rendalli* than in the associated mouthbrooding *Sarotherodon*. Studies in the equatorial Lake George (Uganda) have shown that the highest yield of tilapia from a natural water body is comparable with yields from unfertilized fishponds. But despite this high sustained yield here, it represents less than 1% of the very high primary production (blue-green algae on which the tilapia feed).

The great plasticity of tilapia growth in natural waters suggests that concentrating research on environmental and behavioral factors affecting growth and the switch to reproduction is likely to be more helpful for tilapia culture than the search for faster-growing genetic strains.

Introduction

For this meeting I had been asked to review work on the behavior of tilapias in natural fish communities and species interactions, to see if any inferences could be drawn from this for tilapia culture in polyculture systems. But as feeding and breeding behavior, microhabitats and the dynamics of spatial segregation have already been reviewed in an earlier paper (Philippart and Ruwet, this volume) the emphasis in this paper has been changed to include data on growth and production of tilapias in natural waters.

From the title one might expect discussion of the role of tilapias in relation to their competitors and predators, and the partitioning of resources where more than one tilapia species is present in a community. There has, however, been little direct study of competition with other species, or of differential predation, though more is known about partitioning of resources.

While reviewing the main sources of information on ecology and behavior of tilapias in natural and semi-natural fish communities (for the species listed in Table 1), the aim here was to consider:

1. The partitioning of resources of (i) food, and (ii) space and time breeding (now dealt with by Philippart and Ruwet).
2. Some of the factors which appear to control tilapia numbers in fish communities, with special reference to the size and age at which different populations switch from growth to reproduction.
3. Growth rates in different water bodies.
4. Production and yield from various natural waters.

Tilapias are preadapted by their breeding and feeding habits to thrive in still water and most of the important tilapia fisheries are in lakes. In rivers, tilapia fisheries are usually important only where there are extensive floodplains, and the fishery is then a seasonal one, geared to the annual flood cycle. In the last two decades the creation of large man-made lakes behind hydroelectric dams on African rivers has stimulated pre- and post-impoundment surveys, which have taught us much about the ecology and behavior of the riverine fish communities and how they adapt to lacustrine conditions.

As we have already seen in the previous papers, riverine tilapias tend to be widely distributed, for example, those of the 'soudanian' region and in the Zambezi system. The main river systems of these regions cross savanna floodplains lying at comparable distances north and south of the equator, where seasonal floods follow the overhead sun, so that they flood in their summer season after temperatures and day lengths have started to increase.

The aquatic life is geared to the well-defined seasonal cycles, the high-water season being the main feeding and growing time for the fishes, including the tilapias, when growth is fast and fat stores are laid down to last through times when food is scarce in the low water season. In these vast river systems floods do not necessarily coincide with local rains, as they may result from rain far upstream arriving downriver after local rains are over. Tilapias are less dependent than are the ostariophysan fishes living with them on the start of the rains, or flood, to initiate spawning. These riverine tilapias generally spawn at the end of the dry season, producing at least one batch of young before the water level rises, and continuing to produce batches in succession through part of the highwater season. The growth checks in the dry season show on the tilapia scales as annual rings, so more is known about ages and growth rates of floodplain tilapias than about those dwelling in large lakes, where more climatically stable conditions throughout the year lead to continued growth, making scale ring analyses very difficult. As the floodplains dry out, the tilapias along with the other fish retreat to the main rivers, or are trapped in pools on the floodplain until released by rising waters the next year. Predation from piscivorous fishes, the numerous aquatic birds and other animals, including man, is most intense as the waters shrink, and also in the dry season when the fish may be trapped in pools and riverbeds where there is very little aquatic vegetation to give cover from predators.

Lacustrine tilapia populations include many of the same species found in the rivers, for example the nilotic species in many East African lakes. Some, like the Luapula/Mweru populations of *S. macrochir*, may make regular movements between river and lake. But in some long-enough isolated lakes endemic tilapia species have evolved: for example the *Sarotherodon* species flocks of Lake Malawi, and tilapia species in small crater lakes in West Africa.

As tilapias support the main commercial fisheries in most African lakes (except Lake Tanganyika) it was predicted that they would become the dominant fish in the new man-made lakes, such as Lakes Kariba and Kafue on the Zambezi system, Lake Volta in Ghana, and Lake Kainji on the Niger. The speed with which the tilapia populations have built up from indigenous riverine populations has depended on factors such as the draw-down for hydroelectric needs (which can destroy tilapia nests in shallow water), and perhaps flow-through/storage ratio, a high flow-through militating against the buildup of planktonic food sources. Tilapias became predominant species within two years in Lake Volta and three years in Lake Kariba (Petr 1969), but took longer to build up in Lake Kainji (Lewis 1974).

Ecological Information from Particular Water Bodies

The tilapias most studied in natural and semi-natural communities, and on which this review is based, are listed in Table 1.

1. Nilotic Tilapias from Rivers and Lakes

Fish communities in the soudanian region include three relatively common tilapias, *Sarotherodon galilaeus* and *S. niloticus*, and *Tilapia zillii*, plus the smaller less common *S. aureus* (syn. *S. monodi)*, which is more abundant in the Nile delta and Israel. These fish communities also include many much larger fishes, some of which may compete for resources with tilapias, such as *Labeo* and *Citharinus* which share the bottom detritus used as food by *S. galilaeus* in Lake Chad (Lauzanne 1972). Others are large piscivores, such as *Lates niloticus*, *Hydrocynus* species and *Gymnarchus niloticus*, all known to include tilapias in their diets. The tilapias are generally restricted to pools, lagoons and edges of the river and floodplain, places where aquatic vegetation provides cover. Daget (1954) pointed out that in the Upper Niger *S. galilaeus* predominates over sandy bottoms, *S. niloticus* over muddy ones (as Holden 1963, also found in Sokoto river pools), while *T. zillii*, an ecologically tolerant species, lives over rock, sand or mud, in still or running water. A macrophyte-feeder, *T. zillii* is generally found near plant stands; riverine *S. galilaeus* are detritivores using bottom algae, and *S. niloticus* takes epiphytic algae, diatoms or plankton according to the conditions.

In the lacustrine conditions of the huge but shallow Lake Chad, *S. galilaeus* is the common species, except in the north and northeast where *S.*

Table 1. *Tilapia* species for which ecological information is available from natural and semi-natural waters. These species may be Indigenous (I), Endemic (E), Stocked (S), or have just Appeared (A), probably escapes from dams or ponds.

	S. alcalicus	*S. andersonii*	*S. aureus*	*S. caroli*	*S. esculentus*	*S. galilaeus*	*S.g. multifasciatus*	*S. jipe*	*S. karomo*	*S. karongae*	*S. leucostictus*	*S. lidole*	*S. linnelli*	*S. lohbergeri*	*S. macrochir*	*S. sp. nov.*	*S. melanotheron*	*S. mortimeri*	*S. mossambicus*	*S. niloticus*	*S. pangani*	*S. rukwaensis*	*S. saka*	*S. shiranus*	*S.s. chilwae*	*S. squamipinnis*	*S. steinbachi*	*S. variabilis*	*T. busumana*	*T. discolor*	*T. rendalli*	*T. sparrmanii*	*T. zillii*
RIVERS																																	
Niger			I			I														I													I
Nile			I			I														I													I
Upper Zambezi		I													I																I	I	
Middle Zambezi		A													S			I													I	I	
Lower Zambezi																			I													I	
Pangani								E													E												
Malagarasi									E							E																	
MAN-MADE LAKES																																	
Volta						I														I													I
Kainji						I														I													I
Nasser						I														I													I
Kariba		A													S			I													I		
Kafue		I													I																I		
McIlwaine															S			S															
Nyumba ya Mungu					S			I													I										S		
LAKES																																	
Kinneret (Israel)			I			I																											I
Chad						I														I													I
Albert						I					I									I													I
Turkana (Rudolf)						I														I													I
George/Edward											I									I													
Victoria					E						S									S								E					S
Malawi										E		E											E	E		E					I		
Chilwa																									E							I	
Mweru															I																		
Rukwa																						E											
Magadi	I																																
Nakuru	S																																
Naivasha											S																						
Bosumtwi							E																						I	E			
Barombi Mbo				E									E	E													E						
W. Afr. lagoons																	I																
Sibaya																			I														
Hong Kong Res.																			A														
Nicaraguan lake																			S														

niloticus predominates, and in the southern deltas of the inflowing rivers where *T. zillii* lives over varied types of bottom (Blache et al. 1964). The daily food consumption of *S. galilaeus* in Lake Chad, studied by Lauzanne (1978a, 1978b), depends on the water temperature; the conversion of the detritus eaten is very low (only 3%). In the new man-made lakes of Volta (Ghana), Kainji on the Niger, and Nasser/Nubia on the Nile, *S. galilaeus* has become the dominant tilapia, though in the Volta lake *S. niloticus* also became fairly abundant on the eastern side of the lake, and *T. zillii* in certain western arms. In Lake Kainji the buildup of tilapia populations was delayed for some years, perhaps because the large annual drawdown (10m) affected potential spawning grounds; furthermore *Citharinus* became very abundant when Kainji first filled, the lake filling coinciding with, or stimulating, a good *Citharinus* spawning year, and conditions for the survival of the young were good as predator populations had not yet built up. In Lakes Turkana (Rudolf) and Albert, which both have nilotic faunas, *S. niloticus* is the larger-growing and more abundant tilapia, feeding mainly on phytoplankton. Though here again *S. galilaeus* is found off sand banks; *T. zillii* is rarely seen in the main lake. Lake Albert also has another inshore-dwelling *Sarotherodon*, *S. leucostictus*. In Lakes George/Edward, which drain into Albert, *S. niloticus* and *S. leucostictus* are abundant (the latter in the inshore zone), but *S. galilaeus* and *T. zillii* are absent from the fauna. Both Albert and Turkana have populations of dwarf *S. niloticus*, living in lagoons along the Albert lake shore, and in crater lakes on Central Island in Lake Turkana.

S. niloticus was also stocked into many lakes and dams in Uganda. In some of these the fish grew very large while the populations were small, but then multiplied rapidly and became dwarfed; in other lakes they apparently never grew large, which may have reflected the paucity of suitable food. Data on *S. niloticus* from many of these lakes and dams, summarized by Lowe (McConnell) (1958), brought out the following points:

1. The distribution of *S. niloticus* within a water body varied with the ecological conditions. Phytoplankton, epiphytic diatoms or bottom algae were used as food according to whatever was available.
2. The size to which the fish grew, and at which they matured, varied greatly; fish from large lakes matured at, and grew to, a larger size than those from lagoons or ponds. (The largest up to 64 cm TL came from Lake Rudolf, Worthington and Ricardo 1936). (Figure 1)
3. In the lakes the males and females did not differ significantly in maturation size or final size, but in small bodies of water where the fish were dwarfed, males were larger than females, and also much more numerous than the females.
4. Fish in poor condition (low weight for length) matured at a much smaller size than fish in good condition (Figure 2). The growth rates and ages of large and small fish at maturation were not known, but in some newly stocked waters *S. niloticus* grew to 350 g (equivalent to about 26 cm TL) in one year, whereas in ponds they grew to 17 cm TL (98 g) and started to breed when 7 to 8 months old.
5. In equatorial waters *S. niloticus* in breeding condition may be found at any time of year, but peak spawning coincides with the two rainy seasons (a finding later confirmed by Gwahaba 1973). At higher

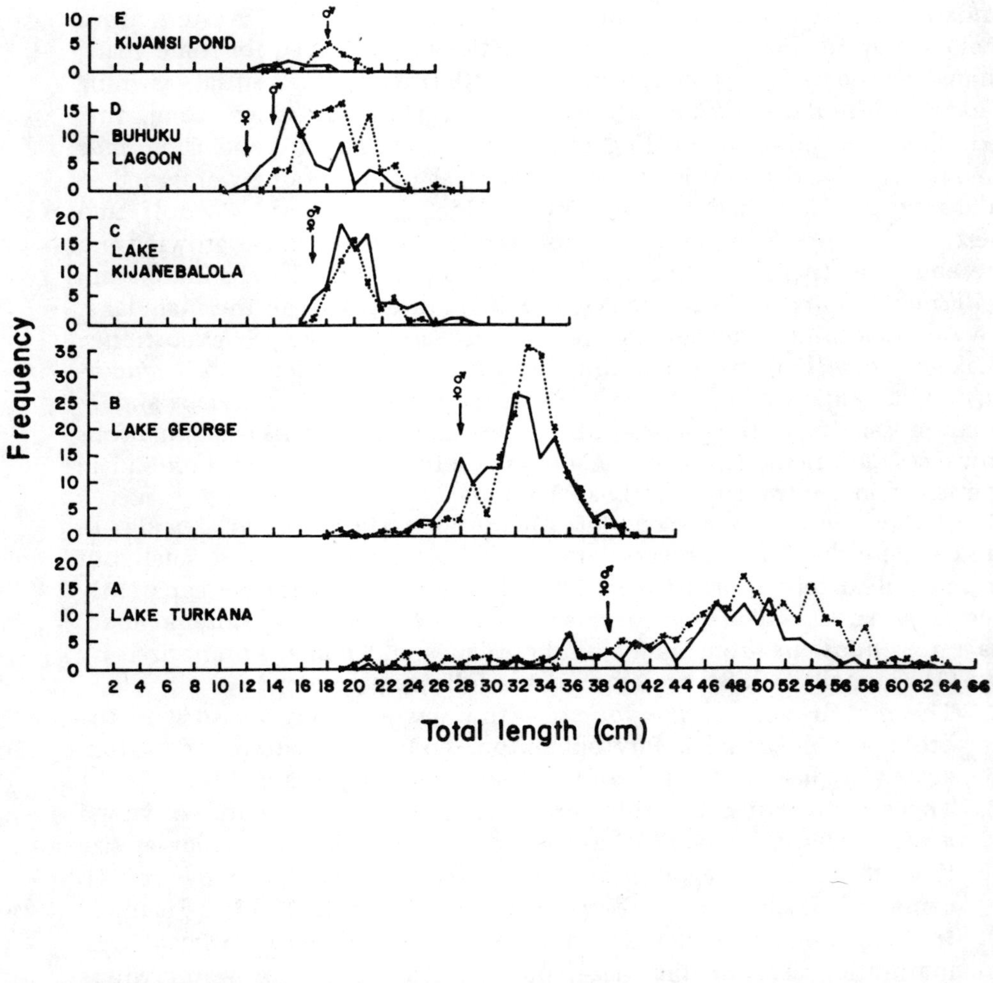

Figure 1. Length frequency distributions of male (x . . . . x) and female (● —— ●) *Sarotherodon niloticus* from various East African waters (after Lowe (McConnell) 1958). A, Lake Turkana, seined fish, maturation size (m.s.) 39 cm TL, both sexes; B, Lake George, gillnetted fish, m.s. 28 cm both sexes; C, Lake Kijanebalola, gillnetted fish, m.s. 17 cm both sexes; D, Buhuku lagoon, Lake Albert, seined fish, m.s. 14 cm male, 12 cm female; E, Kijansi pond, Uganda, fish examined alive when pond drained, males of 18 cm in breeding colors.

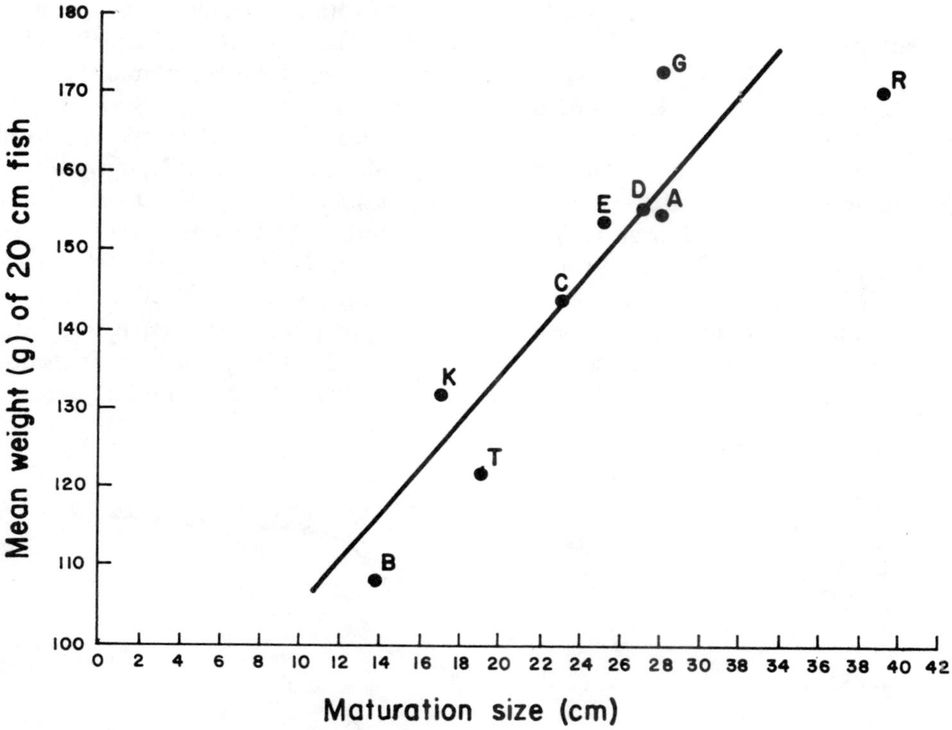

Figure 2. The relationship between weight for length ('condition') and maturation size of *S. niloticus* populations in various East African waters (after Lowe (McConnell) 1958). Mean weight of 20 cm TL fish plotted against maturation size of population for fish from Lake Turkana (R), Lake Albert (A), Lake George (G), Lake Edward (E), Lake Katinda (D), Lake Chanagwora (C), Lake Kijanebalola (K), Tonya lagoon of Lake Albert (T), Buhuku lagoon of Lake Albert (B). (Calculated means are used for Lakes Turkana, Albert and Chanagwora as samples from these waters had no 20 cm TL fish).

latitudes this species breeds mainly in the rainy season.

6. The dwarfed *S. niloticus* from shallow waters were often heavily parasitized, those from the large lakes relatively free of parasites.
7. Some hybridization had occurred between *S. niloticus* and *S. spilurus niger* both stocked in Lake Bunyoni, and possibly between *S. niloticus* and *S. esculentus* both stocked in an Ankole crater lake.

During the intensive International Biological Programme studies on the ecology and production of *S. niloticus* in the equatorial Lake George (Uganda), Gwahaba (1973) found that the maturation size of these fish had diminished since earlier studies (Figure 3). This he attributed to the effects of intensive fishing. Early in the history of the fishery, 50% of the females were mature at 27.5 cm TL, by 1960 this had fallen to 24.5 cm, and in 1972 was 20 cm. Data on *S. leucostictus* in Lakes George/Edward and Albert, where it is indigenous, and from numerous lakes, dams and ponds where it has been introduced, summarized by Lowe (McConnell) (1957), show that, in this species too, maturation size varies greatly with the conditions (from 26 cm TL in Lake George to 7 cm TL in ponds). And about twenty years later Gwahaba (1973) found the maturation size in Lake George had declined to 14 cm TL, though the reason for this decline was not clear. In certain waters male *S. leucostictus* matured at, and grew to a larger size than the females; in ponds this species bred when only three months old. In some ponds immature males were known to have grown faster than the females.

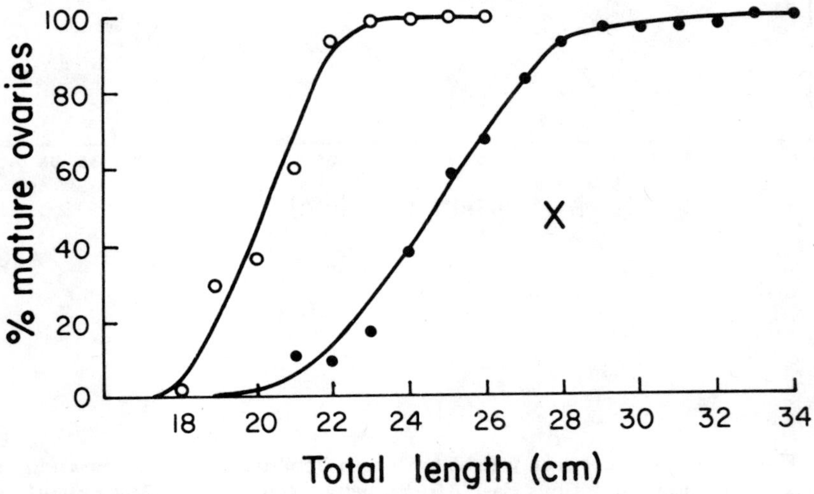

Figure 3. The decline in maturation size of female *S. niloticus* in Lake George, Uganda. Gwahaba's 1971-72 maturity curve (o) compared with Fry and Kimsey's 1960 curve (●) and Lowe (McConnell's) (1958) mean maturation size (x) (after Gwahaba 1973).

In natural waters *S. leucostictus* occupies an inshore zone and is common in lagoons. In equatorial lakes, as in ponds, breeding occurs throughout the year, without well-defined seasons. Egg numbers increased about 100 in a 7 cm female to roughly 1,000 in a 26 cm female, a relationship later studied by Welcomme (1967a). Small females tend to produce smaller, less robust fry than large females. Eggs hatch in about six days after fertilization and young are brooded for a further three weeks, though for the last week only at night. Lake George individuals caught in midlake showed a preponderance of blue-green algae in the gut, whereas fish in the inshore regions had ingested detritus; since this species is more abundant inshore, it is essentially a detritivore (Moriarty et al. 1973).

Introductions of *S. leucostictus* showed how rapidly tilapias can be established in new waters from very few fry. All the *S. leucostictus* in Kenya waters appear to have originated from six specimens between 6 and 11 cm TL taken to Kisumu ponds from Uganda. In the Teso dams (Uganda) where *T. zillii* and *S. leucostictus* were stocked simultaneously, *T. zillii* predominated in the first few years after stocking, but were gradually overtaken in numbers by *S. leucostictus*, although this species lays fewer eggs at a time (Lowe (McConnell) 1955b). Such changes may have been related to food supply; in one of the dams *T. zillii* predominated while dense stands of macrophytes existed, but as it cleared these from the dam, *S. leucostictus* became the predominant species.

S. leucostictus introduced into Lake Naivasha, Kenya, in 1956 hybridized with another introduced species *S. spilurus niger*. Hybrids were abundant in 1961, but had disappeared by 1974 when Siddiqui (1977b) studied the lake. He found *S. leucostictus* then to be the dominant species, and to spawn throughout the year, though Hyder (1970a) had previously maintained they had a relatively quiescent period here from July to September. Males matured at 18 cm TL, females at 16 cm. Fecundity ranged from 320 to 1,328 eggs in 16 to 32 cm females. Males predominated in the catches.

2. Lake Victoria

The nilotic *S. niloticus* and *T. zillii*, together with *S. leucostictus*, have been studied in Lake Victoria to see how they have fitted in with, and taken over from, the indigenous *S. esculentus* and *S. variabilis*. Earlier studies (Lowe (McConnell) 1956a; Fryer 1961a) had indicated that these two indigenous tilapias were ecologically complementary species with little or no competition between them; *S. esculentus* predominated in land-locked bays where the bottom was of soft flocculent mud from phytoplankton deposition, while *S. variabilis* was most abundant off more exposed shores, in water lily lagoons and in the outflowing River Nile. Both species occupied different biotopes according to the size and sexual condition of the fish. *S. variabilis* spawned in shallower water than *S. esculentus* (spawning behavior observed in a water lily swamp, Lowe (McConnell) 1956a), and juveniles occurred in different nursery zones.

In an attempt to boost the tilapia fisheries, *T. zillii* from L. Albert was introduced to Lake Victoria in 1954, and along with it came *S. niloticus* and *S. leucostictus*. Tilapias in ponds draining the lake may also have gained access to the lake around this time. *T. zillii* appeared in gillnet catches by 1956, *S. niloticus* and *S. leucostictus* both appeared in commercial records for the first time in 1960 (Welcomme 1967a). Some suspected hybrids between (i) *T. zillii* and *T. melanopleura* (from pond escapes), and (ii) *S. variabilis* and *S. niloticus*, were described by Welcomme (1967b). The spread of the introduced species was described by Welcomme (1966): *T. zillii* spread rapidly at the northern end of the lake, appeared in the southern and eastern parts in 1960, and was abundant by 1964; it later became the dominant species in the former *S. variabilis* areas. Meanwhile *S. niloticus* and *S. leucostictus*, both rarely seen in commercial catches before 1962, became much more abundant from 1964 onwards, *S. leucostictus* becoming the dominant species in lagoons around the lake and near the papyrus and in the shallows of muddy bays, *S. niloticus* on the former *S. esculentus* grounds, where they were caught both in surface and bottom set gillnets. *S. niloticus* from offshore areas grew much larger than the indigenous tilapias (personal observation c.1970). The *T. zillii* lived in shallow marginal waters all round the lake, fry and juveniles (to 5 cm TL) inhabiting sandy shelving rock shores in sheltered places, situations favored as *S. variabilis* nurseries.

Welcomme considered possible competition between juveniles, and also between adults for feeding and breeding places. There appeared to be little competition for food between adults: *T. zillii* taking higher plant material, *S. esculentus* planktonic diatoms, while *S. niloticus* and *S. variabilis* had more flexible diets, feeding either on the bottom or on epiphytic or planktonic diatoms depending on the habitat. *S. leucostictus* seemed to feed exclusively on bottom material (Welcomme 1966, 1967b). Spatial segregation of feeding individuals helped to reduce any competition, *S. leucostictus* being confined to shallow margins of the lake near, and often behind the papyrus fringe (in water 30 cm to 5 m deep), a habitat shared with young *S. esculentus* (of 13 to 18 cm TL), while *T. zillii* and *S. variabilis* tended to spread into harder-bottomed habitats and more exposed waters. The larger *S. esculentus* remained in mud-bottomed bays 5 to 10 m deep. *S. niloticus* was found in most of the habitats. All the introduced species grew to a large size in the lake and were in good condition (high weight for length).

Competition for breeding grounds was more apparent. The possible lack of suitable areas for spawning, and of nursery grounds for the very young, appeared to be a factor limiting population expansion. This view was supported by the dramatic increase in *S. esculentus* catches in 1964/65 following the unprecedented increase in lake level (1.4 m above the previous highest recorded level) in 1961-62. This flooded lagoons behind the papyrus fringe, places then used as *S. esculentus* nurseries (Welcomme 1970). Clean, firmer substrates were needed on which to spawn by the guarder species *T. zillii*, and by the very small *S. variabilis*, and the ousting of *S. variabilis* by *T. zillii* appeared to be due to competition between them for such places for breeding and nursery grounds. *S. leucostictus* spawned in shallow (30 cm deep) water in mud-bottomed areas at the edges of lagoons; here the males in breeding colors guarded their nests at sites not frequented by other species. *S. esculentus* evidently spawned deeper, where it could find firm enough bottom in the sheltered gulfs, but the juveniles lived in the lagoons at the lake edge. *S. niloticus* appeared more catholic in its habits and could be found in association with any of the other species; its preference for hard substrates was reflected in some hybridization with *S. variabilis*, but later it thrived in *S. esculentus* areas.

Welcomme's (1964) study of juveniles on nursery beaches showed that juvenile tilapia were abundant on gently sloping 'gradient' beaches, *Haplo-*

chromis off more steeply shelving 'non-gradient' beaches. The small tilapias could withstand temperatures up to 38°C, lethal to the *Haplochromis*. Tilapias moved on and off these gradient beaches in accordance with the diurnal temperature fluctuations, the shallows cooling rapidly at night and the fish then moving into deeper water. An approximately linear relationship was found between the size of fish and depth of water. This enables small tilapia to feed without disturbance from larger individuals. Also they are then out of reach of the piscivorous catfishes, *Clarias mossambicus* and *Bagrus docmac*, often caught just off the tilapia beaches. The nursery beaches, though sharing high temperatures, differed in their dissolved oxygen tensions (DO), for which the different species had characteristic, though to some extent overlapping, preferences. *S. leucostictus* young were confined to areas of low DO (0.6 to 2.0 ppm), *S. esculentus* and *S. niloticus* lived in slightly better aerated water (1.0 to 3.0 ppm and 2.0 to 4.0 ppm respectively), while *T. zillii* and *S. variabilis* needed well-aerated slightly exposed areas. Competition for nursery areas appeared greatest between *T. zillii* and *S. variabilis*.

The introduced *Lates niloticus* fed mainly on cichlids in Lake Victoria and included some tilapia in its diet (Gee 1968, 1969).

3. Zambezi System

Zambezi system tilapias have been much studied as part of the pre- and post-impoundment surveys of hydroelectric schemes at Kariba and Cabora Bassa on the Middle Zambezi and on the Kafue tributary (see Balon and Coche 1974; Bowmaker et al. 1978; JFRO Annual Reports, and for information on the Barotse floodplain on the Upper Zambezi, Duerre 1969). Tilapias flourish better in the more lake-like 'reservoir rivers', which retain water and a good growth of aquatic vegetation throughout the year, than in the 'sandbank rivers', which dry out into pools devoid of plant cover in the dry season. Throughout the Zambezi system the presence of large piscivores, especially *Hydrocynus vittatus (Hepsetus odoe* in the Kafue), appear to control the ecology of the tilapias. The Upper Zambezi and Kafue have two indigenous species of *Sarotherodon*, *S. macrochir* and *S. andersonii*, and two *Tilapia*, *T. rendalli* (which greatly resembles *T. zillii* in the northern soudanian floodplains) and the smaller *T. sparrmanii*. In the Middle Zambezi *S. mortimeri* (which greatly resembles *S. mossambicus* in the Lower Zambezi) is the dominant species, though *S. macrochir* is also present; *S. andersonii* had not been recorded from the Middle Zambezi until 1971, after Lake Kariba had been established for twelve years. This example stresses that tilapias may move about more than realized, but strays do not get established unless they find suitable ecological conditions, in this case due to a man-made change.

Summarizing information on Upper Zambezi tilapias, Bell-Cross (1974, 1976) described *S. macrochir* there (the subspecies with a 'volcano' shaped nest) as being well distributed and abundant throughout the Upper Zambezi and its larger tributaries, growing to 1.8 kg. It feeds mainly on periphyton and detritus, though other algae are taken in summer months when they are abundant. Breeding starts early in the summer, probably triggered by increasing water temperature and/or light periodicity; there appears to be a correlation between onset of rains and breeding activity, in the river at least one brood of young is produced before the floods, and one or more during the highwater season (in ponds four broods may be produced between September and February). The female mouthbroods the young. Once juveniles leave the floodplain, where they grow fast, they are restricted to lagoons and backwaters until they reach a size large enough to coexist with piscivorous *Hydrocynus.*

The Mweru/Luapula system has a subspecies of *S. macrochir* which has a characteristic star-shaped nest; this undertakes seasonal migrations from the deeper north end of Lake Mweru to the shallower southern end and up the Luapula River during the period July to October (it breeds September to March), with a reverse movement northwards in February-May; it appears to need shallow water in which to spawn, and cover for the young. Juvenile *S. macrochir* have a higher tolerance of low DO than adults, and remain on the spawning grounds up to a length of 18 to 20 cm. Marshall (1979b) gives egg numbers ranging from 1,000 to 1,500 in 16 to 27 cm SL female *S. macrochir.*

S. andersonii, a larger-growing species (males up to 3.2 kg) occurs widely except in small streams and is found in all types of habitat, but it is a shy fish which likes to hide in deep pools. Though it takes the same food as *S. macrochir*, it appears to be less specialized and will also eat aquatic and terrestrial insects and small Crustacea. The female mouthbroods, and seems to produce only one brood a year, between November and January. The eggs are large (374 to 593 eggs in females of 17 to 25 cm TL; Mortimer 1960).

T. rendalli, widely distributed and abundant throughout the Upper Zambezi, moves up tributary rivers and onto the floodplain, but spawns once in the dry season before leaving the river as well as on the floodplain; juveniles in tributaries continue to move upstream. This species prefers water with dense growths of aquatic plants. In addition to macrophytes and detritus, it eats aquatic and terrestrial insects. It grows to 1.4 kg. A multiple spawner, it breeds during the rains between October and February; the nest is a series of small holes made in shallow water where both parents guard eggs and young.

The smaller *T. sparrmanii* (growing to 0.25 kg in rivers—the largest, up to 25 cm in Lake Young) is the commonest cichlid in the Upper Zambezi and distributed throughout the system; it prefers quiet water with good weed cover, in riverine, swamp and lake habitats. It is a fairly omnivorous species, feeding primarily on periphyton, aquatic plants, detritus and algae, but Crustacea, insects (terrestrial and aquatic) and even worms and fish may be eaten. A multiple spawner, batches of eggs are produced between October and February, both parents guarding the eggs and young. In Lake Young Ricardo-Bertram (1943), who described the spawning behavior there in shallow water, considered that suitable breeding grounds with sandy bottom were very limited, and there was great intraspecific competition for nest sites, the nests almost touching one another.

Predators on these Upper Zambezi cichlids include piscivorous cichlids, *Serranochromis* spp. (though these feed mainly on cyprinids and cyprinodonts). The characoid *Hydrocynus vittatus* takes many cichlids and its presence appears to be a factor limiting cichlid populations. Underwater observations have confirmed that cichlid distribution in the openwaters of the main rivers is limited to those tilapia which have managed to reach a size large enough (c.19 cm TL) to withstand predation by *Hydrocynus*. In the Kafue *Hepsetus odoe* is the common openwater predator, but in the Upper Zambezi *Hepsetus* is apparently restricted by competition to habitats not frequented by *Hydrocynus*, such as lagoons, backwaters and smaller tributaries. Bell-Cross (1974) commented that in years of poor rainfall, when the main river is confined between sandy banks for most of the year and the isolated floodplain pools, which serve as reservoirs for tilapia stocks, dry up, the predation on juvenile and small adult cichlids must be extremely severe. Preimpoundment studies on the Kafue floodplain suggested that predation is particularly high at two periods a year: for tilapia in the rivers in the dry season, and for juvenile fishes as the floodplains dry up, a time when piscivorous fishes *(Hepsetus* and *Clarias)* and the numerous birds and other animals can catch them very easily (Williams 1971).

The preimpoundment survey for Lake Kariba in the Middle Zambezi showed *S. mortimeri* (then called *S. mossambicus*) to be the commonest tilapia, but not nearly as abundant as *Labeo*, *Distichodus*, *Hydrocynus* and *Alestes*. The tilapia hugged the shore or lived in vegetation rather than in openwater, the juveniles in very shallow water. Some *T. rendalli* (then called *T. melanopleura*) lived in weedy shallows. In an attempt to provide a more openwater-living species the new Kariba lake was stocked with *S. macrochir*, known in Lake Mweru to live in openwater, but in the event stock was taken from the nearby Kafue, so was what was later known to be the 'volcano-nest' subspecies, not the 'star-nest' one used to openwater conditions. In the four months after dam closure in 1959 over 11,000 fingerlings of *S. macrochir* and *T. rendalli* were stocked. For many years these *S. macrochir* were not seen; many of the stocked fish were apparently consumed by *Hydrocynus* in the new lake, a piscivore not previously encountered by the Kafue fishes (Van der Lingen 1973). It was the indigenous *S. mortimeri* which thrived and underwent a population explosion in the new lake, becoming very abundant within three years. Recently *S. macrochir* have reappeared and are slowly increasing in numbers (Bowmaker et al. 1978). In Lake Kariba, after the flooded terrestrial vegetation rotted (by 1963) there was a hiatus before the establishment of rooted macrophytes, and tilapia numbers then fell until these were established (Bowmaker et al. 1978). Perhaps these recently found *S. macrochir* came from upstream, for *S. andersonii* also appeared for the first time (in 1971) when conditions had been altered by the new lake and there was cover for the fish.

In Lake Kariba, Donnelly (1969) found that juveniles less than 10 cm TL of both *T. rendalli* and *S. mortimeri* lived in water less than 30 cm deep, the 'primary nurseries' on gently sloping shorelines. Fish of 11 to 19 cm TL (age one year) tend to leave these for adult habitats, though they are not sexually mature till at least a year later. *S. mortimeri* over 20 cm TL live in a variety of habitats and in water up to 15 m deep. *T. rendalli* is confined to macrophyte beds, though adults return to shallows to breed, or during a rise in lake level to feed on decomposing grasses. These Kariba tilapias breed throughout the year, but with spawning peaks from October to

December. Their nesting sites are ecologically distinct. *T. rendalli* nests along the lake margins in water 60 to 100 cm deep, sometimes in colonies, the parents depositing the alevins in pits. The *S. mortimeri* nests appear to be in water 4 m deep. The juveniles move from the nurseries into deeper water at night, but in Lake Kariba tilapia keep near weed cover till 17 to 19 cm TL (though in dams which lack *Hydrocynus*, tilapia schools of all sizes mingle at varying depths). The largest tilapia in *Hydrocynus* stomachs were 18 cm TL. In Kariba *Hydrocynus* seldom enter water shallower than three times their own length; small *Hydrocynus* taken in the primary nursery areas were still feeding on invertebrates. Other piscivores here are the catfishes *Clarias gariepinus* and *Eutropius depressirostris; Mormyrus deliciosus* also forage in the deep nurseries. The combination of active predation (mainly by *Hydrocynus)* and fluctuating water levels has enhanced the importance of submerged plants in providing cover for tilapias in Lake Kariba.

The Lower Zambezi is the type habitat for *S. mossambicus*, now a circumtropical pond fish. Bell-Cross (1976) described this species as an omnivorous feeder, taking plankton, algae, aquatic and terrestrial insects, shrimps and vegetable matter. A polygamous mouthbrooder, several broods are produced during the summer (starting about September but varying with water temperature). Some move up tributaries to spawn. The most thorough ecological studies of this species in natural waters are those made in Lake Sibaya (Kwazulu, South Africa).

4. LAKE SIBAYA

S. mossambicus lives in Lake Sibaya with *T. rendalli swierstrae* and *T. sparrmanii* (see Bruton 1979). *S. mossambicus* nests on terraces in open, shallower water than that used by the other two species, whose nests are among macrophytes. Male *S. mossambicus* move in to establish territories in September, when the water temperature first exceeds 20°C. There may be up to 13 nests per 1,000 m^2. There is intense intraspecific conflict between males prior to excavating nests. Small males (10 to 13 cm SL) are repulsed and returned to deeper water; they may establish nests if the lake level rises, but not at low lake level. After a brief active courtship the female leaves the nest and mouthbroods eggs and fry for 14 to 22 days in the macrophyte beds. The fry when about 9 mm SL are released in very shallow water (10 cm deep) along barren sandy shores or in marginal pools. They occupy progressively deeper water by day as they grow larger, but all move into deeper, warmer water at night, where they lie motionless. The juveniles, which return to the shallows when the temperatures first exceed those of the main lake in the morning, form small schools (up to 20 fishes) on terraces by day for feeding; the composition and size of the feeding groups change constantly.

Scale studies have shown that males grow faster than females after the first year, as discussed below. This Lake Sibaya *S. mossambicus* population is breeding precociously, then growth deceleration occurs and the final size is smaller than in other natural systems in southern Africa, for reasons discussed by Bowen (this volume).

The large adult catfish *Clarias gariepinus* feeds almost exclusively on *S. mossambicus* in this lake (as studied in some detail by Bruton), and both juveniles and adults are preyed on by the numerous birds: cormorants, darters, herons, kingfishers, fish eagles and ospreys.

5. Pangani River System

Another east-flowing river, the Pangani in Tanzania, has two endemic tilapias, *S. pangani* with a subspecies *S. pangani girigan* in Lake Jipe where it is sympatric with *S. jipe*. In Lake Jipe *S.p. girigan* feeds on water weeds *(Najas)*; the former pharyngeal teeth of *S. jipe* suggest it consumes smaller particles (Lowe (McConnell) 1955a). A new man-made lake on this Pangani system, Nyumba ya Mungu reservoir completed in 1965, studied by Bailey and Denny (1978) and Bailey et al. (1978) has three species of *Sarotherodon* (the indigenous *S. pangani* and *S. jipe*, introduced *S. esculentus*) and *T. rendalli*. These indigenous *Sarotherodon* grew to a large size, about 50 cm TL and 1.75 kg (much larger than previously recorded from Lake Jipe). They were found mainly in the littoral areas of the southern part of the lake, where *S. jipe* clearly predominated. The *S. esculentus* lived in the southern deeper offshore waters, where it consumed phytoplankton. *T. rendalli* found in shallows close inshore to fringing reed swamp was surprisingly rare in the north of the lake which has marked development of macrophytes. *S. jipe* and *S. pangani* were chiefly browsers of periphyton and bottom deposits (phytobenthos). In both these species males grew to a much larger size than females (*S. jipe* males to 48 cm, females to 27.5 cm; *S. pangani* males to 47 cm, females 31.5 cm TL). In Lake Jipe and the Pangani River, where males were much more numerous than females in catches of these species, male maturation and maximum sizes were also larger than in females (Lowe (McConnell) 1955a).

6. Malagarasi Swamps

Two endemic species of *Sarotherodon* live sympatrically in the Malagarasi swamps (over 1,100 km^2) which drain westwards into Lake Tanganyika: *S. karomo*, whose breeding behavior in the swamps was studied by Lowe (McConnell) (1956a), and a new species (then referred to as a form of *T. nilotica*). These breed at different times of year. *S. karomo's* teeth appear beautifully adapted for rasping epiphytic algae off the abundant water lilies and other macrophytes.

7. Lake Malawi

Lake Malawi has a unique species flock of four endemic *Sarotherodon* species adapted for openwater life: *S. squamipinnis*, *S. saka* and *S. lidole* distributed round the lake where there are areas of water less than 40 m deep, and *S. karongae* apparently confined to the northern end of the lake (Lowe 1952, 1953). These fish, often in mixed schools living in midwater, feed on phytoplankton. *S. lidole*, the most openwater-living, may move considerable distances to do so (80+ km), the others may turn to bottom feeding if phytoplankton is scarce. These tilapias have evolved relatively short annual breeding seasons at staggered times. Season and place of spawning, reinforced by differences in colors of the breeding males, keep the species distinct. *S. saka*, which has a black breeding male, spawns from August to November, the hot weather before the rains in relatively shallow water (4 m deep) near macrophyte beds, off weedy or sandy shores; brooding females move into weed or reed beds. *S. squamipinnis*, the non-breeding fish of which are scarcely distinguishable from *S. saka*, spawns in the rainy season, December to February, in deeper water (about 16 m), generally off more sandy, open shores; brooding females also retire into the weed and reed beds; the breeding male is sky blue with a white head which is clearly visible to the ripe females which swim above the nesting arenas (see Berns et al. 1978). *S. lidole*, like *S. saka*, also has a black breeding male and breeds before the rains, but in much deeper water or off open sandy, clean weedless beaches.

Fecundity appears to be much reduced in these Malawi *Sarotherodon*, by late maturation (not spawning till three years old), the production of but one brood a year (rarely two?), and of relatively few but large ova (370 to 549/fish in *S. squamipinnis*), the young being brooded to a larger size than in other tilapias (to 30 mm TL in *S. squamipinnis*, 52 mm in *S. lidole)*. In Lake Malawi, *S. saka* and *S. squamipinnis* mature when about 25 cm TL, *S. lidole* at 28 cm TL, with no size difference between the sexes; in aquaria they can mature at a smaller size (10 cm SL—for details of aquarium behavior see Berns et al. 1978).

S. shiranus, of different stock, lives in a more inshore zone of Lake Malawi and has a more extended breeding season, but mainly December-January. *T. rendalli* appears to be confined to vegetated lagoons round the lake.

Malawi tilapias show adaptations to inhabit increasingly openwater conditions, from the inshore-dwelling *S. shiranus* to the most offshore living *S. lidole*, viz: (a) the food contains a higher proportion of phytoplankton rather than bottom diatoms and bottom debris; (b) the tendency to collect in small, close schools increases; (c) the amount of movement increases, *S. shiranus* populations being very localized and *S. lidole* far-ranging fish; (d) the young are brooded to an increasingly large size; (e) the rate of growth in length appears to increase, but growth in weight for length decreases, giving a more streamlined form in openwater species. Each species spawns in the clearer water end of its range, but broods young in the richer, greener water. The brooding females carry the eggs and young inshore, from spawning to brooding or nursery grounds. It is possible to tell from the characteristics of the water, such as clarity, color, and whether the shore is of mud, silt, sand or rock, shore vegetation, etc., which species are likely to predominate off a particular beach, though mixed catches are often taken. *S. shiranus* predominates in muddy swampy areas, *S. saka* off sandbanks near weedbeds close to swampy *(Papyrus)* and reedy *(Phragmites)* shores, *S. squamipinnis* off more open beaches with *Phragmites*, and *S. lidole* off open, clean, clearwater beaches or where the bottom shelves steeply, and out into openwater.

8. LAKE CHILWA

Lake Chilwa, lying southeast of Lake Malawi, has a subspecies of *S. shiranus*, *S. s. chilwae* as a predominant member of its very small fish fauna (which includes *T. sparrmanii*, one species of *Barbus*, one of *Haplochromis* and two of *Clarias*). Mass mortalities of the fish occurred when this lake virtually dried out in 1965-68 (see Furse et al. 1979). Tilapia populations started to decline ahead of the main drying out—possibly due to sandy shallows used as spawning grounds being reduced in area (which seems to happen about every six years or so in this lake). Populations of dwarfed tilapia survived in the permanent streams of the basin and pools in the marshes; these fish were able to repopulate the lake when it refilled. But aided by stocking, it still took three years before tilapia numbers were high again. This subspecies is smaller in adult size than *S. shiranus* in Lake Malawi, matures earlier, produces smaller and more numerous eggs, and has two main spawning periods in an extended breeding season. It also matures more quickly in freshwater when rains come again. In Lake Rukwa in Tanzania, Ricardo (1939) had found *S. rukwaensis* to have dwarf populations in river pools near the lake after it dried out. These appeared adapted to repopulate the lake after it refilled.

9. WEST AFRICAN CRATER LAKES

Endemic lacustrine tilapias have also evolved in small crater lakes in west Africa. Lake Bosumtwi (Ghana), whose fish communities have been studied very comprehensively by Whyte (1975), has three tilapias, *Sarotherodon galilaeus multifasciatus* (an endemic subspecies), *Tilapia discolor* (endemic and very like *T. zillii)*, and *T. busumana*. The adult *S. g. multifasciatus* live in midwater feeding on phytoplankton, here mainly blue-green algae and desmids. Unlike other tilapias (and *S. galilaeus* in Lake Chad) this species is said to feed at night, between 6:00 P.M. and 5:00 A.M. undergoing a diurnal vertical migration to do so; sinking to 30 m by day and rising to surface waters at dusk to feed. Adults of both *Tilapia* species are bottom dwellers. *T. busumana* feeds mainly on aufwuchs on stony and sandy substrates by day and moves into deeper water (3 to 7 m) by night. *T. discolor* feeds mainly on periphyton on reeds *(Typha)* and water plants, and on aufwuchs, but feeds mostly by night, moving into deeper water (5 to 10 m) by day. The juveniles of all three species are surface dwellers round the lake edge, using different foods from their adults; they all share the allochthonous insects on the surface, eaten together with phytoplankton in juvenile *S. g. multifasciatus* and *T. discolor*, while *T. busumana* juveniles take aufwuchs instead of phytoplankton. There is thus separation in feeding habits between species, between juveniles and their adults, and also in vertical use of space, juveniles living in surface waters, *S. g. multifasciatus* in midwater, both *Tilapia* species at the bottom, one feeding by day the other by night.

Other cichlids in Lake Bosumtwi are *Chromidotilapia guentheri*, a benthos feeder in which the juveniles take the same food sources as the adults, and *Hemichromis fasciatus*, the juveniles of which share the allochthonous fauna at the lake surface with juvenile tilapia, as do the juveniles of the catfish *Heterobranchus isopterus*. Adult *Hemichromis fasciatus* and *Heterobranchus isopterus* are both piscivorous, eating juvenile tilapias. Once the tilapias reach a certain size they are relatively safe from predation.

The breeding places and habits of these three tilapias also segregate them. Most *S. g. multifasciatus* spawn between November and April (though some ripe fish may be found at any time of year). Nests are made at 1 to 3 m in the littoral zone over a *sandy* substrate, among reed beds *(Typha australis)* or in openwater; eggs are brooded by both parents and brooding fish may be found 5 to 20 m from shore. Of the two *Tilapia* species, *T. discolor* has a protracted breeding season from December to May, making nests with axillary pits at 1 to 7 m in the littoral zone over *muddy* bottoms (240 to 700 eggs per nest). *T. busumana* peaks from October to February (though some are ripe at any time of year), making a bowl nest in permanent streams and the littoral zone over *sandy* or stony shores in 10 cm to 1 m water (60 to 280 eggs per nest). The nests are of very different form: the *T. busumana* nest is a simple bowl; the *T. discolor* nest has axillary pits around the main nest. These tilapias do not seem to grow very large.

In west Cameroons the oligotrophic Lake Barombi Mbo has four endemic species of *Sarotherodon*, of which two species, the *galilaeus*-like *S. steinbachi* and *S. lohbergeri* share organic debris obtained in shallow water, from sandy areas by *S. steinbachi* and aufwuchs sucked off rocks by *S. lohbergeri*. Both species also probably take some food when skittering at the surface (Green et al. 1973). The other two species (*S. linelli* and *S. caroli* probably related to the *S. melanotheron* group) share phytoplankton-feeding a meter or so below the lake surface; possibly differences in breeding seasons help to keep these two species apart (Trewavas et al. 1972). Another Cameroon crater lake, Lake Kotto is eutrophic with a dense bloom of blue-green algae, food of three tilapias, *S. galilaeus*, *T. mariae* and the endemic *T. kottae*. The two *Tilapia* species both take some bottom food (invertebrates in *T. kottae*) as well. This lake lacks the aufwuchs-feeding niche found in the oligotrophic Barombi Mbo. The phytoplankton eaten by adults is so abundant in Lake Kotto that tilapia numbers are more likely to be limited by availability of food and feeding habitats for the young stages which live and feed inshore (Corbet et al. 1973).

10. Kenya Rift Soda Lakes

Lake Magadi in the Kenya rift, a very alkaline lake (pH 10.5), with hot springs (43°C), and a high specific gravity (1.01 to 1.03) is inhabited by a little *Sarotherodon*, *S. alcalicus grahami* (formerly known as *S. grahami*) whose ecology and behavior was studied by Coe (1966). These fish browse on blue-green algae on the stony bottom, together with small crustaceans (copepods), and dipteran larvae, in the warm shallows near the hot springs, moving into deeper water at night when the surface temperature begins to fall. High temperatures (above 32°C) appear necessary for the males to assume breeding colors in these fish, and at high temperatures they breed almost continuously. Females move to brooding areas, and somehow manage to take some food while brooding; they release the brooded young in shallow nursery areas. Here the young escape predation by adult fish, but are open to attack from numerous kinds of birds, especially pelicans. There is also some cannibalism of eggs.

This species was introduced into Lake Nakuru around 1960, where it is now one of the main herbivores and has altered the whole ecology of the lake (Vareschi 1979). Fish distribution here is very patchy. At noon they concentrate near the shore, at night offshore, a migration pattern probably reflecting preference for high temperatures. They feed primarily on the very dense cyanophyte *Spirulina platensis*, 80% of the tilapia living in the top 100 cm where they feed. The total ichthyomass of *S. alcalicus* in this lake was estimated to increase from 90 t (dry weight) in 1972 to 400 t in 1973 (2.1 g/m^2 to 10.2 g/m^2). The main impact of this introduction on this lake has been to increase diversity by extending food chains to fish-eating birds. Before the introduction such birds only occasionally visited the lake, now over 50 species are recorded there, of which the great white pelican *(Pelecanus onocrotalus roseum)* predominates. Adult pelicans are estimated to eat 1,330 g (fresh weight) of tilapia a day, the young some 770 g/day from hatching to fledging. Calculations suggest pelicans remove 16 to 20,000 kg fresh weight of fish per day from the lake. Since they nest on a neighboring lake, taking fish there to feed their young, they have been calculated to export some 13 t phosphorus a year from the Nakuru system.

11. West African Coastal Lagoons

In west African coastal lagoons *Sarotherodon melanotheron*, a species which can withstand high and changing salinities, has been studied as a possible candidate for fish culture with mullet *(Mugil* sp.) by Pauly (1976). This tilapia only grows to 25 cm TL, the adults feeding on bottom mud (30% organic content, 1.2 Kcal/g calorific content). Daily food intake of a 20 g fish was estimated to be 1.5 g (dry weight) and the calorific assimilation 900 cal/day. Juveniles take zooplankton and phytoplankton. *S. melanotheron* is a male mouthbrooder (whose behavior has been much studied in aquaria under the name *T. macrocephala* or *T. heudelotii*, e.g., papers by Aronson 1951). In the field this species breeds throughout the year with equinoxial peaks. Parasitism is low, attributed to the varying salinity in the lagoon.

12. Exotic Communities

One of the best-studied examples of the role of a tilapia in an exotic fish community is that of *S. mossambicus* in Plover Cove Reservoir, Hong Kong, where an arm of the sea has been turned into a freshwater storage reservoir (Hodgkiss and Man 1977a, 1977b; Man and Hodgkiss 1977a, 1977b).

An initial twenty species in the reservoir fish fauna included marine species which survived but did not reproduce, rice paddy and riverine fish from the drainage basin, and Chinese carps stocked in 1967-71. The *S. mossambicus* were escaped pond fish which established themselves and soon became the dominant species in gillnet catches. Chiefly a detritus-feeder, the tilapia here had the most diverse food spectrum of any species studied, thus proving to be a good reservoir fish since almost all items were exploited, and when certain items were scarce a change to others was possible. Tilapia also helped to control algae and chironomids, and to remove nutrients and decaying organic detritus from the mud, thus indirectly helping to reduce algal growth. Earlier studies on *S. mossambicus* in Indonesia had provided much basic information about this species in Asian waters (Vaas and Hofstede 1952).

In Lake Moyua in Nicaragua (Central America) the stocked *S. mossambicus* matures between 12 and 14 cm (90 to 100 g) when 5 to 6 months old. This lake lacks piscivorous fishes, but numerous water birds feed on the tilapia fingerlings.

13. Hybridization

Where several tilapia species share a water body, nesting areas are usually distinct in place (depth or type of bottom) or time. In substrate-spawners the elaborate courtship behavior probably helps to prevent any hybridization. Courtship behavior is generally much briefer in the mouthbrooders. Cases of hybridization, or suspected hybridization in natural waters, due to introductions include those of:

- *S. niloticus* x *S. spilurus niger* in Lake Bunyoni (both introduced)(Lowe-McConell 1958)
- *S. niloticus* x *S. esculentus* in Lake Nkugute (both introduced) (Lowe-McConnell 1958)
- *S. niloticus* (introduced) x *S. variabilis* (indigenous) in Lake Victoria (Welcomme 1966)
- *T. zillii* x *T. rendalli* (both introduced) in Lake Victoria (Welcomme 1966)
- *S. amphimelas* (indigenous) x *S. esculentus* (introduced) in Lake Kitangiri (Trewavas and Fryer 1965)
- *S. s. niger* x *S. leucostictus* (both introduced) in Lake Naivasha (Elder et al. 1971, who considered how these hybrids fitted into the ecological zonation in the lake). These hybrids, abundant in 1961, following introductions in 1956, had disappeared by 1974 (Siddiqui 1977b).

Tilapia Growth Rates in Natural Waters

The most reliable growth rate data come from waters with marked seasonal changes which affect the biology of the fish, leaving marks on scales, bones, otoliths, etc., and where seasonal spawning allows length-frequency mode progressions to be traced in the young fish. Examples are given in Figure 4, based on data from nilotic tilapias by Blache et al. (1964), Ben-Tuvia (1959), Daget (1956), Jensen (1957), Lauzanne (1978), Payne and Collinson (in

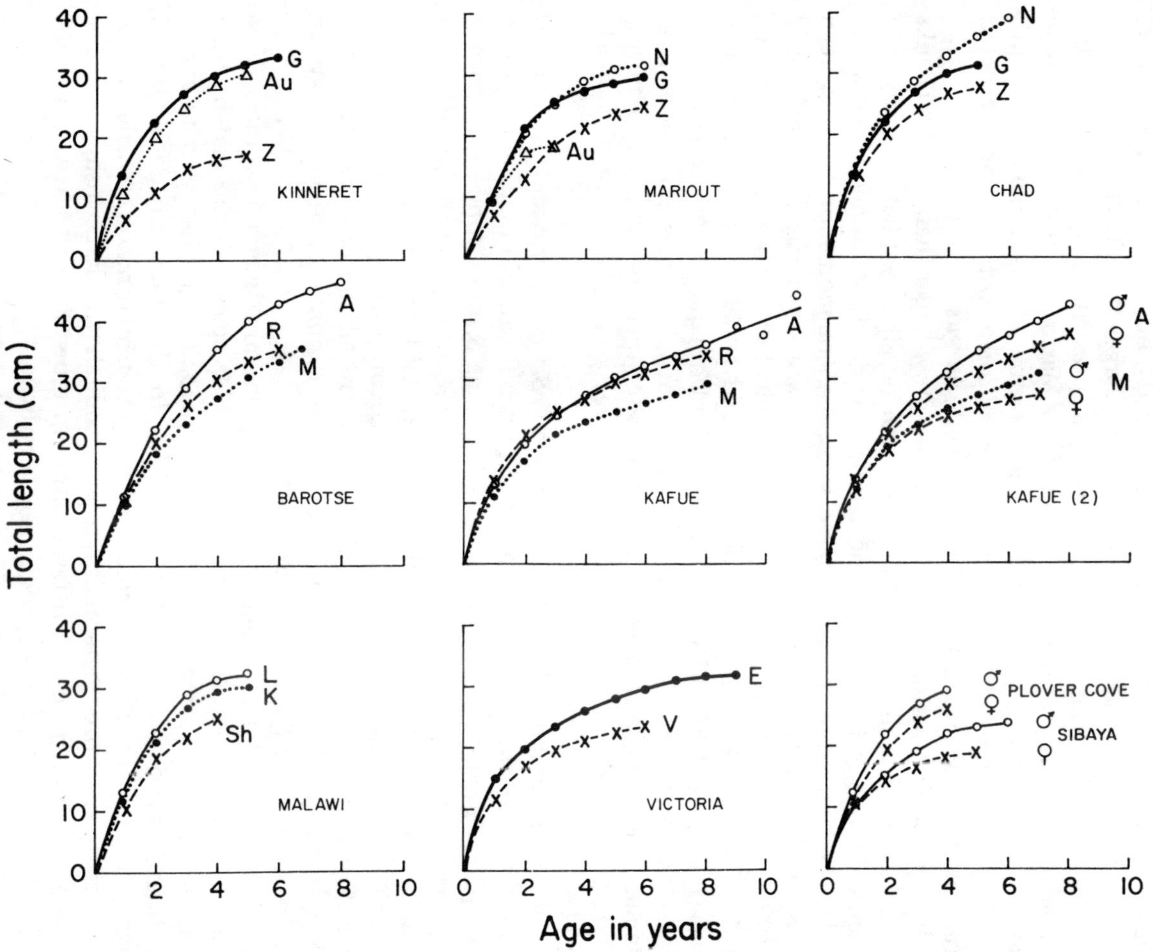

Figure 4. Comparison of tilapia growth rates in various natural waters (for sources of data see text). Nilotic species: *Sarotherodon niloticus* (N), *S. galilaeus* (G), *S. aureus* (Au), *Tilapia zillii* (Z), from Lakes Kinneret (Israel), Mariout (Egypt) and Chad. Zambezi species: *S. andersonii* (A), *S. macrochir* (M), *T. rendalli* (R), from Barotse and Kafue floodplains. Lacustrine species: *S. lidole* (L), *S. saka* (K), *S. shiranus* (Sh) from Lake Malawi; *S. esculentus* (E), *S. variabilis* (V) from Lake Victoria. Male and female curves are shown separately for Kafue (2) and for *S. mossambicus* from Lake Sibaya (S. Africa) and Plover Cove Reservoir (Hong Kong).

press), from Zambezi tilapias by Duerre (1969), Kapetsky (1974), Dudley (1974), for *S. mossambicus* by Bruton and Allanson (1974), Hodgkiss and Man (1977b) and from the Great Lakes from Lowe (1952) and as summarized by Fryer and Iles (1972, Table 14). Total lengths were calculated from standard length data for Lakes Chad and Sibaya.

Growth data for the nilotic tilapias all indicate that within a water body *S. niloticus* grows to a larger size than *S. galilaeus* (though *S. galilaeus* is deeper bodied so weight differences are not so marked): both grow faster and larger than *T. zillii*. *S. aureus* also grows faster and larger than *T. zillii*, but not as large as *S. galilaeus*. Payne and Collinson (in press) have suggested that the extended breeding season of *S. aureus* in Lake Mariout (Egypt) slows somatic growth, for here *S. niloticus* and *S. aureus* grow at comparable rates until they mature at one year old, after which *S. niloticus* with its more restricted breeding season grows faster and larger than *S. aureus*.

In riverine populations *S. niloticus* males grow larger than females (Banks et al. 1965). This also occurs in lagoons and ponds as we have already seen (Figure 1), but in large lakes, where growth is good, males and females may mature at, and grow to, comparable sizes. This phenomenon has also been noted in *S. esculentus*, where growth of the two sexes was not significantly different in lake populations (Lowe (McConnell) 1956b; Garrod 1959). Again, in *T. zillii* in Lake Kinneret, Israel, males and females were of comparable sizes (maturing at 13 to 14 cm) though in the Dor aquaculture station ponds, where they mature at a smaller size than in the lake, males grow faster than females, even though the pond fish grow faster than the lake fish, achieving 18 to 20 cm in one year instead of two years as in the lake (Chervinski 1971).

All three Zambezi tilapias *(S. andersonii, S. macrochir* and *T. rendalli)* grow faster and to a larger size on the Barotse floodplain than on the Kafue floodplain (Figure 4). But on both floodplains *S. andersonii* grows fastest and to the largest size, followed by *T. rendalli*, while *S. macrochir* does not grow as fast or as large. It is interesting that in this case the *Tilapia (T. rendalli)* grows better than one of the accompanying *Sarotherodon (S. macrochir)*, unlike the soudanian region where all the *Sarotherodon* grow better than the *Tilapia (T. zillii)*. The Zambezi region does, however, have an additional smaller *Tilapia (T. sparrmanii)*.

In natural waters, *S. andersonii* rarely matures when less than three years old (at 27 cm TL); a few *S. macrochir* spawn at the end of their first year (when 13 to 16 cm TL). In *T. rendalli* some fish start to spawn in their second year (when 17 cm TL). In the Zambezi, *Sarotherodon* sexual differences in size become marked after maturation, female growth slowing down more than male growth, and all the large fish are males. *T. rendalli* showed no sex differentiation in growth for the first two years. After the sixth year these tilapias grew very little.

From the Kafue, tilapia scale rings permit growth to be backcalculated enabling variations in growth from year to year in the different age groups to be examined in relation to environmental factors such as water temperatures and flood levels. Such data for fifteen years examined by Dudley (1974, 1979) indicated that: a) there are significant variations in growth rate from year to year for most age groups of all three species and b) growth of the young and extent of flooding are positively correlated, suggesting that high flooding would increase growth and survival (Dudley 1974). Later studies, however (Dudley 1979), indicated that year-classes are larger following low-flood years. These *Sarotherodon* will not spawn if the water is too deep, but Dudley considered that high year-classes following low-flood years might be due to the *Sarotherodon* entering a stressed condition and breeding at a smaller size in dry, low-flood years. In support of this view he found ripe females of *S. andersonii* of 18 cm TL in 'dry' years, compared with a minimum breeding size of 26 cm TL in high-flood years (Dudley 1979).

The relative growth rates of young and older fish may differ in different lakes. For example, *T. rendalli's* first year growth appeared to be slower in Lake Kariba than on the Kafue floodplain, but the Lake Kariba population nearly caught up in weight in the fourth year and in length in the fifth year (Kapetsky 1974).

For *S. mossambicus* in Lake Sibaya, South Africa, growth rates were lower and fell off more rapidly (Bruton and Allanson 1974) than for this species in Plover Cove Reservoir, Hong Kong (Hodgkiss and Man 1977b) (Figure 4). The precocious breeding in Lake Sibaya has been explained in terms of food quality (Bowen, this volume), the adults being in poor condition as the food available to the deeper-living adults is less nutritious than that available to the juveniles living in the warm shallows. Female growth rates fell off more rapidly than male growth rates in both places. In Lake Sibaya, females may breed at one year (8 to 10 cm SL), males when one or two years old (12 cm SL); maximum age was 7 to 8 years, the largest males 23/29 cm SL/TL, the largest females 18/22 cm (Bruton 1979).

Tilapia Production In, and Yield From, Natural Waters

For natural waters we have to distinguish clearly between the biological production, i.e., the total elaboration of fish tissue during a given time interval (generally taken as a year), and the yield, i.e., the harvestable part of the production. The catch, sometimes loosely called 'fish production,' is really a yield. Ecologists considering energy transfers from one trophic level to another are concerned with total production (all the fish tissue produced in the time interval, whether or not it survives to the end of that time), whereas fishery biologists are more interested in the available production (the amount of fish tissue surviving at the end of that time). In natural waters the yield (catch) is generally only a small proportion of the available production, depending on the selectivity and efficiency of the fishing method (both the gear used and how much the fish move about). Also, due to the many causes of mortality in natural waters, the available production may be much less than the total production (Table 2 and Figure 5 of data from the Kafue floodplain show this well). In ponds, where fish are cropped by draining the pond, the yield reflects the available production much more closely, and, if predation and diseases are controlled, available production may be much nearer to total production than it can be in natural waters.

Analyses of catch statistics are used for determining whether a fishery is declining, but for computations of production in natural waters growth rate and biomass data are essential. The relatively few comprehensive studies of tilapia production yet made in natural waters include those for the Kafue floodplain and Lake Kariba, both in the Zambezi system, and for the equatorial Lake George in Uganda. For these sites the computed abundances of the fish, their biomasses, production, yield and turnover are indicated in Table 2.

Table 2. Examples of tilapia *(Tilapia* and *Sarotherodon)* production, yield and turnover rates in natural waters.

	Abundance n/ha	Biomass kg/ha B	Production kg/ha/yr Total A	Production kg/ha/yr Available P	Yield kg/ha/yr Y_A	Yield kg/ha/yr Y_P	Turnover A/B
Kafue eastern floodplain							
lagoons (31 spp.)	11,053	680					
channels (24 spp.)	3,894	333					
weighted mean (all spp.)	8,191	541		c. 1,000			
3 spp. tilapia lagoons	2,282	364					
3 spp. tilapia channels	2,351	246					
3 spp. tilapia mean		317					
T. rendalli (lagoons:channels)	706: 187	125	198	110	18	8	1.59
S. macrochir	737:1,327	145	145	96	39	23	1.0
S. andersonii	839: 837	147	119	92	23	15	0.75
Lake Kariba							
20 spp. (revised estimate)	—	827	1,224	720	400	202	
of these 'preferred' spp.							0.7
'accompanying' spp.							2.3
mean turnover ratio (revised)							1.48
T. rendalli (revised estimate)	957	38	44	30	19	12	1.15
S. mortimeri (revised estimate)	2,122	216	178	133	139	103	0.83
Lake George (Uganda)							
All (10+) spp. mean		220 (60-900 inshore)			catch 137		
S. niloticus (offshore:inshore)	63:1,094	37(?) mean			catch 111		
S. leucostictus (offshore: inshore)	7: 649	5(?) mean					

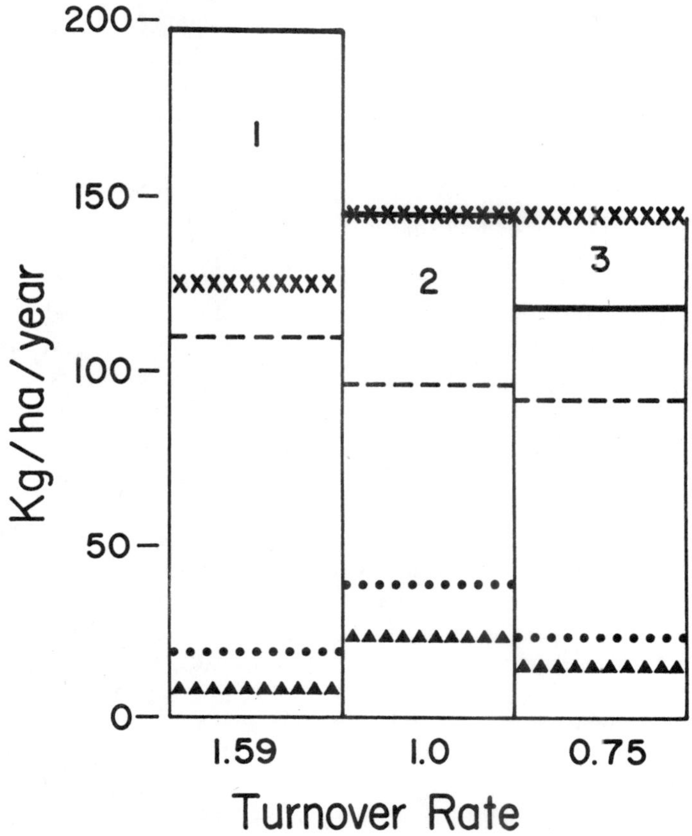

Figure 5. Kafue floodplain production of: 1, *T. rendalli*; 2, *S. macrochir* and 3. *S. andersonii.* Total production (———), Biomass (XXXXX), Available production(— — —), Yield (available) (● ● ● ●), Yield (production) (▲▲▲▲▲): see Table 2 and text (data from Kapetsky 1974).

The Kafue eastern floodplain (38,880 ha) is part of the 6,000 km^2 Kafue floodplain lying at 15° 30′S, 900 m, which produces an annual crop of 5,000 t of fish, over half of the catch (59%) being three tilapia species: *S. andersonii*, *S. macrochir* and *T. rendalli.* The building of the Kafue Gorge hydroelectric dam, closed in 1971, stimulated pre- and post-impoundment studies. Here fish biomasses were determined by chemofishing 25 sample areas (Lagler et al. 1971; Kapetsky 1974). First estimates ranged from 64 kg/ha in grass marsh, 337 kg/ha in open lagoons, 337 kg/ha in river channels, to 2,682 kg/ha in vegetated waters. The areas of different types of habitat, at low and high water, were calculated from planimeter measurements on 1:50,000 maps. Meanwhile tilapia growth rates were determined by scale ring analyses (Chapman et al. 1971; Kapetsky 1974). *S. macrochir* and *T. rendalli* were considered harvestable at the end of the second year, *S. andersonii* end of third year.

The resulting computed biomass, production (total and available), yields and turnover rates of the tilapia populations on the eastern floodplain (Table 2) are shown in Figure 5. This brings out clearly that the total production is considerably greater than the biomass present in *T. rendalli* (turnover rate A/B, 1.59), equal to the biomass in *S. macrochir* (A/B, 1.0), and less than the biomass in *S. andersonii* (A/B, 0.75).

Kapetsky then computed a yield for the whole Kafue floodplain (121,000 ha), weighting for areas of lagoons and channels. This gave an estimated total tilapia production of 70,000 t/yr from the whole floodplain, total yield 13,000 t/yr and available yield 7,000 t/yr. Present catch is 5,000 t/yr of all species (Dudley and Scully 1980), so if computations are realistic it looks as though only about half the possible tilapia crop is being caught (my interpretation of the findings).

The Kariba dam across the Zambezi was closed in December 1958. Here chemofishing in lake coves was used to determine fish biomasses (Balon 1974). The calculations of production etc., for each species and the whole lake had, however, to be revised (Mahon and Balon 1977, whose revised estimates for total production, available production, total yield and available yield, all species, were 38%, 105%, 67% and 107% respectively of the original estimates). Tilapias were a much less important part of the catch (20 species considered) than on the Kafue floodplain. However, *S. mortimeri* (then called *S. mossambicus)*, though present at a 'low population density' (2,122 fish per inhabited ha), had a mean biomass of 215.7 kg/ha and was estimated to give the third highest total production (178.5 kg/ha/yr) (after a mormyrid and *Alestes* sp.), and the highest total yield (139 kg/ha/yr) of the nine economically preferred species. *T. rendalli* was much less abundant, but had a higher turnover rate than *S. mortimeri* (1.15 compared with 0.83).

Lake George (Uganda). In an attempt to determine how this shallow equatorial lake, mostly less than 3 m deep (250 km^2) continues to produce such good catches of tilapia *(S. niloticus*, which forms 80% of the catch, and *S. leucostictus)* an International Biological Programme team spent six years studying production at different trophic levels (Dunn 1972; Burgis et al. 1973; Burgis and Dunn 1978; Burgis 1978; Gwahaba 1975, 1978). Fish biomasses (of 10 species) were determined using openwater seines; they were computed to range from 60 kg/ha in openwater to 900 kg/ha inshore (mean 220 kg/ha). The determination of fish growth rates proved very difficult in this equatorial lake, but for *S. niloticus* was indicated by length frequency analysis, of a particular cohort of young fish, to be about 1.2 cm/month (Gwahaba 1978). The fish yield based on catch statistics over twenty years fluctuated around a mean of 3,461 t/yr for all (10) species, 2,790 t/yr for *S. niloticus*, equivalent to 137 kg/ha/yr for all species, 111 kg/ha/yr for *S. niloticus*. The weight of fish landed had been maintained but, as already discussed above, the maturation size of *S. niloticus* had declined (i.e., the numbers had risen). (This high, maintained catch suggests to me that the biomasses to produce these fish must have been considerably higher than recorded here.) This yield is very high for a natural water body, comparable with unfertilized fish ponds.

Lake George was found to have a very high gross primary production (82,400 kJ/m^2/yr) made up primarily of the bluegreen alga *Microcystis*, consumed and digested by the tilapia. (Net primary production proved very difficult to determine; respiration rates of microorganisms were very high at the prevailing high temperatures.) The production of the main herbivores (*S. niloticus* and *S. leucostictus* which also took some bottom debris, *Haplochromis nigripinnis* and the cyclopoid copepod *Thermocylops hyalinus*) was computed to be equivalent to 650 kJ/m^2/yr (Burgis and Dunn 1978; Burgis 1978), a mere 0.8% of the estimated net primary production. Burgis and Dunn compared this figure with Prowse's (1972) finding that fish production was 1.02 to 1.79% of net primary production in Malacca fish ponds containing tilapia and grass carp. Thus, despite the sustained high fish yield, the Lake George system appears to be inefficient as so little of the high initial primary production is passed on to fish flesh. The reason for this is not clear; it was suggested that perhaps the lake has too few herbivores; fish numbers do not seem to be limited by food here, but possibly they are by availability of suitable spawning grounds in this soft-bottomed lake.

Discussion: Factors Controlling Tilapia Numbers in Natural Conditions

1. Physico-Chemical Factors

Physico-chemical events, such as drying up of lakes (as has occurred for Lake Rukwa, Lake Chilwa and much of Lake Chad) have decimated tilapia populations from time to time. The role of deoxygenation, silt stress, alkalinity stress, heat stress and effects of drying out the shallow spawning grounds in the decline of the Lake Chilwa fishes were studied by Furse et al. (1979). Deoxygenation when bottom water was brought to the surface sometimes causes massive fish kills (as in Lake George). Increased alkalinity may kill fishes in the soda lakes and lagoons. Winter temperature drops also kill or stress the fish at high latitudes (as in Hong Kong, Israel). On floodplains, the variations in flooding from year to year affect tilapia survival and growth rates, as has become clear from scale studies backcalculating growth in previous years (Dudley for Kafue fish, Duerre for Barotse fish). But in large lakes biotic pressures may be relatively more important than physico-chemical ones in controlling tilapia numbers.

2. Growth and Reproduction

Tilapia numbers are greatly affected by their ability to switch from growth to reproduction in unfavorable conditions. Though there appear to be some limitations (presumably genetic) to growth possibilities (for instance

T. sparrmanii is basically a small species, *S. andersonii* a large one), in all the tilapias investigated growth varies with the conditions. Certain 'stress factors' evidently switch the physiological state of the fish from growth to reproduction at a smaller size ('dwarfing', 'nanism') and at a younger age than in normal populations, irrespective of the rate of growth to achieve this size. The mechanisms for this are not yet known, but are reflected in the condition (weight for length) of the fish. Pond experiments at Malacca (Chen and Prowse 1964) demonstrated that the size of the inhabited water body influences maturation size and final size in some as yet unexplained way.

Is this ability more marked in species living in habitats likely to dry up? Even species endemic to the Great Lakes of Africa, such as *S. esculentus*, evidently have this ability. In Lake Victoria, *S. esculentus* only matures when two or three years old and at least 20 cm TL, but it can mature at five months old when only 10 cm TL in aquaria (Cridland 1961). The mechanisms for such a switch need investigation. In aquaria fast growth and precocious spawning in *T. zillii* was stimulated by keeping the fish at a high temperature (up to 31°C); changes in illumination had little effect (Cridland 1962).

Field studies have presented a number of clues about the switch: (i) tilapia in poor condition (low weight for length) switch to reproduction at a smaller size than those in better condition (see Figure 2). (ii) In Lake George, Uganda, marked declines in maturation sizes both of *S. niloticus* and of *S. leucostictus* were recorded after twenty years intensive fishing had reduced the general size of fish caught. (iii) In the Kafue, Dudley (1979) found some evidence that *S. andersonii* matures at a smaller size in dry years than in years when the water level remains high at the end of the dry season. (iv) In Lake Sibaya precocious breeding in *S. mossambicus* is associated with decline in food value of the food available to the fishes as they grow and move into deeper water (as discussed by Bowen, this volume). But quite possibly the tilapias are responding to some factors that we are not yet considering.

In addition to (1) the size and age at which the fish mature, tilapia populations are also affected by: (2) the length of the breeding season and number of broods produced a year. This may vary from only one (rarely two) as in Malawi *Sarotherodon* of the *S. squamipinnis* flock (Lowe 1952; Berns et al. 1978) and *S. andersonii* (according to Mortimer 1960), to a succession of broods, as seems more usual. These may be produced either through a defined breeding season, where environmental conditions are more seasonal, or throughout the year as in some equatorial waters; (3) The number of eggs produced at a time ('fecundity') which increases with the size (weight) of the female fish; (4) The egg size which is species-specific, substrate-spawners producing smaller and more numerous eggs than mouthbrooders (Lowe 1955b listed increasing egg size in *T. zillii, S. galilaeus, S. leucostictus, S. niloticus, S. variabilis, S. esculentus, S. karomo, S. lidole* and *S. saka*). Egg size shows some latitude, however, as females of dwarf populations produce somewhat smaller eggs than normal sized populations (Lowe (McConnell) 1957, Peters 1963), which helps to counteract the effect of fewer eggs per female in populations with dwarf females; (5) The number of young that can be mouthbrooded, which tends to increase to a maximum then decrease with increasing size of fish. Thus the 'brooding efficiency' (number of young brooded/number of eggs produced) falls as females increase above a certain size: this aspect has been studied in most detail in *S. leucostictus* by Welcomme 1967a).

The survival of the young depends on many factors. Great variations in survival from year to year have been found in the Kafue populations, which have obscured any effects due to the change from riverine to lacustrine conditions as the new lake filled (Dudley 1979). Under the more climatically stable conditions in the Great Lakes, biotic pressures including predation, competition for suitable spawning grounds, or nursery sites, or food among the juvenile fishes, may be particularly important.

3. PREDATION

Though there has been much discussion on the effects of predators on cichlid speciation (see Fryer 1965; Lowe-McConnell 1975), facts concerning differential predation are much needed. Tilapias, particularly juveniles, are known to be eaten by *Prototopterus* (lungfish), *Lates*, *Hydrocynus*, *Hepsetus*, catfishes such as *Clarias* (see Bruton 1979), *Bagrus* and *Eutropius* and also by piscivorous cichlids in the Great Lakes (e.g., *Haplochromis squamipinnis* in Lake George) and *Serranochromis* spp. in the Zambezi system. Some of these piscivores are diurnal feeders, but the catfishes are nocturnal, taking sleeping tilapias. The reversal of the feeding rhythm in Lake Bosumtwi tilapia was thought by Whyte (1975) to be an anti-predation measure.

The numerous kinds of fish-eating birds that will take tilapias include pelicans, cormorants (though these 'prefer' cylindrical fish) darters, herons, kingfishers, gulls and waders in the shallows (see lists in Coe 1966; Vareschi 1979 and JFRO reports).

4. RIVERINE AND LACUSTRINE TILAPIA POPULATIONS— 'r' AND 'K' STRATEGISTS?

Ability to breed at a dwarf size is shown by all tilapias tested, but in natural waters we find a range of adaptations. Floodplain species tend to have seasonally fast growth, early maturity (1 or 2 yr), high fecundity producing numerous rather small eggs, several broods in succession; rather generalized bottom-feeding habits. Many of these characteristics are typical of 'r strategists' found in pioneer habitats (as indeed the floodplains are—new habitats opened up each year). Lacustrine species in large lakes (Malawi, Victoria), on the other hand, tend to grow more slowly, have delayed maturity (2 to 4 yr) so the maturity/final size ratio is higher (see Iles 1971; Fryer and Iles 1972), reduced fecundity, producing only one (rarely two) brood a year in Malawi species, with relatively few, large ova, and they brood the young to a correspondingly large size. They tend to be specialized phytoplankton-feeders. (Thus they appear to have many features of 'K-strategists'). *S. andersonii* also has some of these features, such as brooding only one brood a year (which suggests that this species may have evolved under more lacustrine conditions).

Substrate-spawners have smaller, more numerous eggs than mouthbrooders, and their populations can build up very quickly in the comparative absence of predators, as, for example, did those of *T. zillii* in Lake Victoria, a species adapted for life with nilotic riverine piscivores, many of which are not found in the lake. The neotropical cichlids are nearly all substrate-spawners, including the *Cichlasoma* adapted for lacustrine life in Central American lakes (papers by McKaye 1977; Perrone 1978; Perrone and Zaret 1979). Lake Tanganyika is the only African lake to include numbers of both substrate-spawners and mouthbrooders among its endemic species (list in Lowe-McConnell 1975); in all other African lakes most cichlids are mouthbrooders. Why should this be so?

5. Polyculture and Introductions

These field studies have shown that it is quite usual for several species of tilapia to share resources in a water body (see Table 1). The main riverine fish communities in the soudanian region and Upper Zambezi each have two *Sarotherodon* species and one or two *Tilapia* species living together in most of the region. The lakes too, unless they dry up (like Lakes Rukwa and Chilwa, which each have one species) tend to have three or more tilapias sharing resources. The experimental addition of more species to the Lake Victoria fauna has led to the decline of the two indigenous species, but five tilapias are still present in this lake. In Lake Malawi the unique *S. squamipinnis* species flock (four species) appears to have evolved in response to the increasingly openwater conditions available in this lake, in one case by the splitting of early and late breeding populations into two species (*S. saka* and *S. squamipinnis*). In the Cameroon crater Lake Barombi Mbo, colonizations from the river may have recurred at different times, and the resultant tilapias have managed to coexist (again possibly by differences in breeding seasons, but more information is needed on this).

In Lake Victoria the coincidental phenomenal rise in lake level appears to have helped the introduced species to get established. In a number of lakes the indigenous tilapias have thrived better than introduced species, for example in Lake Kariba, where despite heavy stockings with *S. macrochir* (albeit from the Kafue, so a riverine form, not the Mweru, lacustrine subspecies intended) it was the indigenous *S. mortimeri* which underwent a population explosion, while *S. macrochir* vanished for many years. However, as ecological conditions change, so may the tilapias; *S. macrochir* is now said to be increasing in Lake Kariba, and *S. andersonii*, previously unknown from the Middle Zambezi, has been recorded here. Again, in the Nyumba ya Mungu reservoir on the Pangani river, it has been the indigenous tilapias, *S. pangani* and *S. jipe* which have grown so well (to 50 cm TL) and dominated catches, not the introduced *S. esculentus* or *T. rendalli.* And in multispecific stocking of Tanzanian dams, Payne (1974) found the indigenous *S. esculentus* to do better than the introduced *S. macrochir; T.*

rendalli was introduced here before *T. zillii* and the *T. zillii* failed to get established.

The establishment and dominance of *S. mossambicus* from pond escapes in Plover Cove Reservoir, Hong Kong, formerly an arm of the sea, stocked with Chinese carps, shows what hardy persistent fish these tilapias are; their ability to use many food sources was thought to have contributed to their success here. This also emphasizes how easily pond escapes occur, and how difficult it may be to eradicate a species once it has gained access to a water body.

6. Conclusions from Growth Studies

Collated information on tilapia growth from many natural waters indicates:

a. That within a water body, the different species tend to grow at different rates, the faster-growing species reaching a larger size (see Figure 4).
b. The same species will grow at different rates in different water bodies, suggesting that environmental differences are more potent than genetic differences in determining maturation and maximum size.
c. Under natural conditions both maturation and maximum size of a species tend to be smaller in small bodies of water than in larger ones (see Figure 1).
d. Populations with low weight for length tend to have a lower maturation size than populations where the tilapia are in better condition (see Figure 2).
e. Maturation size of a species can change within a water body, e.g., rising as lagoon fish escape to the main lake, or falling after intensive fishing (as in Lake George), or in low-flood years (as on Kafue flats).
f. In riverine populations (also in lagoon and aquarium kept-fish) males grow larger than mature females, but in lakes males and females of the same species may mature at, and grow to, comparable sizes.
g. The relative growth rates of young and old fish may be different in different water bodies.
h. An extended reproductive season may slow somatic growth.
i. Both Kafue and Kariba studies indicate that *T. rendalli* has a higher production/biomass ratio than the associated *Sarotherodon* species.

Finally, the great plasticity of growth in natural waters shows that tilapia growth rates are very greatly affected by environmental and behavioral conditions. This suggests that concentrating research on the environmental and behavioral factors affecting growth and the switch to reproduction is likely to be more helpful for fish culture than a search for faster-growing genetic strains.

Acknowledgments

My special thanks go to Dr. E. Trewavas for introducing me to the tilapias, encouraging the field studies through the years, and for drawing my attention to recent papers and for innumerable stimulating discussions. I am also very grateful to the Trustees of the British Museum (Natural History) for working space and excellent library facilities, and to the members of the Fish Section who have helped in so many ways.

SESSION 2: PHYSIOLOGY

Chairman's Overview

D. J. W. MORIARTY

Tilapias are mainly lacustrine fish and are well adapted to enclosed waters. They produce high yields and thus are an important source of protein in many tropical countries. But what are the biological attributes which suit them for this role? What is known of the biology of tilapias that would improve cultural practices and what gaps are there in our knowledge which limit successful culture? The ability of tilapias to utilize blue-green algae has been cited as a major reason for their high yields in shallow tropical lakes, where blue-green algae are often abundant. I think that other factors are also involved, but food is obviously important.

Dr. Bowen examines the nature and quality of the food of tilapias and their digestive physiology. All large juveniles and adults feed on plant material or detritus. (Zooplankton feeding has been described for *S. aureus* in Lake Kinneret; Spataru and Zorn 1978—Editors.) Some species feed on macrophytes, but many feed on phytoplankton or detritus. Most of the important, cultured tilapias feed on bacteria, especially the blue-green algae (or cyanobacteria as they are now commonly called). These microorganisms are lysed by stomach acid and subsequently digested in the long intestine. Tilapias which feed on macrophytes make use of their especially adapted pharyngeal teeth to break open cells, and then digest them by similar methods.

Tilapias are generally slow moving, and do not need as much energy for movement as predatory fish. For rapid synthesis of body protein they require a relatively high protein content in their diet. They would get this from blue-green algae or bacteria, which have C:N ratios from about 4:1 to 10:1. Tilapia fry feed on zooplankton and zoobenthos as well as algae and bacteria, which they collect around shallow lake margins. It is not known whether they can utilize animal protein more efficiently at this stage. As they grow larger, they move into deeper water and their preferred diet changes to phytoplankton (or macrophytes or detritus in some species).

Dr. Bowen's work with *Sarotherodon mossambicus* shows that the fish select food with a high nitrogen content. The requirement for nitrogen and its supply in the diet is one aspect of feeding biology which needs to be investigated in more detail. Tilapias feed at or near the base of the food web in natural systems. This fact could be exploited in tilapia culture and it may be considered economically wasteful to supply manufactured food to tilapias. Many of the phytoplankton-feeding species may not be able to utilize readily other forms of food.

In studying feeding, digestion and growth, temperature is an important factor to consider. Dr. Caulton has shown how *Tilapia rendalli* utilizes temperature variations in its lake environment to maximize growth. By constructing careful energy budgets, he has demonstrated that faster growth rates were achieved when fish moved into warm inshore areas during the day (where feeding and digestion rates were faster) and then retreated to cool, deeper areas at night where energy demands were less, than if the fish remained at a constant temperature. The conclusion is that yields will be much higher in shallow waters where diel temperature changes are more pronounced. This temperature effect helps explain some of the distribution of tilapias in lakes, but it is probably not the only reason why the fish in Lake George, Uganda, are concentrated around the edge. This shallow lake stratifies thermally each day, so the fish could obtain more than a 10°C variation in temperature by varying their position in the water column. Other environmental factors may therefore be involved.

Temperature also limits the distributon of tilapias. Their inability to withstand temperatures much below 16°C confines them to tropical or warm temperate regions. For culturing tilapias near the limits of their range, an optimum water body size and depth is needed, which ideally allows the diurnal temperature variation that Caulton has shown and yet which does not cool down too much in winter. Many tilapias are euryhaline, which increases the potential areas of water available for their culture. Oxygen requirements are also important, but we need more data relating oxygen levels and temperature to the energetics of metabolism and the switch from growth to reproduction. Where there are dense blooms of algae, oxygen may be depleted at night or below the photic zone, to the point where respiration of the fish is affected. Perhaps an effect of this nature might be part of the explanation for preference of tilapias in Lake George for the lake edges, where phytoplankton densities are much lower than in the middle of the lake. The combination of high temperature and low oxygen levels would be stressful and, as Caulton has shown with *Tilapia rendalli*, there is likely to be a fine balance between energy supply in the diet and energy losses for maintenance. If metabolic losses are too high, or low oxygen availability decreases assimilation efficiency and limits energy supply, then growth and reproduction will be affected. There is evidence that growth is rapid and the onset of sexual maturity is delayed in highly oxygenated water.

In addition to food, the high reproductive potential of tilapia is an important factor governing their success in tropical lakes. This also leads to problems, however, because stunting may occur, particularly in overcrowded conditions, with attainment of sexual maturity at an early age. Like other fish, gametogenesis in tilapias is regulated by the influence of external and internal events on the nervous system and by a complex interaction of hormones from the hypothalamus, pituitary and the gonads. We need to know a lot more about the biochemistry of the hormones and the physiology of reproduction before practical benefits can flow to culturists. An example of the influence of social factors on breeding may be seen in Lake George, Uganda. As fishing pressure increased, the minimum size at which females reached sexual maturity decreased from 28 cm to 20 cm. As breeding sites are limited (because most of the substrate is too soft and flocculent for nest-building), there may be competition for sites and the large fish would be dominant. Thus, it may be that occupancy of breeding sites by large fish inhibits the onset of sexual maturity, and as the large fish are removed by fishing, smaller fish become mature.

Environmental Physiology of Tilapias

J. CHERVINSKI

Fish and Aquaculture Research Station
Dor, D.N.
Hof Hacarmel, Israel

CHERVINSKI, J. 1982. Environmental physiology of tilapias, p. 119-128. *In* R.S.V. Pullin and R.H. Lowe-McConnell (eds.) The biology and culture of tilapias. ICLARM Conference Proceedings 7, 432 p. International Center for Living Aquatic Resources Management, Manila, Philippines.

Tilapias in general do not grow at water temperatures below 16°C and are not able to survive water temperatures below 10°C for than a few days. Low water temperature tolerance limits are: *Tilapia sparrmanii* 7°C, *T. rendalli* 11°C, *Sarotherodon aureus* 8-8.5°C, *S. vulcani* 11-13°C and *S. mossambicus* 8-10°C. Hybrids between *S. niloticus, S. vulcani, S. hornorum* and *S. aureus* are similar to *S. aureus* in their low temperature tolerance. Low temperature tolerance is affected by prior acclimation temperatures. Tilapias are tolerant to high water temperatures, e.g., 42°C for *S. alcalicus grahami* and 41°C for *S. aureus*. Tilapias are euryhaline and are able to survive, grow and some species even reproduce in sea water up to 40‰ salinity. The lowest dissolved oxygen concentrations survived for short periods by tilapias are 0.1 ppm for *S. mossambicus* and 0.2 ppm for *S. aureus*. Tilapias are able to tolerate a pH range of 5 to 11. *S. aureus* tolerates a maximum of 2.4 mg/l un-ionized ammonia but after exposure to sublethal concentrations can raise this limit to 3.4 mg/l.

Introduction

Fish of the genera *Tilapia* and *Sarotherodon* are native to Africa and Israel, the northern limit of their distribution. The tropical origin of tilapias is clearly expressed in their ecological physiology, especially in their temperature preference during their reproductive period.

Tilapias become inactive at water temperatures below 16°C, which is the minimal temperature for normal growth. Reproduction occurs above 22°C. Their adaptation to stable ambient temperature regimes has limited their natural distribution to tropical areas.

In recent years tilapias have been distributed all over the world where temperatures are suitable for their growth and reproduction. In many parts of the world tilapias have been introduced for vegetation control, pond culture and recreational fishing. They have become established in numerous lakes in Florida, California and Texas where winter temperatures are not limiting.

They also occur in other regions of the U.S.A. in water bodies which are warmed above normal ambient temperature during the winter by geothermal water sources or artificial heating in conjunction with the operation of power plants. Some tilapias have excellent aquaculture potential because of their fast growth, herbivorous or omnivorous feeding habits, high food conversion efficiency, high tolerance to low water quality, ease of spawning, ease of handling, resistance to disease and parasites and good consumer acceptance.

Temperature

One of the problems of using tilapia for pond culture is their inability, in general, to survive water temperatures below 10°C for more than a few days. The influence of temperature on survival and growth has been studied through field observations, laboratory investigations and a few physiological experiments designed specifically to determine thermal effects. There is a large variation in the reports of thermal tolerance in tilapias, stemming from a lack of standardized methods. Other variations in results might be attributed to differences in: 1) acclimation time, 2) water quality such as total dissolved oxygen (DO), total dissolved solids and salinity, 3) age, size, sex and health of the fish, and 4) duration of the drop in temperature.

Activity and feeding of tilapias become reduced below 20°C and feeding stops completely around 16°C. Although tilapias may be able to resist short-term exposure (a few hours) to temperatures of 7 to 10°C, death can occur (in some species) at temperatures as high as 12°C after long-term exposure.

Some tilapias are more tolerant to low temperatures than others. *Tilapia sparrmanii* is a hardy fish, capable of withstanding much lower temperatures than those tolerated by other species. The lowest water temperature tolerance recorded for this species was 7°C in Zambia (Maar et al. 1966) and the same temperature was recorded in South Africa (Hofstede 1955). *T. rendalli* was able to survive a temperature of 11°C in Zambia (Sklower 1955).

Of the mouthbrooding fish, *Sarotherodon aureus* seems to be the most resistant to low temperature. Yashouv (1960) found that at temperatures below 10°C, *S. aureus* (reported as *T. nilotica*) ceases all motion, while at 6 to 7°C it loses its ability to maintain body position. However, when exposed to low temperatures for only a few hours the fish recovers. Sarig (1969) found that local *S. aureus* in Israel was able to tolerate temperatures of 8.0 to 8.5°C. *S. vulcani*, which was introduced to Israel from Lake Rudolf, dies at temperatures of 11 to 13°C. On the other hand, its hybrid with *S. aureus* had a lower temperature tolerance limit of 8.0 to 9.0°C. Denzer (1968) recorded a temperature of 11°C as the lower lethal limit of *S. niloticus*. Chervinski and Lahav (1976) showed that the hybrids between *S. niloticus* ♀ x *S. aureus* ♂ and *S. vulcani* ♀ x *S. aureus* ♂ have temperature limits similar to *S. aureus*. Similar experiments conducted by Lee (1979) showed that *S. aureus* is more tolerant to low temperature (6.7°C) (criteria, 50% of the fish lost equilibrium) than *S. hornorum* (10.0°C) and *S. niloticus* (7.8°C). The crosses between *S. niloticus* ♀ x *S. aureus* ♂ and *S. hornorum* ♀ x *S. aureus* ♂ were similar to *S. aureus* in their low temperature tolerance.

Another factor that affects low temperature tolerance of tilapias is the thermal history (acclimation) before exposure to low temperature. Chervinski and Lahav (1976) found that *S. aureus* acclimated to 28°C for two weeks began to die at 11°C while those acclimated to 18°C (for two weeks) began to die only at 9°C.

S. mossambicus is killed between 8 and 10°C (Chimits 1957). The minimum temperature at which the fish ceases to feed is 15.6°C and 100% mortality occurs at 8.3 to 9.4°C (Kelly 1956). Lower temperature tolerance (5.5°C) was reported for *S. mossambicus* in Hanoi, Viet Nam (Li et al. 1961). The influence of salinity on the lower temperature tolerance limits of *S. mossambicus* was investigated by Allanson et al. (1971). They found that *S. mossambicus* tolerated 11°C in 5‰ saline water but when in freshwater it could not survive at that temperature. It was suggested by Allanson et al. (1971) that the ability of *S. mossambicus* to withstand low temperature is correlated with the maintenance of high plasma sodium and chloride concentrations.

In contrast to their limited low temperature tolerance, tilapias are very tolerant to high water temperatures. *S. alcalicus grahami*, which lives in the hot springs of Lake Magadi, Kenya, tolerated temperatures up to 42°C (Coe 1966). A similar resistance to high temperatures (up to 42°C) was found in *S. shiranus chilwae* (Morgan 1972). Gleastine (1974) reported a low level of mortality at 41°C for *S. aureus*.The upper lethal limit for *S. niloticus* was also determined to be 42°C (Denzer 1968), while the upper median lethal temperature was 38.2°C for *S. mossambicus* (Allanson and Noble 1964).

Experiments conducted by Beamish (1970), using *S. niloticus*, showed that the temperature preferendum was 30°C when the fish was acclimated to temperatures between 15 and 30°C. A lower preferendum of 28°C was found when the fish was acclimated to 35°C.

Salinity

It is assumed that tilapias evolved from a marine ancestor and that their penetration to fresh water is secondary (Myers 1938; Steinitz 1954). This may account for marked euryhalinity of certain species (Chervinski 1961a).

S. aureus is able to survive direct transfer from freshwater to 60 to 70% sea water (20.2 to 25.0‰ salinity) and through gradual adaptation is able to withstand up to 150% sea water (Lotan 1960). The growth and survival of *S. aureus* and *S. galilaeus* in brackishwater have been studied in small-scale experiments in aquaria, concrete tanks and earth ponds (Chervinski 1961a, 1961b, 1961c, 1966). The growth of *S. aureus* in seawater ponds investigated by Chervinski and Yashouv (1971) was found not to differ significantly from that in freshwater. Good growth of F_1 hybrid offspring of the cross *S. niloticus* ♀ x *S. aureus* ♂ in brackishwater ponds was found by Fishelson and Popper (1968) and Loya and Fishelson (1969).

S. mossambicus is euryhaline and grows and reproduces in fresh, brackish and seawater. According to Vaas and Hofstede (1952) reproduction does not occur in salinities between 30 and 40‰. However, Hora and Pillay (1962) found that reproduction occurred in seawater up to a salinity of 35‰. Recent observations by Popper and Lichatowich (1975) showed that in seawater ponds in Fiji. *S. mossambicus* is able to reproduce at salinities of 49‰. Fry of *S. mossambicus* were found to live and be in good healthy condition at salinities of 69‰ (Potts et al. 1967). These differences may be attributed to different races of *S. mossambicus*.

Tilapia zillii was found to reproduce in Lake Qarun (Egypt) in salinities between 10 to 26‰ (El Zarka et al. 1970a). *T. zillii* was found in the Red Sea at salinities of 42.7‰ (Bayoumi 1969) and in the hypersaline Bardawil Lagoon at salinities of 41 to 45‰ (Chervinski 1972).

Neither *S. aureus* nor *T. zillii* reproduces in sea water. In addition no nest building occurs and the gonadosomatic index dropped in *S. aureus* (Chervinski and Yashouv 1971; Chervinski and Zorn 1974). This fact can be used to control wild spawning of these species. The high tolerance of *S. aureus*, *S. mossambicus*, *S. galilaeus* and *T. zillii* for seawater was attributed by Morgan (1972) to their natural habitat in estuaries and the lower reaches of rivers.

Other Water Quality Parameters

DISSOLVED OXYGEN

Due to their tolerance to poor water quality, tilapias are found in habitats which most other fish genera are unable to inhabit. Even under conditions of heavy feeding, fertilization and manuring no mortality occurs.

The lowest short-term DO limit recorded for a tilapia is 0.1 ppm DO for *S. mossambicus* (Maruyama 1958) and *S. niloticus* (Magid and Babiker 1975). *S. niloticus* ♀ x *S. hornorum* ♂ hybrids tolerate 0.3 ppm DO (Lovshin et al. 1974). Experiments conducted at Texas A&M University using fresh chicken manure to fertilize ponds at a rate of 2,760 hens per hectare of pond water showed DO's at dawn: 0.4 to 0.8 ppm. A correlation was found between the low DO measurements at dawn and reduced growth in *S. aureus* (McGeachin pers. comm.).

Studies by Job (1969a, 1969b) showed that the respiration of *S. mossambicus* was independent of DO at temperatures between 15 and 30°C until the partial pressure of oxygen dropped to 50 mm Hg, equivalent to 32% saturation. Below this level the metabolic rate became dependent on available oxygen. Similar results were shown by Rappaport et al. (1976) who found that the growth of tilapia and carp is reduced below 25% oxygen saturation. Mortality occurs when oxygen remains below 20% saturation for more than 2 to 3 days.

It seems that tilapias are able to tolerate DO's as low as 1 ppm. Below this level they may utilize atmospheric oxygen. Stickney et al. (1977) reported that *S. aureus* reared in ponds receiving swine wastes experienced heavy mortality when the pond surface became covered with duckweed, possibly because its ability to utilize atmospheric oxygen was restricted. A well-known phenomenon in harvesting tilapia ponds is the fact that tilapias can survive for several days in small mud puddles with little ill effect.

pH

Most pH values associated with tilapia growth come from pond observations. Swingle (1961) summarized the relationship between pH and fish culture. He found that the lethal acidic limit is approximately pH 4 and the alkaline limit is pH 11. Huet (1971) recommended pH 7 to 8 for culture. Bardach et al. (1972) stated that tilapias did not grow in the acid waters of West Congo. *S. alcalicus grahami* in Lake Magadi, Kenya was found to tolerate pH 5 to 11 without any adverse effect. Lovshin et al. (1977) reported a pH range in culture ponds in Brazil from a minimum of pH 7.7 in the morning to over pH 10 in the afternoon. Experiments conducted at the Aquaculture Research Center of Texas A&M University, showed that *S. aureus* tolerated a pH range of 7.5 to 10.2 in tanks receiving chicken manure (McGeachin pers. comm.). Also in ponds receiving chicken manure (100 kg/ha/day of dry matter) the pH ranged between 7.2 and 9.3 at 11:00 A.M. (Burns and Stickney 1980).

AMMONIA

Fish excrete most of their nitrogenous waste through the gills in the form of ammonia. Excreted ammonia exists in water in equilibrium between the un-ionized NH_3 (toxic to fish) and ammonium ions NH_4^+ which are not toxic. The toxicity of un-ionized ammonia depends on the DO. When the DO is low un-ionized ammonia is toxic at a lower concentration. The toxicity of NH_3 also decreases with increasing CO_2; this depresses the pH which shifts the NH_3/NH_4^+ equilibrium.

The influence of un-ionized NH_3 on *S. aureus* was investigated by Redner and Stickney (1979). The 48-hour median lethal concentration (LC_{50}) was 2.4 ppm. When fish were acclimated to sublethal concentrations (0.43 to 0.53 ppm) for 35 days, a concentration as high as 3.4 mg/l caused no mortality within 48 hours. This pattern is important when heavy feeding, fertilization and manuring are being applied in intensive pond culture. The maximum total ammonia tolerated by *S. aureus* in experimental ponds receiving fresh chicken manure from 2,760 hens per hectare was 11 ppm at pH 8 and 27°C. The amount of the un-ionized ammonia present was 0.75 ppm (McGeachin pers. comm.). Burns and Stickney (1980) reported the total ammonia level recorded with 4,000 hens per hectare to be 2.4 ppm, which appears low.

Table 1. Salinity tolerance of *Sarotherodon* and *Tilapia* species.

Salinity (‰)	Comments	Source
	S. aureus	
6	Grew and reproduced in brackishwater ponds	Chervinski (1966)
10	Growth was nearly equal to that in freshwater ponds; greater mortalities in brackishwater	,,
18.9	Reproduced in brackishwater	Chervinski (1961b)
20-25	Survived direct transfer from freshwater	Lotan (1960)
36.6-44.6	Grew well; failed to reproduce	Chervinski and Yashouv (1971)
53.5	Able to survive through gradual adaptation	Lotan (1960)
	S. galilaeus	
10-26	Thrived and bred naturally	El-Zarka et al. (1970a)
19.5	Thrived and grew after direct transfer from freshwater	Chervinski (1961c)
	S. mossambicus	
30	Grew well and reproduced in ponds	Chimits (1957)
30-40	No reproduction	Vaas and Hofstede (1952)
35	Reproduced	Hora and Pillay (1962)
49	Reproduced	Popper and Lichatowich (1975)
	T. zillii	
11-29	Thrived and reproduced; survived better than *S. aureus* and *S. niloticus*	El-Zarka et al. (1970a)
23.4-27.3	Maximum salinity tolerated after direct transfer from freshwater	Chervinski and Hering (1973)
39	Was able to tolerate this through gradual adaptation	,,
38.8-43.7	Acclimated to this salinity; Grew better than *S. aureus;* Did not reproduce	Chervinski and Zorn (1974)
42.8	Found in the Red Sea	Bayoumi (1969)
41-45	Found in the hypersaline Bardawil Lagoon	Chervinski and Hering (1973)

Table 2. Lower temperature limits (°C) recorded for *Sarotherodon* species.

Temperature (°C)	Comments	Source
	S. aureus	
5	Survived in ponds when temperature dropped for a short time	Yashouv (1960)
6-7	Survived in ponds when temperature dropped for a short time	Chervinski (unpublished)
8	Survived this temperature for a short time in laboratory	Yashouv (1960)
8-8.5	Died under experimental conditions	Sarig (1969)
9	Began to die: previously acclimated at 18°C for two weeks	Chervinski and Lahav (1976)
11	Began to die: previously acclimated at 28°C for two weeks	,,
11	Higher survival when kept in 5‰ salinity than in fresh water (0.4‰)	,,
5.6	Died: previously acclimated to 21°C for 2 days. (Drop in temperature 0.8°C/1 hour)	Lee (1979)
	S. galilaeus	
8	Survived in laboratory when exposed to this temperature for a short time. Lost equilibrium when exposed to 7.5°C for a short time.	Yashouv (1960)
	S. mossambicus	
5.5		Li et al. (1961)
8-10	Tolerated	Chimits (1957)
8.3-9.4	Total mortality	Kelly (1956)
9-12	Did not survive these temperatures in fresh water; at 11°C, disorientation; survived 11°C in 5‰ salinity	Allanson et al. (1971)
	S. niloticus	
6.7	Determined experimentally	Lee (1979)
11	Determined experimentally	Denzer (1968); Chervinski and Lahav (1976)

Discussion

LOVSHIN: My experience of tilapias is that they do much better in waters with a high pH. At pH 5 to 6 or below, they become sick and lethargic and do not grow very well. Tilapias are found in the Zaïre Congo Basin anyway, where the pH of some tributaries can get down to around 5; tilapias can live in waters of low pH, and I would like to know if anyone has any information on tilapias in low pH waters. My observations suggest that the sarotherodons are alkaline water fish and if we attempt to raise them in acid waters they will not respond well, quite apart from the low productivity. Are there tilapias which do well and spawn in low pH situations? [Tilapias are generally absent from most of the central basin of the Congo, the forested part (Trewavas, pers. comm.), but Dubois (1959) reported *T. congica* living in Lake Tumba, Zaïre, where the pH is 4.5 to 5.0 (see Philippart, this volume)].

LOWE-McCONNELL: I do not think they get into very low pH, but I have pH's for *S. niloticus*. Some of the lagoons have pH's around 9.

ROBERTS: We keep our experimental populations in Scotland in tap water at pH's of 6.4 to 6.8.

HEPHER: What about high pH's? For example, 10.2 was mentioned. What is the source of this high pH? Is it caused by a fluctuation in the bi-carbonate cycle or by deposits of sodium carbonate? A pH as high as 10.2 may affect tilapias.

CHERVINSKI: In fact we are talking about fluctuations of pH.

HEPHER: That is all right. Most fish will tolerate those levels. The main question is, can you construct tilapia ponds in a swampy area which has acid soils?

LOWE-McCONNELL: Dr. Prowse at Malacca had tilapia in ponds at very low pH's; I think around 2 to 4 or something of that order. If you look in the Malacca reports, you will find all the data. These were acid lands which were of little use for anything else but to grow fish. He was liming them. I don't know what the pH was after he limed them.

CAULTON: I would like to comment on an unusual aspect of environmental adaptation shown by *S. mossambicus*. I have been informed by a reliable source of three reports of live *S. mossambicus* being dug out of apparently dry river beds in Zimbabwe. These reports have never been scientifically verified, but I am quite happy with their reliability. In all instances, no surface water was present yet the fish survived in the damp subsurface moisture as deep as 50 cm below the surface. The fish have obviously adapted to a severely modified microenvironment and appear to remain alive under extremely adverse conditions. All attempts at recreating these conditions in the laboratory have, however, resulted in total mortality.

CHERVINSKI: It is known by fish culturists that when you drain ponds, tilapias can withstand very low oxygen levels in the mud. After one or two days, when you put the water back again, they will revive.

HEPHER: But that is not dry.

CHERVINSKI: No, but still the oxygen is still very low, and it may be that tilapias are able to use atmospheric oxygen. We see them in ponds going up to the surface and swallowing air. We have done some small experiments and showed that when tilapias cannot reach the surface, there is a lot of mortality.

ROBERTS: I would like to comment on the pH tolerances. In Mombasa, Haller has stocks of *Sarotherodon alcalicus grahami*. They will not survive in any pH below 9 and live at pH 10 or 12 without any difficulty. Also, I cannot agree with Dr. Chervinski that levels of CO_2 are unimportant. We have a very considerable problem in several places where high CO_2 water supplies cause nephrocalcinosis.

CHERVINSKI: I agree that CO_2 can be very important, but in the work I was discussing, lowered O_2 is more important. The fish can avoid high CO_2 waters.

GUERRERO: I would like to give some information about tilapia culture in the Philippines where people are working a great deal in brackishwater. At salinities between 15 and 30‰, *S. niloticus* do not reproduce. They seem to grow better at 15‰.

HEPHER: But *S. mossambicus* will spawn in seawater.

CHERVINSKI: *Tilapia zillii, S. aureus* and *S. niloticus* are all very salt tolerant.

TREWAVAS: It is interesting that *T. zillii* can tolerate high salinities in, for example. Lake Qarun (Egypt), but appears unable to do so in West Africa where it is replaced by *T. guineensis.*

MIRES: North of Eilat in Israel, hybridization between *S. niloticus* females and *S. aureus* males is very difficult in salinities of 20‰ and above.

AVTALION: Is there any evidence that environmental parameters can affect the sex ratio of the progeny?

MIRES: Not to my knowledge. Temperature, at least, appears to have no influence.

HENDERSON: I would be suspicious of drawing any general conclusions from the effects of varying salinity on reproduction. There are so many physiological factors involved.

PULLIN: I was interested to hear you talk about oxygen tolerance. I believe you said that when a culturist who subjected his fish to low oxygen conditions such as early morning dissolved oxygen (DO) of around 0.1 ppm the fish suffered growth depression, and could not make this up even if the oxygen levels went back to normal during the day. We sometimes get DO's as low or lower than this for several hours in manured ponds in the Philippines.

CHERVINSKI: With experiments conducted with *Sarotherodon aureus* it has been shown that there is a correlation between low oxygen in the early morning and growth.

PULLIN: Twenty-four hour records of oxygen levels are essential in experimental work.

MORIARTY: The important point to consider here is, I think, the interaction between temperature and oxygen. As the temperature goes up, the fish are going to have a higher oxygen demand and be much more stressed. We did have a fish kill some time ago in Lake George, Uganda, but it was in the center of the lake where the plankton densities were highest, not around the edges. It happened at night. The tilapias could have come to the surface and breathed air. None of the air breathing species died. In Lake George, the temperatures are very high. The lethal stress was therefore the high temperature and the high oxygen demand when the oxygen levels are already low.

CAULTON: I would like to make one point which is not often considered. The oxygen stored in the swimbladder can, in *S. mossambicus,* sustain metabolism for some time. My estimates are that a fish of 400 g may survive for as long as 30 minutes using this oxygen at 25°C. Obviously, the greater the depth that the fish is found in, the greater the volume of oxygen available. Similarly, the lower the temperature and the larger the fish, the longer the supply could last.

LOVSHIN: I used to think that tilapias utilized atmospheric oxygen a lot but now I doubt it. I think rather that they come up to the surface film where the dissolved oxygen is high and flush their gills with this water to get enough oxygen to survive. If they do not have access to the surface or are crowded in cages, they die very quickly at low DO's. I think I have read somewhere that tilapia have the ability to lower their metabolism when DO's go down to a certain level. I have seen this in the basic studies we have carried out in Brazil. The fish have the appearance of being dead, but they are not, and the amount of oxygen in the surface layers is enough to keep them going as long as they have contact with it.

Reproductive Physiology in Cichlid Fishes, with Particular Reference to *Tilapia* and *Sarotherodon*

B. JALABERT AND Y. ZOHAR

Institut National de la Recherche Agronomique
Laboratoire de Physiologie des Poissons
Campus de Beaulieu 35042
Rennes-Cedex, France

JALABERT, B. AND Y. ZOHAR. 1982. Reproductive physiology in cichlid fishes, with particular reference to *Tilapia* and *Sarotherodon*, p. 129-140. *In* R.S.V. Pullin and R.H. Lowe-McConnell (eds.) The biology and culture of tilapias. ICLARM Conference Proceedings 7, 432 p. International Center for Living Aquatic Resources Management, Manila, Philippines.

Works dealing with the external and internal factors which regulate reproductive efficiency in *Tilapia* and *Sarotherodon* are reviewed. Gametogenesis, although presenting the same general features as in most other teleosts is characterized by a low production of gametes related to the high efficiency of parental care: in substrate-spawners *(Tilapia)* the number of small sticky eggs is approximately related to the cube of body length, whereas in mouthbrooders *(Sarotherodon)*, eggs are bigger, not sticky, and their number is related to the square of body length.

Both groups seem to exhibit a capacity for precocious sexual maturation, the reason for which is not clear. There is also unlimited successive breeding through whole populations in equatorial areas and an increased tendency for seasonal breeding with increasing latitude, with maximum activity during maximum temperature and light intensity. When temperatures exhibit seasonal variations of large magnitude, low temperature (15 to 22°C) inhibits reproduction during part of the year. Social factors have been shown to enhance spawning frequency and might also influence sexual precocity through mechanisms which need further work.

Despite some attempts, specific tilapia gonadotropins have not yet been satisfactorily purified; the ubiquity of specific prolactins in controlling both osmoregulation and parental care behavior must be confirmed, and their possible inhibitory action on gametogenesis need investigation. The precise nature of steroids which mediate the pituitary action during different phases of the reproductive cycle also needs more research.

Introduction

The efficiency of reproduction in *Tilapia* and *Sarotherodon* has paradoxical consequences: on one hand, this aptitude which allows easy and rapid propagation in various tropical and subtropical environments, partially explains the economic interest in these species for fish culture; on the other hand this reproductive efficiency can be a source of problems because uncontrolled multiplication within a limited environment in situations of food competition is liable to produce dwarf fish populations, of little value.

The behavioral patterns which occur after spawning and which characterize substrate-spawners *(Tilapia)* and mouthbrooders *(Sarotherodon)* have been described and discussed by numerous authors (Lowe-McConnell 1959; Perrone and Zaret 1979). Whatever the role of each sex in brood care, which differs among species, this care provides an efficient protection for eggs and fry against predators, and contributes greatly to the reproductive efficiency of these species. However, the physiological mechanisms which control parental care behaviour are poorly understood.

Another aspect of the reproductive efficiency of tilapias is precocious sexual maturation which can occur as early as 3 months in some species (McBay 1961; Arrignon 1969) and depends probably, in addition to genetic factors, on environmental factors like temperature (Hyder 1970a; Siddiqui 1979a), food availability, social factors, etc. Precise data based on experimentation are generally lacking.

As soon as sexual maturity is attained, and provided temperature is suitable, most cichlid females are able to undergo successive breeding cycles, producing new broods at 4 to 6 week intervals. This usually results in a continuous production of fry throughout a population, with the exception of certain environments subject to substantial seasonal variations (Moreau 1979). But the relative asynchrony between the sexual cycles of individual females can be a problem when mass production of homogenous fry is required for intensive fish farming.

Thus, for practical reasons dictated by fish farming conditions, it would often be advantageous either to inhibit or delay sexual maturation, or in some cases to favor synchronous spawning and breeding for mass production of fry.

This paper reviews the external and internal factors which seem to be involved in the control of different stages of the reproductive cycles of cichlids, to suggest practical means for artificial control. As literature in the field of cichlid reproductive physiology is scattered, references will be made, when necessary, to the present state of knowledge in other teleosts. For more detailed information concerning reproductive physiology and endocrinology in fish, see recent reviews by Dodd (1975), Fontaine (1976), Jalabert (1976), Olivereau (1977), Callard et al. (1978), Peter (1978), Billard et al. (1978) and Breton et al. (1980).

General Characteristics of Gametogenesis

Sexual differentiation of the gonad into a juvenile testis or ovary with a characteristic morphology occurs very early in *Tilapia* (Yoshikawa and Oguri 1978) and in *Sarotherodon* (Nakamura and Takahashi 1973), around 15 to 30 days after fertilization (at 23 to 25°C). Sexual maturity can be then completed after a few months.

Gametogenesis in cichlids appears to present the same general features as in most other teleosts and lower vertebrates, whether in males or in females (Barr 1968; Hoar 1969). Thus some stages in the following description are taken from noncichlids. All the available information suggests that this is valid (Dadzie 1969; Hyder 1970a, 1970b; Polder 1971; Von Kraft and Peters 1963; Hodgkiss and Man 1978; Babiker and Ibrahim 1979; Moreau 1979b).

In males, the testis possesses a stock of undifferentiated gonial cells (type A) which originates from the primordial germ cells of the embryo through mitotic divisions. Active spermatogenesis begins with the isolation of type A spermatogonia, each one surrounded by a few somatic cells, followed by successive synchronous mitotic divisions of spermatogonia (type B), while surrounding somatic cells divide to form a continuous layer of cells, called "Sertoli cells". This process results in numerous cysts of cells throughout the testis. After an unknown number of spermatogonial mitotic divisions, meiosis occurs synchronously for each cyst, each spermatogonium then producing four spermatids which will differentiate into spermatozoa (spermiogenesis). Cyst evolution occurs along testicular lobules which are separated from each other by fibroblast cells and interstitial tissue. The latter is particularly well developed in cichlids in comparison with some other teleosts. Spermiation occurs when spermatozoa are released from cysts after separation of the Sertoli cells, first into the lobules and then into the vas deferens.

In female teleosts, the ovaries contain a stock of undifferentiated oogonia which seems to be renewed by mitotic divisions throughout life unlike higher vertebrates. For example, unilateral ovariectomy stimulates oogonial mitosis in the remnant ovary of *Sarotherodon aureus* (Dadzie and Hyder 1976). Some oogonia begin a meiotic division but remain at arrested prophase (primary oocytes). Each primary oocyte increases in size and is progressively surrounded by layers of somatic cells: inner granulosa cells which form a monolayer directly in contact with the thickening outer oocyte envelope (the zona radiata, i.e., the future chorion) and an outer theca, made of several layers of fibroblasts penetrated by capillaries. The theca forms the external cellular envelope of the ovarian follicle. The morphology of ovarian follicles in *Tilapia* and *Sarotherodon* has been well described by Von Kraft and Peters (1963).

Oocyte growth can be divided into two main phases: first, the previtellogenic phase when size increases (up to 0.6 to 0.9 mm diameter in *S. niloticus*) which is considered to be the result of synthesis occurring mainly within the oocytes (endogenous vitellogenesis) and second, the vitellogenic phase which results from the rapid accumulation, after minor biochemical modifications, of "vitellogenin": a lipophosphoprotein which is synthesized in the liver, released into the blood and incorporated into the oocytes by micropinocytosis (exogenous vitellogenesis or yolk deposition). At the end of vitellogenesis, meiosis resumes: a general process called oocyte maturation, characterized by completion of the first meiotic division, with first polar body emission, which is accompanied by important changes in the gross morphology of yolk and cytoplasm just before ovulation (expulsion of mature secondary oocyte from the follicle). Oocyte maturation ends only after sperm penetration with the second meiotic division and second polar body emission.

In contrast to the spherical shape of the mature egg in most teleosts, *Tilapia* and *Sarotherodon* eggs are ovoid. In substrate-spawners *(Tilapia)*, ovulation is accompanied by the production of sticky material (probably mucopolysaccharides), which seems to be secreted by the granulosa, from light microscope evidence (Von Kraft and Peters 1953) and electron microscope observations (Nicholls and Maple 1972). This sticky substance is deposited around special threads emerging from the zona radiata of oocytes (Von Kraft and Peters 1963), and glues all the eggs into one mass which sticks to the substrate. In mouthbrooders *(Sarotherodon)* eggs are usually not embedded in these sticky threads, with the exception of *S. galilaeus* which leaves the eggs some minutes on the substrate before commencing mouthbrooding (Fishelson 1966b).

As discussed by Perrone and Zaret (1979), egg size and fecundity in fish are strongly related to parental care patterns. In *Tilapia* and *Sarotherodon*, whose behavior provides a high level of parental care, production of gametes is rather low. In males, testis weight is very low compared to that of other teleosts without parental care (Peters 1971), whereas in females the number of eggs per spawning is of the order of hundreds in *Sarotherodon*, and a few thousand in *Tilapia* (Peters 1963): very different from the millions of pelagic eggs produced by some species without parental care. In tilapias, as in other teleosts, fecundity, egg size and egg weight usually increase with female size: egg production increases approximatively in relation to the square of body length in *Sarotherodon* and in relation to the cube of the length in *Tilapia zillii* as discussed by Welcomme (1967b).

The temporal patterns and rhythmicity of gametogenesis have been much studied in *Tilapia* and *Sarotherodon*. Individual females spawn successively, either during a defined breeding season or year-round, with a few weeks interval between spawnings (Moreau 1979): this requires either rapid or continuous gametogenesis. In both sexes successive waves of gametogenesis have been demonstrated (Von Kraft and Peters 1963; Peters 1963; Dadzie 1969; Moreau 1971, 1979; Hyder 1970a; Bruton and Boltt 1975; Siddiqui 1977b, 1979a; Babiker and Ibrahim 1979).

In females some doubt remains as to the stage from which a new wave of oocytes develops to prepare for the next spawning. Some authors (Von Kraft and Peters 1963; Peters 1963; Hyder 1970a) have reported that the next wave of oocytes is already in the process of active vitellogenesis in spent fish just after spawning, but some other evidence suggests that each new batch arises from a stock of previtellogenic oocytes (Silverman 1978a, 1978b; Moreau 1979). This dubious point may be due to imprecisions in the exact definition of active vitellogenesis (or exogenous vitellogenesis, i.e., under pituitary control), to differences in species, and to enviromental differences. For example, Moreau (1979) showed that the mean interval which separates the last two successive waves of oocytes in *Tilapia rendalli* is different in two different lakes. This point, however, would need more precise observation and experimentation to understand the underlying endocrinological mechanisms (see below), particularly in species where the female exhibits brood care behavior after spawning, during which ovarian growth might be inhibited.

In males, spermatogenesis is reported to occur continuously, the testis containing cysts at all different stages, in both *Sarotherodon* (Dadzie 1969; Hyder 1970a, 1970b; Moreau 1979) and *Tilapia* (Moreau 1979), but quantification reveals that substantial variations can occur in the intensity of spermatogenesis in marginal temperature conditions (Moreau 1979). Methods of quantification (Billard et al. 1974) should be used more extensively to assess spermatogenetic continuity. As in the case of females, this would allow investigation of the occurrence of any post-spawning inhibition in species where the male exhibits brood care behavior.

The Role of Environmental Factors in the Regulation of Reproductive Activity

In most teleosts, spawning periods appear to be adjusted to (and by) environmental factors (photoperiod, temperature, salinity, rainfall, etc.) so that they are suitable for rearing offspring (de Vlaming 1974). Fish of temperate zones, where photoperiod and temperature variations are of great magnitude, spawn during a limited period of the year, and only once in most cases. In equatorial and tropical regions, where these variations are more limited, temperature is rarely a limiting factor, but considerable environmental changes can occur which might inhibit or favor offspring survival, favoring related adaptations (for example rainfall and cloud cover during the rainy season).

Tilapia and *Sarotherodon* species are abundant in both equatorial regions and subtropical regions as well as the mediterranean and have even been introduced into environments which can be considered as marginal, for at least a part of the year, especially with respect to low temperature tolerance. This is the case in lakes at high altitude in Madagascar (Moreau 1971, 1979). Here the low temperatures encountered are very different from those in the original habitats of tilapias and the response in breeding patterns may be considered as the limit of a tendency. *Sarotherodon* and *Tilapia* extend their breeding seasons for as long as temperature is favorable. During the cold season spermatogenesis is greatly retarded but all its developmental stages remain present in the testis. In females, on the contrary, exogenous vitellogenesis seems to be completely inhibited by low temperatures and all yolk-laden oocytes disappear.

In more appropriate environmental conditions, such as equatorial lakes and ponds, *S. niloticus* was found to breed throughout the year, though the number of breeding fish was slightly higher during the wet season (Lowe (McConnell) 1958). In areas distant from the equator, the same species exhibits a well defined breeding season, spawning mainly during the warmest and most sunny season (Lowe (McConnell) 1958). Among *Tilapia*, *T. zillii* in equatorial lakes shows no reproductive seasonality and individual fish spawn successively year-round (Siddiqui 1979a). On the other hand, the same species in northern areas presents a definite seasonal breeding, spawning during the period of maximum water temperature and maximum light (Ben Tuvia 1959; Fishelson 1966a, 1966b; Siddiqui 1977b).

However, the assumption that no reproductive seasonality occurs in equatorial areas might even be questioned in regard to the precision and the validity of the method of appreciation, if we consider the contradictions between Hyder (1969, 1970a) and Siddiqui (1977b) concerning *S. leucostictus* in the same equatorial lake: the first studies show a seasonal variation in breeding activity, while the latter denies this finding and concludes that breeding is non-seasonal.

It seems therefore that *Tilapia* and *Sarotherodon* have a capacity for unlimited successive breeding (at least when the whole population is considered) in equatorial regions and an increased tendency for seasonality with increasing latitude, with maximum reproductive activity during periods of maximum temperature and day light. The effects of seasonal rainfall cannot, however, be excluded (Aronson 1957; Lowe (McConnell) 1958; Lowe-McConnell 1959; Hyder 1969, 1970a; Marshall 1979a, 1979b; Moreau 1979) but these remain controversial, probably because of differences between species.

It remains difficult to assess which are the key factors to stimulate, inhibit, or exert any regulation on the various stages of the breeding cycles of *Tilapia* and *Sarotherodon*. This uncertainty is due in part to the absence of any experimental study to dissociate the role of separate environmental factors, and also part to the imprecise methods which have been used to determine the different parameters relative to the intensity of reproductive activity.

The Role of Social Factors in the Regulation of Reproductive Activity

Social interactions are known to influence some parameters of reproduction, particularly the timing of first sexual maturation, spawning frequency and fecundity. In cichlids, spawning frequency is increased by different sensory stimulations coming from conspecific fishes, e.g., visual stimuli, sound production, lateral line contacts and probably chemical communication (Aronson 1945, 1951; Polder 1971; Marshall 1972; Chien 1973).

In some *Sarotherodon* species the female is able to spawn regularly even when isolated, but shows increased interspawning intervals compared to non-isolated females (Aronson 1945; Marshall 1972; Silverman 1978a, 1979b). In *S. mossambicus* Silverman (1978a, 1978b) was able to dissociate the effects of different levels of social contact on separate parameters of the interspawning interval. He distinguished between unlimited contact (several fish in the same aquarium), medium contact (adjacent aquaria each containing one fish so that each fish can see into the other tank) and low contact (visually isolated fish in different aquaria) and showed that visual stimuli hastened mainly ovulation with little influence on oogenesis but that other non-visual stimuli (e.g., tactile or chemical) in unlimited-contact females were able, in addition, to advance yolk deposition by about 7 days. Males and females were shown to be equally effective as stimulus animals. This may reflect some "general conspecific effect" related to gregariousness of *S. mossambicus*, a species where "presence" within a social group is important, according to Silverman. He suggested that regular spawning in isolation would most likely occur in species where males with ripe gonads are continuously available waiting for new females to encounter their nests. Although very interesting, these experiments concerning *S. mossambicus* should be performed on other species, and efforts should be made to look for the existence of pheromone-like substances which have been shown in other fish (Solomon 1977).

The Role of Internal Factors in the Regulation of Reproductive Activity

Numerous organs, endocrine glands, and hormones are involved, directly or indirectly, in the regulation of reproduction in fish as in other vertebrates (Fontaine 1976; Olivereau 1977). This complex control system is directed by the nervous system which integrates external stimuli together with ontogenetic and physiological constraints, and exerts its control mainly through the hypothalamo-pituitary-gonad axis. The following section will deal essentially with the functional modalities of this axis. Points which are still very controversial in other teleosts, and which have not even been studied in cichlids to our knowledge will be omitted: for example, the role of the pineal organ.

The Role of the Pituitary

Many works reviewed by Pickford and Atz (1957) have demonstrated how important the role of the pituitary is in the control of reproduction in fish. This role was confirmed in cichlids *(Sarotherodon spilurus)* by means of chemical hypophysectomy using methallibure and replacement therapy (Hyder 1972; Hyder et al. 1974; Hyder et al. 1979). Methallibure, which was shown to inhibit gonadotropin secretion by the pituitary in fish (Breton et al. 1973) as it is known to do in mammals, induces an effective gonadal regression in both male and female *S. spilurus* after 4 to 5 weeks of treatment with daily doses of 1 mg/l of aquarium water (Hyder 1972). This treatment results in an extensive resorption of yolk-laden oocytes and a complete inhibition of vitellogenesis in females and in the complete inhibition of two steps of spermatogenesis in males: the step between spermatogonia and spermatocytes (meiosis) and final spermiation. A similar effect was confirmed by Lanzing (1978).

The present state of knowledge of teleost fish reproduction tends to support the hypothesis that two different pituitary gonadotropins are involved in the control of gametogenesis. One purified at first from carp, *Cyprinus carpio* (Burzawa-Gerard 1971), trout, *Salmo gairdneri* (Breton et al. 1976) and in salmon of the genus *Oncorhynchus* (Donaldson et al. 1972) is characterized by its control over final intrafollicular oocyte maturation (Jalabert et al. 1974) and its effect upon spermatogenesis (Billard et al. 1970). A second kind of gonadotropin which seems necessary for vitellogenesis was recently purified from carp and salmon pituitaries (Idler and Bun Ng 1979).

In cichlids, some attempts have been made to purify and characterize pituitary gonadotropins. Farmer and Papkoff (1977) obtained two preparations from *S. mossambicus* which exhibited some biochemical characteristics similar to luteinizing hormone (LH) and follicular stimulating hormone (FSH) from higher vertebrates. The first preparation stimulated testosterone production in isolated rat testis Leydig cells, an assay which is considered highly sensitive and specific for mammalian LH. Neither of the two preparations was assayed for its activity on *Sarotherodon* gonads. More recently, Hyder et al. (1979) observed enhanced gonadotropic activity (as judged by testis stimulation in methallibure-treated *S. spilurus*) after chromatographic separation of a glycoprotein fraction from *S. niloticus.* The fraction was the same as that known to contain FSH when the same biochemical procedure is applied to pituitary preparations from higher vertebrates. Results are thus still confusing, and more work remains to be done in order to find specific bioassays for the gonadotropins of cichlids.

Another pituitary hormone, characterized as a "prolactin-like" hormone, is believed to play an interesting role in the reproductive physiology of cichlids, in relation to the regulation of breeding activity and parental care Blüm and Fiedler (1964) found that injection of ovine prolactin induced behavioral and histiotropic effects in *Symphysodon aequifasciata axelrodi.* The behavioral reactions were of fanning movements with the pectoral fins orientated to a distinct reference point (normally associated with parental care towards the brood) and suppression of feeding and fighting tendencies. The histiotropic effect was an increase in mucous cell production on the body surface which normally serves to nourish the young in that species (Hildemann 1959), thus exhibiting a curious analogy with the secretion of crop-milk in pigeons and milk production in mammals.

In other cichlids, e.g., *Pterophyllum scalare*, *Aequidens latifrons*, *Cichlasoma severum* and *Astronotus ocellatus* (Blüm and Fiedler 1965) ovine prolactin induces similar effects, including increase in mucous cell production although to a lesser extent than in *Symphysodon* and here this is not related to nursing of the young in that species. It is possible that mucus secretion can be related to the osmoregulatory properties of prolactin in fish (Blüm 1973; Bern 1975). Isolation and purification of *Sarotherodon* prolactin was successively reported by Blüm (1973) using behavioral and histiotropic effects in *Symphysodon* and by Farmer et al. (1977) using specific assays for osmoregulation. Considering the pleiotropic action of prolactin, isolating the same molecule through different kinds of bioassays is not surprising, but a comparison of different preparations using the same tests would be of interest in assessing the reality of this pleiotropic role for the same hormone. Among the alleged multiple actions of prolactin in fish, one could be of interest for *Tilapia* and *Sarotherodon* culture if its mechanism could be elucidated: the inhibition of gonad development (Blüm 1976). In natural conditions, gonadal development after spawning might be inhibited during parental care, which is precisely the period when prolactin is supposed to be acting. In addition, some antagonism between prolactin and gonadotropins was found to occur regarding spawning behavior (Blüm and Fiedler 1965).

In addition to the central role of gametogenesis the gonads possess complex endocrine properties which, under pituitary control, contribute to the regulation of reproductive cycles by direct action on gamete differentiation, by controlling the activity of different organs and tissues involved in reproduction (such as the liver, fat and bone mineral stores) and also by controlling the development of secondary sexual characteristics. Finally the endocrine secretions of the gonads participate in the regulation of pituitary activity (a feed-back mechanism) and also act on the central nervous system, allowing different kinds of behavioral patterns to occur during the successive periods of the sexual cycles.

Sexual steroids are the main hormones produced by the gonads, but other compounds, the occurrence of which have not been investigated in cichlids, must certainly be secreted (Breton et al. 1980). Steroids are generally considered to be produced by specialized cells presenting histochemical and morphological features common to all vertebrates (Hoar and Nagahama 1978).

In males these cells form a typical interstitial tissue which has been characterized as steroidogenic by demonstration of 3β-hydroxy-steroidogenase (Yaron 1966) and from ultrastructural evidence (Nicholls and Graham 1972). They exhibit increased activity (number, size, lipid concentration) during spermatogenetic progression, with maximum activity during spermiation, when there is also rapid development of sexual coloration, nest building activity and territoriality (Hyder 1970b), and when the testis contains high levels of testosterone (Hyder and Kirschner 1969). The fact that interstitial tissue activity is under pituitary control can be demonstrated by several methods. Treatment with human chorionic gonadotropin (HCG), which exhibits gonadotropic activity in some fishes, induces both stimulation of interstitial tissue and an increase in the testosterone content of the testes (Hyder et al. 1970). Conversely, methallibure inhibition of pituitary gonadotropin secretion lowers interstitial cell activity, an effect which can be overcome by administration of HCG or *Sarotherodon* pituitary extracts (Hyder 1972; Hyder et al. 1974). But testosterone is probably neither the only nor the main steroid mediator produced in the gonad, as testosterone propionate administered to methallibure-treated *Sarotherodon* failed to restore spermatogenesis, producing only some spermiation at high doses (Hyder et al. 1974). As a range of androgenic steroids has already been found in fish, like 11-ketotestosterone in female *S. aureus* (Eckstein 1970), further studies would be of great interest in attempting to identify the major active steroids in *Tilapia* and *Sarotherodon*, and to understand their action on the different steps of spermatogenesis, sexual behavior and secondary sexual characteristics: for example, the genital tassel of male *S. macrochir* which seems to be an important signal in the spawning behavior of that species (Wickler 1965, 1966b).

In female cichlids both histochemistry (Livni 1971; Yaron 1971) and ultrastructural morphology (Nicholls and Maple 1972) show that most ovarian steroidogenesis is probably located in two kinds of follicular cells: the granulosa cells (although the ultrastructural evidence is here somewhat equivocal) and special theca cells, located mainly close to capillaries.

Estradiol-17-β (E2) seems to be of general occurrence in fish. It is mainly involved in the control of vitellogenin synthesis in the liver and of mobilization of mineral and fat stores (Olivereau and Olivereau 1979; Mugiya 1978). This seems to be true also in *S. aureus* where estradiol-17β was identified in the ovary by Katz et al. (1971). Later a positive correlation was found between plasma E2 concentration and ovarian weight, i.e., the stage of vitellogenesis (Yaron et al. 1977; Terkatin-Shimony and Yaron 1978). In the same species, some ovarian steroidogenic pathways were investigated by Eckstein (1970) and Eckstein and Katz (1971), leading to the identification of new steroid metabolites like 11-ketotestosterone and 11-hydroxytestosterone among others which are more classical (testosterone and progesterone). Their exact function, although related to sexual activity (Katz and Eckstein 1974) is unknown. Deoxycorticosterone, a corticosteroid which can be synthesized either in the interrenal tissue or in the teleost ovary (Colombo et al. 1973) was found to increase 38-fold in the blood of *S. aureus* between a sexual resting phase at 18 to 20°C and sexual activity at 30°C. As this steroid is assumed to be one of those mediating gonadotropin action on final maturation and ovulation in catfish, *Heteropneustes fossilis* (Sundararaj and Goswami 1977), it is tempting to propose a similar model for final maturation control in *Sarotherodon*. However, the state of "sexual activity" (stage of ovarian development) at which the blood of female *Sarotherodon* was collected by Katz and Eckstein (1974) was not known precisely and further work is needed to find out which steroids mediate pituitary gonadotropin control over final oocyte maturation and their mechanism: for reviews on this problem in other teleosts, see Jalabert (1976) and Sundararaj and Goswami (1977).

Prolactin also has some control over ovarian steroidegenesis as can be inferred from the observation of Blüm and Weber (1968) that ovine prolactin greatly stimulates the activity of 3β-steroid dehydrogenase in ovaries of the cichlid *Aequidens pulcher.* This observation suggests that inhibition of ovarian growth by prolactin is mediated through steroid action. An understanding of the precise mechanism for this action would be of great interest.

Regulation of Pituitary Activity

As in higher vertebrates, gonadotropin secretion by the pituitary appears to be regulated by the nervous system through the hypothalamus which secretes a releasing factor first indicated in fish *(Cyprinus carpio)* by Breton et al. (1971) and Breton and Weil (1973). This factor is probably not very different from mammalian luteinizing hormone-releasing hormone (LH-RH) (Breton et al. 1972, 1975a) at least as far as the biologically active part of the molecule is concerned. Recent studies show that LH-RH's from lower vertebrates, including *S. mossambicus*, differ from those of higher vertebrates both in immunological and biochemical characteristics (King and Millar 1979). Other experimental data in fish support the hypothesis that factors inhibiting pituitary gonadotropin secretion could be present in the fish hypothalamus (Peter 1978).

Ovarian endocrine secretions, particularly some steroids, seem to be able to exert positive or negative feed-back action upon the activity of the hypothalamo-pituitary complex in fish (Billard and Peter 1977; Breton et al. 1975b; Billard 1978; Jalabert et al. 1980) as they do in mammals.

A better understanding of all these controls over pituitary activity could provide practical means to manipulate some aspects of reproduction in cultured fish.

Conclusion

Some *Tilapia* and *Sarotherodon* species are well-suited for experimental work in reproductive physiology. They perform gametogenesis and regular spawning in aquaria, where external factors (light, photoperiod, temperature) can be easily controlled and they are of a convenient size for endocrinological studies involving blood sampling on living cannulated fish at different stages of the sexual cycle. Experimental data are rather scarce, however, making it very difficult to propose any techniques immediately applicable in tilapia culture. Thus, experimental research is still greatly needed to determine the respective roles of internal and external factors on first sexual maturation in the different steps of gametogenesis and on spawning frequency. Concerning the role of hormones, important results may come from work on other groups of fish and other vertebrates, concerning, for example, the hypothalamic regulation of gonadotropin secretion and non-specific means to control this activity. The development of specific methods to manipulate reproduction in tilapias requires better knowledge of their specific protein and steroid hormones involved in the control of sexual differentiation, gametogenesis and sexual and brooding behavior.

Discussion

NASH: I agree with you on the importance of egg quality. For example, salmon hatcheries tend to discard the first and last eggs available in a spawning season as they are invariably of poorer quality than those taken in the middle. Do you recommend for describing oogenesis in tilapias your six-part scale or Yamazaki's five-part scale? In your six-part scale, stages 3, 4 and 5 appear to be stages IIIa, IIIb and IIIc of the five-part scale. I assume that your stage 6 is atresia?

JALABERT: The important point is that we find oocytes at all stages in just about every fish in the population.

HEPHER: In temperate regions, it is found that vitellogenesis in fish is affected very much by day length. You haven't mentioned this at all. Is this because tilapias are from the tropical regions where there is little change in day length?

JALABERT: Some people say that there is a daylength effect even in equatorial regions, others say there is none. In tilapias, there is no clear evidence for such an effect and it would need very precise criteria to assess small variations in the intensity of gametogenesis to investigate this. I remember for instance the contradictions between Hyder (1970a) and Siddiqui (1977b) concerning the seasonality of *S. leucostictus*. Such contradictions probably indicate the lack of precise criteria.

HEPHER: The works I know show that, for fish in general, day length affects vitellogenesis more than the other stages of oogenesis, but since you didn't mention this effect at all, may I assume that it does not exist in tilapias?

JALABERT: I don't know. We have to make controlled environment experiments to investigate this. The same is true for the effects of salinity on reproduction and growth. The evidence seems to be that in tilapias, responses to environmental changes are very variable indeed and may differ even within the same species. We conducted some recent experiments with trout to try to select for early and late reproduction to spread availability of eggs. We observed a shift in the timing of reproduction after only one generation. For tilapias, therefore, we can expect different responses for different species and strains.

NASH: We cannot generalize about reactions to environmental stimuli with subtropical and tropical fish. In mullet it is the light stimulus that triggers oogenesis and then the temperature takes control. And, although you cannot suppress development completely, you can delay it by temperature control even after triggering by light. There are groups of fish for which the opposite is the case. The temperature acts as the trigger and the light control acts as the monitor.

MORIARTY: I would like to ask a question from the point of view of the fish culturist. Is it going to be of any practical benefit to have a much more detailed knowledge of the reproductive hormones of these fish, or do you think that this is so complex that we would be better advised to study environmental and social factors?

JALABERT: Although the practical applications are not evident now, we do need more work on the reproductive hormones of fish and on their modes of action, in order to understand how environmental and social factors interact to modulate sexual activity. I believe that management decisions could be more efficient if based on such basic knowledge. But, from a strategical point of view, I feel that some fundamental questions are closer to application than some others; for example, are pheromones important in tilapia reproduction?

GUERRERO: In one experiment you described the separation of the two sexes of tilapia. You say that there was no pheromone action because the fish still spawned.

JALABERT: In this experiment the isolated females spawned anyway, or, more exactly, released their eggs, but the interval between successive spawnings was increased. When fish were together, this interval was reduced. We cannot rule out pheromones here, among the other factors. On the other hand, spawning in artificial environments can be affected by many stress factors which are not necessarily specific.

Feeding, Digestion and Growth—Qualitative Considerations

S. H. BOWEN

Department of Biological Sciences
Michigan Technological University
Houghton, Michigan 49931
U.S.A.

BOWEN, S.H. 1982. Feeding, digestion and growth—qualitative considerations, p. 141-156. *In* R.S.V. Pullin and R.H. Lowe-McConnell (eds.) The biology and culture of tilapias. ICLARM Conference Proceedings 7, 432 p. International Center for Living Aquatic Resources Management, Manila, Philippines.

Tilapias feed as herbivores and detritivores. Food particle size is reduced in the pharyngeal mill which facilitates peristaltic mixing and increases surface area for exposure to digestive fluids in the gut. Gastric acid secreted to pH values < 2.0 lyses cell walls and cell membranes of bacteria and algae. Subsequent intestinal digestion occurs gradually as the food passes down the extremely long intestine. In addition to microorganisms, detritus is digested and is a major nutritional resource for detritivorous tilapias. Whole diet assimilation efficiencies are lower for tilapias than for carnivorous fishes. Food quality appears to limit the growth of tilapias in natural populations. The limited data available indicate that within a given water body, tilapias select precisely the food that will maximize growth. The combined abilities for cell wall lysis and selective feeding suggest that tilapias hold considerable promise for low technology, protein efficient aquaculture.

Introduction

Animal growth rate is determined through the combined effects of food quantity and food quality. The quantity of food consumed is regulated through appetite to satisfy the animal's energy requirements (Rozin and Mayer 1961, 1964). Limited food availability that does not allow full appetite satisfaction results in growth rates below the maximum potential. Food quality depends on the composition of the diet, and the extent to which the components are digested and assimilated. Quality is rarely a limiting factor in the growth of carnivores since their diet is consistently of very high quality, but the quality of diets consumed by herbivores and detritivores is extremely variable and plays a major role in control of growth at these trophic levels. This contribution reviews recent scientific research on aspects of the diet, digestion, assimilation and growth of the tilapias (family Cichlidae; genera *Tilapia* and *Sarotherodon)*, which are herbivorous and detritivorous, to assess our current understanding of how food quality influences the growth of these animals in natural populations.

The Feeding Apparatus and Digestive Tract in Tilapias

It is important to begin with consideration of the feeding apparatus and digestive tract since these structures limit the range of potential food items that can be consumed and digested efficiently. Compared to the haplochromis cichlids, the feeding apparatus of tilapias is simple and unspecialized (Fryer and Iles 1972). The jaw teeth are small unicuspid, bicuspid or tricuspid structures that occur in one to five rows (Plate 1). In those species for which descriptions are available, the jaw teeth are flattened distally to form blades that appear to be useful as scrapers (Fryer and Iles 1972; Lanzing and Higginbotham 1976). Neither the gill rakers nor the buccal cavity appears to be specialized for feeding, but considerable specialization is evident in the dentition of the pharyngeal bones (Plate 2). These teeth range from fine, thin, hooked structures on the pharyngeal bones of *Sarotherodon esculentus*, a phytoplankton consumer, to the coarse, robust structures on the pharyngeal bones of *T. rendalli*, a macrophyte consumer (Caulton 1976). Mechanical and myological details of the cichlid pharyngeal apparatus are described by Fryer and Iles (1972) and Liem (1973).

The role of the pharyngeal apparatus is to prepare food for digestion. In many species this is done by breaking or cutting the food into smaller sized units. In *S. esculentus*, filamentous and large colonial phytoplankton may be broken into smaller units. In *S. mossambicus* and *S. melanotheron*, detrital aggregate is broken into finer fragments (Bowen 1976a; Pauly 1976) and in *T. sparrmanii* long filamentous periphyton is shredded to short segments of uniform length. The advantages of reduced particle size include a greatly increased surface to volume ratio that facilitates enzyme-substrate interaction, and reduced resistance to peristaltic mixing. In addition, mechanical disruption of macrophyte cell walls by the pharyngeal apparatus in *T. rendalli* increases the efficiency with which this food is digested (Caulton 1976).

The esophagus is short with a small diameter and leads to a small sac-like stomach. Some investigators have questioned the identification of this latter structure (Kamal Pasha 1964; Man and Hodgkiss 1977b) but its separation from the intestine by a sphincter, the low pH of the fluid it contains (Moriarty 1973; Bowen 1976b; Caulton 1976) and the pH optima of proteases extracted from its mucosa (Fish 1960; Nagase 1964; Moriarty 1973) all attest to its gastric function. Immediately behind the pyloric sphincter, the intestine receives a common bile duct. The first, short intestinal segment is thin-walled and of greater diameter than the remainder. Perhaps the most striking feature of the digestive tract of tilapias is the exceptional length of the intestine. Quantitative data for *T. rendalli*, *S. melanotheron* and *S. mossambicus* show that the ratio of intestinal length to fish standard length is between 7:1 and 10:1 (Caulton 1976; Pauly 1976; Bowen, unpublished). Other non-quantitative observations reported in the literature suggest these ratios are representative of the tilapias as a group. By comparison, Fryer and Iles (1972, p. 41) report that only three of 106 cichlids collected from Lake Tanganyika and its affluents had gut length: total length ratios greater than 6:1 and these three may have been tilapias. The intestine ends in an anal sphincter.

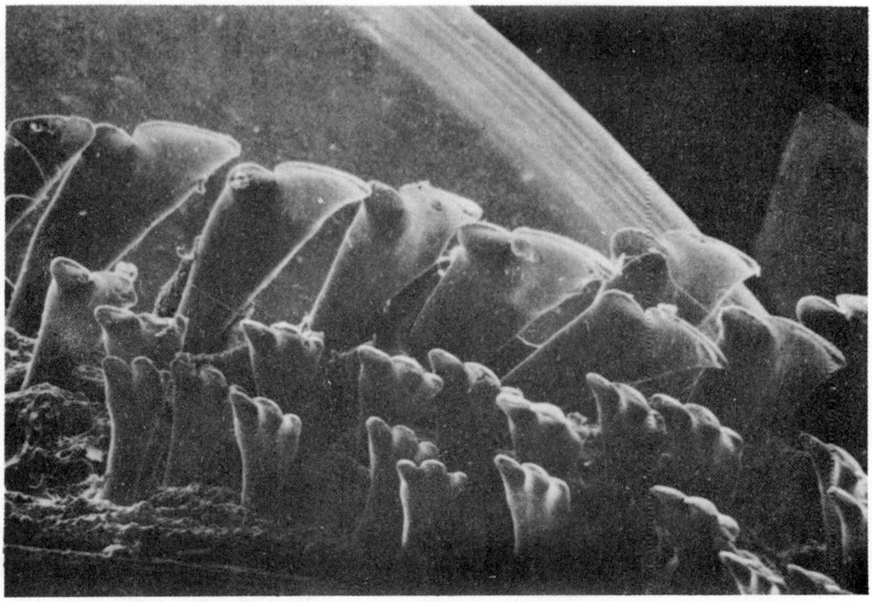

Plate 1. Jaw teeth of an 8 cm standard length *Sarotherodon mossambicus* from Lake Valencia, Venezuela.

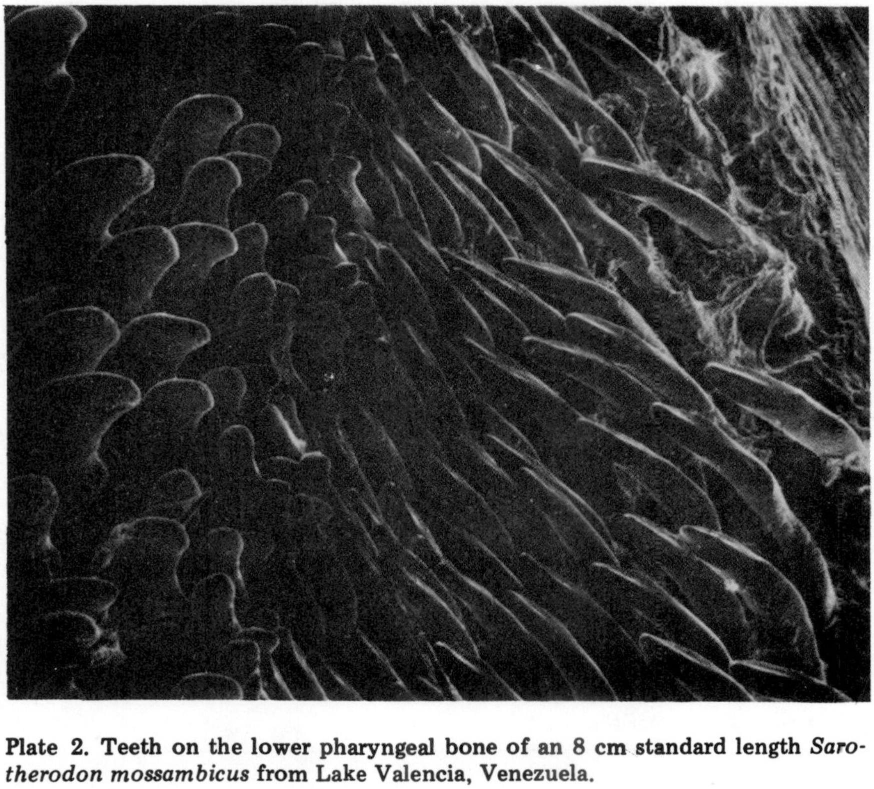

Plate 2. Teeth on the lower pharyngeal bone of an 8 cm standard length *Sarotherodon mossambicus* from Lake Valencia, Venezuela.

The Diet of Tilapias

In common with early developmental stages of nearly all fishes, the larvae, fry and early juvenile tilapias feed on small invertebrates, especially Crustacea (Le Roux 1956). The transition from an invertebrate diet to the typical adult diet is usually abrupt (Bowen 1976a; Moriarty et al. 1973) but in some cases it may occur gradually over the period of a year or more (Whitfield and Blaber 1978).

The diets of adult tilapias have been reported for more than 17 species sampled on three continents and a number of island systems (Table 1). Practically every aquatic animal, vegetable and mineral small enough to pass through the esophagus has been found in the guts of these fish. Some of this variety must be attributed to items that are occasionally abundant in the diet but are rare or absent most of the time and have no long-term significance in the fish's nutrition. For example, Bruton found that *S. mossambicus* in Lake Sibaya fed intensively on formicids for the few days when these were abundant at the lake surface during an annual swarming period, but they did not occur in the diet at other times of the year (Bruton pers. comm.). To describe the characteristic diet of a tilapia species, it is necessary to identify those food items that are consistently present in the diet over long periods of time.

Table 1. Diets reported for adult tilapias in natural habitats.

Species	Diet	Authority
Sarotherodon shiranus	macrophytes, algae, zooplankton	Bourn 1974
S. pangani	periphyton	Denny et al. 1978
S. jipe	periphyton	Denny et al. 1978
S. esculentus	phytoplankton	Denny et al. 1978; Fish 1951, 1955
T. rendalli	macrophytes, attached periphyton	Caulton 1976, 1977b; Denny et al. 1978
S. mossambicus	macrophytes, benthic algae, phytoplankton, periphyton, zooplankton, fish larvae, fish eggs, detritus	Bowen 1979, 1980b; Man and Hodgkiss 1977b; Munro 1967; Naik 1973; Weatherley and Cogger 1977
S. aureus	phytoplankton, zooplankton	Fish 1955; Spataru and Zorn 1976, 1978
S. niloticus	phytoplankton	Moriarty and Moriarty 1973a
T. kottae	phytoplankton, detritus, invertebrates	Corbet et al. 1973
T. mariae	phytoplankton, invertebrates	Corbet et al. 1973
S. galilaeus	phytoplankton	Corbet et al. 1973; Spataru and Zorn 1978
T. zillii	macrophytes, benthic invertebrates	Abdel-Malek 1972; Buddington 1979
T. guineensis	algae, detritus, sand, invertebrates	Fagade 1971
S. melanotheron	algae, detritus, sand, invertebrates	Fagade 1971
S. variabilis	algae	Fish 1955
S. leucostictus	phytoplankton, detritus	Moriarty et al. 1973
T. sparrmanii	periphyton	Bowen unpublished

The characteristic diet of adult tilapias is plant matter and/or detritus of plant origin. Blue-green and green algae, diatoms, macrophytes and amorphous detritus are all common constituents of adult tilapia diets (Plate 3). Bacteria are also present and may be very important for some species. Although tilapias collect their diets from such diverse substrata as macrophytes, lake bottoms, rock surfaces and from suspension, it is the uniformity of their diets that is especially noteworthy. Comparisons of diet composition for tilapias reveal extensive overlap. Tilapias that feed on macrophytes also ingest the attached algae, bacteria and detritus. This attached material is likely to be an important component of the diet. Although quantitative data are lacking for macrophytes consumed by tilapias per se, the dry weight of attached material was found in one study to exceed 25% of the total supporting macrophyte dry weight (Bowen 1980a) and could comprise an even greater percentage of the weight of leaves actually consumed by tilapias. Tilapias that feed on epiphytes also frequently ingest some of the supporting macrophyte (Munro 1967). Deposit feeders that feed in the littoral zone ingest a mixture of algae, detritus and bacteria that is essentially indistinguishable from the material attached to macrophytes. Deposit feeders that feed in deeper water ingest a mixture of sedimented phytoplankton and detritus essentially indistinguishable from the diets of suspension feeders. These various sources provide diets that are so similar that tilapias may switch from one source to another with little change in diet composition (Moriarty et al. 1973; Bowen, unpublished). Thus, despite the diversity of food resources exploited by tilapias, their diets are qualitatively very similar mixtures of algae, bacteria and detritus with or without macrophytes.

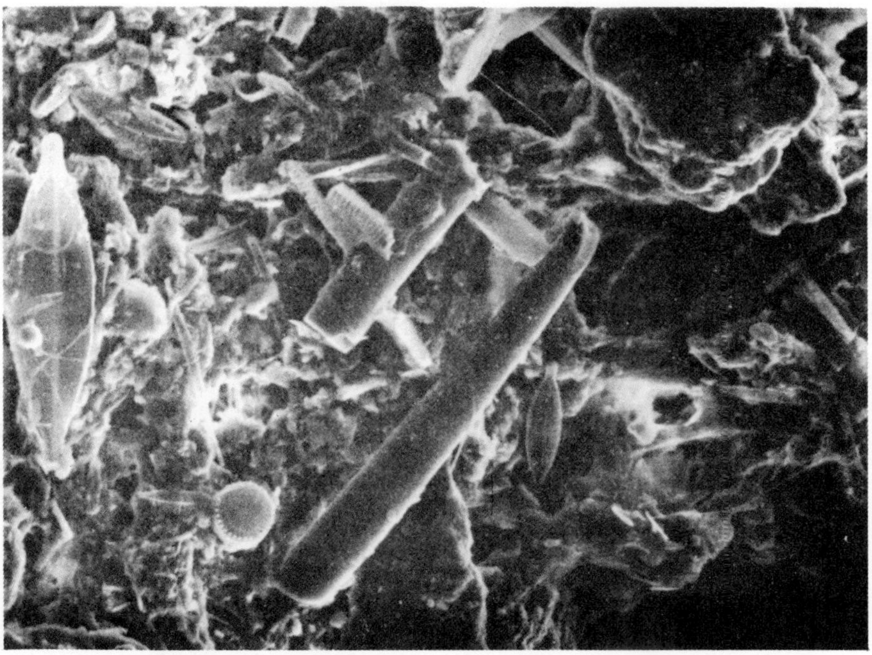

Plate 3. Scanning electron micrograph of typical stomach contents from *Sarotherodon mossambicus* in Lake Valencia, Venezuela.

The role of animal found in the diets of tilapias is, at present, an enigma. In some studies, animal remains are rare or absent from gut contents even when hundreds of specimens were examined. In most cases animal remains are present in low numbers and investigators have concluded these were ingested incidentally, either as whole invertebrates or as fragments of dead invertebrates, while the fish fed on other more typical foods. But in some cases, invertebrates clearly make up a significant proportion of the diet and are probably ingested intentionally (Abdel-Malek 1972; Spataru and Zorn 1978; Whitfield and Blaber 1978). Inclusion of invertebrates in the diet may be an important variable in the feeding strategy of tilapias, but until more quantitative data are available to describe circumstances under which animal prey are selected and the significance of such selection for nutrition, this aspect of the diet will remain an enigma.

The Process of Digestion in Tilapias

Digestion in tilapias is a two-step process with distinct gastric and intestinal components. The mechanism for gastric digestion found in tilapias appears to be unique among animals. In other animals, the pH of fluids in an actively digesting stomach ranges from about 2.0 to 2.2 (Barrington 1957). This is the pH at which vertebrate gastric digestive enzymes, including those of tilapias (Fish 1960; Nagase 1964; Moriarty 1973) show maximum activity. In contrast, the pH of stomach fluid in actively digesting tilapias is frequently as low as 1.25 (Moriarty 1973; Bowen 1976; Caulton 1976) and values as low as 1.0 have been recorded (Payne 1978). Moriarty (1973) was the first to describe the role of gastric acid in digestion by tilapias. He studied the population of *S. niloticus* in Lake George, Uganda, which feeds on phytoplankton dominated by colonial and filamentous blue-green species. As the algae pass from the esophagus they may travel either of two routes through the stomach. Peristaltic movement carries the food in a posterior direction along the ventral wall, up the posterior wall and back along the top to the pyloric sphincter. Algae that travel this route are exposed to progressively lower pH as HCl is secreted by the gastric mucosa. Exposure to acid at pH 1.8 or lower decomposes the algal chlorophyll to phaeophytin and thus a gradual change in color from green to brown is seen as the algae move along the gut wall. Some algae take a shorter route passing across the anterior of the stomach from the esophagus directly to the pyloric sphincter and are not exposed to acid concentrations below pH 2.0. These cells remain green. Moriarty demonstrated that acid not only decomposes chlorophyll but actually lyses blue-green cell walls. This makes subsequent intestinal digestion possible by providing intestinal enzymes access to the algal cytoplasm. Following Moriarty's discovery, it was found that the same mechanism allows *S. mossambicus* to digest bacteria associated with detritus in its diet (Bowen 1976a). These observations are significant since vertebrates lack gastric enzymes capable of attacking the prokaryotic cell wall. Development of a special mechanism for lysis of blue-green algae and bacteria allows the tilapias special access to a relatively protein-rich (about 50%) food resource for which there is little, if any, vertebrate competition.

Gastric acid is similarly important in digestion of some eukaryotic cells. The efficiency of diatom digestion increases with lower pH to pH 2.5 (Bowen 1976b). Tilapias that consume macrophytes also have low gastric pH when feeding and there is some indication that gastric acid may facilitate digestion of their food as well (Caulton 1976).

Details of gastric digestion have been studied for *S. mossambicus* in Lake Valencia, Venezuela (Bowen 1981). Comparison of food and stomach contents for juveniles fed periphytic detrital aggregate (PDA) in aquarium experiments showed that gastric acid decomposed much of the mineral component of the diet (Table 2). In effect, this increased the concentration of organic matter available for the fishes' nutrition by a factor of 1.5. Measurable protein values increased by a factor of 1.6. Although a small fraction of this increase may be gastric enzymes, *in vitro* simulation of gastric acidification showed most of the increase was due to the effect of acid in facilitating protein extraction for subsequent assay (Table 3). Experimental data indicate that about 20% of the dietary carbohydrate is decomposed by gastric acid.

Table 2. Results of aquarium experiments to assess effects of gastric digestion on periphytic detrital aggregate ingested by *Sarotherodon mossambicus*. Samples for which the coefficient of variation (CV) is not given were pooled. Total organic values are % of sample weight; protein and carbohydrate values are % of organic weight.

Food Component	Experiment number	Food intake (%)	Stomach contents (%)	CV (%)	Stomach (%) / food (%)
Total Organic	1	35.7	55.8	12.7	1.56
	2	32.1	61.2	12.4	1.91
	3	43.7	64.9	4.1	1.49
	4	56.1	62.8		1.11
	5	44.7	56.4		1.26
	6	40.2	53.4		1.33
	7	38.3	63.1		1.65
					$\bar{x}$ = 1.47
Protein	4	1.42	1.97		1.39
	5	1.79	3.35		1.87
	6	0.84	1.19		1.42
	7	1.19	2.00		1.68
					$\bar{x}$ = 1.59
Carbohydrate*	4	17.4	14.2		0.82
	5	16.7	13.1		0.79
	6	18.5	14.0		0.76
					$\bar{x}$ = 0.79

*Mean of three determinations

Table 3. Effect of gastric acid on the food of *Sarotherodon mossambicus* in Lake Valencia, Venezuela simulated *in vitro* ([a]% sample weight, [b]% organic weight).

pH	Organic[a]	Protein[b]	Carbohydrate[b]	Carbonate[a]
7.0	41.0	1.52	11.8	11.9
3.0	47.6	2.57	10.6	11.0
2.5	48.5	2.63	10.6	10.1
2.0	49.7	2.47	10.1	9.5
1.5	49.9	2.79	9.8	9.8

The importance of gastric enzymes in digestion by tilapias is not clear and may vary with species or with diet for a single species. Although a protease with pH optimum around 2.1 is present in the gastric mucosa of *S. mossambicus* (Fish 1960; Bowen 1976a) and *S. niloticus* (Moriarty 1973), it has not been detected in the gastric fluids (Moriarty 1973; Bowen 1976a). This enzyme has a greatly reduced reaction rate at pH values below 1.5, and it is possible that it is not secreted when gastric acid is used to lyse cells.

Secretion of gastric acid stops at the end of the daily feeding period and stomach fluids return to pH 5 to 7. In *S. mossambicus* (Bowen 1976b) and *S. niloticus* (Moriarty 1973) the stomach is completely empty during this resting phase but in *T. rendalli* a small amount of food remains (Caulton 1976). At the start of the next day's feeding, acid secretion begins only as the stomach is filled with the result that the first food to pass into the intestine is not exposed to strong acid. In those species whose diet is rich in chlorophyll, the food ingested at the start of a feeding period remains green and undigested as it moves along the length of the intestine (Moriarty 1973; Caulton 1976). Similarly the first diatoms to be ingested by *S. mossambicus* remain undigested (Bowen 1976b). This explains why early workers who compared stomach and posterior intestine samples sometimes reported that algae were not digested.

The second step in the digestive process occurs in the intestine. The common bile duct which opens into the intestine on the back of the pyloric sphincter adds bile salts that maintain the pH between 6.8 and 8.8 (Fish 1960; Nagase 1964). Trypsin, chymotrypsin, amylase and esterase activity have all been identified in intestinal fluid (Fish 1960; Nagase 1964; Moriarty 1973).

Details of intestinal digestion have been studied for *S. mossambicus* in Lake Valencia, Venezuela. Contents of digestive tracts removed from freshly caught specimens were divided into five subsamples: stomach contents, and the contents of the first, second, third and fourth quarters of the intestine. Each subsample was analyzed for organic matter, carbohydrate, protein and total amino acids (Bowen 1980b, 1981). Values from these analyses were expressed in units of weight per weight of hydrolysis resistant sample ash (HRA). This indigenous component of the diet is used as an undigested reference compound relative to which digestion and assimilation are measured (Conover 1966; Bowen 1981). The results show total organic matter and carbohydrate are digested rapidly in the first and second quarters of the intestine with no evidence of digestion in the third and fourth quarters

(Figure 1). Digestion of protein is complete in the first quarter. This component of the diet is thought to be microbial protein present in living cells and protein that may be associated with detritus. The results indicate that microbial cells lysed in the stomach are digested immediately in the intestine while those not lysed are not digested at any point. Very different results were obtained for total amino acids (Figure 2). This class of compounds is comprised of amino acids bound into the complex chemical milieu of the detritus in addition to amino acids that make up microbial proteins. Whether detrital amino acids are single amino acids, oligopeptides or polypeptides is not yet known. As a group, the total amino acids are digested gradually as food passes down the intestine and maximum digestion is not achieved until the end of the digestive tract.

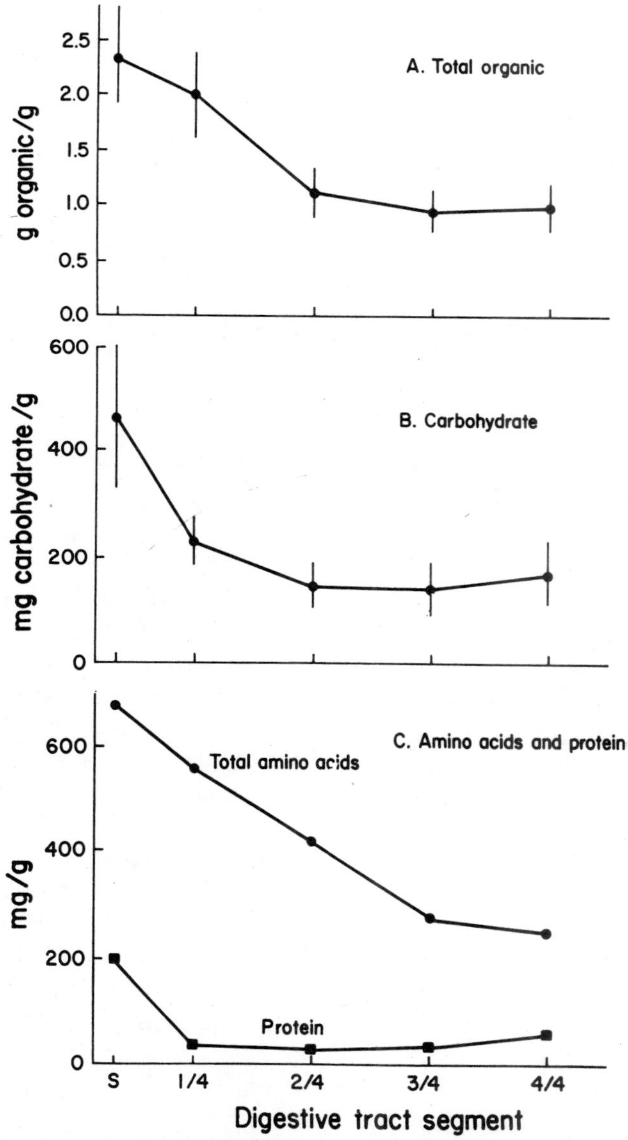

Figure 1. Comparisons of samples from the stomach (S) and the four quarters of the digestive tract of *Sarotherodon mossambicus* from Lake Valencia, Venezuela, to show the course of digestion. Means ± 95% confidence limits, n = 55 (after Bowen 1980).

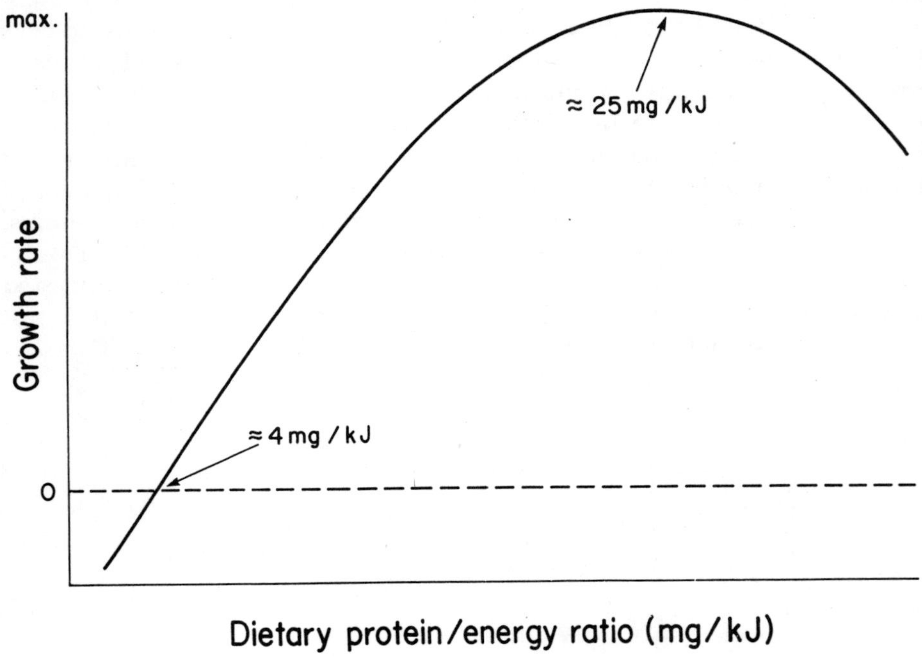

Figure 2. Graphic description of the relationship between dietary protein/energy ratios and growth rate in *Sarotherodon mossambicus*.

Assimilation of Digested Food

Digestion breaks macromolecules in the diet into progressively smaller subunits. Assimilation occurs when these subunits are small enough to pass through the gut wall. Because assimilation is the point in the nutritional process at which energy and materials gathered from the habitat are actually incorporated into and become an integrated part of the consumer, a measure of assimilation is the single most valuable quantitative descriptor of the diet.

Assimilation is usually quantified in terms of assimilation efficiency (AE): (amount assimilated ÷ amount ingested) x 100. AE's of tilapia feeding on various diets have been determined by two techniques. The mass-balance technique estimates the amount assimilated as amount ingested minus the amount defecated. A disadvantage of this approach is that precise estimates of ingestion and defecation are often difficult to obtain. In addition, this approach can only be used under controlled experimental conditions that necessarily differ from the natural feeding habitat. Since secretion of gastric acid is very sensitive to disturbance (Fish 1960; Moriarty 1973), efficiency of assimilation can be altered by experimental conditions. Nonetheless, the mass-balance technique has produced valuable results when used with fish acclimated to laboratory conditions and precise quantification that is possible using ^{14}C-labeled algae.

The second technique for determination of AE uses some refractile compound in the diet as an undigested reference. Some workers have added a reference compound such as Cr_2O_3 (Furukawa and Tsukahara 1966; Bowen

1978), while others have utilized refractile indigenous components such as cellulose for macrophyte grazers (Buddington 1979; Buddington 1980) and ash or HRA for detritivores (Bowen 1979, 1980b, 1981). AE is calculated by comparison of food:reference, and feces:reference ratios using the formula developed by Conover (1966). The advantage of the refractile reference technique is that it may be applied to analysis of tilapias taken directly from their feeding habitat. Since it appears there is no assimilation through the stomach, samples of stomach contents and posterior intestinal contents from freshly caught specimens may be compared for the purpose of estimating AE. This means determinations may be made for fish feeding under totally natural conditions and all handling effects are eliminated.

The few data available for assimilation efficiency of tilapia feeding on natural diets are summarized in Table 4. Two important generalizations can be drawn. Firstly, AE values are high for blue-green algae and diatoms, but are low for green algae and macrophytes. This is not surprising since the gastric acid that lyses blue-green cell walls and denatures diatom cell membranes would not be expected to have a direct effect on the cellulose cell walls of green algae and higher plants. Secondly, whole-diet AE estimates for these four tilapias are low in comparison to the average of 85% for a variety of carnivorous species (Winberg 1956). Again, this is not surprising when we consider that macrophytes and detritus contain abundant refractile compounds including cellulose and lignin that are not vulnerable to vertebrate digestive enzymes. A similarly low AE is reported for *Ctenopharyngodon idella*, a cyprinid macrophyte grazer (Hickling 1966). The low total AE for *S. niloticus* fed suspended matter > 100 μm filtered from Lake George appears to be the result of the long time interval between the start of feeding and the point at which digestion reaches peak efficiency in this species (Moriarty 1973).

Another interesting aspect of the data in Table 4 is that protein and lipid are assimilated much more efficiently than total organic matter or food energy. This same result was reported for tilapia fed pelleted algae that had been dried and ground to a powder such that all natural impediments to enzyme-substrate interaction had been removed (Kirilenko et al. 1975). Thus, the efficiency with which macromolecular groups are digested is related not to selective disruption of cell walls or membranes, but rather to basic characteristics of the digestive process itself.

Food Quality and Growth

The food quality of a given diet is directly proportional to its ability to support growth. The dietary component most important in limiting the growth of herbivorous and detritivorous fishes is protein. There are 10 to 13 amino acids that are essential for building new tissues (Cowey and Sargent 1972). These amino acids cannot be synthesized by fish and must be obtained from the diet. Carnivorous fish consume prey that may be > 80% protein by dry weight, but the diets of tilapias range from about 50% to < 1% protein. Values below 15% are most common (Boyd and Goodyear 1971; Caulton 1978a; Bowen 1979, 1980a, 1980b).

Table 4. Assimilation efficiency (AE%) estimates for tilapias feeding on natural diets.

Species	Experimental diet	Component of diet	AE%	Technique	Reference
Sarotherodon niloticus (phytoplankton grazer)	*Microcystis*	^{14}C	70	mass balance	Moriarty and Moriarty 1973a, b
	Anabaena	^{14}C	75		
	Nitzschia	^{14}C	79		
	Chlorella	^{14}C	49		
	Lake George suspended matter > 100 μm	total C	43		
Tilapia zillii (macrophyte grazer)	*Najas guadalupensis*	dry weight	29	reference (cellulose)	Buddington 1979
		organic matter	32		
		non-cellulose organic matter	56		
		protein	75		
		lipid	76		
		energy	45		
Sarotherodon mossambicus (detritivore)	Benthic detrital aggregate	organic matter	38	reference (ash)	Bowen 1979
		protein	46		
		carbohydrate	35		
		energy	42		
	Periphytic detrital aggregate	organic matter	63	reference (HRA)	S. Bowen 1981
		carbohydrate	63		
		protein	77		Bowen 1980b
		detrital amino acids	64		Bowen 1980b
Tilapia rendalli (macrophyte grazer)	*Ceratophyllum demersum* (aplical segments only)	dry weight	53-60	mass balance	Caulton 1978
		protein	80		
		energy	48-58		

To describe food quality, protein levels are frequently expressed as mg assimilable protein per kJ assimilable food energy. This reflects the fact that appetite controls protein intake indirectly through its response to food energy levels, and consequently the amount of protein assimilated depends on the amount associated with assimilable food energy (Harper 1967; Russell-Hunter 1970). A minimum of 4 mg protein per kJ is required for maintenance (Figure 2). Growth increases with increasing protein levels up to a maximum of about 25 mg/kJ. Protein at higher levels is in excess of the animal's ability to utilize it anabolically and thus growth decreases presumably as a result of the energetic cost of protein catabolism.

Protein-energy ratios have been used to assess food quality in two studies of the feeding of *S. mossambicus*. One study was conducted at Lake Sibaya, South Africa (Bowen 1979). The Lake Sibaya population was especially interesting because its juveniles grew well and were in good condition but the adults were stunted and showed extensive marasmas indicative of malnutrition (Bruton and Allanson 1974; Bruton and Boltt 1975). Although both juveniles and adults fed on benthic detrital aggregate, they collected their food from different parts of the lake. Juveniles fed near shore, frequently at depths less than 30 cm, while adults fed in water 3 to 5 m deep. Analyses of samples from the lake bottom showed that protein-energy ratios were high in shallow, near-shore waters, but dropped rapidly with increasing depth. Analyses of stomach contents confirmed that juvenile diets contained considerably more protein that adult diets. When the protein/energy criterion was applied, these results showed that juveniles have enough protein in their diets to produce good growth, but the adult diet lacks adequate protein for maintenance and the observed malnutrition would be expected (Figure 3). Reasons why

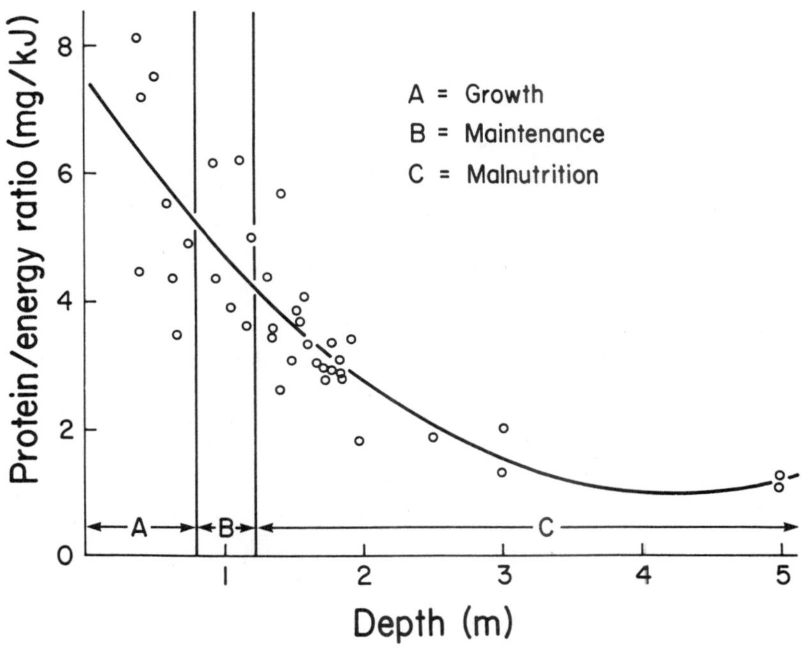

Figure 3. Relationship between protein/energy ratios and depth for benthic detrital aggregate consumed by *Sarotherodon mossambicus* in Lake Sibaya, South Africa, showing expected nutritional significance of feeding at different depths. Curve fitted by parabolic regression (after Bowen 1979).

adults do not feed in the shallows are discussed elsewhere (Bowen 1976b; Bruton 1979). With regard to the food quality problem, the important findings of the Lake Sibaya study are: 1) food quality varies significantly between and according to a predictable pattern and 2) juvenile *S. mossambicus* risk the perils of the near-shore shallows (predators, heavy-wave action) to utilize selectively food of the highest available quality.

Similar results were obtained in a study of *S. mossambicus* adults in Lake Valencia, Venezuela (Bowen 1980a, 1980b, 1981). These fish fed on periphytic detrital aggregate attached to *Potamogeton* spp. The protein content of this food resource varied from site to site and was directly proportional to the slope of the littoral zone bottom. Adult *S. mossambicus* fed selectively in the steeply sloped littoral zone (Figure 4). Some difficulty arose in interpretation of these results because assimilated protein:assimilated energy ratios indicated that protein levels were too low to support growth, but the fish were in excellent condition and reached a large maximum size. Further study revealed that in addition to protein, the diet contained detrital amino acids bound into the amorphous detritus. Detrital amino acids were most abundant in the steeply sloped littoral zone where the fish feed (Figure 5) and were assimilated with an efficiency of 64%. With this source of amino acids taken into consideration, adult *S. mossambicus* in Lake Valencia assimilated the equivalent of 14.4 mg of protein per kJ of assimilated food energy: an amount expected to produce good growth. As with the Lake Sibaya study, these results show *S. mossambicus* in Lake Valencia selected high quality food that maximized growth.

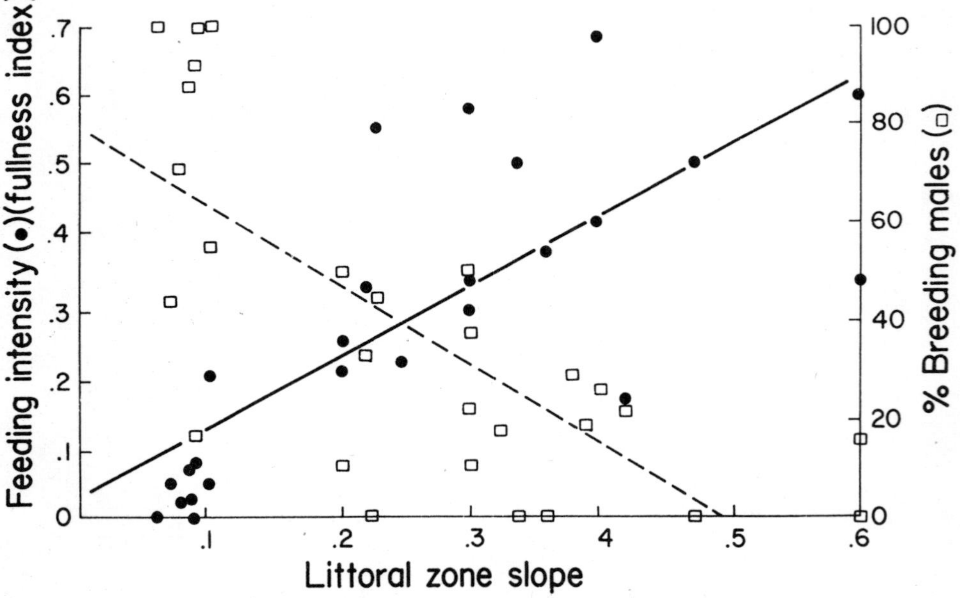

Figure 4. Habitat partitioning by male and female *Sarotherodon mossambicus* in Lake Valencia, Venezuela. Males in breeding coloration build nests and defend territories in the gently sloped littoral while females and non-breeding males feed in steeply sloped littoral where food quality is highest. Lines fitted by least-squares. (Reproduced by permission of the Ecological Society of America.)

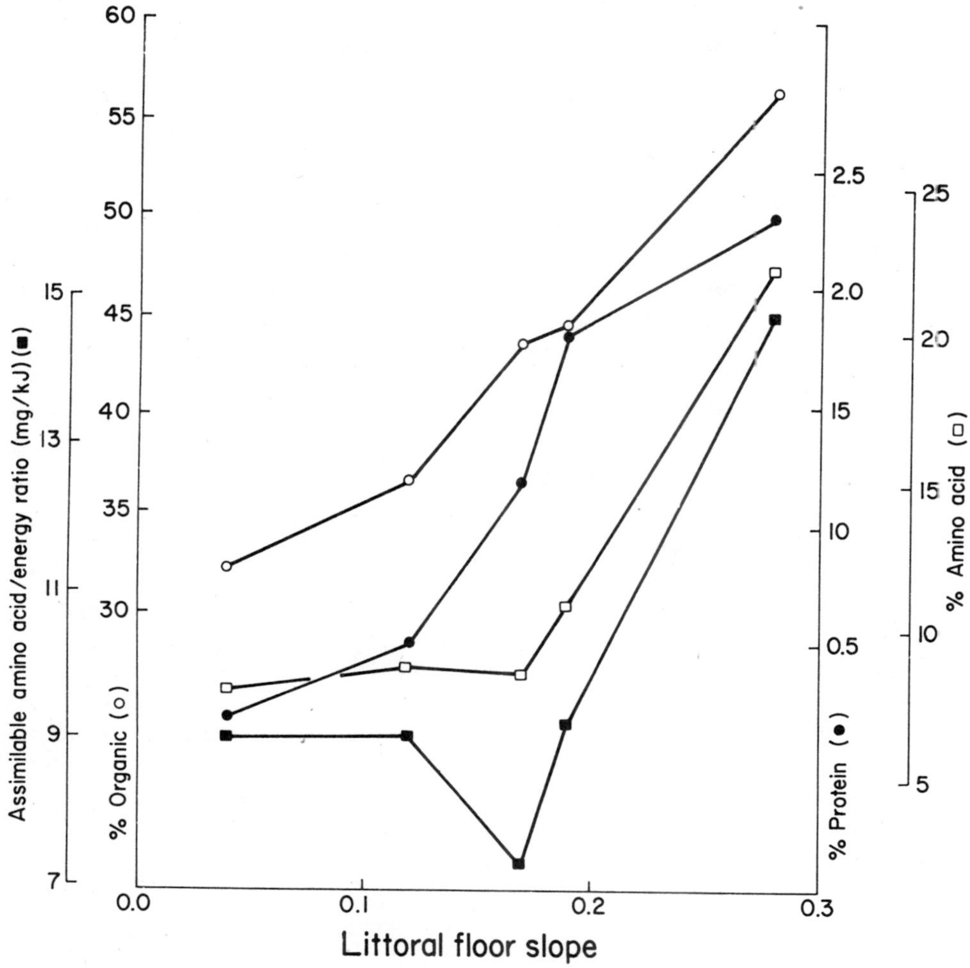

Figure 5. Relationship between organic content, protein content, total amino acid content and assimilable amino acid/energy ratios in the diet and slope of the littoral floor in Lake Valencia, Venezuela. Adult *S. mossambicus* feed selectively in the steeply sloped areas. (Reproduced by permission of the American Association for the Advancement of Science.)

In view of the abundance of detrital energy and the limited availability of protein-rich detrital food, both the Lake Sibaya and the Lake Valencia populations provide excellent examples of the way in which "ecosystems function to expend readily available energy to minimize the constraints imposed by limiting nutrients" (Reichle et al. 1975).

Summary and Conclusions

The characteristic diet of adult tilapias is a mixture of algae, detritus, bacteria, and, in some cases, macrophytes. Tilapias possess morphological and physiological adaptations for utilization of this diet. Pharyngeal teeth break food components into smaller units for easier peristaltic mixing and increased exposure to digestive enzymes. Gastric acid secreted to an unusually low pH, frequently below pH 1.5, lyses prokaryotic cell walls and denatures prokaryotic and eukaryotic cell membranes to expose the cytoplasm therein to intestinal enzymes. The extreme length of the intestine is essential for efficient digestion and assimilation of some components of the diet.

Food quality, quantified in terms of mg assimilable protein / kJ assimilable food energy, is potentially a limiting factor in the growth of tilapias. Studies of the detritivorous *S. mossambicus* show that the quality of available food resources varies widely, and that the fish utilize selectively the resource that produces maximum growth. Future studies of tilapia that feed on macrophytes and suspended matter are also likely to reveal selective feeding for growth maximization.

These trophic abilities of tilapia make them ideally suited for aquaculture. Since they feed at the base of the food chain, they are energetically very efficient and ecosystem carrying capacity should be high for these organisms. Their uncommon ability to exploit prokaryotic algae and bacteria as sources of protein makes them unusually protein efficient since they do not rely on invertebrate intermediates. In addition, the ability to utilize prokaryotic protein may open new doors to cost-efficient aquaculture. A principal economic obstacle in culture of fish is the cost of protein in culture diets (Weatherley and Cogger 1977). It is likely that bacteria and blue-green algae can be grown either directly in culture ponds or in adjacent facilities using locally available agricultural waste products and low-technology manipulations (Schroeder 1978). The protein/energy criterion could be used to optimize the quality of the food produced. Although systematic evaluation of this approach has not yet been attempted, what we have learned about the feeding of tilapias in natural systems suggests that it is a very promising possibility for the future.

Acknowledgments

I am grateful to Dr. D.J.W. Moriarty for helpful discussion of the manuscript. Plates 1 and 2 were produced by the Electron Microscopy Unit, Rhodes University, Grahamstown, South Africa.

Discussion

LOVSHIN: Dr. Bowen, I have one question, just a yes or no will do. Did you study the density of fish per unit area or the predator populations within the two lakes? Was there some sort of control on the effects of tilapia density between the two lakes, because I think your results could be affected tremendously by differences in density.

BOWEN: In Lake Sibaya the densities of all fish were very low and there were a lot of predators for tilapia. In Lake Valencia the tilapias were very dense, but still in very good condition. (See Caulton, this volume, for further discussion—Editors.)

Feeding, Metabolism and Growth of Tilapias: Some Quantitative Considerations

M. S. CAULTON
King Prawns (Pvt.) Ltd.
P. O. Box 135
Kariba
Republic of Zimbabwe

CAULTON, M.S. 1982. Feeding, metabolism and growth of tilapias: some quantitative considerations, p. 157-180. *In* R.S.V. Pullin and R.H. Lowe-McConnell (eds.) The biology and culture of tilapias. ICLARM Conference Proceedings 7, 432 p. International Center for Living Aquatic Resources Management, Manila, Philippines.

Tilapias are strongly thermophilic with an eccritic thermal range varying from 30°C to 36°C depending on species. Natural populations of tilapias often show a diel movement which mirrors the diel thermal fluctuation of shallow inshore waters of lakes, pans or inundated river flood plains. The movement of young tilapias is shoreward, toward warm water, during the day and away from the shore during the night as the shallows cool rapidly. Metabolic energy demand, being a function of both temperature and mass, is strongly influenced by the thermophilic behavior of tilapias living in natural waters. Tilapias that feed on aquatic macrophytes demonstrate that food intake is related to both body mass and environmental temperature. Assimilation efficiency increases with increasing temperature. The cost of food processing is a function of both food quality and quantity but is also influenced by environmental temperature. The increased net gain in assimilated energy at high temperatures may be almost offset by the increased cost of food processing at comparable temperatures.

When considering these factors and formulating a simple energy budget it is clear that high daytime temperatures coupled with lower nighttime temperatures lead to an optimal use of energy resources and results in increased growth potential. Aquaculturists should take note of such physiological optima and with careful design of ponds and water management could create a diel temperature variation resulting in enhanced growth potential and decreased food conversion ratios.

Introduction

Rapid growth rates, good productivity per unit volume of water and economic, efficient food conversion make the tilapias a suitable fish for the needs of the modern aquaculturist. This group of fish made an early appearance in the field of aquaculture, but the rather premature enthusiasm waned somewhat when many of the earlier trials failed to come up to expectations. Today, however, significant advances have been made in the basic understanding of the biology of these fish, and controlled production is becoming increasingly successful. Notwithstanding this recent upsurge in the importance of these fish to aquaculture, there still remains a paucity of information which would contribute to our understanding of their physiological and related environmental requirements. The principal environmental factor influencing physiology, i.e., temperature, will be the theme of this paper which is structured to include some aspects of feeding, metabolism and growth and to demonstrate the inter-relationships of these functions through the influence of temperature in the development of a simple energy budget.

Historically, the spread of tilapias from the hot equatorial inland lakes and rivers of Africa immediately suggests that these fish are well adapted to the prevailing climatic conditions of these areas. Just how well these fish have adapted to their environment may be of importance in our attempts to domesticate these fish. The survival of tilapias is generally limited to waters with temperatures warmer than 10°C to 12°C, although under exceptional circumstances (e.g., in areas of increased saline content such as estuaries - Allanson and Noble 1964), thermal tolerance of lower temperature may be marginally increased. It is not perhaps the tolerance to temperature that is important to most wild populations, but rather the effect that temperature, by its influence on the physiology, may have on growth.

Temperature and Metabolism

It is generally accepted that tilapias cease growing significantly at temperatures below about 20°C, but at the same time constant warm temperatures may not, by themselves, be adequate for optimal growth. Figure 1 illustrates the distribution of two species when subjected experimentally to a thermal gradient in a horizontal test tank. In each instance, the fish respond positively to the warm water, orientate rapidly and swim actively toward water which is only marginally cooler than their upper lethal temperature tolerance.

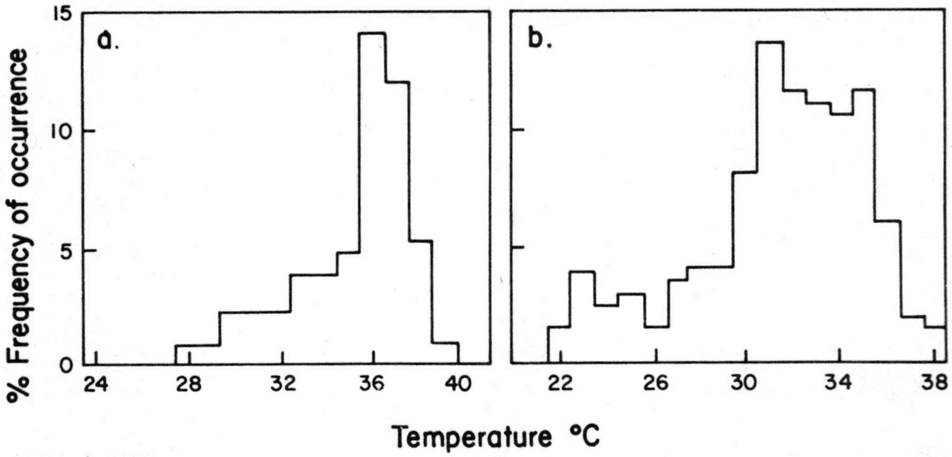

Figure 1. Temperature 'selection' by juvenile a) *Tilapia rendalli* and b) *Sarotherodon niloticus* in a thermal gradient test tank (after Caulton 1979).

Such an intense behavioral response to temperature clearly indicates that these fish are strongly thermophilic and in some instances, this response is so strong that individuals briefly enter water warmer than that in which they could normally survive. The eccritic temperatures demonstrated by the species tested (*Sarotherodon mossambicus, S. niloticus, S. macrochir* and *Tilapia rendalli*) varied between 30°C and 36°C. Temperatures of this magnitude may initially appear extraordinarily high, but they are not uncommon in the shallow marginal waters of tropical and equatorial lakes, lagoons and river pools of Africa. These high, temperatures, however, are generally not stable throughout the day but, due to the shallowness of the water, are subject to diel oscillations often in excess of 15°C per day. It has been noted by many authors (e.g. Welcomme 1964; Donnelly 1969; Fryer and Iles 1972; Bruton and Boltt 1975; Caulton 1975) that a variety of tilapias react to such temperature oscillations by moving inshore during the day and offshore at night. A graphical example of such a movement by juvenile *T. rendalli* is illustrated in Figure 2.

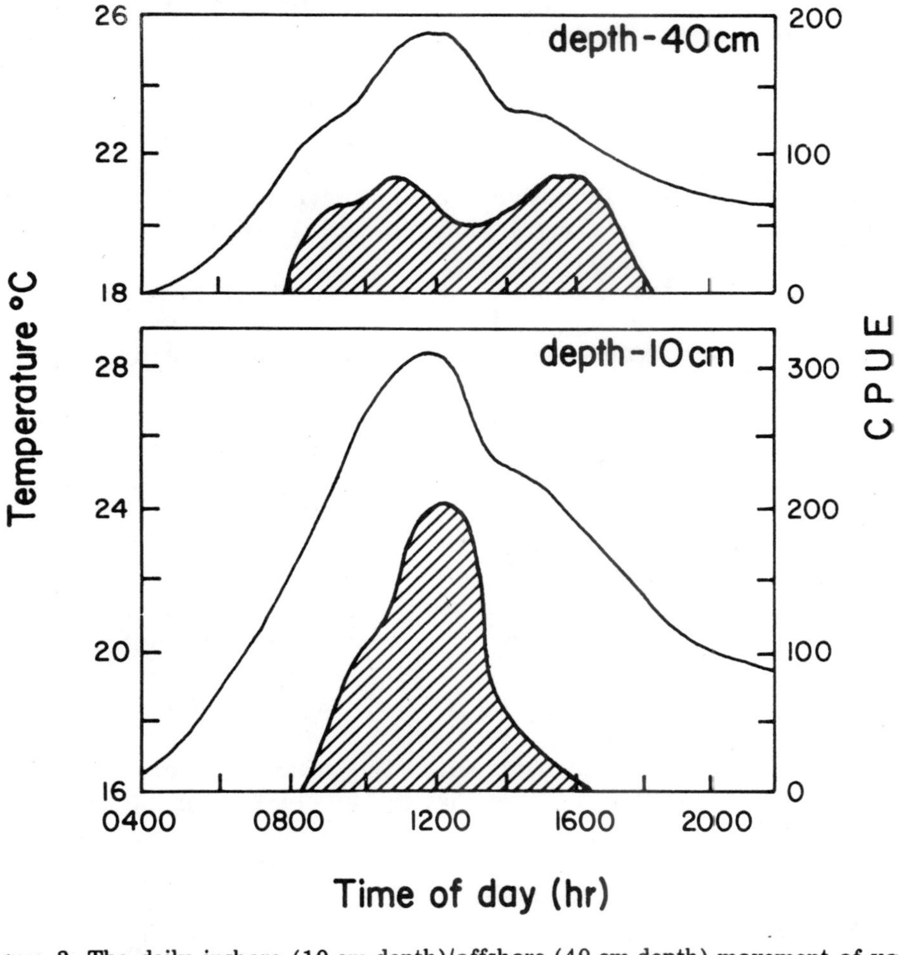

Figure 2. The daily inshore (10 cm depth)/offshore (40 cm depth) movement of young *Tilapia rendalli* in Lake McIlwaine, Zimbabwe. (1.10.74). Catch per unit effort (CPUE) is given as the number of fish caught over a 15 min time period using a 2 m cast net. The numbers of fish caught are shown by the cross hatch areas. Upper lines show the change in temperature (after Caulton 1975a).

With the knowledge that tilapias are strongly thermophilic, it may not be surprising that such daily inshore/offshore movements do occur. In the example illustrated, the movement of young *T. rendalli* (up to 10 cm SL) can be followed from the deep (> 1.5 m) homothermal water, where they stay overnight, into water 40 cm deep, beginning at 8:00 A.M. or when the temperature at this depth exceeds that of the homothermal waters. The inshore movement of juvenile fish continues throughout the morning and eventually large schools of juveniles are found in shallow warm water at midday. A reverse movement is noted in the late afternoon when the fish, again responding to temperature, leave the now cooling margins to return to the relatively warmer, homothermal deeper waters.

Such diel movements are commonly encountered in shallow gradient shoreline areas, while fish habitually favoring deeper open waters may occasionally show a less spectacular but nevertheless visible vertical movement for a similar reason. These patterns of diel movement can however be easily disrupted; for example, during periods of even moderate winds the wave wash along the shore is sufficient to restrict movement into very shallow water. Similarly, during exceptionally warm periods, or in areas where excessive shoreline vegetation assists in insulating the warm daytime water in the shallow margins, a nocturnal presence of juveniles is not uncommon. When in areas of food paucity, fish often have to remain in the shallows at night in order to feed. The presence of predators in deep water is also a strong factor which will seriously disrupt any set daily pattern of movement although avian predation along the margins during the day does not seem to influence the movements very strongly.

Notwithstanding the possibility of such numerous disruptions, many young tilapias are often subjected to a daily thermal variance of between 10°C and 15°C and, since fish are poikilothermic, such thermal variations will understandably have an important bearing on almost all of the animals' physiological functions. Of these functions, metabolism would be expected to be most markedly affected. In broad theoretical terms, metabolic energy demands are expected to approximately double with every 10°C rise in body temperature. Thus a tilapia moving into the warm shallows during the day must expend considerably more energy than it would during the night in the relatively cooler deeper water. It may be expected that as a consequence of this fluctuating energy demand, fish living under these conditions of thermal fluctuation would show unnecessarily high expenditure of energy during the day, which could otherwise have contributed to either storage or growth.

To investigate this, it is necessary first to look in some detail at the effect temperature has on metabolism. Metabolism, as a function of aerobic respiration, can be quantitatively equated to the uptake of molecular oxygen which in turn is a simple measure that can be obtained in the laboratory. Many extrinsic variables can modify the basic metabolic rate or oxygen consumption of a fish in nature, but, suppressing most of the variables, with the exception of temperature, relationships demonstrating the influence that temperature has on metabolism can be investigated. Obviously, a laboratory test animal out of its natural environment is subjected to a number of unnatural stresses which are capable of either increasing or suppressing normal oxygen consumption but, in general, laboratory acclimated juvenile tilapias are reasonably suited to respirometry and show few visible signs of the stress so often encountered in the apparently more nervous cyprinids and other fish.

Using a simple continuous flow, continuous recording respirometer as described by Caulton (1975a), a satisfactory measure of routine metabolism in both *T. rendalli* and *S. mossambicus* has been obtained. Routine metabolism by definition can be described as the energy required by an unfed fish exhibiting spontaneous rather than directed movement. Such a measure of metabolism is particularly suited to tilapias, since movement even in their natural habitat is very seldom vigorously directed but can more often be described as spontaneous fin generated movements. Obviously, exceptions do occur, e.g., during intense feeding activity, migrations or predator avoidance. This suggestion is borne out to a certain extent by the fact that schools of juvenile *S. mossambicus* in a large communal respirometer all displaying normal swimming, schooling behavior and feeding movements (into sterile sand) showed an increased metabolic demand 25% greater than that required by the same fish at night, when they were almost stationary, or that obtained from the summation of individual oxygen requirements in a normal single chamber respirometer. This contrasts markedly with generally faster moving, continuously swimming fish species, especially predators, which may require at least twice as much energy during normal active periods than during inactive periods.

The conversion for equating oxygen consumption to the amount of metabolic energy liberated by herbivorous tilapias was calculated as 13.68 J/mg O_2 (Caulton 1977a), and thus with the availability of such a conversion factor, it is possible to obtain more quantitatively meaningful information from respirometry. The routine metabolic energy demand by *S. mossambicus* can be satisfactorily described by the equation:

$$R_r = 0.0086\, t^{2.0783}\; M^{0.652}$$

where R_r is the energy of metabolism expressed in J/hr, 't' the temperature in °C and M the fresh mass of the fish in grams (restricted to fish less than 150 g, the maximum size used in the trials). The full details of the derivation of this formula are given in Caulton (1978c).

A similar relationship between temperature and metabolism, but for a single size range of fish, is shown graphically in Figure 3, where the metabolism

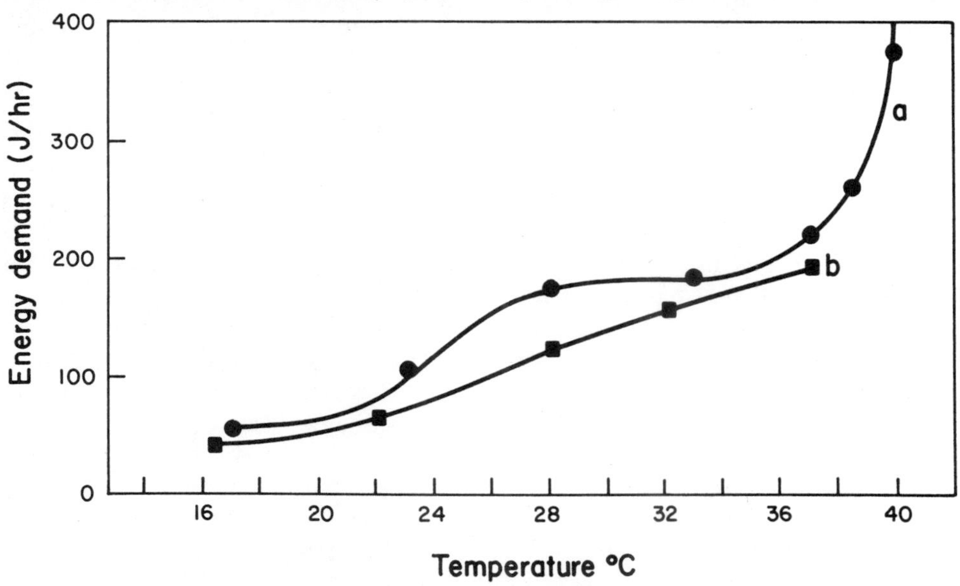

Figure 3. The influence of temperature on metabolic energy demand in a) *Tilapia rendalli* and b) *Sarotherodon mossambicus* (modified after Caulton 1977a, 1978c).

of *T. rendalli* and *S. mossambicus* is compared. Unlike *S. mossambicus*, *T. rendalli* shows an unexpected three-phase energy demand in which the distinctive plateau phase, extending from approximately 28°C to 37°C, is an unusual characteristic. Over this temperature range the metabolic demand of *S. mossambicus* increases by 62% whereas the metabolic energy demand by *T. rendalli* over the same temperature range increases by only 22%. Similarly structured, unusual metabolic curves have been reported for other fish species (Schmein-Engberding 1953; Sullivan 1954; Job 1969a, 1969b and Fry 1971 for *S. mossambicus*). Denzer (1968) reports a similar function in *S. niloticus*. Exactly how important, or in fact how real, such thermal homeostasis is, is difficult to determine and respirometry technique (static versus flowing systems) must be considered when appraising results. Obviously any function that can appreciably depress an expected rise in energy demand must also benefit the overall energy balance in consequence.

The metabolic energy required to sustain routine maintenance by a 50 g *T. rendalli* at, for example, 28°C is approximately 175 J/hr (Figure 3) but to appreciate this requirement in terms of the utilization of storage tissue, it may be pertinent to consider which catabolic fuels are responsible for the supply of energy. Glycogen or the major carbohydrate fraction is an energy source usually stored in the liver but, in *T. rendalli* at least, is not of great significance as a sustained catabolic energy source. A 50 g *T. rendalli* in good condition has a maximum glycogen content of 0.46% by mass (about 230 mg) and of this amount 144 mg are stored in the liver while the remainder is present in the muscle. On starvation approximately 100 mg of glycogen is utilized within the first few days, thereafter the level remains relatively constant even to the state of near death from starvation. This 100 mg could supply sufficient energy to maintain a 50 g fish at 28°C for less than 12 hours and thus cannot be classified as of great importance as a storage fuel to a starving fish, which is thus reliant mainly on lipids and protein as a source of catabolic energy. The importance of condition is immediately evident and a close relationship does exist between condition and the type of catabolic fuel mobilized during routine maintenance. The mobilization of lipids is always associated with some protein mobilization (and vice versa) and follows a pattern closely linked with condition.

Condition in the present context is described as a morphometric interpretation of a fish's plumpness as compared to the population mean. As such, condition is an extremely useful measure of assessing the physiological state of a fish and more cognizance of this measure should be taken by the fish farmer. The classical measurement of relative condition as described by Le Cren (1951) using the formula $Kn = 100\ W/L^b$ where Kn is the relative condition, W the fresh mass in grams, L the standard length in cm and 'b' the length/mass regression exponent, provides an adequate quantitative assessment of condition if the basic measurements are made carefully. In *T. rendalli* close relationships do exist between condition and the proportionate mobilization of fat/protein during metabolism but extreme care must be exercised when interpreting these relationships (see Caulton and Bursell, 1977, for a discussion of these problems, especially the use of percentages when measuring or interpreting results).

The mass of all major body constituents, with the exception of the inorganic compounds, must, by definition, decrease with a decline in condition, but it is the relative pattern of decline of each constituent that is important. The lipid content (predominantly the triglyceride groups) declines rapidly with decreasing condition (Figure 4) indicating a preferential mobilization of triglyceride lipids during routine catabolism in fish of good condition. When compared with protein mobilization, the initial preferential utilization of storage fats as a catabolic fuel becomes even more evident (Figure 5). Tilapia in good condition derive nearly 3 times as much metabolic energy from fat reserves than from protein while fish in poor condition come to rely almost exclusively on protein as a catabolic fuel when triglyceride lipid has declined to a level which may be critical or almost non-existent (Figure 4). Since lipid is a high energy compound containing 1.7 times more energy per unit mass than protein, it follows that as the fish become more and more reliant on a protein source for catabolism, so the rate of decline in mass or condition accelerates. This may be one of the reasons why a decline in condition so often appears to lag behind an expected decrease due either to a predicted environmental stress or to some expected physiological stress at the end of a period of good growth.

The preferential mobilization of either lipid or protein as the major metabolic energy source has little effect on oxygen consumption since 13.72 J of energy is liberated from lipid for every mg of oxygen utilized while only marginally less energy, 13.39 J, is liberated per mg of oxygen if protein is the prime energy source. Thus, although fish that show rapidly declining condition will be largely utilizing somatic protein as a catabolic fuel, oxygen consumption basically remains a function of the mass of tissue and only in extremely accurate work is the proportionate composition of the fish important.

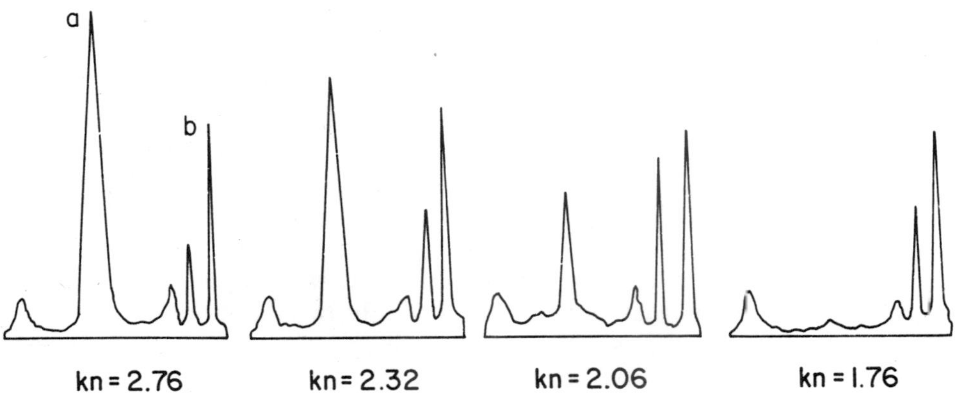

Figure 4. Semi-quantitative analysis of the various lipid fractions from *Tilapia rendalli* of varying condition (a = triglyceride, b = phospholipid, kn = relative condition factor).

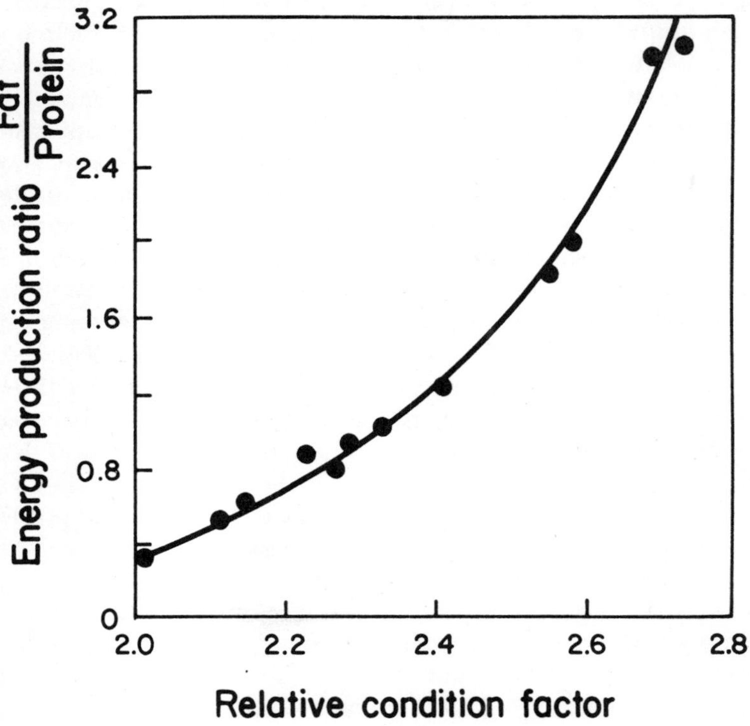

Figure 5. The energy production ratio of fats and proteins during change in relative condition factor for *Tilapia rendalli* (after Caulton 1978b).

The problem of accurately predicting the tissue mass of a live fish prior to respirometry is difficult because the relationship between tissue mass, water content and condition is continually changing. The proportionate relationship between tissue (lipid and/or protein) and water is seldom a simple inverse linear function, as so often suggested, but changes with changing condition. The ratio of water and tissue in *T. rendalli*, for example, is lowest in fish of good condition (Figure 6), highest in medium condition fish and marginally declines in fish of poor condition. This complicating factor will certainly have a small but important effect on oxygen consumption when it is related to the fresh mass of the fish and thus for precise measurement of metabolism, oxygen uptake or the metabolic demand should be related rather to the dry tissue mass or the energy content of the fish and not simply the fresh mass unless the fish are all in a similar condition.

The information already discussed can be correlated and cross-referenced to metabolism: an exercise that gives confidence to some of the results discussed. For example, a morphometrically similar group of young *T. rendalli* with an average mass of 42.63 g and having an average condition of 2.23 were maintained without food for a period of ten days at a constant temperature of 18°C. The average mass loss over this period was measured as 2,020 mg per individual. From our knowledge of oxygen consumption and energy utilization by *T. rendalli* (Figures 3 and 5), it is calculated that over the trial period 11.7 kJ of energy was released to sustain routine metabolism. From the information given in Figure 5, a fish of condition 2.23 would be expected to derive 43.8% of its energy from fat catabolism and 56.2% of its energy from a protein source. Thus of the 11.7 kJ of energy required for maintenance, 5.1 kJ would be supplied from a lipid source and 6.6 kJ from a protein source. This, in terms of total tissue mass, represents 130 mg. of lipid and 280 mg of protein or a total of 410 mg of tissue. It can be estimated that approximately 1,760 mg. of water associated with the tissue would be simultaneously lost (Figure 6) thus giving a total fresh mass loss of 2.17 g: a reasonable comparison to the measured loss of 2.02 g. Similar comparative estimates for various other temperatures are given in Table 1. The reasonably close values demonstrate that tissue composition as derived from careful condition factor analysis is a useful and fairly accurate predictive method to determine composition for detailed respirometry or to establish the pattern of utilization of catabolic fuels during metabolism without killing the experimental fish.

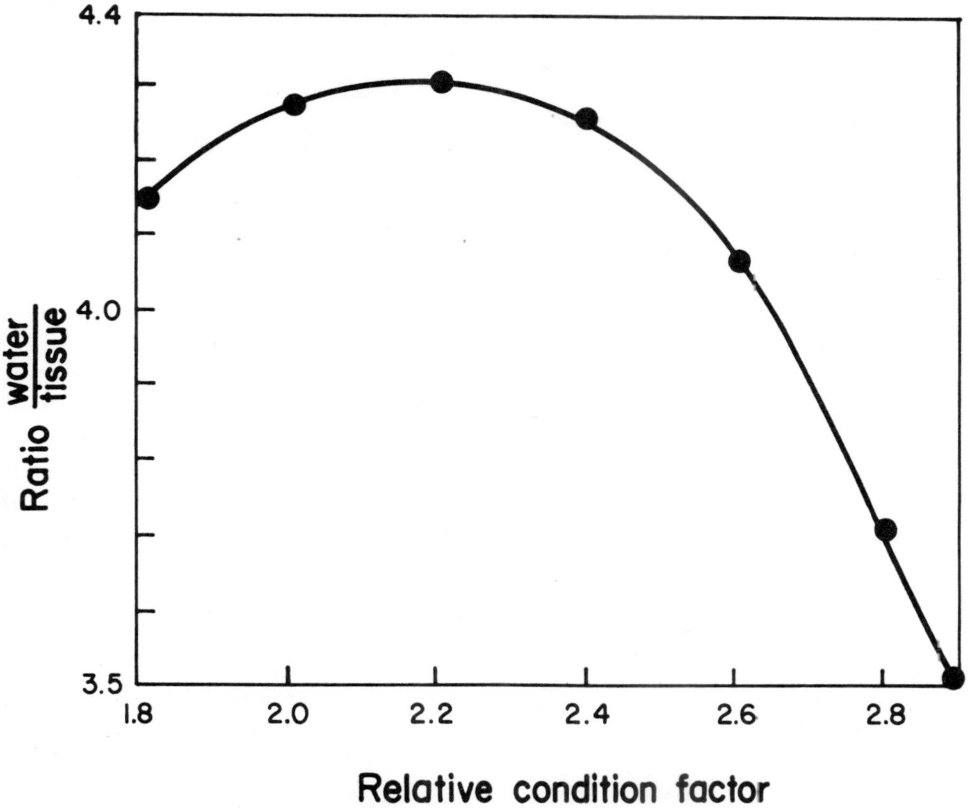

Figure 6. The relationship between water content and tissue content for *Tilapia rendalli* of varying relative condition factor (after Caulton 1978b).

Table 1. Comparison of measured and estimated change in mass during starvation of *Tilapia rendalli* at six experimental temperatures (after Caulton 1977c).

Temperature °C	18	20	21	24	30	34
Time of starvation (d)	10	6.25	6.75	6	12	8
Mean condition factor (kn)	2.23	2.42	2.51	2.68	2.58	2.60
Number of fish	19	19	19	8	14	18
Total measured mass lost (mg)	2020	1260	1940	3282	3641	4196
Estimated mass of tissue used (mg)	409	268	373	607	779	802
Estimated mass of water lost (mg)	1759	1140	1563	2398	3101	3272
Total estimated mass lost (mg)	2168	1408	1936	3005	3880	4074

Feeding and Growth

Sub-adult and adult *T. rendalli* are essentially macrophagous plant feeders in areas of abundant aquatic vegetation, a feature which lends this species most favorably to quantitative feeding experiments. The ubiquitous plant *Ceratophyllum demersum* heads a list of preferred food types eaten by *T. rendalli* when available, and it was for this reason that this plant was chosen for the work to be described. Like almost all species of tilapia, *T. rendalli* generally restricts its feeding to daylight hours (Figure 7) and thus to maintain some level of simulated natural conditions laboratory feeding trials were conducted over the same time period.

C. demersum growing shoots (terminal 1-7 g) were used throughout the trials and the following necessary facts about the plant were established. To convert fresh mass (limited centrifugation to remove surface moisture) of *C. demersum* to dry mass equivalents the following relationship was used:

$D = 0.0839\ W - 0.7845$ ($r = 0.986$, $n = 50$, SE 'b' = 0.003) and for converting dry mass to fresh mass, $W = 11.5941D + 26.7257$ ($r = 0.986$, $n = 50$, SE 'b' = 0.478) where W is the fresh mass and D the dry mass both expressed in milligrams. The average energy value of the dry shoots was determined as 17.9477 kJ g^{-1} ($n = 20$, SE = 0.0236).

Laboratory maintained fish fed to satiation during a prescribed daylight period of 12 hours showed a linear relationship between food consumed and fish mass (Table 2), a feature also reported for wild fish (Moriarty and Moriarty 1973a; Caulton 1977a) but in reality, over the range juvenile to large adult, it may be expected that as growth rates decline so less food per unit mass of fish would be consumed. Using the data presented in Table 2 for consumption rates at various temperatures, the relationship between food consumed and temperature, for a given size fish, is found to be curvilinear showing increasing food intake with increasing temperature (Figure 8). This relationship holds over the range 18°C to 30°C, but between 30°C and 35°C consumption is little affected by temperature, while at temperatures in excess of 35°C, food intake declines and ceases at about 37 or 38°C.

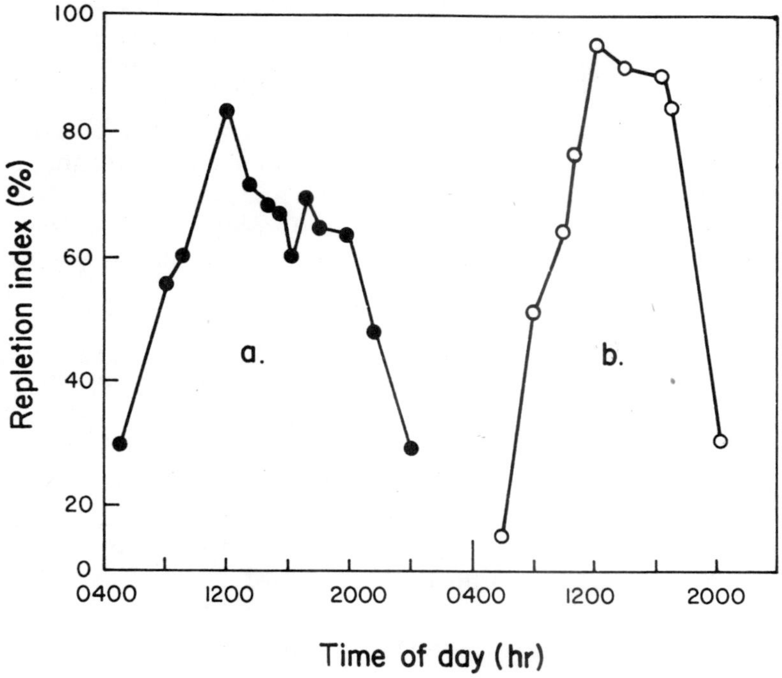

Figure 7. The daily feeding cycle of *Tilapia rendalli* in Lake McIlwaine, Zimbabwe (a, 1.9.75; b, 30.10.75).

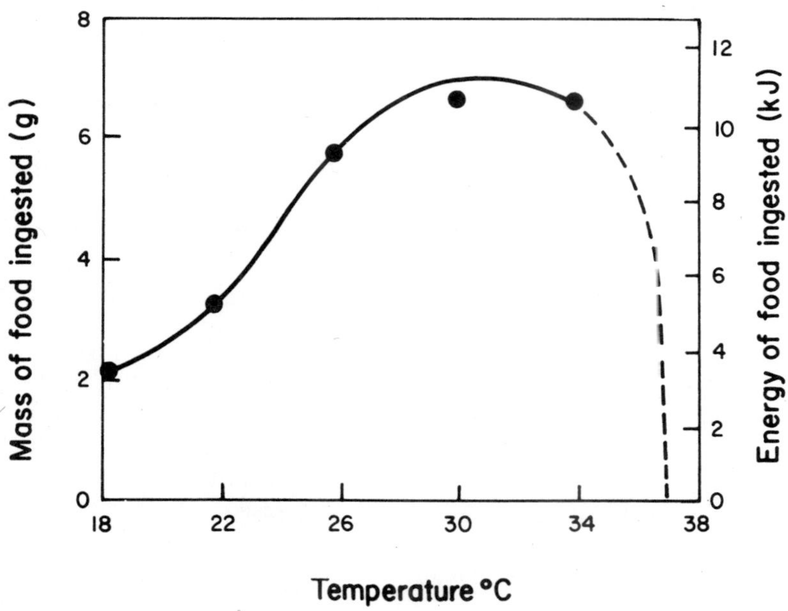

Figure 8. The effect of temperature on the daily food ingestion of laboratory-maintained juvenile (50 g) *Tilapia rendalli*. Dotted line extrapolated to show decline and cessation of feeding above 35°C (see text).

It is well known that many species of fish, especially the tilapias, will regulate their food intake in confined captivity. Thus, it is extremely difficult to relate quantitatively the patterns of food intake shown in Figure 8 and Table 2 to wild populations. Estimates of food ingestion in some wild populations have, however, been calculated using a most useful technique initiated by Moriarty and Moriarty (1973a). This particular technique is suited for use in some algal feeding tilapias, but it is a technique possibly even more suited to macrophagous feeders, such as *T. rendalli* when feeding on *C. demersum*: a food with a characteristic green color. Moriarty and Moriarty (1973a) estimated, for example, that a 200 g *S. niloticus* in Lake George would consume just less than 3 g (dry mass equivalent) of algae per day while a *T. rendalli* of equal proportion in Lake Kariba would consume about 3.3 g (dry mass equivalent) of *Panicum repens* per day (Caulton 1977b). Similarly, a 100 g *T. rendalli* in Lake Kariba would consume about 2.2 g (dry mass) of food per day, yet a laboratory maintained fish of 100 g at 28°C (roughly equal temperatures) consumes 12 g of fresh *C. demersum* per day (Table 2) or an equivalent of 1 g dry mass of food. Thus, it would appear that laboratory satiation is equivalent to about half the daily food intake of wild, free-living fish. A similar observation published by Moriarty and Moriarty (1973b) shows that this function is not restricted to macrophagous tilapias, but is also a feature of phytoplankton-feeding species.

Table 2. Linear regressions describing the amount of *Ceratophyllum demersum* growing shoots ingested by young *Tilapia rendalli* at various temperatures when fed *ad lib* in laboratory. (Feeding period = 12 hours; C is the fresh mass of food ingested and M the mass of the fish) (both measured in grams: after Caulton 1978b).

$C_{18°C} = 0.0667M - 1.061$	(n = 37, r = 0.975, S. E. 'b' = 0.002)
$C_{22°C} = 0.0709M - 0.261$	(n = 28, r = 0.946, S. E. 'b' = 0.001)
$C_{26°C} = 0.1097M + 0.278$	(n = 30, r = 0.974, S. E. 'b' = 0.004)
$C_{30°C} = 0.1169M + 0.683$	(n = 23, r = 0.970, S. E. 'b' = 0.003)
$C_{34°C} = 0.1205M + 0.868$	(n = 19, 5 = 0.986, S. E. 'b' = 0.005)

Quantitative and accurate measurements of the assimilatory potential of *T. rendalli* fed on *C. demersum* in the laboratory are possible (Caulton 1978a). Table 3 summarizes the results obtained for such experiments over a temperature range 18°C to 34°C. A definite relationship between assimilation efficiency and temperature indicates an increased efficiency of assimilation with increasing temperature. A relative increase in efficiency of 18.6% between 18°C and 34°C is a feature that may be expected to have some favorable effect on the energy balance resulting in better growth. Not only is more nutrient mass being extracted from the food at higher temperatures but also more energy per unit mass of food is being assimilated. This feature is reflected in the relative decline in the energy content of the feces with increasing temperature. It is also noted that the inorganic mineral content of the feces recovered from a given meal of *C. demersum* varies from 4% to 15% less than that consumed, but this removal of minerals snows no relationship to, or definite pattern with, either assimilatory efficiencies, meal size or temperature. In the results presented in this paper cognizance is taken of this fact and the results given are corrected accordingly. A further feature of note is that the relative energy content of the feces is greater (gram for gram, mineral corrected) than the food. This is a common feature, but such differences are often misinterpreted when investigating assimilation.

Table 3. The effect of temperature on assimilation by juvenile *Tilapia rendalli* when fed *ad lib* on *C. demersum* (17.96 kJ/g). (Number of determinations (n) = 28 per temperature).

Temperature °C	Mass (mg) of feces recovered per 1,000 mg dry *C. demersum* consumed	Energy (J) of feces recovered per kJ *C. demersum* consumed	Mean energy (J) assimilated per kJ *C. demersum* consumed	Mean energy (kJ/g) content of feces	(2 x SE)
34	399.2	413.0	587.9	18.6	(0.29)
30	394.6	417.8	582.2	18.9	(0.32)
26	413.5	448.5	551.6	19.42	(0.21)
22	449.8	470.0	530.0	19.34	(0.41)
18	470.5	522.1	477.6	19.92	(0.30)

An assimilation efficiency ranging from 47.8% to 58.7% may be regarded as being very good for a primary macrophagous herbivore. This efficient utilization of food can be attributed largely to successful primary food trituration and efficient pre-assimilatory processing of the food (Caulton 1976). The breakdown of resilient cell wall structures is the key factor and as such, assimilatory capabilities will vary considerably,depending on the species and composition of the plants eaten.

The mechanical trituration of the food, like digestion, absorption and transportation of primary nutrients, also requires an energy input by the fish. Such energy-demanding functions can be collectively termed 'apparent specific dynamic action' (S.D.A.) or the calorigenic cost of food processing. The additional collective energy required for these processes can be measured by feeding fish in a respirometer and measuring the post-feeding increase in oxygen consumption and equating any increase in oxygen uptake to the level of energy input through feeding. The energy costs of feeding *C. demersum* to *T. rendalli* in a respirometer are shown in Figure 9 and Table 4.

From the data presented it is apparent that meal size and processing costs are related in a linear fashion with greater processing costs (relative per unit food consumed) being required to synthesize larger meals. A linear trend of increasing processing costs with increasing meal size may be expected and has been described for a variety of fish (mainly predatory species—Edwards et al. 1972;Hamada and Ida 1973; Beamish 1974), but exponential relationships have also been described for some species (Tandler and Beamish 1979) so there appears to be no standard function to describe this relationship. A second noticeable feature of the results presented in Table 4 is that in *T. rendalli* food processing costs also increase with increasing temperature irrespective of the amount of ingested energy. As previously shown from results of experiments to determine assimilation efficiency, the amount of energy liberated per unit mass or energy of food ingested increases with increasing temperature thus it may not be surprising that the processing costs of food preparation increase proportionately. The increase in food processing costs with the increasing temperature can, at warm temperatures, seriously deplete the favorable gain in energy due to the greater assimilation efficiency at those temperatures. For example, a fish that consumed 1 kJ of food at 18°C would have available 478 J of post-assimilatory energy from which a further 20 J of energy must be deducted for processing costs leaving approximately 458 J of free energy available for metabolism and growth. Likewise, a fish at 30°C would have 582 J, less a calorigenic cost of 109 J, or 473 J of energy available. The difference in assimilatory efficiency between fish feeding at 18°C and those feeding at 30°C results in a gain of some 104 J, but when the cost of food processing is considered this difference is reduced to an almost insignificant 15 J. This clearly demonstrates just how energy demanding food processing is at higher temperatures. This example also serves to illustrate that digestibility alone, as a comparative measure used to forecast growth potential attributable to various diets, may not always be a suitable measure unless the associated food processing costs are also considered. This is especially important to the users of pelleted rations and fish farmers should not be misled by manufacturers' claims of good digestibility and consequently optimistic growth forecasts, as deduced from digestibility trials alone.

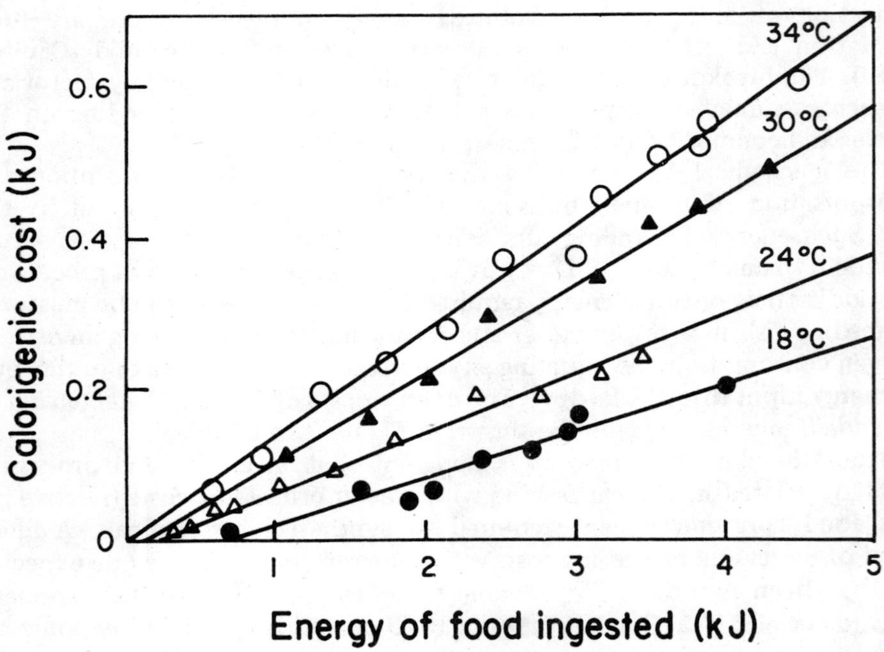

Figure 9. The effect of temperature on the calorigenic cost of food processing shown by *Tilapia rendalli* feeding on *Ceratophyllum demersum* (n = 39, after Caulton 1978a).

Excretory products, predominantly in the form of ammonia, are largely derived as a by-product of protein deamination and as such will reflect the effect of temperature on assimilation. Ammonia has an energy content of 20.4 kJ g^{-1} (Elliott and Davison 1975) and thus by determining the rate of either protein catabolism or protein assimilation, an estimate of the loss of excretory energy can be calculated. Table 5 gives an estimate of the rate of excretory energy loss when *T. rendalli* is feeding on *C. demersum*.

Table 4. Linear regressions describing the effect of temperature on the calorigenic cost of food processing. *Ceratophyllum demersum* by *Tilapia rendalli* (R_s = cost of food processing expressed in joules, C = the energy of *C. demersum* consumed expressed in joules and S. E. 'b' is the standard error of regression slope. (After Caulton 1978b).

Temperature	Regression
18°C	$R_s = 0.063C - 43.41$ (r = 0.94, S. E. 'b' = 0.010)
22°C	$R_s = 0.074C - 26.69$ (r = 0.96, S. E. 'b' = 0.008)
24°C	$R_s = 0.082C - 16.95$ (r = 0.98, S. E. 'b' = 0.006)
26°C	$R_s = 0.094C - 7.99$
30°C	$R_s = 0.119C - 10.10$ (r = 0.99, S. E. 'b' = 0.004)
34°C	$R_s = 0.142C - 10.75$ (r = 0.99, S. E. 'b' = 0.007)

Table 5. An estimate of the expected loss of ammoniacal energy per unit energy of food consumed for *Tilapia rendalli* feeding on *Ceratophyllum demersum*.

Temperature °C	34	30	26	22	18
Joules of NH_4 excreted per joule *C. demersum* consumed	0.028	0.028	0.027	0.026	0.023

The collective data already provided, in various forms of energy gains or losses, can now be integrated to provide some information pertaining to the growth and optimization of productivity by tilapias. For example, a simplified model of physiological growth can be derived from the use of so-called balanced energy equation (Davis and Warren 1971). This equation is derived from the balance between energy inputs and energy outputs and from such an equation one can determine, for example, growth potential as well as some of the ways in which temperature can affect growth. The input of energy in the form of the amount of food energy consumed (C) can be balanced against the various energy losses through egesta (F = feces plus U = excretory products), as heat by routine metabolism (Rr), through the cost of food processing (Rs), by movement (Rm) as well as that energy stored as fat, protein or glycogen, which together represents the energy of growth or productivity (ΔB). Thus, the basic equation can be given as:

$$C = F + U + Rr + Rs + Rm + \Delta B.$$

or on re-arrangement, growth potential can be derived from:

$$\Delta B = C - (F + U + Rr + Rs + Rm).$$

Substitution of the results given in the appropriate figures and tables into the equation results in the derivation of a simple formula that can describe growth in *T. rendalli.* For example, at 18°C a fish of M grams consuming C joules of *C. demersum* per day would have a daily growth potential of:

$$\Delta B_{18^\circ C} = C - (0.522C + 0.023C + 27.53M + 0.063C - 43.41)$$

or on re-arranging the equation:

$$\Delta B_{18^\circ C} = 0.392C + 43.41 - 27.53M$$

Similar formulae can be derived for the other experimental temperatures and the resultant equations obtained are listed in Table 6. (Energy of movement is not included in these formulae since its omission does not affect the following discussion).

Utilizing the equations an estimate of the food intake required for maintenance, i.e., when ΔB = 0, can be obtained. At 18°C, for example, maintenance energy for a 50 g *T. rendalli* would approximate 3.4 kJ/d (Table 6) which, in terms of *C. demersum* intake, would be equivalent to 189 mg dry mass or 2.2 g of fresh plant material. When compared to the amount of food ingested *ad lib* over a 12-hour laboratory feeding period at 18°C (Table 2, Figure 8), the level of so-called satiation is almost exactly the same as the estimated energy required for maintenance. This similarity in the level of food intake and maintenance energy requirement not only applies to low temperatures but to all the temperatures investigated (Table 7). This phenomenon illustrates the problems associated with attempting to determine growth rates under restricted laboratory conditions, but although field related growth patterns cannot be simulated under restricted conditions, some useful principles concerning growth may still be derived from laboratory work.

T. rendalli fry, like many other tilapia fry, will grow in 80-liter aquaria to about 30 or 40 g, when growth generally decelerates and then almost stops at about 80 g—even under conditions of abundant and nutritious food availability. In aquaria, 10 g *T. rendalli* fry maintained at 18°C showed negligible growth, but a comparable group maintained at 30°C showed a fairly good rate of growth with the latter fry reaching 40 g in six to eight weeks.

At the beginning of this paper, it was suggested that many marginal dwelling tilapias are, in their natural environment, seldom restricted to a

Table 6. Theoretical growth (ΔB) equations to describe the growth potential of young *Tilapia rendalli* at various temperatures. ΔB is expressed in joules per day, C is the energy (J) of food consumed per day and M the fresh mass of the fish, weighing between 40 and 60 g.

$$\Delta B_{18°C} = 0.392C + 43.41 - 27.53M$$

$$\Delta B_{22°C} = 0.428C + 26.69 - 39.66M$$

$$\Delta B_{26°C} = 0.430C + 7.99 - 69.87M$$

$$\Delta B_{30°C} = 0.437C + 10.10 - 88.70M$$

$$\Delta B_{34°C} = 0.422C + 10.75 - 91.34M$$

Table 7. A comparison of the energy required for satiation in the laboratory (see Table 2) and the estimated energy required for maintenance as derived from Table 6 when $\Delta B = 0$, for *Tilapia rendalli* feeding on *Ceratophyllum demersum.*

Temperature °C	Laboratory satiation ($kJ\ d^{-1}$)	Maintenance energy requirement ($kJ\ d^{-1}$)
18	3.43	3.40
22	4.95	4.57
26	8.70	8.11
30	9.84	10.13
34	10.42	10.80

constant temperature but are usually subjected to varying degrees of daily thermal fluctuations. If fry (10 g) restricted in aquaria, were subjected to a thermal oscillation (18° C night, 30° C day), then their growth rate was found to be very similar to that shown by a control group of fry maintained at 30° C. The major difference between the two groups was found, however, in their feeding, with those fry maintained at a constant temperature of 30° C consuming up to *twice* as much food as those fry subjected to daily temperature changes. Both groups of fry grew from a mean standard length of about 6.5 cm to a maximum of 10 cm over a 45 day period. This growth represents an increase from 10 g to about 40 g in mass. Although the two groups showed very similar growth, their food intake was very different. An explanation for this difference can be obtained using the growth equations of Table 6. For example, a 50 g fish maintained at a constant temperature of 30° C and having a growth rate of 1,000 J/d (196 mg fresh mass per day) would be expected to consume:

$$((88.7 \times 50) - 10.1 + 1000)/0.437 = 12414 \text{ J/d}$$

A similar fish subjected to a 12-hour daytime temperature of 30°C and with a nocturnal non-feeding temperature of 18°C would, in order to grow by the same 1,000 J/d, have to consume:

$$((88.7 \times 50 \times 0.5) - 10.1 + 1000/0.437) + (27.53 \times 50 \times 0.5) = 8028 \text{ J/d}$$

Thus, for individuals from either of the two groups to grow by equal amounts, those fish maintained at a constant temperature require 1.6 times more food than the group subjected to a thermal oscillation of 30°C to 18°C. On testing a variety of combinations in a similar manner, it becomes evident that as the temperature oscillations become less marked, so the ratio between the energy of growth and the energy of food consumed (Δ B/C, Odum's ecological growth efficiency) decreases, resulting in a decline in the efficiency of production.

Estimates of food consumption conducted in (optimal?) *T. rendalli* habitats at Lake Kariba during summer indicate that a 50 g fish consumed an average of 12.6 kJ/d of food, which is equivalent to about 8.2 g/d of fresh food (maximum measured was 10 g/d). Growth estimates obtained from the same wild population by tagging indicate an instantaneous growth capability of 690 mg/d or 3.6 kJ/d. Calculating the ecological growth efficiency (using energy values) of this population, a value of 0.29 is obtained. If the same size fish consumed the same amount of food energy but was maintained at 28°C (mean maximum and minimum of midsummer Kariba temperature oscillation) then growth would be calculated to be 1.3 kJ/d (derived graphically from Table 6) with an ecological growth efficiency of 0.10 or approximately half that calculated for the wild population.

The key factor in this change of growth efficiency is the metabolic energy demand, since, if food is not limiting, the other parameters of the energy balance do not vary in sufficient magnitude to seriously affect the balance at any given temperature. The relationship between food intake and temperature has, in the laboratory, been shown to be strongly temperature related but indications are that in the field, a less marked influence is found, although fish do feed more actively at higher temperatures. This being the case, then, some influence will be reflected in the growth efficiency, but it still remains more likely that metabolism is the governing factor. The pattern of change in metabolic demand with changing temperature may be a complicating factor, but, on the whole, relationships of the type shown for *T. rendalli* or *S. mossambicus* (Figure 3) result in only very subtle variations in the final result. More important is the effect of fish size on metabolism. The smaller the fish the greater the benefit of thermal oscillations may be since the metabolic demand by small fish is greater, per unit mass, than larger fish. Thus, there may well be a limit whereby larger fish no longer benefit from diel migrations to warm shallow water and this, combined with food distribution, may be a reason why juvenile fish are far more abundant in shallow marginal areas of gradient shorelines.

One is able to draw on limited examples from wild populations to support the theory that thermal oscillations are beneficial to the growth of juvenile cichlids. The work of Coe (1966, 1967) showed that the resident population of *S. alcalicus (T. grahami)* in the thermal springs of Lake Magadi (Kenya) showed stunting and poor growth. These fish were generally located close to the 'eye' of the spring (35°C) at night (response to thermophilia) but during the day fed in the cooler (28°C) periphery—the only source of food. This movement is a complete reversal of the diel migrations previously discussed and it may be expected that as a consequence growth efficiency would be extremely low. Notwithstanding the possibility of poor food quality or abundance or possible problems associated with water quality in this environment, all indications are that the thermal peculiarities of the system are themselves responsible to some degree for the poor growth. These fish do not show any genetic malfunction since on transference to a more 'normal' habitat, growth resumes a normal pattern (Fryer and Iles 1972).

Similarly, an early record by van Someren and Whitehead (1959b), who reported that the growth rates of *S. spilurus niger (T. nigra)* in ponds was better in shallow, 30-cm deep ponds than in 60-cm deep ponds, may indicate another example of growth enhancement due to thermal oscillation. Ponds of 30-cm deep would be expected to exhibit a larger diel variation in temperature and tilapia that are restricted to such an environment would be expected to show growth superiority or, at least a better growth efficiency, if food was not limiting. Thus it may not be surprising that the shallow ponds (caused by sedimentation) from which van Someren and Whitehead made the observations showed a good production. My own observations on pond trials tend to confirm this observation but the elimination of the numerous variables in trials of this type is extremely difficult.

A number of documented examples from various African lakes show that tilapias, especially juveniles, have a preference for shallow inshore waters. Recent examples of this distribution preferenda include the tilapias present in Lake Sibaya (Bruton and Boltt 1975), Lake Kariba (Donnelly 1969; Caulton 1977b) and Lake George (Burgis et al. 1973; Moriarty et al. 1973; Gwahaba 1975). In the two former examples food distribution and/or predator pressure may well influence the inshore distribution of young tilapias but in the Lake George example food is more abundant in the deeper areas of the lake yet tilapias are far more abundant in the inshore areas (Gwahaba 1975).

There are obviously a variety of factors contributing to the distribution of tilapia in any particular water body, but indications are that a diel thermal fluctuation may be an important factor to consider. With the exception of the deep rift valley lakes, natural inland waters in Africa are generally composed of shallow river floodplains, seasonal pans, lagoons and marshes as well as large areas of permanent swamp. These areas are sites of tilapia radiation and, prior to the advent of man-made impoundments, constituted the major tilapia habitat. All these areas are subjected to daily temperature fluctuations often in excess of 10°C and thus it stands to reason that the tilapias would have evolved various physiological functions suited to optimizing such conditions.

In concluding this presentation, it is clear that more applied physiological work is urgently required in order to determine the functioning of the various species of tilapia in their natural environment and only after determining the optimum levels of physiological efficiency can we begin to optimize on productivity both in natural water fisheries and in pond culture conditions.

Discussion

NOAKES: A comment more than a question. One of your points is that temperature is very significant for these fish. Their thermal preferanda should therefore change with changing conditions, for example, fed vs. unfed. Is this so?

CAULTON: No, not as far as I know. If you feed the fish in a gradient tank, it makes no difference where the food is, they still go to the same temperature.

NOAKES: Dr. Caulton, you were saying though that different physiological functions have different optima. If this is the case, then it would be reasonable to assume that the fish, when it is undergoing certain physiological functions, would choose the corresponding optimum temperature.

CAULTON: Although this is a reasonable assumption, I do feel that certain physiological optima are sacrificed for others. Certain physiological processes do certainly function most efficiently at a prescribed temperature, yet metabolism, for example, is efficient, in terms of energy saving, at low temperatures yet the overall physiological optimum for a feeding tilapia is at a temperature of about 30°C.

NOAKES: It has been suggested in some temperate regions where there are thermoclines in the summertime that some of the large predator species move above the thermocline in the evenings to feed at a higher temperature where they can move around more efficiently. Then they will go down into colder water where they can digest their food and grow very efficiently.

BOWEN: The thermal preferendum of *Sarotherodon mossambicus* does change with age. It gets lower as the animal gets larger. But even the thermal preferanda of the largest fish are well above any temperatures normally achieved in the natural environment. So, in many cases, they would be expected to migrate back and forth between the thermal gradients along the shores.

HEPHER: Dr. Caulton, what was the average weight of your fish?

CAULTON: We worked with juvenile fish from about 20 to 50 g, but for a lot of the preferenda work, we used fish up to 300 or 400 g. For respirometry work we used a large range of sizes, up to 200 or 300 g. Basically, however, we were concerned with modelling on juvenile fish.

MIRES: Dr. Caulton, from what part of the fish did you take the fat content tissue sample? From muscle?

CAULTON: We used the whole fish.

GUERRERO: Can tilapias get their essential amino acids from blue-green algae and bacterial protein as they can from animal protein?

BOWEN: Yes, it is good quality protein.

MORIARTY: In Lake George, the diet of *Sarotherodon niloticus* consists of about 70% *Microcystis* with a few other blue-greens and diatoms and a few percent bacteria. This was obviously a good food.

GUERRERO: Dr. Bowen, have you studied the amino acid make-up?

BOWEN: Yes I have and it is not significantly different from the amino acid composition of plant proteins in general, including those from macrophytes.

GUERRERO: How does it compare with fish meal?

BOWEN: I have not compared it with fish meal, but I have compared it with the essential amino acid requirements of warmwater fish given in the literature and it is more than adequate to meet these.

MORIARTY: Another point on that of course is that the protein content of microalgae and bacteria is very high compared with that of other plant material. It is around 50 to 60% of the dry weight. This varies of course.

BOWEN: Yes, there are very few storage compounds, such as carbohydrates, in blue-green algae and bacteria and instead of storing energy, these cells continue to reproduce, making more protein.

HENDERSON: You mentioned that there are data for assimilation efficiencies of grass carp feeding on plant protein.

BOWEN: Yes, these are around 50%.

JALABERT: Is there any evidence of feeding rhythms?

MORIARTY: My experience of the Lake George fish is that they feed during daylight hours only. They start feeding at dawn or just before. (Editor's note: Whyte (1975) reported that *Sarotherodon galilaeus multifasciatus* and *Tilapia busumana* in Lake Bosumtwi, Ghana, both feed at night. The water of this lake is very clear.)

JALABERT: Does the efficiency of the food utilization vary according to the time at which it is eaten during the day? For example it has been shown in some fish, particularly catfish, that the efficiency of food utilization is not the same in the morning as in the afternoon.

CAULTON: Yes, we have shown that with the acid secretion cycle in the tilapia. Until you have acid production in the stomach sufficient to lyse the food items, you will not get any release of nutrients. So the first portion of the meal ingested in the morning is hardly digested at all, and there is a large increase in efficiency as the day progresses.

BOWEN: I have actually worked that out by counting the number of empty diatom frustules at different locations in the gut. There is a group of undigested diatoms which is derived from the first part of the daily meal but this is not a very significant portion of the diet and the digestive system becomes efficient very rapidly. So that there is 10% or less of the food that is not fully digested and the remainder is very effectively digested.

NASH: In this detritus, is there any gelatinous material? There seems to be a nice parallel between this freshwater situation and that sometimes found in saline waters. For example, there are some excellent feeding grounds for marine shrimp in Brazil in the Cabo Frio area, and excellent feeding grounds for milkfish in many of the Pacific low atolls where the substrate is a deposit of blue-green algae and detritus, but both are characterized by being very gelatinous. The substrate is similar to the lab-lab prepared in the Philippine milkfish ponds which again has a very high protein level.

BOWEN: Is this gelatinous appearance the result of a large mass of blue-green algae?

NASH: I don't know what it is.

LOVSHIN: There are some blue-greens which form gelatinous masses.

BOWEN: I know that some diatoms secrete gelatinous strands which tend to bind them together into a fine substrate, but I don't know if blue-greens do that.

MORIARTY: Many benthic and planktonic blue-green algae secrete large amounts of slime. Its nutritive value for fish has not been assessed, but it may well be utilized.

COCHE: There is one factor which has not been mentioned from the fish culture point of view. Is the amount of food available affecting growth efficiency? It is well known in fish culture, for instance, that if you feed tilapia *ad libitum* their assimilation efficiency will decrease. That is why the ration becomes so important for a culturist.

HEPHER: There is another point which we should stress more and that is the effect of temperature, particularly at night, on growth rate. I do not know if you hold a fish for long periods at 18°C whether it will grow as well as if you hold it at higher temperatures.

BOWEN: With regard to Dr. Coche's comment, unlike carnivorous and omnivorous animals, the herbivores and detritivores that I am aware of do not in the natural environment ever confront a shortage of food. So quantity to my knowledge is never a limiting factor. It is quality and if I understand Dr. Caulton's contribution correctly, what he has done is to look at how the fish interact with the variables in their environment and try to maximize the amount of food that they are capable of consuming in the absence of any limit on availability. So this does not apply immediately to the culture pond situation where you are trying to maximize the efficiency of food conversion.

LOVSHIN: This is just a comment on the migrations to different temperatures. At Auburn we have a lot of small pool units from which we pick small fry. There are very definite migrations towards the surface when the fish are in small schools. The fry are much easier to collect between 11:00 AM and 2:00 PM when the temperatures are high. There is a very definite time of day when the schools are much more visible and higher up in the water.

CAULTON: Thermal preferanda could be responsible for both a horizontal and a vertical migrations—see text of my paper. In natural tilapia systems, juveniles invariably show a horizontal migration since they are predominantly marginal in habit while adults, present over deeper water, often show a vertical movement. Both environments are affected by daily temperature inversions, hence, the movement of fish.

MORIARTY: Dr. Caulton, did you do any temperature preference studies at night as well as by day?

CAULTON: Yes. The results of these experiments demonstrated that there was no significant difference between day/night thermal preferenda. It also made no difference if the fish were fed or unfed males or females or introduced singly or in shoals.

PULLIN: I would like to comment on the questions raised by Dr. Coche. The amount of food given in a culture situation is obviously very important. I find Dr. Bowen's view that herbivores and detritivores in their natural environment are never likely to encounter shortages of food quantity very significant as some of the detritivores we are trying to grow in fertilized ponds may be encountering a shortage of food. There may well be ways which we can manipulate and improve this kind of culture environment. This may not be as easy as regulating supplemental food, but has anyone any ideas on how detrital feeding can be improved in pond culture?

HENDERSON: The quality of the food is most important because ingestion by the fish is partly related to how much time they spend for the process. If they are processing poor detritus and using a lot of energy and time in the process, they are going to get a lot less benefit than from feeding on blue-greens. But then again, a high quality detritus system may be an improvement.

PULLIN: Yes, and a fertilized pond culture system using organic manures would have a head start over one using inorganic fertilizer.

MORIARTY: The form in which detritus occurs is also important. We have heard mainly about macrophyte feeders and detritus feeders and not very much on the phytoplankton feeders which are also used in pond systems. For the detritus feeders their food must be in a form which they can pick up easily: if it is more or less soluble or in the form of free swimming bacteria then it may be of little use to the fish. It has to be precipitated in a detrital aggregate or has to be taken up by microorganisms which themselves grow in clumps and form larger aggregates for the fish to ingest. Whether this is in fact detritus *per se*, i.e., non-living material, or whether it is aggregations of living microorganisms is another matter. Detrital foods in ponds must be of the right consistency to be available to fish as well as having the right protein and carbohydrate makeup.

BOWEN: In this regard, it is interesting that a study of manured pond culture of tilapia published recently by Schroeder in Israel looked at microbial production and estimated it. He found that even when he considered that source of nutrition for the tilapia, he could not account for all the production of fish. It remains a great mystery how those fish grew so well, because even if they were cropping 100% of the microbial production, he could not account for all the growth of the fish. He suspects that there must have been a non-living source of nutrients, i.e., organic matter derived from the manure.

MORIARTY: There are a lot of problems in relating microbial production to fish growth.

LOVSHIN: We have talked about various forms of detritus, and I am interested in knowing whether detrital matter has to stay in the water and be worked on by bacteria for some time before it is usable or whether the tilapias can use it fresh, like fresh cut grass?

BOWEN: There are two main categories of organic compounds to consider here with respect to macrophyte material such as palm leaves, grass clippings etc.: structural components that are insoluble in water, and non-structural components which are largely soluble. It is the soluble compounds which are valuable for the nutrition of fish. Cellulose and lignin are not used directly by fish, but soluble proteins, soluble carbohydrates, and lipids, which will go in solution soon after waste plant material is put into the pond will all have a real impact on the fish's nutrition.

LOVSHIN: What type of detrital matter are you talking about here?

BOWEN: I distinguish between morphous detritus and amorphous detritus. Amorphous detritus is the material which has been formed from dissolved organic matter. Its formation is not by fragmentation of leaves, for example, but it is via the dissolved organic components. There are many ways in which you get from dissolved organic to particulate amorphous organic matter. Microbial action is one. Interface reactions either at the air-water interface or the water-mineral interface are others. Simple chemical polymerization in solution is another. UV light also has a physical effect on polymerization. All these factors play a role, but the importance of each of them is not understood. The material which I have found in my studies which plays an important role in the nutrition of *S. mossambicus* is the amorphous organic matter—the amorphous detritus.

LOVSHIN: I think it would be a very important and very interesting line of study to see what material we could throw in the fish pond, or what combinations of materials, to get from these processes the maximum amount of detrital matter which could then be utilized by the fish.

MORIARTY: That is a subject that really does need to be examined in much more detail because detritus is used as such a broad term that in different situations you can have different processes happening.

BOWEN: Schroeder's work shows (or strongly suggests) that an approach to improve detrital feeding would be a very productive one. There is a huge body of literature now on decomposition processes in aquatic environments, which gives some very good ideas of where to begin searching for the manipulation which would simulate detritus formation and give a product that has a high nutritional value.

MORIARTY: Going back to Dr. Chervinski's paper, one thought which occurred to me is that the influence of temperature at the limits of distribution of tilapias, particularly where low temperatures occur, is going to be affected by the surface area and volume of the culture water body. Dr. Caulton's work suggests the use of a shallow pond which will have high temperatures during the day and low temperatures at night, but extreme low temperatures must be avoided. [Editor's note: In Transvaal, dams are stocked with *S. mossambicus.* The fishes survive winter frosts only if they can retire to a part of the dam that is at least 2 m deep (Lombard 1959).] In summarizing the work of Drs. Caulton and Bowen, the quantity and quality of food that tilapias eat need to be considered in relation to environmental factors such as temperature and oxygen. The fish are all considered to be herbivorous, although one group is better termed detritivorous. The origin of detritus is not necessarily from plants, but the role of micro-organisms is obviously very important in the decay of materials such as grass clippings etc. The available protein at the start of decay is not going to be very high and Dr. Bowen's work has shown that it is the total available nitrogen that is important.

Their nitrogen, therefore, has to be in a form that is digestible by proteases, i.e., protein or polypeptides. Amino sugars and amino acids in slime layers may also be utilized, but further work is necessary to study digestibility.

The fish have to use their digestive enzymes to digest detritus. The macrophyte feeders have a very different task. They have thick cell walls to get through and therefore the pharyngeal teeth of the fish are very important. We have said little about the phytoplankton feeders. A lot of the fish that are used for culture feed on blue-green algae which are the most common algae in lakes in the tropics. The work that we did in Lake George showed that the blue-greens were a very nutritious source of food for the fish. The fish had rapid growth rates. The actual assimilation efficiencies were between 30 and 65%: about 45% on average. The maximum assimilation efficiency was 80%. The variability is due to the physiology of digestion and not because there are refractory compounds present in the food. The fish have to secrete gastric acid, and the pH has to fall to about 1.5 for effective digestion to occur. This takes five to six hours—half-way through the daytime feeding period, which accounts for the low assimilation efficiency averaged over a 24-hr period. Even though they assimilate only 45%, our work on the amount that the fish ate per day and their growth rates showed that the amount that they assimilated was more than adequate to support the growth rates obtained and their reproduction. Finally, I would like to stress the importance of Dr. Caulton's work on diurnal migrations between different water masses at different temperatures. Attention to this and to temperature manipulation could be of direct benefit to tilapia culturists.

SESSION 3: CULTURE

Chairman's Overview

H. F. HENDERSON

One of the major difficulties in comparing different culture systems is the lack of an appropriate yardstick for performance. One of the more obvious examples is the expression of the productivity of a system as the biomass per unit of time *per unit area.* In extensive aquaculture in large lacustrine systems, this unit would seem to have the same sort of validity as in comparative studies of the ecology of aquatic ecosystems in general. In pond systems it still seems a useful figure for judging the most economical use of space, but has somewhat doubtful validity in comparing biological efficiencies unless depth is brought into the picture.

As soon as we move into cage and pen systems, however, one hardly bothers to calculate such figures. From the point of view of the culturist, input-output information is of more interest. Unfortunately the commonly quoted feed conversion ratio is not a very satisfactory biological index unless the water content of the feed is taken into consideration. Growth rate and production are of direct interest to both the scientists and the culturists, but are difficult to compare for different lengths of growing season. Dr. Coche gives special attention to these problems, providing several different measures of performance to compare cage culture systems. Recognizing that it may not be very useful to compare the efficiency of the use of space, say, between pond and pen culture, it does nevertheless seem worthwhile to define standard performance criteria for culture systems. These criteria should obviously extend to economic as well as biological factors.

The reviews prepared by Dr. Coche and Drs. Hepher and Pruginin suggest that pen and cage culture of tilapias have advantages for the developing countries while pond culture seems to require rather complex management for success, and may be better suited to the developed countries. It is clear, however, that the main technical and managerial complexities lie in the production of seed rather than in the growout phase, particularly the hybrid seed which seems to be required for profitable production in temperate climates. The question of suitable pond and cage culture systems should also be considered in both high and low technology contexts. Offhand it would seem to me that cage design and construction would offer more problems in developing countries than pond construction. However, inexpensive pre-fabricated cages could conceivably be developed.

Dr. Coche remarks that keeping cage floors well above the bottom of the water body seems to reduce the incidence of disease. He also states, however, that caged fish may be under greater stress than pond fish from crowding and suggests that they may be more readily attacked by disease and parasites.

Manures and fertilizers seem to be more effectively utilized in pond culture systems. In tropical countries tilapia culture is particularly attractive when it is combined with pig or duck farming. Drs. Hepher and Pruginin point out in their paper that organic manures, by directly stimulating heterotrophic food-chains, are better than inorganic fertilizers for the microphagous tilapias.

Only pond and cage culture systems have been reviewed. It is also valuable to refer to two other systems. One is the very intensive culture of tilapias in tanks and in raceways, systems which are particularly appropriate in utilization of waste thermal waters. The other is extensive culture in small reservoirs, natural lakes and ponds. The highly intensive culture systems pose special problems of providing complete diets as well as removal of wastes from the system. Management of extensive systems, on the other hand, should be of interest in relation to the extent to which cage or pen culture can or should be combined with harvest of the more free-living stocks. This is an important question in, for example, the improvement of the fisheries of the 10,000 or so small reservoirs of Sri Lanka. In small reservoirs, there are some interesting possibilities for the control of reproduction of tilapias through water level control: for example, by exposing nests at critical periods.

Where cage culture has prospered, local administrators have often been unprepared for the resulting legal and environmental problems. Dr. Coche cites several cases in the Philippines where this has happened and I have been told of similar sorts of problems in Polish trout culture. It is, however, not only administrators who have been caught unprepared. There is very little information available on the effects of cage culture in lakes, ponds or bays upon which to base predictions of permissible levels of occupancy. At high levels of occupancy, reduced production in each cage unit becomes obvious to the operators. As in capture fisheries, however, the overutilization of water bodies for cage culture is not likely to be reduced to optimal levels voluntarily. Unfortunately, we know little about predicting permissible levels of occupancy before production begins to drop, nor, to my knowledge, has there been any attempt to define effluent standards for cage culture, analogous to those which have been adopted in some places for pond culture.

Speaking of environmental matters, Dr. Coche also notes that cage culturists and others should take care that inappropriate species are excluded from waters to be used for cage culture. Competition for food between *S. aureus* and wild *S. mossambicus* in cages in Puerto Rico has been cited as an example of this problem. There, the wild *mossambicus* were also thought to be a source of parasites and disease for the cultured *aureus*.

There is a problem not mentioned explicitly in any of the papers, but which will emerge as tilapia culture becomes more intensive and more highly selected strains are used for culture. This is the increased movement of both broodstock and seedstock across national boundaries and the attendant risks of transfer of parasites and disease, and of contamination of stocks with genes of related species. In most countries, there is as yet very little control on either the species, or more importantly, the health of the fish transported.

Roberts and Sommerville note that the diseases of tilapias have not been as well studied as those of other cultured species, both because their culture has only recently been developed to an intensive level where disease problems are more evident, and because diagnostic facilities are, in general, poorly developed in areas where tilapia culture is most common.

Tilapia polyculture systems are discussed in some detail by Hepher and Pruginin, and Coche mentions one instance, in Lake Victoria, of *S. esculentus* and *T. zillii* being cultured together in cages, and two instances of *S. niloticus* being cultured with common carp. It would seem that there is less reason for polyculture in cage systems than in ponds, though others may recognize the availability of distinct niches in cages. In ponds, Hepher and Pruginin suggest that in Israel polyculture is especially advantageous when low to moderate stocking densities of young-of-the-year tilapia are used. At high stocking densities, which are more typical of intensive production systems with hybrid fish, the natural food component is negligible, and hence the feeding synergisms are less important. The use of predators in tilapia culture is perhaps a special case as such use is primarily for controlling recruitment. Hepher and Pruginin suggest that common carp may carry out this latter function to some degree, as well as that of increasing the utilization of natural food.

Dr. Roberts draws attention to the recent demonstration of a biochemical compound in water in which *S. mossambicus* had been intensively cultured, which produces a kind of "shock syndrome" in this and other tilapias. The production of such substances by fish has been frequently referred to elsewhere, but seems not to have been given much attention by culturists. There is some evidence that such substances may be important in spawning behavior.

Tilapia Culture in Ponds under Controlled Conditions

B. HEPHER

Fish and Aquaculture Research Station
Dor, D.N. Hof-Hacarmel
Israel

AND

Y. PRUGININ

Givat-Shapira, Israel

HEPHER, B. AND Y. PRUGININ. 1982. Tilapia culture in ponds under controlled conditions, p. 185-203. *In* R.S.V. Pullin and R.H. Lowe-McConnell (eds.) The biology and culture of tilapias. ICLARM Conference Proceedings 7, 432 p. International Center for Living Aquatic Resources Management, Manila, Philippines.

Tilapias are valuable pond fish. They are appreciated by consumers in many countries and can produce high yields on relatively low inputs. However, tilapias pose a special management problem. They breed in production ponds when still young and small, greatly increasing the population. This results in competition and stunting due to lack of food. Only by controlling reproduction in production ponds can fish of marketable size and high yields per unit pond area be obtained. Two management methods to overcome this problem are reviewed: (a) rearing a mixed male-female population of young of the year before they attain sexual maturity and (b) rearing an all-male population. Factors affecting the choice of one of these methods are discussed as well as the techniques involved in each method. The effects of chemical fertilization, organic manuring and feeding on the production of tilapia in ponds are also discussed.

Introduction

Tilapias have become increasingly important in fish culture, especially in warm climates. According to FAO (1978), the total world production of tilapias (both *Tilapia* and *Sarotherodon* species, but excluding other cichlids) reached 197,000 t in 1977. Only a part of this production is obtained through aquaculture, but that portion is increasing steadily.

The number of tilapia species cultured in ponds, both experimentally and on a commercial scale, is quite large. Huet (1970) mentions 16 species. Balarin and Hatton (1979) give a list of 23 species which have been cultured at some stage. Of these, however, only two *Tilapia (T. rendalli* and *T. zillii)* and three *Sarotherodon* species *(S. mossambicus, S. niloticus* and *S. aureus)* have seen widespread use. The plant feeder *T. rendalli* is cultured to some extent in the Malagasy Republic and some Latin American countries, e.g., Brazil and Mexico (Balarin and Hatton 1979). *Tilapia zillii* is cultured in some east African countries. It seems that there is considerable variation in the performance of cultured *T. zillii* since this species can reach a good marketable size in certain east African countries, while in other countries, such as Israel, it is considered a pest due to its small size at maturation. The major production of tilapias in ponds is derived from the three *Sarotherodon* species and most of the references in this paper will deal with these.

Tilapias, though valuable pond fish, pose a special problem when grown in ponds, whether in polyculture or in monoculture. Their early maturation and prolific "wild" spawning produce such large numbers of small fry as to cause stunting of the entire tilapia population and often of other species present in the pond. In some regions it is customary to harvest tilapias a number of times during the growing season. Each harvest removes the larger fish, leaving the smaller ones. When this practice is employed with fish that do not spawn freely in ponds, it usually results in a larger cumulative yield by providing more food for the remaining fish (Van der Lingen 1959a). With tilapia, however, this practice usually lengthens the culture period and results in heavy "wild" spawning. The longer the culture period, the more fry are produced and the population becomes stunted. The number of market-sized fish and the yield at each harvest becomes progressively smaller. In commercial farms, which have to supply relatively large quantities of market-sized fish regularly, such sequential harvesting is not practical and reduces profitability. Silliman (1975) points to another disadvantage of this practice in that the remaining fish which spawn in the pond and from which fry are usually taken for further rearing, are actually selected for slower growth, and this trait may be inherited by their offspring.

There is no doubt that early and prolific breeding under tropical pond conditions is perhaps the greatest disadvantage in the use of indigenous cichlid species for aquaculture in many parts of the world (Okorie 1975; Pillay 1979). It is obvious that Huet (1970) was correct in recommending the culture of tilapia during a definite period, with complete draining of the pond between cycles to control populations. Even then, special measures must be taken to overcome "wild" spawning. The two most common ways to overcome this problem are: culturing a mixed sex population and harvesting before they attain sexual maturity or culturing an all-male population. Two main factors should be taken into account when deciding which method to employ: the size preference of the market and the species used.

If the market accepts small fish (up to 150 to 300 g) and if these can be obtained before they reach sexual maturity, the first method is usually preferred since it saves effort, fingerlings and space. If, however, the market accepts only large fish which reach maturity before marketing, it will be better to use only male fry and thus avoid "wild" spawning and stunted populations.

In some cases the market accepts both sizes but prices are higher for the larger fish. Careful economic calculations should then be made, weighing the higher income for producing larger fish against the extra costs of obtaining male fry and wastage of females. The calculations should also take into account the higher yield expected from an all-male pond as compared to a mixed sex one, since in most tilapia species the males grow faster than the females (Fryer and Iles 1972).

Some species, such as *Tilapia zillii* or *Sarotherodon mossambicus*, breed when they are young and small and cannot reach market-size before maturation. Monosex populations have been obtained, mainly by manual sexing, but this can only be done reliably when the fish have reached a size of 20 to 50 g, so they must first be nursed to at least this size. In many cases these species will breed at an even smaller size. Pruginin and Arad (1977) report that in Malawi, *S. mossambicus* bred and growth stopped due to stunting when the fish reached 30 g. As a consequence the yield after 100 to 150 days culture did not exceed 300 kg/ha. It may be advisable, therefore, to choose species which spawn at an older age, such as *S. niloticus*, *S. aureus* or others. Pruginin (1965) found that while *S. hornorum* in Uganda reached not more than 150 g in about one year, *S. niloticus* usually reached 250 g in 5 to 6 months by the time they reached sexual maturation.

Culture techniques, such as pre-nursing, choice of size at stocking, stocking density, fertilization, feeding, etc., are affected to a large extent by the choice of management approach: mixed sex or all male. However, other factors such as climate and the type of culture system (monoculture or polyculture) have also to be considered.

Young-of-the-Year Culture

For unsexed young-of-the-year populations, the period before they reach sexual maturity is short: about 3 to 6 months. Thus the management method is based on having 2 or 3 cycles/yr, with complete drainage between cycles. In order to ensure rapid growth to the desired market weight during this short period, the stocking density is usually relatively low and depends on the inputs used, e.g., fertilization, manuring and feeding.

Van der Lingen (1959a) has found that with increasing levels of management, the maximum standing crop (carrying capacity) which can be sustained in the pond increases. He gives the following carrying capacities for *S. mossambicus*: natural feeding (no fertilization), 840 kg/ha; high level fertilization, 2,466 kg/ha and high level supplementary feeding with fertilization, 6,165 kg/ha. Van der Lingen (1959a) further states that over the same culture period and under the same conditions the yield per unit area at any level of nutrition is dependent upon the initial stocking biomass per unit area and not on the number of fish per unit area. He therefore recommends stocking by weight according to the expected carrying capacity. However, this statement is based on only a few experimental observations. Since the relative growth rate (i.e., relative to unit body weight) of small fish is higher than that of larger fish (Ghosh 1974; Hepher 1978), it is usually found in practice that for a given biomass of fish, the higher the density the higher the yield, provided enough food is available. Moreover, the practice of stocking by weight may lead to stunting of the population if the average weight of the stocked fry is low, since the number of fry per unit of stocked weight will be large and they will reach the carrying capacity while still small. Van der Lingen (1959a, 1959b) stocked fingerlings of 21 to 56 g average weight. His recommended stocking weights and densities relative to the carrying capacities for different inputs are given in Table 1. The number of fish stocked

Table 1. Stocking biomass recommended by Van der Lingen (1959a, b) for tilapia cultured in ponds with different inputs and the calculated fish densities (assuming an average weight of 38 g).

Inputs	Stocking biomass (kg/ha)	Calculated density (fish/ha)
None (natural feeding, no fertilization)	56-112	1,475-2,950
Fertilized ponds	112-224	2,950-5,900
Fertilized ponds with supplementary feeding	560-1,680	14,750-44,200

were calculated, taking into account an average weight of 38 g. If all the fingerlings stocked according to these recommendations survive by the time carrying capacity is attained and no "wild" spawning occurs, the average individual weight at harvest (carrying capacity/density) will be 139 to 417 g, which is a good market weight. However, if fry of 1 to 2 g are stocked, as is common in many countries, the average weight at harvest under the same conditions will be only 7 to 22 g. In Van der Lingen's (1959b) experiments only 15 to 30% of the harvested fish reached a market size of over 224 g. The densities given in Table 1 seem therefore to be too high since they will not permit rapid growth. In Israeli ponds, where protein-rich pellets are fed, tilapia are stocked at densities of 3,000-5,000/ha.

Allison et al. (1979) experimentally cultured unsexed *S. aureus* in concrete tanks (surface area 0.002 ha) at very high densities: 50,000, 100,000, 200,000/ha. "Wild" spawning decreased with increasing density (from 222,900/ha in the lowest to 38,380/ha in the highest density) and yield was high (from 1 t/ha in the lowest to 17.3 t/ha in the highest density). Allison et al. (1979) do not give the rate of growth but calculating from the yield and the density, it seems to be very low. The extrapolation of such data for application in commercial fish ponds is very questionable.

Yashouv (1969) has demonstrated that in a polyculture of *S. aureus*, common carp *(Cyprinus carpio)* and grey mullet *(Mugil cephalus)*, the tilapia (if stocked below 5,000/ha) do not affect the growth of the carp but in many cases even stimulate it. The growth of the tilapia is also not affected by the presence of the common carp or mullet at densities up to 2,500 to 3,000/ha. These synergistic effects may be explained by the increased amounts of detritus through the presence of the common carp and its consumption by the tilapia. The detritus carries dense populations of bacteria and protozoa (Odum 1968) and constitutes a nutritious food for the tilapia. On the other hand, the consumption of the detritus by the tilapia improves the oxygen regime for the benefit of the common carp. The increased yield in the polyculture pond explains why most, if not all, young-of-the-year tilapia culture in Israel is done in polyculture systems.

Polyculture may have an additional advantage. Since common carp and grass carp *(Ctenopharyngodon idella)*, when large enough, can prey to some extent on tilapia fry spawned in the pond (Spataru and Hepher 1977), they can, therefore, help to alleviate the problem of "wild" spawning that may develop towards the end of the culture period. This problem can, however, be solved much more effectively by including a predator in the polyculture, such as the Nile perch *(Lates niloticus)* in Africa (Pruginin 1965, 1967; Meschkat 1967), the mud fish *(Channa striata)* in the Philippines (E.M. Cruz, pers. comm.) and Thailand (Chimits 1957), *Cichlasoma managuense* in Central and South America (Dunseth and Bayne 1978); and the sea bass *(Dicentrarchus labrax* or *D. punctatus)* in Israel (Chervinski 1974, 1975). The last two species are marine fish which can adapt to freshwater. The predators are in most cases stocked at about 10% of the tilapia population. The use of predators in tilapia ponds, despite its promise, has not received wide application and has been practiced mainly on an experimental scale, partly because of the difficulty in obtaining predator fry. This is true for both the Nile perch and the sea bass. Their inability to breed spontaneously in the pond can be considered an advantage but a source of fry must be found for commercial application.

In the subtropics the winter is usually too cold for growth of tilapia and the growing season is restricted to the summer. Balarin and Hatton (1979) quote Bishai (1965) who gives the range of 17.2 to 19.6°C, below which the growth rate of most tilapias decreases. For spawning to take place, temperatures must also be higher than this. The lowest temperature for spawning is 20 to 23°C (Huet 1970; Uchida and King 1962). This restricts the length of the season for culture of young-of-the-year even more. In Israel spawning starts only in May, and fry of approximately 1 g for stocking growout ponds are not available before June. This limits the season for rearing young-of-the-year from June to October so that only one cycle can be carried out. The fry are transferred from the spawning pond into the growout ponds when they are over 1 g. The culture period is sufficient to bring them to a market size of 200 to 300 g.

Since tilapias grow well above 18°C, the early part of the summer (March to June) can also be utilized for culture but the only fry then available are those hatched the previous year. If a mixed sex population is cultured they will spawn in the ponds when the temperature reaches 20 to 21°C in May. Since this occurs shortly before harvesting, the fry thus produced do not reach a size which can cause much harm in the pond and the stocked tilapia reach market size unaffected. Halevy (1979) reported that unsexed fry hatched in the previous year could be cultured during the first cycle and young-of-the-year during the second. A typical example of one growing season during 1977 is given in Table 2. In this case it can be seen that the cool spring weather is an advantage and high yields of market-sized tilapia can be obtained.

It is obvious that fish size, the time of stocking and especially the time of harvesting may be crucial to the success of culturing the previous year's fry in spring. The later the stocking and the smaller the fry, the smaller the fish at harvest time and the later the harvesting, and the more troublesome is the "wild" spawning. This is a considerable drawback in large farms where, due to market demand, the stocking and harvesting times are usually phased so that some ponds are stocked in late spring and harvested in mid-summer. These cannot, therefore, be used for early spring stocking and growout.

Table 2. Polyculture of tilapia *(Sarotherodon aureus)* with common, silver and grass carp in a 1.4-ha pond at Dor, Israel (after Halevy 1979).

Species	First culture period: February 6 to May 30 — Stocked: Density (fish/ha)	Stocked: Average weight (g)	Harvested: Density (fish/ha)	Harvested: Average weight (g)	Gain: Average daily weight gain/fish (g)	Gain: Average daily yield (kg/ha)	Total yield (kg/ha)
Tilapia *(Sarotherodon aureus)*	5,000	6	2,140	200	1.3	2.8	398
Common carp *(Cyprinus carpio)*	3,000	5	2,960	642	4.4	13.0	1,885
Silver carp *(Hypophthalmichthys molitrix)*	930	378	890	2,000	11.3	10.2	1,428

Species	Second culture period: July 3 to November 22 — Stocked: Density (fish/ha)	Stocked: Average weight (g)	Harvested: Density (fish/ha)	Harvested: Average weight (g)	Gain: Average daily weight gain/fish (g)	Gain: Average daily yield (g)	Total yield (kg/ha)	Annual total yield (kg/ha)
Tilapia *(Sarotherodon aureus)*	5,000	0.5	4,560	224	1.6	7.3	1,018	1,416
Common carp *(Cyprinus carpio)*	7,170	125	4,450	533	2.9	12.9	1,776	3,661
Silver carp *(Hypophthalmichthys molitrix)*	1,070	750	1,070	2,220	10.3	11.0	1,572	3,000
Grass carp *(Ctenopharyngodon idella)*	1,430	10	1,290	150	1.0	1.3	180	180
							Grand total annual yield	7,957

There is another serious drawback to mixed sex spring culture which is economic in nature. The cost of overwintering tilapia fingerlings is high and the rate of growth of unsexed fingerlings in spring is lower than that of all-male fingerlings. This is because males grow faster and because large fish grow better than small ones in culture ponds. At much the same cost, therefore, one can raise overwintered all-male fingerlings which have been nursed to a larger average weight. This is the reason why most tilapia culture in Israel uses all-male populations.

When rearing young-of-the-year in warmer tropical climates, it should be remembered that their age is a most important factor affecting sexual maturity. Stunted fry cannot be cultured because they will breed early in the ponds. It is important therefore to use recently hatched fry. It is also important to completely drain the rearing pond between cycles and eliminate all the remaining fish, if necessary by poisoning. The intrusion of fish from outside the pond should be prevented by screening the water inlets.

All-Male Culture

A monosex male population can be obtained in three ways: manual sexing; crossing two species of *Sarotherodon* to produce all-male or a high percentage (90% and over) of male hybrids and sex-reversal at an early age by incorporating hormones in the feed.

While much experimental work is being carried out on hybridization and sex-reversal and these methods appear promising for the production of all-male populations, it is only very recently that commercial use has been made of monosex hybrids and no commercial application has yet been made of sex-reversal. The main method used today to achieve an all-male population is still manual sexing. This is a relatively simple procedure. In many tilapia species the sexes can be distinguished by the genital papilla which has one orifice in the male as compared to two orifices in the female. The female often also has a smaller genital papilla.

It is important to sex the fish carefully. The less errors in the sexing, the less "wild" spawning occurs. The earlier the sexing is done the better since the females are then discarded. Early sexing thus saves space which can be used for rearing of the males and minimizes feed wastage on unwanted females. There is, however, a certain minimum size of fish for sexing with an acceptable degree of confidence. In field conditions the optimum size for sexing in most tilapia species is 50 to 70 g. This means that the fry have to be nursed to at least this size before growout. The hybrid crosses that produce high percentages of males are advantageous even if they do not produce 100% all-male population. Using such hybrids, fewer females are discarded, space is saved and sexing can be done on large fingerlings with a greater degree of confidence.

Here again, climate is an important consideration. Where a cold winter exists, nursing is done in the summer and growout is usually postponed to the following year. The final weight of the nursed fingerlings will depend to a large extent on the length of the nursing period. Those which hatched early and were stocked in the nursing ponds early in the season (end of

May-beginning of June) can reach a final weight of 100 g and over by autumn. However, those hatched later and stocked in the nursing ponds in July-August will only reach a weight of 40 to 60 g. The minimum weight for overwintering fingerlings is about 20 g. With smaller fingerlings the survival rate during winter is very low. The stocking density preceding overwintering is adjusted according to these expected final weights. When fingerlings are expected to reach a final weight of 100 g, the stocking density is about 50,000/ha. In order to better utilize the natural productivity of the pond, fry are often stocked at a density of 100,000/ha for the first part of the nursing period until they become 50 g fingerlings. They are then thinned out to 50,000/ha. Fry of later spawning, which can reach a final weight of 50 g before overwintering are stocked at a density of 100,000/ha.

One of the most important considerations in determining the desirable final weight of the nursed fingerlings in regions with cold winters is the available overwintering capacity. If special facilities and large investments are required for retaining a higher than ambient water temperature in the overwintering ponds (e.g., covering of the pond and/or warming the water), then this usually means a restricted overwintering pond area and, therefore, a limitation of the standing crop of fish that can be held over the winter. A certain number of fingerlings are required for stocking the culture ponds in spring and the maximum average weight of these can be obtained by dividing the standing crop which can be overwintered by the required number of fingerlings. Smaller fingerlings can be obtained by increasing the fry stocking density in nursing ponds relative to the desired final fingerling weight. These fingerlings can be nursed again in spring if necessary.

Nursing can also be done in polyculture ponds while rearing other fish (sometimes including large male tilapia) to market weight. In such cases the density is sometimes lower than for nursing ponds. Table 3 gives an example of such nursing on an Israeli farm where tilapia were introduced in late summer (August 9).

The culture of an all-male population removes the restrictions on final age and thus on weight at harvest associated with mixed sex culture. The fish can therefore be cultured to a much larger final weight, usually 400 to 600 g, even though this takes longer. Growout can be done either in polyculture or in monoculture systems. The advantages of polyculture have already been discussed above. The stocking density used in polyculture is not much different to that for young-of-the-year culture, but since males grow better and the growing period is longer, they usually attain a much larger final weight.

Here again, the density depends on the levels of inputs (fertilization, manuring and feeding). Pruginin (1967) stocked all-male hybrids *(S. niloticus* x *S. hornorum)* of 50 g average weight at densities of 1,000 to 1,500/ha. The daily weight gain was 1.5 to 3.0 g/fish, and the average weight at harvest was 200 to 450 g after a culture period of only 100 to 150 days. No information was given on fertilization or feeding rates. Shell (1968) conducted an experiment to study the stocking of male *S. niloticus* using supplementary feeding (pelleted Auburn No. 2 fish feed, rich in protein). The growth rate was independent of fish density up to 5,000 fish/ha at 2.1 to 2.3 g/fish/d. Rearing at this density gave the highest yield of 1.6 t/ha/163 d. The feed conversion

Table 3. Nursing tilapia fingerlings in a 5.7-ha polyculture pond with other species grown to market weight at Hama'apil, Israel.

Species	Stocking date	Harvesting date	Culture period (days)*	Average stocking weight (g)	Density at harvest (fish/ha)	Average weight at harvest (g)	Yield (kg/ha)
Common carp *(Cyprinus carpio)*	Mar. 12	Dec. 29	245	29	4,540	1,106	4,889
Mullet *(Mugil cephalus)*	Mar. 17	Dec. 29	244	156	1,750	500	563
Silver carp *(Hypophthalmichthys molitrix)*	Mar. 7	Dec. 29	245	195	1,400	2,268	2,980
Tilapia *(Sarotherodon aureus)*	Aug. 9	Dec. 29	98	13	9,120	100	835
					Annual yield (245 days)		9,267

*Taken as within the period March 15 to November 15, when the temperature is suitable for fish growth.

Table 4. Culture of all-male tilapia hybrids *(Sarotherodon niloticus* x *S. aureus)* in a 4.2-ha polyculture pond at Gan-Shmuel, Israel, over two culture periods.

Species	Stocking date	Harvesting date	Culture period (days)*	Average stocking weight (g)	Density at harvest (fish/ha)	Average weight at harvest (g)	Yield (kg/ha)
Common carp *(Cyprinus carpio)* I	Feb. 24	Aug. 6	143	50	3,000	942	2,670
Common carp *(Cyprinus carpio)* II	Apr. 6	Aug. 6	122	15	3,444	230	730
Tilapia	Apr. 5	Aug. 6	123	148	2,020	487	680
Silver carp *(Hypophthalmichthys molitrix)*	Feb. 24	Aug. 6	143	200	1,720	1,311	1,910
		Totals for first culture period			10,184		5,990
Common carp *(Cyprinus carpio)* I	Aug. 8	Dec. 12	99	387	3,230	1,018	2,040
Common carp *(Cyprinus carpio)* II	Aug. 11	Dec. 12	96	10	3,120	235	700
Tilapia	Aug. 8	Dec. 12	99	48	2,850	288	680
Mullet *(Mugil cephalus)*	Aug. 22	Dec. 12	85	160	1,180	430	320
Silver carp *(Hypophthalmichthys molitrix)*	Aug. 22	Dec. 12	85	416	1,600	1,000	930
		Totals for second culture period			11,980		4,670
		Total annual yield (242 days)					10,660

***Taken as within the period March 15 to November 15, when the temperature is suitable for fish growth.**

ratio (FCR) was 2.31.

In Israel the stocking densities for all-male tilapias in polyculture systems are usually 3,000 to 5,000/ha, when the fish are fed supplementary feed (25% protein pellets). Table 4 presents results obtained from a well-managed polyculture pond in Israel which included sexed all-male tilapia hybrids *(S. niloticus* x *S. aureus)*. The tilapia stocked in spring were nursed the previous year and then overwintered. Those stocked in the second cycle in August were young-of-the-year, which were nursed to 48 g and then sexed. The total annual yield of tilapia was 1.36 t/ha: over 10% of the total pond yield.

The most economic stocking density is not necessarily that which results in the highest average growth rate, but rather that which results in the highest yield per unit area. Up to a certain density, an increase in stocking density does not depress growth rates proportionally and the yield per unit area increases. This effect is quite pronounced with tilapia. At densities of 3,000 to 4,000/ha, when fed protein-rich pellets, each fish can gain 3 to 5 g/day. The daily weight gain drops when the stocking density is increased, but at 20,000/ha it is still 1.5 to 2 g/day, which results in a high yield. The standing crop of such densely stocked ponds can reach 25 t/ha. This is only possible, however, when no restrictions are imposed on the length of the culture period and the fish can be given longer period to reach market weight: as in all-male culture.

Lovshin et al. (1977) cultured all-male *S. niloticus* and all-male hybrids of *S. niloticus* x *S. hornorum* at densities of 10,000/ha. The tilapia were stocked at 60 to 63 g and harvested at an average weight of about 240 g. The yields after a 180-day culture period were 2.8 and 3.2 t/ha for the *S. niloticus* and hybrids, respectively.

Sanchez (1974, cited by Lovshin and Da Silva 1975) stocked male *S. aureus* at 40,000/ha with an average weight of 113 g. After 60 days he harvested the fish at an average weight of 143 g. Though the yield was high (2.1 t /ha), the growth was very low (0.5 g/fish/day). This may have been due to the feed which contained 30% coffee pulp.

Sarig and Marek (1974) stocked male *S. aureus* at even higher densities: up to 60,000/ha. The fish were fed a pelleted diet containing 25% protein and in spite of the high density gained 2.39 g/day. The mean feed conversion ratio (FCR) was somewhat high (3.4) but the total yield during the 67-day culture period was 15.6 t/ha.

When the stocking rate is so high, the existing natural food resources of the pond are divided between a great number of fish, and the role of natural food and productivity in the overall nutrition of the fish decreases. The advantages of polyculture are then very limited and the extra work involved in sorting the different species of fish at harvest becomes a burden. Monoculture is therefore more rational at high stocking densities. This can, however, be modified by environmental factors. In Israel filamentous algae develop early in the growing season and become a nuisance when the fish are harvested. In order to control this a number of common carp are introduced into the ponds (stocked at up to 20% of the tilapia density). By burrowing in the mud, the common carp prevent the development of the filamentous algae. The monoculture thus becomes a duoculture, with the tilapia as the dominant member.

Table 5. Yields of all-male tilapia hybrids *(Sarotherodon niloticus* x *S. aureus)* cultured alone and with common carp *(Cyprinus carpio)* in two Israeli fish farms.

1. Monoculture at the Ginossar technological experiments station (U. Rappaport, pers. comm.)
Pond area: 0.02 ha

Species	Stocking date	Harvesting date	Rearing period (d)	Density at harvest (fish/ha)	Average stocking weight (g)	Average weight at harvest (g)	Yield (kg/ha)	Daily weight gain (kg/ha)
Tilapia hybrid	Aug. 20	Nov. 11	76	30,000	217	380	4,890	64.3

2. Duoculture at a farm at Nir David.

Species	Stocking date	Harvesting date	Rearing period (d)	Density at harvest (fish/ha)	Average stocking weight (g)	Average weight at harvest (g)	Yield (kg/ha)	Daily weight gain (kg/ha)
Tilapia hybrid	July 1	Sept. 1	62	18,400	250	450	3,680	
Common carp	July 1	Sept. 1	62	1,000	250	600	350	
							4.030	65.0

Lovshin et al. (1977) experimented with a duoculture of 8,960/ha all-male hybrid tilapia *(S. niloticus* x *S. hornorum)* and 1,785/ha common carp. They found that the total yield was somewhat lower than that for tilapia hybrid monoculture. However, the feed added to the duoculture pond was much lower than that added to the tilapia monoculture pond.

In ponds with very high stocking densities, oxygen usually becomes a limiting factor. Aeration is therefore mandatory, at least during the night. Other drawbacks of the system are the accumulation of metabolites in the water and organic matter in the bottom mud. The latter causes a reduction in the redox potential and the appearance of H_2S which is toxic to fish and fish food organisms. These two factors curtail the growing season to about 100 days, after which the water has to be changed.

In densely-stocked ponds, fish are fed protein-rich pellets which seem to be sufficient to sustain their growth at 1.5 to 2 g/d. In spite of the high density, the FCR usually remains relatively low. Table 5 presents the yield in two ponds under such a culture system.

Inorganic Fertilization and Manuring

The tilapia species dealt with here are all microphagous: feeding either on plankton or detritus. When reared at low densities they can obtain a major part of their nutritional requirements from natural sources. Increasing the production of this natural food by inorganic fertilization and/or manuring, when coupled with increased stocking density, usually results in a considerable increase in yield. Van der Lingen (1959a) found that the yield of *S. mossambicus* could be increased 2.5 times over that from natural production alone by means of inorganic fertilization. George (1975) conducted an experiment on the effect of fertilizers on the yield of *S. niloticus* in ponds of 0.02 and 0.15 ha. Although no replications were carried out for treatments, the effect of both inorganic fertilizers (triple superphosphate) and organic manures (cow and chicken) were clearly demonstrated. Superphosphate increased production 3.4-fold, cow manure 1.7-fold and poultry manure 3.3-fold over natural production. The highest yield was obtained using a combination of superphosphate and cow manure, which increased the yield 5-fold.

Organic manures stimulate food production in a different way to inorganic fertilizers. They add detritus which stimulates the heterotrophic food chains, producing more bacteria and zooplankton (Schroeder 1978). The microphagous tilapias are very responsive to such treatment and yields can be increased considerably. Chimits (1955) describes a method of rearing of *S. mossambicus* in Thailand where considerable amounts of manure (pig, cow and chicken) were added to ponds. The annual yield reported from these ponds was about 4 t/ha. Moav et al. (1977) experimented with the effect of liquid cow manure on fish in polycultures which included tilapia. They found that high yields of tilapia could be obtained by manuring, even at high stocking densities, thus replacing supplementary feed to a large extent. Feeding tilapia under these conditions did not have much additional effect on their growth rate. Table 6 gives the relevant results.

Table 6. Yields of tilapia *(Sarotherodon aureus)* from ponds receiving cow manure with and without supplementary feeding with grains (after Moav et al. 1977).

	Stocking density (fish/ha)	Average initial weight (g)	Average final weight (g)	Daily weight gain (g/fish)	Survival (%)	Daily yield (kg/ha)
Ponds receiving only cow manure						
Tilapia 1	3,320	97*	430	2.6	91.5	8.9
Tilapia 2	1,680	21	221	1.6		
Ponds receiving cow manure + supplementary feed						
Tilapia 1	3,320	97*	416	2.5	88.5	8.5
Tilapia 2	1,680	21	220	1.6		

*Hand-sexed, all-male

Lovshin and Da Silva (1975) report that ponds stocked with all-male hybrid tilapia at a density of 8,000/ha (average weight 25 g), when manured with 500 kg/ha/wk chicken manure gave an average yield of 1.35 t/ha after 189 days and an average weight at harvest of 186 g.

Fresh manure disintegrates in water into colloidal particles which are attacked by bacteria and readily incorporated in the food web. Integrated farming systems where animals are kept over fish ponds and their wastes fall directly into the pond usually result in high fish yields. Here again, tilapias are very responsive. Van der Lingen (1959c) cultured ducks (1 duck/8.3m^2) on a pond stocked with tilapias *(S. mossambicus, S. macrochir* and *T. rendalli)*. He obtained a yield of 3.48 t/ha, of which more than 40% were over 225 g. Culture experiments carried out in Israel also showed increased yields of *S. aureus* in a polyculture (H. Barash pers. comm.).

The integration of tilapia culture and pig fattening has also given good results. Lovshin and Da Silva (1975) constructed pigsties on the borders of 0.01 ha ponds to give manure loadings of 70 pigs/ha of pond. The pig manure was washed daily into the ponds. The ponds were stocked with 8,000/ha tilapia (all-male hybrids of 25 g) and were harvested after 189 days at an average weight of 205 g. The fish yield was 1.5 t/ha. The only supplementary feed was that supplied to the pigs.

Feeding

Tilapias larger than 4 to 5 cm take supplementary feed readily (Le Roux 1956; Bishai 1962; Huet 1970). Meschkat (1967) lists many feedstuffs used in tilapia culture such as plants, copra wastes, cotton seeds, etc. No information is given, however, on the effectiveness of these feedstuffs. It seems that some feeds are less effective than others. Experience in Israel has shown that whole sorghum grain does not affect the growth of *S. aureus* much, either because they are not eaten or have a low nutritional value. A possible explanation is that the tilapia cannot crush the hard grain of sorghum as well as can the common carp. The latter have molar-like pharyngeal teeth suitable for such a task.

Stickney and Simmons (1977) incorporated dried poultry waste into pelleted trout feed and fed this to *S. aureus*. At levels of incorporation of 20 and 30%, a considerable negative effect on the growth rate of the fish was noticed. Supplementary feeding with a proper diet can however increase yields very considerably. Huet (1970) suggests that yields of tilapias with supplementary feeding can be increased 2- to 10-fold over yields from non-fed ponds. The effect of supplementary feed is emphasized at higher stocking densities. Lovshin et al. (1977) conducted a feeding experiment at two stocking densities (5,600 and 8,960/ha) with all-male hybrid tilapia (*S. niloticus* x *S. hornorum*). They fed a mixture of 50% wheat chaff and 50% castor bean meal. The diet contained 25% protein and was fed at 3% of body weight. Taking the net yield in the control ponds as 100% (= 288 kg/ha at 5,600 fish/ha and 179 kg/ha at 8,960 fish/ha) the following increased yields were found: at 5,600/ha with organic fertilizer, 265% and with supplementary feeding, 326%; at 8,960/ha with organic fertilizer, 518% and with supplementary feeding, 938%.

The FCR can serve, to a certain extent, as an indicator of the nutritional value of a feed, although feed conversion is also affected by other factors such as the physiological state of the fish, environmental conditions, the amount of available natural food and the amount of feed consumed. Balarin and Hatton (1979) give a feed conversion table for various supplementary feedstuffs. The following FCR values for *S. niloticus* will illustrate their wide variability with different feeds: groundnut cake, 3.6; cottonseed cake, 4.8; pelleted chicken feed + 10% fresh fish equivalent, 1.8 to 6.5; brewery waste, 12.6; cottonseed crush, 18.9.

Natural food in ponds contains about 55% protein on a dry weight basis and can therefore be supplemented, to a certain extent, by carbohydrate-rich feeds such as rice bran. Tilapias seem to utilize such carbohydrates well. E.M. Cruz (pers. comm.) conducted experiments in the Philippines to determine the effect of feeding rice bran and copra meal on the production of *S. niloticus* and common carp. The addition of these feedstuffs increased the yield of the tilapia by about 50% over that of fertilized ponds with no supplementary feeding.

With increased standing crops, the quality and quantity of dietary protein become more important. Inclusion of protein in diet of *S. niloticus* reduced feed conversion considerably (de Kimpe 1971). In Israel it has been observed that feeding pellets of 25% protein, which include 10 to 15% fishmeal, has a pronounced effect on the growth of tilapia hybrids (*S. niloticus* x *S. aureus*), especially at high densities (Piperno 1970a, 1970b; Marek 1970).

Not many systematic studies have been done on rates of feeding. Since natural food constitutes an important part of the nutrition of tilapias, the amount of supplementary feed given is usually lower than that for common carp. Shell (1967) shows that the best FCR for protein-rich pellets (Auburn No. 2) by *S. mossambicus* was when fed at a rate of about 2% of its body weight per day. For *S. niloticus* the best was obtained at 4% of body weight per day. Marek (1975) developed a feeding chart for tilapias in Israel

Table 7. A feeding chart for the culture of tilapias in Israel (after Marek 1975). The daily feeding rates are expressed in g/fish and % body weight.

Fish weight (g)	Daily feeding rates: For polyculture with carp (g)	(%)	For monoculture (g)	(%)	Fish weight (g)	Daily feeding rates: For polyculture with carp (g)	(%)	For monoculture (g)	(%)
5-10	0.4	5.3	0.5	6.7	100-150	2.2	1.8	2.7	2.2
10-20	0.6	4.0	0.8	5.3	150-200	2.5	1.4	3.0	1.7
20-50	1.3	3.7	1.6	4.6	200-300	3.0	1.2	3.7	1.5
50-70	1.6	2.7	2.0	3.3	300-400	3.6	1.0	4.5	1.3
70-100	1.9	2.2	2.4	2.8	400-500	4.2	0.9	5.2	1.2
					500-600	4.8	0.9	6.0	1.1

(Table 7). He considers that tilapia in polyculture gain some natural food from association with common carp and other fish species and therefore the amounts of supplementary food he recommends are somewhat lower than for monoculture. In both cases higher rations are given when the tilapia are small and the ration decreases with increase in body weight. Differences in response of tilapia to supplementary food in polyculture and monoculture were also found by Lovshin et al. (1977). In a polyculture pond the FCR for tilapia and carp combined was lower than that for tilapia alone in a monoculture. It was concluded that less feed was required to raise a given biomass of hybrid tilapia and common carp than was needed to raise the same biomass of hybrids cultured alone.

Conclusion

From this review it can be seen that very high yields of tilapia can be obtained with relatively low inputs. This, however, requires complete control of reproduction and the choice of proper methods according to existing conditions. It is doubtful whether this can be done in small homestead ponds, but with sufficient know-how it can be achieved in small or large commercial ponds.

Discussion

HENDERSON: Can you give us some idea of the relative price structure for fish sold from these pond culture systems? What percentage of the profit of the farm comes from farming tilapias and what percentage from the other species?

HEPHER: I am afraid I cannot. Maybe Mr. Mires can help?

MIRES: On a national basis, we produce per year about 7,000 to 8,000 t of common carp, about 1,000 t of silver carp and about 3,000 t of tilapias. The prices are: common carp, about \$2/kg; tilapias also about \$2/kg and silver carp about \$1.50/kg. We also produce about 700 to 800 t of mullet which fetches about \$4/kg and miscellaneous species—grass carp, big headed carp, etc. totalling about 200 t. There is a price differential between small and large tilapia. A 200 g fish costs about $1.20 and the bigger ones about $2.

NASH: Dr. Hepher, can I ask if you tried the same stocking formula for polyculture in different sized ponds, and do you get the same results operating at say, 1 ha and ½ an acre.

HEPHER: Yes, we do. Once you have prevented wild spawning or uncontrolled reproduction, the results are the same irrespective of pond size.

LOVSHIN: I would like to suggest another method which I think is valid for raising tilapias. Dr. Hepher has been talking largely about the Israeli situation, but elsewhere, a predator to control unwanted recruitment can be very useful if a good native species is available. In Israel, they do not have a predator but there are many places in the tropics where native predators are available and can be reared easily. This is a good system.

HEPHER: I agree. The only problem is that most information on the use of such predators is from experimental work not commercial culture. I think the main reason is the great difficulty in obtaining the fry of some of these predators. Take for instance the Nile perch which when grown in combination with tilapias will prey on the small tilapia. You cannot get large quantities of Nile perch fry very easily. It will not spawn readily in ponds. I think though that you have used predators successfully.

LOVSHIN: We have used *Cichla ocellaris*, but mostly for experimental work. The mass rearing of this species is no problem. I do know also of small scale commercial use of a *Cichlasoma* sp. in El Salvador.

MIRES: What is its native name?

LOVSHIN: It is a cichlid. *Cichlasoma managuense*.

HEPHER: Although it has not been used commercially, I can give you another example of a good predator—the seabass species, *Dicentrarchus labrax*, or *D. punctatus*. Again, you face the same problem. They are wonderful fish, but you cannot get the fry.

LOVSHIN: In Taiwan, I know that the snakehead *Channa striata* is used commercially They reproduce it and they put it in the tilapia ponds, and it works well as a controlling predator. What I am saying is that there are some commercial examples; not as much as the use of monosex culture, but still enough to indicate a valid system. I think as tilapia culture expands, we are going to see more and more use made of predators.

HEPHER: I agree.

PULLIN: In the Philippines, the ICLARM and Central Luzon State University cooperative project on integrated animal-fish culture uses snakehead as a controlling predator for tilapia culture in manured ponds. We have found, however, that simple predator:prey ratios are not adequate as management guidelines for different lengths of culture period. You have to adjust to the different recruitment loads.

LOWE-MCCONNELL- Dr. Hepher, is sexing done by eye, because I know that some tilapias are much more difficult to sex than *Sarotherodon mossambicus*?

HEPHER: We do not sex *Sarotherodon mossambicus*, but we do sex *S. aureus* and hybrids. It is quite easy. At Dor, all we have is about a 3 to 4% error. It should be done when the fish are over 50 g because then the differences are more marked.

GUERRERO: In the Philippines, we are experimenting with polyculture of tilapias with invertebrates, with freshwater shrimps and prawns. We think there is commercial potential for these systems, particularly in developing countries.

COCHE: In relation to the two previous questions, the prey-predator relationship and sexing, at the pilot commercial fish farm in the Ivory Coast sexing of *S. niloticus* is done by hand on 25 g fish with about 10% error. Then to control this error, a small predator *(Hemichromis fasciatus)* is used in the ponds without any calculation of prey:predator ratio. One kilogram of predators is used per 1,000 m^2 pond. They remain very small and can be reused several times. When the pond is drained, the predators are taken out and transferred to another pond. When some tilapia recruitment is needed, the predators can be removed and tilapias can be bred in the ponds. This combination of predators and early hand sexing seems to work well.

HEPHER: May I comment on the response of tilapias to pond fertilization? Fertilization may be organic or inorganic. Tilapias are much more responsive than other fish to organic fertilization. This may be because of their feeding habits, for example, consumption of detritus. A wide variety of different manures are readily divided into small colloidal particles, for instance, liquid cattle or chicken manure, and even the sludge remaining after processing manure for bio-gas. Tilapias are very responsive to all of these. We manure ponds every day, in some cases in very big quantities, up to 150 to 180 kg dry matter/ha/day. With respect to feeding, we have found that the feeding rate of tilapias is almost half that of carps. Using feeding charts, we can feed half the amount of supplemental feed for the same standing crop of tilapias as for carps.

MORIARTY: You are using *S. aureus* and a *niloticus* x *aureus* hybrid. Presumably, *S. aureus* feeds on phytoplankton in nature. Are you encouraging blooms of algae in the ponds?

HEPHER: No, more detrital feeding.

MORIARTY: Are they really feeding on detritus? Has anyone actually looked at the contents of the stomach?

HEPHER: Yes, Spataru has analyzed the stomach contents of *S. aureus* and *S. galilaeus* from Lake Kinneret. The majority of the contents is detritus. This is probably why they are so responsive to manures.

MORIARTY: The manures will also encourage the algal blooms as well.

HEPHER: Yes.

MIRES: Our carp ponds always suffered very much from blue-green algae blooms in the past. Then, with the production of tilapias and with a combination of tilapia and silver carp, this problem has been totally eliminated. Although what is found today in the stomach of tilapias may not suggest that they consume much blue-green algae, we have had the impression in the past that they can clear these algae from the pond.

CHERVINSKI: We experimented at Texas A&M University growing *Macrobrachium rosenbergii* by itself or in combination with *S. aureus.* They don't suffer from blooms of blue-green algae in the tilapia pond, but they do suffer from these when cultured alone. We also tried culturing *S. galilaeus* in dirt ponds, but they did not perform well because they are almost exclusively plankton feeders whereas *S. aureus, S. niloticus* and hybrids will bottom-feed as well.

BOWEN: There appear to be two feeding categories amongst the tilapias and sarotherodons. Those adapted to feeding on coarse material, like macrophytes, and the microphagous group. The comments made here emphasize the fact that the microphagous group is highly flexible in its feeding mode. They can feed from suspension or they can feed from substrates, depending entirely on food availability. Certainly, there are some species which tend to feed either always from suspension or always from the bottom in the natural environment, but even in the natural environment, some species are flexible.

There is a good example from Lake George where one species *(S. leucostictus)* is a filter feeder in the open water, but feeds on the bottom near shore. From my own experiments, I know that *S. mossambicus*, which almost always feeds from a substrate in the natural environment, when confined in an aquarium and excluded from any substrate will filter feed on plankton and suspended detritus. So, they are highly flexible and I do not think an absolute distinction between filter feeders versus substrate feeders is valid, especially when we are talking about aquarium or aquaculture situations.

Cage Culture of Tilapias

A. G. COCHE

Senior Fishery Resources Officer
(Aquaculture)
Inland Water Resources and Aquaculture Service
Fishery Resources and Environment Division
Food and Agriculture Organization
Rome 00100, Italy

COCHE, A.G. 1982. Cage culture of tilapias, p. 205-246. *In* R.S.V. Pullin and R.H. Lowe-McConnell (eds.) The biology and culture of tilapias. ICLARM Conference Proceedings 7, 432 p. International Center for Living Aquatic Resources Management, Manila, Philippines.

This study presents a world-review of tilapia cage culture which is today practiced in an increasing number of countries, mostly in tropical freshwaters. Most culture systems are on an experimental scale, and use *Sarotherodon niloticus, S. mossambicus* or *S. aureus.* Among the technological aspects of tilapia cage culture, construction and design of cages, site selection criteria, and management of tilapia stocks are discussed briefly before presenting the various technologies used for the production of either juveniles or food-fish in cages. On the basis of feeding practices, adult tilapia may be raised in extensive, semi-intensive or intensive cultural systems. Extensive systems based on natural feeding are described from eutrophic lakes and fertilized water bodies. Semi-intensive systems (where low-cost and low-protein diets are fed) have been successfully developed on an experimental scale either combining phytophagous tilapia and a vegetable diet or utilizing agricultural byproducts in the presence of algal blooms. As feed quality improves, the cultural system gradually intensifies and feeding aspects become more important from the economic viewpoint. These aspects are therefore discussed thoroughly before reviewing the available data pertinent to intensive tilapia cage culture. Finally, the advantages, constraints, research needs and prospects of this particular technology are discussed. It is concluded that some tilapias present good production potentials, particularly in cultural systems with low energy inputs.

Introduction

Fish cage culture has been defined as the rearing of fish stocks, generally from juvenile to market size, in a totally enclosed water volume through which a free water circulation is maintained.

General reviews of fish cage culture as practiced in inland waters have been published earlier (Coche 1978, 1979). The objective of the present review is to present a synthesis of the information available on the application of this culture technique to the production of tilapias.

Tilapia cage culture has a relatively short history. The first scientific experimentation started around 1970 at Auburn University, Alabama with the rearing of *Sarotherodon aureus* in cages placed in fish ponds (Armbrester

Table 1. Geographical distribution of cage culture of tilapias (*Sarotherodon* and *Tilapia*): in fresh waters, either experimental (EF) or commercial (CF) and in brackish waters similarly (EB or CB).

Country	*S. mossambicus*	*S. niloticus*	*S. aureus*	*S. esculentus*	*S. galilaeus*	*S. melanotheron (T. heudeloti)*	*S. niloticus* x *S. mossambicus**	*T. rendalli*	*T. guineensis*	*T. zillii*	Reference
Europe											
Belgium		EF							EF		Philippart et al. 1979
North America											
Alabama		(EF)	EF								Armbrester 1972; Pagán 1973, 1975; Suwanasart 1972; Anon. 1979c
Africa											
General		EF									Coche 1979; ADCP 1980
Central Africa		EF									N'Zimasse 1979
Ivory Coast		EF									Coche 1977; Campbell 1978a, 1978b
————		EB									De Kimpe 1978
————		CB				CB					Campbell (pers. comm.)
Nigeria		EF			EF					EF	Ita 1976
Tanzania				EF						EF	Ibrahim et al. 1975
Asia											
China	(EF)						EF				dela Cruz 1979; Song (pers. comm.)
Indonesia	EF	EF					(EF)				Rifai 1979, 1980; Pedini (pers. comm.)
Japan	EF										
Philippines	EF	EF	EF				EF			EF	FAO 1977a; Guerrero 1975, 1979a, 1980a, 1980b; Pantastico and Baldia 1979
————	CF										Garcia 1979; Guerrero 1979a; Radan 1979
Sri Lanka	EF										Sollows (pers. comm.)
Thailand	EF										Sollows (pers. comm.)
Latin America											
General	EF	EF	EF					EF			ADCP 1978
Brazil								EF			ADCP 1978
Colombia	EF							EF			Corredor 1978; McLarney 1978
Costa Rica		EB	EB								Nanne 1979
Cuba	EF	EF									Rodriquez 1976
El Salvador			EF								Godinez and Jose 1976a
————			EF								Godinez and Jose 1976b; Hughes 1977; Street 1978
Guatemala	EF										Bardach et al. 1972
Puerto Rico			EF								Jordan and Pagán 1973
————			EB								Miller and Pagán 1973; Miller and Balantine 1974

*hybrid

1972; Suwanasart 1972). Since then, the technique has spread progressively to several other regions of the world (Table 1).

Because of some of its inherent advantages, such as the possibility of using existing tropical water bodies to produce a fast growing and well accepted food fish, tilapia cage culture is raising more and more interest, particularly in tropical, developing countries. It is hoped that this review will contribute not only to the justification of such interest but also to an improved technology and its wise application.

The Present Status of Tilapia Cage Culture

Although a relatively new development, tilapia cage culture is now found in several tropical countries of Africa, Asia and Latin America (Table 1). In North America it remains confined to the state of Alabama, where it has been studied for about 10 years. In Europe, tilapia cage culture is practiced only in Belgium in the thermal effluent of a nuclear power station (Philippart et al. 1979).

Most tilapia cage culture is done on an experimental scale and in fresh waters. The few exceptions are as follows: experimental/brackishwater in the Ivory Coast, Costa Rica and Puerto Rico; commercial/freshwater in the Philippines and El Salvador and commercial/brackishwater in the Ivory Coast.

The tilapias most commonly used are *S. mossambicus* (in Asia and Latin America), *S. niloticus* (especially in Africa) and *S. aureus* (especially in North and Latin America). Six other species are also listed (Table 1), but of these only *Tilapia rendalli*, as a phytophagous fish in Latin America and *S. melanotheron* (formerly *T. heudelotii*) as a brackishwater species in the Ivory Coast are likely to become important. The hybrid *S. niloticus* x *S. mossambicus* is being tried in a few Asian countries where it may have better qualities than *S. mossambicus* for cage culture.

There is a definite potential for tilapia cage culture in the rivers and lakes of many Latin American countries (ADCP 1978) and Africa (Coche 1979; ADCP 1980), both on a commercial scale and as subsistence fish culture for local people. Plans for expansion of cage culture are contemplated in several Asian countries including China (Song pers. comm.) and the Philippines (Guerrero 1979a and Castro pers. comm.). It looks, therefore, as if tilapia cage culture has a bright future, particularly in tropical, developing countries. In colder climates, the utilization of thermal effluents may also lead to the development of tilapia cage culture (Philippart et al. 1979).

Technological Aspects of Tilapia Cage Culture

The general technological aspects of cage culture have been discussed elsewhere (Coche 1978, 1979). The purpose here therefore is to stress those aspects which are particularly relevant to tilapia cage culture, although some repetition appears unavoidable.

1. Type and Size of Cages

Surface standing cages, resting on the bottom, are used in shallow water bodies such as ponds and streams. Floating cages are preferable, however, wherever the water depth permits such as in lakes and rivers. In all such cases, the floor of the cage should be kept at least 0.5 to 1.0 m above the bottom sediment, where wastes may accumulate and dissolved oxygen (DO) is lowered. A water depth of 5 to 10 m is recommended to reduce parasitism and disease outbreaks.

The size of cages varies for different operations. Breeding cages and fingerling production cages are smaller than growout cages. Experimental cages do not generally exceed a few cubic metres until the pilot-scale stage is reached. At the subsistence level relatively small cages are also preferred. For commercial exploitation, medium-sized cages (6 to 20 m^3) should first be used at the artisanal level while larger cages (50 to 100 m^3) may be envisaged for the industrial level. Very large cages (1,000 m^3 or more) have also been used (Table 2).

Table 2. Construction costs for tilapia cages in the Philippines and Ivory Coast, 1976 to 1978.

			US$
(a)	Philippines, 1978 (Guerrero 1979a):		
	1)	Experimental cage, capacity 1 m^3, wooden frame with polyethylene netting of 25 mm mesh using styrofoam floats	10
	2)	Commercial cage, capacity 6,250 m^3 (50 x 25 x 5 m), bamboo and wood with nylon netting bag of 12.7 mm mesh	2,000
(b)	Ivory Coast, 1976 (De Kimpe 1978):		
	1)	Experimental cage, capacity 20 m^3, wooden surface structure with nylon netting bag of 14 mm mesh and 200 l metal drums as floats	185
(c)	Ivory Coast, 1978 (Campbell 1978a):		
	1)	Experimental cage, capacity 1 m^3, wooden frame with plastic netting of 8 mm mesh and styrofoam floats	55
	2)	Experimental cage, capacity 6 m^3, floating wooden frame with plastic netting bag of 25 mm mesh and 20 l plastic barrels as floats	100
	3)	Experimental cage, capacity 20 m^3, floating wooden frame with nylon netting bag of 14 mm mesh in 210/18 twine mounted 33 per hundred meshes and 60 l plastic barrels as floats	170

The size chosen for cages should reflect the level of technology available. In principle very large cages can result in the loss of several inherent advantages of cage culture, mainly flexibility and maneuverability. With tilapias however, a relatively large cage environment results in better growth rates, in reduced feed losses and in improved survival at very low DO's (Campbell 1978a). It seems that there is also a minimum cage size for guaranteeing a good feed conversion ratio (FCR). Cages have to be sufficiently large to reduce feed losses through the walls from the turbulence created as the tilapias feed voraciously.

The water depth in the cages has been shown to influence growth and reproduction (Maruyama and Ishida 1976). When comparing the growth of *S. mossambicus* in water depths ranging from 0.5 to 1.5 m in square cages of 6 m side, the best growth and the highest fry production were observed in the deepest cage. A depth of at least 0.75 m was recommended.

For *S. niloticus* in the Ivory Coast, Campbell (1978a) recommends the simultaneous use of various sizes of floating cage. For fingerling production, 0.5 and 1.0 m^3 cages should be successively used as the fish grow. For market fish production at the artisanal level 20 m^3 cages (3 x 3 x 2.5 m) are the most suitable. He believes that the maximum cage sizes above which handling becomes a problem without special equipment are 22.5 m^3 (3 x 3 x 2.5 m) for plastic netting and 30 m^3 (3.5 x 3.5 x 2.5 m) for synthetic-fibre netting.

2. Cage Construction

Standing cages have a supporting frame extending 0.2 to 0.3 m below the cage floor to keep it away from the bottom sediment. The mesh walls are attached to the upper part of this frame. Floating cages are made of two components. The surface structure consists of a floating rigid frame and the subsurface structure of either a rigid frame with mesh walls or a flexible mesh bag designed to retain a rectangular shape. In the presence of strong water currents (above 20 to 30 cm/sec), a rigid construction is preferred over a flexible bag with heavy corner anchors. The choice of materials is important (see below), but the mesh size of the walls remains the most important factor. This should be as large as possible, according to the size of the fish being raised, to allow a free circulation of water through the cage at all times.

Accessory items are used according to each particular situation. For *S. niloticus* culture, Campbell (1978a) observed no advantage in using cages with either a solid bottom or an opaque top cover. A feeding ring is also unnecessary under normal circumstances. Against bird predation, a light covering net should be used. If the subsurface structure includes fibre netting (rather than more resistant plastic or metal netting), additional protection against aquatic predators may be required either as a second stronger fibre net with larger meshes added to each cage or a larger anti-predator net around the culture site.

There are definite advantages in grouping several cages (e.g., four to six) together with a stable working platform, to form a raft. When poaching is a major problem, a watchman can be housed on this, as is commonly done in Asia. As the structure increases in size, however, more attention should be paid to the design of an effective anchoring system, adapted to the local water conditions.

3. Materials and the Working Life of Cages

The choice of materials for the construction of cages varies greatly from country to country. Local materials such as wood and bamboo may be used, but generally their working life is short when continuously submerged. Boring insects such as *Povilla adusta* in the Ivory Coast (Coche 1979) readily attack light wooden frames. Mahogany frames are more resistant but rather heavy. Some bamboos are more resistant to attack than others.

For the subsurface structure, there is a tendency to eliminate supports and to use a net bag either of synthetic fibre or plastic.

Fibre netting with a nylon twine size R 470 tex (e.g., 210/18) has been successfully used in the Ivory Coast, mounted on nylon ropes at 33 per hundred meshes and spread at the bottom by a steel frame (Campbell 1978a). Knotless netting with square rather than diamond-shape meshes is preferred. In brackishwater, considerable damage may be experienced due to crabs. Chua and Teng (1977) recommend in such a case an R 1150 to 1300 tex polyethylene netting (21 to 24-ply threads). Compared to nylon netting such material is not only much cheaper but it also appears to be able to better withstand the tropical sun for a considerable period of time. Treating the netting with tar may also increase its resistance and reduce fouling (Coche 1979).

Plastic netting combines the advantages of being light and durable with some extra rigidity. It should, however, contain ultra-violet stabilizers for longer-lasting performance. Its only drawback might be its price, especially in countries where it has to be imported. Even then, careful consideration should be given to plastic netting because of its inherent advantages, particularly durability.

The working life of cages and their depreciation period vary greatly with the materials used for their construction and the local conditions, e.g., climate, limnology, handling, maintenance, etc. Under careful management the components of floating cages used in Lake Kossou (Ivory Coast) had the following estimated working life (Campbell 1978a): surface floating frame, (ordinary wood, 6 x 6 cm) 5 years; floats, (empty plastic barrels, 20 to 60 l) 3 years; subsurface wooden frames (mahogany, 5 x 5 cm) 3 years; subsurface wooden frames (ordinary wood) 1 year; nylon fibre netting (210/18 twine, 14 mm mesh) 3 to 5 years and plastic netting (8 and 25 mm mesh, Netlon) 5 to 10 years.

4. Cage Prices

The prices of some cages being used for tilapia culture illustrate the magnitude of the initial investment to be made (Table 2). The average cost/m^3 varies from US$55 to US$8.50 according to the size of the cage and the material used. The larger the size, the cheaper the unit volume cost but also the lower the recommended (see below) fish density. The average fish production/m^3 therefore decreases as the cage volume increases.

Campbell (1978a) therefore recommends the following cage dimensions for artisanal tilapia culture in the Ivory Coast: for 15 to 30 g fingerlings successively use (a) small cylindrical 0.5 m^3 cages made entirely of 4 mm mesh plastic netting hanging from a small, rigid, floating frame; (b) 1 m^3 cubic cages of the same design with 8 mm mesh plastic netting; for 200 to 300 g growout, use medium size cages of about 20 m^3 with either nylon fibre netting (20 mm mesh; R470 tex twine) or plastic netting (18 to 25 mm mesh) as in Table 2.

Cage Culture: General Considerations

1. Site Selection

Good water circulation and adequate protection against floating debris and wave action are normally the two essential requirements for a culture site. Other factors, e.g., water quality, site accessibility, security and distance to markets, are also important. A water depth sufficient to place the cages at least 2 to 5 m above the bottom sediments is preferred.

In lake environments, wind-induced surface currents and fish movements should provide a continuous exchange of the water in the cages, keeping the DO high and removing wastes. Knowledge of seasonal limnological cycles will help to identify any critical periods during which cage culture might have to be discontinued. Such periods correspond to the seasonal turnover of the water mass, when the thermal stratification breaks down. Deep deoxygenated water is suddenly brought to the surface where it may cause heavy fish mortalities in the cages. A period follows during which very low DO's may persist for several days.

Tilapias are relatively tolerant of low DO. Caged *S. niloticus* in the Ivory Coast have survived concentrations as low as 0.7 mg/l or 9% saturation for several days (Coche 1977). In 1976, however, 64% of the adult fish and all the fingerlings suddenly died within 3 days, when the DO dropped below 0.5 mg/l (Traore and Campbell 1976). From recent observations during similar periods, it would appear that a DO of 3 mg/l should be considered the limit in cage culture below which adverse effects begin to appear: feed digestion stops, the growth rate decreases sharply and stress intensifies.

In the presence of such critical periods, it might be advisable to suspend culture if no adequate artificial aeration can be supplied to the caged fish. If such a suspension is not feasible, the following practices are suggested to minimize the mortality risks: space the cages further apart, several meters away from each other; use larger cages providing a relatively larger air/water interface; use lower fish stocking rates, never exceeding 15 to 20 kg/m^3 and low densities not exceeding 200 to 250 fish/m^3; stop feeding the fish (Coche 1977, Traore and Campbell 1976).

2. Management of Caged Tilapia Stocks

The management of caged fish stocks encompasses the stocking of juvenile fish, feeding during growout and cage maintenance. During a cage production cycle, several interacting factors have to be optimized to give maximum production efficiency.

The supply of seed (fry and/or fingerlings), must be adequate both in quantity and in quality and available as and when required. There should be as little as possible variation in size. This remains the major limitation to the further development of tilapia cage culture. Existing breeding practices are inadequate and a new technology should be developed for selected tilapias (Coche 1977, 1979).

Adequate feeding is essential for growth and survival. In some cases, natural food (plankton, aufwuchs or benthos) may be sufficient and no supplementary feeding is required. When natural food is insufficient but still an important part of the diet, supplementary feeding is practiced with relatively low-cost ingredients. Under intensive rearing conditions at high stocking densities, the natural food available for the caged fish becomes insignificant and a complete artificial food is required.

The choice of the daily feed ration which will optimize the utilization efficiency particularly for costly food is of crucial importance: much more so in cages than in earth pond culture. Very little information exists on the relationship between feeding rate and FCR for a particular type of feed as a function of interacting factors such as species, fish size, fish density and water quality.

Cage maintenance should be regular and geared towards increasing the working life of equipment, maintaining the water quality by fouling control and reducing fish losses from escapes, mortalities and predation. Both the environment and the fish stock should regularly be monitored.

3. Growth and Production

The biomass of the caged fish per unit volume (B kg/m^3) is related to: 1. the individual growth rate which tends to decrease as the biomass increases; 2. the average monthly production (MP kg/m^3) which increases as the biomass increases until the carrying capacity of the cage is reached and decreases thereafter and 3. the FCR which decreases as the biomass increases until an optimum biomass is reached, above which FCR increases. The average individual weight of the caged fish (Pm) is related to:

1. the individual growth rate which decreases as Pm increases and
2. the FCR which, for a constant daily feed ration, increases as Pm increases.

The maximum carrying capacity of a cage (MCC in kg/m^3) is mainly determined by the DO throughout the cage. This varies with the mesh size and cage size. Campbell (1978a) has defined the MCC's (in kg/m^3) for *S. niloticus* reared in well-oxygenated water with good circulation (at least 2 cm/sec) as follows: with plastic netting of 25 mm mesh, 90 for 1 m^3, 70 for 6 m^3 and 40 for 20 m^3; with nylon fibre netting of 14 mm mesh, 40 for 20 m^3. Therefore, as cage size increases, MCC decreases. In practice, it is always safer to stock below the MCC, as the risks of diseases and mortalities greatly increase as the MCC is approached. For 1 m^3 cages, the recommended safe limit is about 73 kg (FAO 1976).

Cage production increases as the initial biomass (B_i kg/m^3) at stocking increases until it reaches an optimum value. For this B_i optimum, the final biomass (B_f) will equal the MCC at the end of the production cycle. In the Ivory Coast, in 1 m^3 cages, the MCC = 90 kg/m^3 was reached in 4 months from B_i of 20 kg/m^3 *S. niloticus* (at 250 to 350 fish/m^3). The corresponding maximum production was about 70 kg/m^3 or a monthly average of 17 kg/m^3 (Coche 1977).

It is well known that male tilapias grow faster than females. Any shift in the sex ratio of the cultivated population towards a male predominance will therefore accelerate production. In cage culture, increases in growth rates and production as well as decreases in FCR's have been recorded for monosex male populations (Coche 1977). Campbell (1978a) has also observed good production with 84% male *S. niloticus* following size grading of the juveniles. In both monosex male and mixed sex *S. niloticus* cage culture at Auburn University, Alabama, male growth rates were about 2.4 times than those of females (Anon. 1979c).

Seed Production

1. Fry and Fingerling Production in Cages

Most tilapia cage culture is concerned with growout to market size and takes advantage of the fact that reproduction is usually suppressed by the cage environment. Tilapias may, however, spawn in cages under certain conditions, depending mainly on the mesh size and on the fish density. For example, in Lake Atitlan (Guatemala), *S. mossambicus* have produced larvae in floating cages (Bardach et al. 1972) and Guerrero (1975) has observed in the Philippines female *S. mossambicus* mouthbrooding in cages with 200 fish/m^3. I have also found mouthbrooding *S. niloticus* in floating cages with 25 mm mesh but these were only a few individuals within a large population and the young fry disappeared rapidly from the cages through the netting. Suwanasart (1972) observed that *S. aureus* spawned successfully at densities of 500 fish/m^3 in cages with a small mesh screen placed on the bottom. In Indonesia, Rifai (1979, 1980) has also bred *S. niloticus* in cages with 3 mm mesh but "the occurrence of reproduction was relatively low"—only 5 out of 27 cages. Therefore to use cages with the definite purpose of producing tilapia fry is unpromising and a better technique is required.

In the Philippines, Guerrero (unpublished data) uses fine mesh nylon or mosquito net cages, termed hapas, to breed *S. niloticus* and *S. niloticus* x *S. mossambicus* hybrids. The broodstock live in these cages, breeding continuously, and the fry produced are collected once a month. These fry are then grown on to fingerling size, either in another cage or in a nursery pond.

The hapas (1.5 x 1 x 1 m) are suspended just above the pond bottom from poles in water about 1 m deep, from a good quality supply. The water depth inside the cages is about 0.6 m giving a rearing volume of about 0.9 m^3. Table 3 summarizes some results of fry production in hapas for 5 weeks. For both crosses, the most advantageous sex ratio was three females to one male and the breeding success was much higher for the intraspecific *S. niloticus* cross. The average monthly production of *S. niloticus* fry was 1,320/cage: equivalent to 880/m^2 or 1,466/m^3.

Table 3. Some results of tilapia fry production in 0.9 m^3 mosquito net cages (hapas) at total stocking rates of 8 to 12 fish of individual weight 90 to 135 g per cage. This stocking rate approximates to 5.3 to 8.0 fish/m^2 or 8.9 to 13.3 fish/m^3 (Guerrero 1979a).

Spawning Cross Male x Female	Sex ratio (Male:Female)	No. of fry produced per cage	No. of fry produced per female	% of spent females
S. niloticus x *S. niloticus*	1:5	1,660	309	53
	1:3	1,647	407	67
S. niloticus x *S. mossambicus*	1:5	509	280	18
	1:3	527	458	19

A similar system has been proposed by the Aquaculture Department of the Southeast Asian Fisheries Development Center (SEAFDEC) for small-scale operators of *S. niloticus* cages in the eutrophic lake Laguna de Bay, Philippines (Radan 1979). Six to seven broodstock/m^3 of water are placed in hapas at the 1:3 sex ratio. Every 3 to 4 weeks, an average of 250 fry/spawner are collected. These are sorted by size and reared in further hapas at densities of either 500/m^3 (no supplementary feeding) or 1,000/m^3 (feeding with algal cake), until ready for transfer to growout cages after 1 to 2 months.

2. Fry Production in Earthen Ponds Followed by Fingerling Production in Cages

At Lake Kossou (Ivory Coast), *S. niloticus* fry were regularly produced in relatively large numbers using two earth ponds, 30 x 20 x 0.4 m deep (Campbell 1978b). Large female fish and smaller males (mean weights 700 and 200 g) were stocked at an average density of 0.5/m^2 with a 1:4 to 5 sex ratio in one pond. Intense supplementary feeding on a diet exceeding 30% protein was provided for one month. The broodstock were then removed with a cast net and transferred to the second pond where the treatment was repeated. Intense supplementary feeding was continued in the first pond for one more month at the end of which about 5,000 3 to 4 cm fry were harvested. This pond was then immediately restocked with the broodstock removed from the second pond. Such a culture system can therefore produce each month an equivalent of 4.2 fry/m^2 pond or 10.4 fry/female, much lower production figures than those reported above from hapas.

The fry produced in the ponds in two months were then transferred to 1 m^3 floating cages with 8 mm mesh at a stocking density of 1,000 fry or more per cage. For two months, a complete feed (25% protein) was given at the daily rate of 10 to 8% of the biomass. A first selection of 20 to 30 g fingerlings was made at the end of the first month and a second one month later. Such selections based on growth rate resulted in populations which were on average 84% males. The slower growing fry, mostly females, were discarded after two months. Up to 50 kg of fingerlings was produced per 1 m^3 cage.

On the basis of this experience Campbell (1978a) has suggested some improvements. The use of smaller mesh (4 mm) cages would allow stocking with 10 to 15 day-old fry, about 1.5 cm long. This would make it possible to remove fry earlier from the ponds and increase their efficiency. With such small mesh, however, the water exchange inside the cages would be greatly reduced. Therefore, smaller cages should be used (0.5 m^3) and the stocking rate should be reduced accordingly. As soon as the fry reach 3 to 4 cm, they should be transferred to 1 m^3 cages with 8 mm mesh as above.

Some experiments towards this were performed in the Philippines (Guerrero 1980a) but at very low fish densities (100 fish/m^3). Fry of mean weight 2.6 g were stocked at 0.26 kg/m^3. They were fed daily for 56 days at 5% of the biomass with a mixture of fish meal and rice bran in the form of a mash. The mean weight at harvest was 15 to 20 g. The best feeding efficiency (FCR=1.7) was obtained with 25% fish meal. Similar results could probably have been obtained with higher densities.

3. Mass Production Technique

Mass production of fry and fingerlings of *S. niloticus* and *S. melanotheron (T. heudelotii)* is now being attempted in slightly brackish (4 to 8‰) water in Jacqueville (Ivory Coast) on the western Ebrie Lagoon (Campbell pers. comm.). The main characteristics of the culture system, which is still in the developing stage, are summarized here.

Four 3 x 18.3 m tanks are used for spawning. These have a water depth of 30 to 40 cm. The water supply (26 to 32°C) is sufficient to give a total exchange every 6 hours. Aeration equipment and automatic feeders are used. Examples of stocking rates for broodstock are as follows: *S. niloticus*, 4.4 to 6.5 200-400 g fish/m^2 with sex ratio 1:5; *S. melanotheron (T. heudelotii)*, 14.5 150-200 g fish/m^2 with sex ratio 1:1.

Every day at noon, about 1,250 to 2,500 fry (still in cloud formation) are harvested from the lower part of each tank and transferred to a plastic tank. The four tanks produce on average 5,000 to 10,000 fry/day.

The fry are grown on in 4 x 4 m plastic tanks with a lateral water inlet and central drain. The water depth is gradually increased as the fry grow over a 2-month period. Intensive hand feeding is first employed followed by automatic feeding. The water exchange rate is high. The fry harvest from one week (35,000 to 70,000) is concentrated at first into one tank. As they grow, they are regularly graded by size and transferred to other tanks. The stocking density therefore gradually decreases. Regular prophylactic treatment is given against parasites. The normal survival rate is 80 to 90%. Up to 30,000 fry, mean weight 2 to 3 g, can be grown in each tank. These are then transferred to fingerling production cages.

The stocking rate for fingerlings varies according to the availability of fry. The mesh size is 8 mm. With intensive hand feeding (5 times daily), 25 g fingerlings are produced in one month.

Extensive Culture

1. Commercial Cage Culture, including Eutrophic Lake Sites

In extensive cage culture there is no supplemental feeding. In most cases, the tilapias feed on the plankton either as natural blooms in eutrophic waters or in fertilized fish ponds.

The species used are microphagous *S. aureus*, *S. niloticus* and *S. mossambicus* whose natural diet normally includes algae.

The only large-scale, extensive, commercial tilapia cage culture in eutrophic waters is that of *S. mossambicus* in the Philippines in eutrophic natural lakes and in reservoirs. Guerrero (1980b) states, however, that *S. niloticus* is now the preferred species in Laguna de Bay where it may be raised at the density of 20 to 25 fish/m^2 without supplementary feeding, from 3 to 4 cm juveniles to 100 g adults in 4 to 5 months.

The eutrophication of these water bodies derives from the richness of local volcanic soils, e.g., Lake Bunot and Lake Sampaloc, or the large inflow of nutrients (N,P), e.g., Laguna de Bay. A rapid expansion of the tilapia cage culture industry is expected (Guerrero 1979a). In Lake Bunot more than 70 commercial-size cages were added from 1975 to 1978. Tilapia cage culture is now practiced also in Laguna de Bay, Lake Calibato, Lake Gunao, Lake Paoay, the Pantabangan Reservoir, Lake Sampaloc and Lake Sebu.

Some examples of the technology used and its results are grouped in Table 4. Very large floating cages (fibre netting) are used: more than 6,000 m^3. Small fingerlings (5 to 10 g) are generally preferred for stocking because of their greater availability and lower price. The stocking rate is relatively low: usually less than 0.5 kg/m^3, less than 70 fish/m^3. The average production rarely exceeds 1 kg/m^3/month, because of the large size of the cages, but up to 10 to 15 t of *S. mossambicus* can be harvested every 6 months, providing the farmer with a net income of more than US$3,000.

The success of the first cage farmers accelerated the development of the industry so much that it got out of control within a few years in most of these eutrophic lakes. In Lake Sampaloc for example, the water surface has become so congested with floating cages that the average original production of 3.8 kg/m^3/yr has drastically dropped today to 0.8 kg/m^3/yr and the growth rate of the tilapias has decreased by a factor of 9. It takes now as long as 12 months to produce 50 to 60 g fishes compared with earlier culture of fish to 150 to 200 g in 4 months. Some control has been introduced and a license is now required for cage culture.

Table 4. Extensive commercial cage culture of *S. mossambicus* in Philippine lakes.

Location environmental conditions; source	Floating cage data	Stocking data			Harvesting data			Culture period (mo)	Remarks
		Mean weight (g)	No. of fish/m^3	Biomass (kg/m^3)	Mean weight (g)	Biomass (kg/m^3)	Production (kg/m^3)		
Laguna de Bay 900 km^2; 2-8 m in depth; salinity up to 3-5 ppt hypereutrophic in the summer Sollows (pers. comm.)	20 x 10 x 2 m deep; 8 mm mesh	5 to 7	25 to 75	0.125 to 0.525	100 to 150	3.7 to 7.5	3.5 to 7.0	4 to 5	2 crops/year
Lake Sampaloc volcanic crater lake Sollows (pers. comm.)	500 m^2 x 4 m deep	25 to 30	25	0.700	150 to 200	4.5	3.8	4	— before congestion due to uncontrolled increase of cage numbers
					50 to 60	1.5	0.8	12	— under congested conditions
Lake Bunot volcanic soil region Garcia 1979 and Guerrero 1979	50 x 25 x 5 m deep	5 to 10	16	0.080 to 0.160	100 to 150	1.6 to 2.4	1.5 to 2.2	6	— occasional supplemental feeding with rice bran and water hyacinth — yield 10-15 tons/cage giving net income eq. US$3,378

2. Experimental Cage Culture

Some experimental results are also available from Laguna de Bay (Pantastico and Baldia 1979) for *S. mossambicus.* Three floating cages of about 9 m^3 capacity each (5.4 mm mesh) were stocked with fingerlings of mean weight 10 g at 50 fish/m^3 and grown for 3 months. The results were as follows: average growth rate of individual fish, 0.36 g/d; average monthly specific growth as a percentage of the original weight at the start of the month, 110%; average monthly production, 0.825 kg/m^3; daily rate of increase of the biomass,* 1.31%. Growth and production were relatively low when compared to those obtained in fertilized ponds with *S. aureus* (Table 5).

3. Cage Culture in Effluents, Fertilized Ponds and Canals

In Tihange (Belgium), a series of experiments were conducted in a pond fed by the heated effluent of a nuclear power plant (25 to 37°C) to quantify the monosex production of male *S. niloticus* in cages (Philippart et al. 1979). Populations of 200 males/m^3 were reared in floating 0.5 m^3 cages for 15 days. Three size classes of fish were used, giving a range of initial biomass (B_i) of 5.1 to 10.4 kg/m^3. The main results are given in Table 6. The quantities of plankton available were not recorded but the water exchange rate was high and it may be assumed that there were no blooms. The food supply was therefore probably inadequate and the results were rather poor, considering that only males were used: see Table 5 for comparison. These data can, however, be taken as a basis for obtaining true FCR's of trials using artificial diets by subtracting the production due to natural feeding.

In Alabama, U.S.A., *S. aureus* were cultured in experimental floating cages in ponds in which other fish were also cultivated. These ponds were fertilized regularly to develop either moderate or dense plankton blooms: average Secchi disc visibility depths were 64 and 37 cm. Table 5 summarizes the data. Dense plankton blooms produce much higher fish biomasses, especially with the small fish. For the larger fish, it appears that the MCC of the cages might have been exceeded (where B_i = 12.5 kg/m^3) which would explain the reduced MSG and DRIB values (compared to where B_i = 5 kg/m^3) but the Pmi difference also has an effect. The smaller the fish, the greater the benefit they are likely to derive from an algal diet. All these results demonstrate the real potential of *S. aureus* cage culture in the presence of plankton blooms.

In West Java (Indonesia), common carp *(Cyprinus carpio)* are cultured in fixed cages placed in streams and canals heavily enriched by domestic and agro-industrial effluents on a commercial scale. This technology is

---

*The daily rate of increase of the biomass is given the acronym DRIB. The compound interest formula describing this is $B_f = B_i\,(1 + i)^n$, where B_f is the biomass at harvest; B_i is the initial biomass at stocking; n is the culture period in days and $i = \frac{\text{DRIB}}{100}$.

Table 5. Cage culture of *S. aureus* in fertilized ponds in Alabama, U.S.A. without supplementary feeding.

Density of plankton	Pm_i (g)	B_i (kg/m^3)	N/m^3	B_f (kg/m^3)	Culture period d	G (g/d)	MSG (%)	MP (kg/m^3)	DRIB (%)	Source
Moderate	2.9	1.7	600	17.9	70	0.39	400	6.9	3.4	Armbrester 1972
Dense	2.9	1.7	600	44.3	70	1.01	1,048	18.3	4.8	(cages 0.25 m^3)
Dense	10	5.0	500	43.1	87	0.78	235	12.7	2.5	Suwanasart 1972
Dense	25	12.5	500	54.8	87	0.72	86	14.1	1.7	(cages 0.21 m^3)

Pm_i : mean weight at stocking

B_i : biomass at stocking

B_f : biomass at harvest

G : average growth rate of individual fish during the culture period

MSG : average monthly specific growth rate as % Pm_i for 30 days

MP : average monthly production: $(B_f - B_i)$ observed and recalculated on a 30-day basis

DRIB : daily rate of increase of the biomass calculated from $B_f = B_i\ (1 + i)^n$ where n is the culture period in days and i is $\frac{DRIB}{100}$

now being applied to tilapias also, although still on a relatively small scale (Rabelahatra 1979). For example, in Cianjur large submerged bamboo cages (5 x 4 x 0.7 m deep) are stocked with *S. mossambicus* fingerlings (mean weight 20 g; stocking rate 0.7 kg/m^3). This yields 50 to 60 kg of fish in six months, which corresponds to an average monthly production of about 0.55 kg/m^3: a little less than the production realized in Laguna de Bay by Pantastico and Baldia (1979), see above, but only about half of that obtained by commercial farmers in eutrophic lakes (see Table 4).

Table 6. Summary of data from the culture of all-male *Sarotherodon niloticus* stocked in 0.5 m^3 cages at 200 fish/m^3 in a pond receiving heated water effluent (25 to 37°C) from a nuclear power plant at Tihange, Belgium (after Philippart et al. 1979). Bracketed values are representative of the range of fish sizes.

Pm_i (g)	B_i (kg/m^3)	B_f (kg/m^3)	G (g/d)	MSG (%)	MP (kg/m^3)	DRIB (%)
25	5.1	6.8	0.6	68	3.5	1.94
32-35*	(6.7)	(7.3)	0.2	13-20	0.8-1.4	0.40-0.64
52*	10.4	(11.0)	—	2-18	0.2-1.9	0.06-0.61

* duplicate experiments

Pm_i : mean weight at stocking
B_i : mean biomass at stocking
B_f : mean biomass at harvest
G : average growth rate of individual fish during the culture period
MSG : average monthly specific growth rate as % Pm_i for 30 days
MP : average monthly production ($B_f - B_i$) observed and recalculated on a 30-day basis.
DRIB : daily rate of increase of the biomass calculated from $B_f = B_i (1 + i)^n$ where n is the culture period in days and i is $\frac{DRIB}{100}$

Semi-Intensive Culture

1. Definition

In semi-intensive operations some supplemental feed is given but in contrast to intensive operations, this feed is usually relatively poor in protein (less than 10% dry weight) and made from local materials readily available at low cost. Semi-intensive cage culture of tilapias uses mainly phytophagous species which receive vegetable material or agricultural by-products as supplementary low-protein diets in the presence of algal blooms.

2. Semi-Intensive Culture using Aquatic and Terrestrial Plants

A caged monosex male population of *T. guineensis* in a thermal effluent was fed at 4.8% of the fish biomass/day on the freshwater algae *Hydrodictyon* sp. (Chlorococcales) and *Cladophora* sp. (Ulotrichales) (Philippart et al. 1979). The average production was about 1 kg/m^3/month and the algae gave an FCR on a dry weight basis of 4.5 which is relatively good (Table 7). This points out the potential of a vegetable protein diet for caged tilapia, in this case 28% wet weight protein. The results are particularly good as *T. guineensis* usually performs less well under culture conditions than *S. niloticus* and *S. aureus.*

The comparative values of three aquatic macrophytes for growing adult *S. niloticus* in cages have been determined in Indonesia (Rifai 1979, 1980): *Hydrilla verticillata*, a submerged perennial growing to as much as 3 m long; *Lemna minor*, which is characterized by small, free-floating thalli and *Chara* sp., a coarse plant, usually coated with precipitated calcium carbonate. *Lemna minor* was preferred by the fish and gave the best growth, although with the lowest FCR (33). Table 7 summarizes the results. These were probably adversely affected by the very small mesh size used (reduced water exchange) and the shallow water depth in the cages.

Cage culture of *T. rendalli* has been proposed for rural areas in Colombia using tropical terrestrial plants as supplemental feed. The required characteristics of such plants are a high protein content in the leaves, edible tubers, vegetative reproduction and good growth even in poor soils. McLarney (1978) citing in part Prof. A.R. Patino's observations, suggests the following: 1) *Manihot esculenta* (Euphorbiaceae) commonly known as cassava which has edible tubers and leaves with 17.2% dry weight of protein; 2) *Alocasia macrorhiza* (Araceae) which has edible tubers and leaves with 23.2% dry weight and 6.25% wet weight of protein and 3) *Colocasia* spp. (Araceae) commonly known as taro which has edible tubers, large leaves and grows well on pond dykes.

Cnidosculus chayamansa, which has leaves with 24.2% dry weight of protein and *Xanthophyllum* spp. with edible tubers have also been recommended.

For tilapia culture, research seems to have concentrated on *Alocasia macrorhiza.* Data from three sets of experiments in Colombia using *T. rendalli ('T. melanopleura')* are summarized in Table 7 (Popma 1978). In one of these when 10 g fish were stocked there was no growth during the first month and for the next 14 weeks the growth was very slow. As a result, it took 7.2 months of feeding to produce 80 to 130 g fish. Ten grams is therefore obviously too small a size for stocking.

The food value of *Alocasia* leaves is better shown by the results for larger individuals (25 to 40 g) with an initial biomass (B_i) not exceeding 3 kg/m^3. For these the average individual growth remains relatively good at about 1 g/d and fishes weighing around 150 g can be produced in 4 to 5 months. When the B_i exceeds 4 kg/m^3 the individual growth rate decreases but the production increases. The best production (3.5 kg/m^3/mo) was obtained with a B_i of 6.6 kg/m^3.

The leaves of *Alocasia macrorhiza* and ipil-ipil *(Leucaena leucocephala*, Leguminoseae) (24.5% dry weight of protein) have also been used in combination with other ingredients, e.g., wheat bran or rice bran, as a source of relatively cheap vegetable protein for caged tilapias (see below).

Table 7. Semi-intensive cage culture of tilapias using aquatic and terrestrial plants. Bracketed values are representative within the size ranges of fish used.

Species and sexes used: country	Supplemental feed	Stocking data Pm_i (g)	N/m^3	B_i (kg/m^3)	Growth data G (g/d)	MSG (%)	DRIB (%)	Harvesting data Pm_f (g)	B_f (kg/m^3)	MP (kg/m^3)	FCR	Culture period (mo)	References and remarks
T. guineensis, males: Belgium	Fresh algae (26% protein), DFR 4.8% B	42	86	3.6	0.4	26	0.82	53	4.6	0.97	46 dry 4.5	1	Philippart et al. 1979; cages, 0.5 m^3; average temperature 25.8°C
S. niloticus mixed sex: Indonesia	*Hydrilla verticillata* *Chara* sp. *Lemna minor* all at or above DFR 30% B	170 (96- 276)	5 or 15 or 45	0.8 to 6.6	0.44 0.43 0.73	—	—	170 to 218	1.1 to 7.6	(0.5) — (1.5)	23 19 33	3	Rifai 1979, 1980; cages, 1 m^3 with 3 mm mesh
T. rendalli, mixed sex: Colombia	*Alocasia macrorhiza* leaves, ad lib.	22.5	100	2.25	0.95	127	1.31	165	15.8	2.7	—	5	McLarney 1978; cages, 1 m^3; survival, 96%
	a. *Alocasia macrorhiza* leaves, DFR 10% to 20% B	10 10 10	50 100 200	0.5 1.0 2.0	0.55 0.46 0.34	164 136 100	1.16 1.09 0.97	128 108 82	6.0 10.5 16.0	0.79 1.33 1.98	10 to 13	7.2	Popma 1978; cages, 1 m^3; survival, 96.4%; a) complete trial with small fishes; b) same fish from week 19 to week 31
	b. *Alocasia macrorhiza* leaves, DFR 10 to 20% B	48 42 33	50 100 200	2.4 4.2 6.6	0.97 0.81 0.59	54 51 48	0.95 0.95 0.92	128 108 82	6.0 10.5 16.0	1.40 2.33 3.48	less than 10	3+	
	Alocasia macrorhiza leaves, ad lib	27.4 26.6	175 225	4.8 6.0	0.41 0.50	45 56	0.83 0.96	76.4 86.4	13.0 18.8	2.0 3.2	—	4	Corredor 1978; 4 cages, 1 m^3; complete trial over 6 mo but difficulties after 4 mo

Pm_i : mean weight at stocking
N : number of fish
B_i : mean biomass at stocking
G : average growth rate of individual fish during the culture period
MSG : average monthly specific growth rate
DRIB : daily rate of increase of the biomass calculated from $B_f = B_i (1 + i)^n$ where n is the culture period in days and i is $\frac{DRIB}{100}$
Pm_f : mean weight at harvest
B_f : mean biomass at harvest
MP : average monthly production: $(B_f - B_i)$ observed and recalculated on a 30-day basis
FCR : feed conversion ratio, on a net weight basis unless otherwise indicated
DFR : daily feeding rate as % B (biomass)
B : biomass

3. Semi-intensive Cage Culture using Low Protein Mixed Feeds

Various feeds based on local ingredients have been used for caged tilapias. In some cases, such as in the eutrophic lake Laguna de Bay (Philippines), such feeds are used to augment natural feeding on well-developed algal populations. In other instances, they may constitute practically the only nutritional source available to the fish. The protein content of these feed is generally less than 10% of their total dry weight. Several ingredients may be combined as diverse as rice bran, snails, plant leaves, brewery waste, oil cake and cattle blood, according to local availability and price. In Lake Ilopango (El Salvador), the Fisherman's Cooperative, with 210 m^3 of cages, harvests about 900 kg of *S. aureus* and private operators sell another 2,500 kg annually (Street 1978) but no details are available on the exact nature of the diet.

Data related to various other tilapias are summarized in Table 8. *S. niloticus* and *S. mossambicus* were reared experimentally in floating cages in Laguna de Bay using rice bran alone (Anon. 1979a). *S. niloticus* gave a much better production (2.3 kg/m^3/month) than *S. mossambicus* (Pantastico and Baldia 1979), even though the latter was fed additional chopped snails (*Stenomelania canalis* and *Melanoides* sp.). The average initial biomass, although not clearly ascertained, was most probably higher than for *S. mossambicus* of which the production potential appears to be lower than that of *S. niloticus*. McLarney (1978) obtained good results in Colombia rearing *T. rendalli* in 1 m^3 cages in fish ponds. Starting with 22.5 g fingerlings and an average B_i of 2.25 kg/m^3, 21.6 kg/m^3 of 200 to 250 g fish were harvested after 5 months. *Alocasia macrorhiza* leaves and wheat bran were fed daily. The average individual growth (1.3 g/day) was good and the monthly production nearly 4 kg/m^3. In fact, these are the best results available for semi-intensive tilapia cage culture (Table 8).

In the first experiment with tilapia cage culture in Africa in 1972, *S. esculentus* and *T. zillii* were reared, either separately or together in Lake Victoria, Tanzania (Ibrahim et al. 1975). Feeding mostly on brewery wastes and some fish meal plus plant leaves (for *T. zillii*), the fish's average B_i's were very low (0.2 to 0.4 kg/m^3). The resulting average growth rates and production were relatively low to medium, probably due in part to poor feed value.

Trials to raise tilapia (probably *S. mossambicus*) in floating cages in the inlet to the Negombo Lagoon, north of Colombo, Sri Lanka were started recently with relatively good results (Sollows pers. comm.). Some water is added to the feed components (see Table 8) to form a mash.

Intensive Cage Culture

1. Feed Composition

As the quality of supplemental feed—in particular its protein content and its nutritional balance—are improved so cage culture of tilapias may be intensified. From the economic point of view, the cost of feeding gains in importance, and can account for more than 50% of the production costs

Table 8. Semi-intensive cage culture of tilapias using mixed feeds. Bracketed entries are representative of the range of fish sizes used.

Species and location	Feed composition and feeding rate	Stocking data Pm_i (g)	Stocking data N/m^3	Stocking data B_i (kg/m^3)	Growth data G (g/d)	Growth data MSG (%)	Growth data DRIB (%)	Harvesting data Pm_f (g)	Harvesting data B_f (kg/m^3)	Harvesting data MP (kg/m^3)	FCR	Culture period (mo)	References and remarks
1. *S. niloticus*, Laguna de Bay (Philippines)	rice bran in the presence of algal blooms	(10)	(100)	(1.0)	(0.8)	—	—	150	15	2.3	—	6	Anon. 1979a; Floating cages; 20 x 25 x 2.5 m deep; 1,000 m^3 capacity
2. *S. mossambicus*, Laguna de Bay (Philippines)	Rice bran, 70%; chopped snails, 30%; equivalent to 41% protein; in the presence of algal blooms	10	50	0.50	0.52	157	1.9	57.1	2.8	0.77	6	3	Pantastico and Baldia 1979; 6 cages (5.4 mm mesh) 25 x 25 x 1.5 m deep; each with 9.4 m^3 capacity
3. *S. aureus*, Lake Ilopango (San Salvador)	Undefined	—	—	—	—	—	—	—	—	2.25	—	—	Street 1978; 7 commercial cages; each with 30 m^3 capacity
4. *T. rendalli*, Ponds (Colombia)	*Alocasia* leaves + 0.5 to 1.0 kg wheat bran daily	22.5	100	2.25	1.3	173	1.5	200 to 250	21.6	3.87	—	5	McLarney 1978; 1 m^3 cages; survival 96%
5. *S. esculentus*, Lake Victoria (Tanzania)	Brewery waste, 99%; fish meal, 1%; DFR = 15 to 30% B	19	18.9	0.36	0.16	25	0.4	46.6	0.70	0.06	—	<6	Ibrahim et al. 1975; cage, 42 m^3; (8 mm mesh)
6. *T. zillii*, Lake Victoria (Tanzania)	as 5. above plus plant leaves	2.56	84.2	0.21	0.14	164	1.6	15.6	0.92	0.23	—	3	as 5. above
7. *S. esculentus* + *T. zillii*, Lake Victoria (Tanzania)	as 6. above	16.3	22.9	0.37	0.45	83	1.1	(80)	1.78	0.30	—	4.7	Ibrahim et al. 1975; cage, 125 m^3 (20 mm mesh)
8. *S. mossambicus* ? Negombo Lagoon inlet (Sri Lanka)	Rice bran, 16/21; coconut oil cake, 4/21; cattle blood, 1/21; plus wheat flour, 450 g/21 kg;	1.5 to 3.0	—	—	—	—	—	80 to 250	—	—	—	6	Sollows pers. comm.: floating cages

Pm_i : mean weight at stocking
N : number of fish
B_i : mean biomass at stocking
G : average growth rate of individual fish during the culture period
MSG : average monthly spcific growth rate
DRIB : daily rate of increase of the biomass calculated from $B_f = B_i (1 + i)^n$ where n is the culture period in days and i is $\frac{DRIB}{100}$
Pm_f : mean weight at harvest
B_f : mean biomass at harvest
MP : average monthly production: $(B_f - B_i)$ observed and recalculated on a 30-day basis
FCR : feed conversion ratio, on a net weight basis unless otherwise indicated

(Coche 1978, 1979). The utilization of a high initial biomass then becomes not only possible but also necessary if production costs are to be minimized and net profits maximized.

Researchers have used either commercial fish feeds or self-made compounded feeds. Commercial fish feeds have a very high protein level (e.g., Purina Trout Chow, 40%; Trouvit, 46%), a balanced composition including essential minerals and vitamins (at least for the target fish species), and a high price. Self-made feeds, however, have used mixtures of locally available ingredients, giving lower protein levels (usually 20 to 30%) and lower costs. Several empirical feed formulations have been evaluated but none has been developed from scientific data.

Table 9 gives examples of artisanal feeds which have been found satisfactory under various local conditions. Their composition is largely based on relatively cheap materials, such as rice bran and cotton seed oil cake, enriched with either fish meal or blood meal. The actual price of the mixed ingredients is now about US$0.15 to 0.20/kg, and the cost of pelleting probably adds another US$0.10/kg. The Campbell formula B4 (Table 9) is particularly attractive because of its relatively high protein content (25%), low cost (currently US$0.20/kg) and high efficiency (FCR = 2.0), at least for *S. niloticus.*

In the Philippines, Pantastico and Baldia (1979) have used a feed combining 20% ipil-ipil leaf meal *(Leucaena leucocephala*, Leguminoseae), 20% fish meal and 60% rice bran for *S. mossambicus* cage culture. This feed contained 27.3% dry weight of protein and cost only US$0.09/kg. Unfortunately the adverse conditions under which these experiments were conducted (small-size mesh, low density, low biomass, high feed ration) prevented a realistic demonstration of the efficiency of this feed (FCR = 4.0).

For *S. niloticus* in particular, an omnivore with a vegetarian tendency in its natural environment, Coche (1977) stressed that the ideal artificial diet *a priori* should contain a relatively high percentage of carbohydrates as an energy source. This appears to have been confirmed by Campbell (1978b) for *S. niloticus* in Lake Kossou, Ivory Coast. In his B4 formulation (Table 9) the 25% protein includes at least some of animal origin (i.e., 4% fish meal). The addition of a mineral/vitamin premix to B4 did not improve the results.

2. Feeding Rate

The feeding rate is usually quantified as the daily feed ration (DFR), the amount of feed (wet or dry) being fed daily (generally six days a week) to the fish, expressed as a percentage of the best available estimate of the fish biomass (%B). Such estimates, based on cage sampling, are usually made every 15 to 30 days.

It is well known in fish husbandry that food requirements per unit weight of fish decrease as the fish increase in size, but for the tilapias there is little information on this. In intensive cage culture of *S. niloticus* in the Ivory Coast, it was shown that the DFR had to be decreased from 6 to

4% B after the fish had reached 40 g average weight, to improve the feeding efficiency (Coche 1977). Since then, this problem has been tackled by other researchers.

Table 9. Satisfactory artisanal feeds for the cage culture of tilapias, using local ingredients, with feed conversion ratios (FCR) and costs where available.

A. Guerrero 1979a and Anon. 1979a for *S. niloticus* and *S. mossambicus* culture in the Philippines

1. Ingredients (% by weight)

Rice bran	77%
Fish meal	23%

2. Cost (1979) US$0.17/kg[a]
3. FCR 2.5

B. Campbell 1978b and pers. comm. for *S. niloticus* culture in the Ivory Coast

	Feed formulation			
1. Ingredients (% by weight)	B1	B2	B3	B4
Misc. carbohydrates[b]	—	—	—	45
Rice polishings	65	61	65	—
Wheat middlings	12	12	12	12
Peanut oil cake	18	18	—	—
Cottonseed oil cake	—	—	18	38
Fish meal	4	8	4	4
Oyster shell	1	1	1	1
Total protein content (as % dry weight)	20	22	20	25
2. Cost (1978) in US$/kg[c]	0.11	0.13	0.07	0.09
3. FCR (approx.)	2.4	2.0	2.2	2.0

C. N'Zimasse 1979 for *S. niloticus* in the Central African Republic

1. Ingredients (% by weight)

Cottonseed oil cake	82%
Wheat flour	8%
Cattle blood meal	8%
Bicalcium phosphate	2%

2. Cost (1979) US$0.17/kg[d]
3. FCR 3.2[e]

[a]Moist pellets have higher costs than this. This is the cost of ingredients only.
[b]E.g., brewery waste, maize and rice bran mixed according to availability.
[c]This gives the cost of ingredients only. The pelleting cost to less than 10% moisture pellets was about US$0.07/kg.
[d]This gives the cost of ingredients only. Dry pellets cost an extra US$0.06/kg.
[e]Poor, due partly to oxygen deficiency.

Campbell (pers. comm.), experimenting empirically with *S. niloticus* and a 25% protein feed, reached the conclusion that DFR's should be adjusted with size as follows: fry/fingerlings less than 25 g, 10 to 8% B; 25 to 150 g, 6 reducing to 4% B; 150 to 200 g, 3% B; and over 200 g, 2% B. He has even reduced the DFR to 1% B for fish larger than 200 g to increase the efficiency of pelleted feeds. Taking a more scientific approach, Mélard and Philippart (1980) estimated the optimum DFR for *S. niloticus* in tanks and cages using a 46% protein commercial feed at 27 to 31°C. Table 10 summarizes their recommendations. In the Philippines, Guerrero (1980a) has also proposed a progressive reduction of DFR with increasing size for *S. niloticus:* from 5% B (fish less than 50 g) to 4% B (50 to 100 g) and 3% B for larger fish. These are smaller DFRs than those in Table 10 and also refer to lower protein feeds.

Table 10. Recommended daily feeding rates (DFR) expressed as percentage fish biomass (% B) for *Sarotherodon niloticus* in tanks and cages at 27 to 31°C, fed a 46% protein commercial fish food (after Mélard and Philippart 1980).

Mean fish weight (g)	DFR (% B)
0 to 5	30 reducing to 20
5 to 20	14 reducing to 12
20 to 40	7 reducing to 6.5
40 to 100	6 reducing to 4.5
100 to 200	4 reducing to 2
200 to 300	1.8 reducing to 1.5

When fixing the DFR for a particular operation, one should also take into account the natural productivity of the environment. In Alabama, *S. aureus* were fed 40% protein floating pellets (DFR, 3% B) for 10 weeks in 0.25 m^3 cages placed in fertilized ponds (Armbrester 1972). Although the feeding efficiency was very good in presence of moderate algal blooms (Net FCR = 1.1 to 1.5), it decreased considerably in ponds with dense algal blooms (Net FCR = 3.3 to 6.3). This was attributed to the abundant availability of natural food.

3. Feeding Techniques

The methods used to distribute feed to caged tilapias may greatly influence the results. This depends on numerous factors including digestive physiology and feeding behavior relative to fish size; the shape and size of the cage; the water circulation through the cage and the stocking density. There are three major variables: feeding frequency, feed presentation and distribution methods.

Increasing feeding frequency by dividing the DFR into several meals may allow increased DFRs and give improved production and feeding efficiency. Although no scientific evidence exists to support such practices, in one commercial cage farm *S. niloticus* are fed their DFR as five meals (Campbell pers. comm.).

4. Feed Presentation

Caged tilapias have been grown using mash or pelleted feeds, the pellets being wet or dry, sinking or floating. Guerrero (1980a) regards *S. niloticus* as an avid surface feeder in cages and advises the use of mash feeds for small fish only. With large fish, the water agitation is so strong that much of the mash feed is washed out of the cage and lost. Pelleted feeds are therefore more efficient for large fish but the simplicity and low cost of mash make it still attractive for artisanal farmers. Rather than pelleting compounded feed, they could prepare meal in advance, in the form of "balls" or mash and air-dry them. Such feed "balls" can be placed in the middle of the cage-covering net, which is then slightly lowered under the surface of the water to give the fish access to the food (Campbell pers. comm.). A similar feeding method can also be used with fresh mash balls, particularly if the area of the cage is of the order of several square meters.

Guerrero (1980a) fed *S. niloticus* moist and dry compounded artisanal feeds (65% rice bran, 25% fish meal, 10% copra meal) at DFR 4% B as two meals per day in 1 m^3 cages (mesh 2.5 cm) placed in a pond for 24 days. The results are summarized in Table 11A. He concluded that moist pellets were better utilized by the fish. The FCR was slightly higher with the moist pellets, but there was no significant statistical difference between the two treatments. Moist pellets were easier and cheaper to produce.

Table 11B gives the results of feeding floating and sinking pellets (40% protein) to *S. aureus* in 0.12 m^3 cages placed in fertilized ponds for 87 days at DFR 3% B for 70 days of this period. The floating pellets gave better growth, higher production and a lower FCR. This holds true particularly for small cages with high fish densities and biomass. The difference may be less for larger cages (several m^3), at lower densities (below 400/m^3) and with a smaller biomass (up to 40 kg/m^3). In such cases, the extra cost of floating pellets might not even prove economical, but good data are missing.

Dry pellets may be distributed in cages either by hand or mechanically. In the latter case, automatic rather than demand feeders are preferable, especially if the fish density is high. Again, economical considerations should guide the choice. If labor is expensive, automatic feeders can reduce costs. They also facilitate dividing the DFR. On the contrary, in artisanal cage farms, hand feeding may be more advantageous. It also gives the opportunity to monitor the fish stocks more closely. No comparative data exist for tilapia cage culture.

5. Feed Conversion

The efficiency of feed conversion is generally quantified as the feed conversion ratio (FCR): the weight of feed required to produce unit weight of live fish during a determined feeding period. Some authors calculate a "Net FCR" where the production due to the natural feed is taken into consideration. In intensive cage culture particularly (Coche 1978, 1979) FCR is determined by the interactions between the fish (individual size, sex ratio and density), the feed (quality, DFR and distribution) and the

Table 11. Summary of experiment trials on the intensive cage culture of *Sarotherodon niloticus* and *S. aureus*: A. comparison of moist and dry pelleted feed (after Guerrero 1978); B. comparison of floating and sinking pellets (after Suwanasart 1972).

A. *S. niloticus* receiving moist and dry feed (65% rice bran, 25% fish meal, 10% copra meal on a dry weight basis) at a daily feeding rate of 4% fish biomass split into 2 meals in 1 m^3 cages (2.5 cm mesh).

Type of pellets	Pm_i (g)	N/m^3	B_i (kg/m^3)	B_f (kg/m^3)	Pm_f (g)	G (g/d)	Survival (%)	MP (kg/m^3)	FCR	Culture period (d)
Moist	96	150	13.8	25.5	179	3.63	95 to 97	14.5	2.7	24
Dry	96	150	15.0	23.4	161	2.54	95 to 97	10.6	2.5	24

B. *S. aureus* receiving floating and sinking pellets (commercial feed, 40% protein) at a daily feeding rate of 3% fish biomass in 0.12 m^3 cages.

Type of pellets	Pm_i (g)	N/m^3	B_i (kg/m^3)	B_f (kg/m^3)	Pm_f (g)	G (g/d)	MP (kg/m^3)	FCR	Culture period (d)
Floating	25	500	12.5	79.5	172	1.7	22.3	1.33	Total 87:fed for 70
Sinking	25	500	12.5	60.6	127	1.2	16.0	1.86	Total 87:fed for 70

Pm_i : mean weight at stocking
N : number of fish
B_i : mean biomass at stocking
B_f : mean biomass at harvest
Pm_f : mean weight at harvest
G : average growth rate of individual fish during the culture period
MP : average monthly production: $(B_f - B_i)$ observed and recalculated on a 30-day basis
FCR : feed conversion ratio, on a net weight basis unless otherwise indicated

rearing environment (cage size, water exchange rate, DO and temperature). To maximize the feeding efficiency, every one of these factors should be maintained at its optimum level for the particular species being cultured. Most of these aspects have already been reviewed above.

Feed losses through the cage walls should be kept to a minimum by adapting feeds and feeding to the particular conditions prevailing. Excessive water movement either caused by the fish themselves or due to a strong water current may wash a great proportion of the feed out of the cages. A dense wild fish population may even learn how to create the water current necessary to achieve this. The fish density should be kept below a threshold value, above which the FCR increases. For *S. niloticus* in 1 m^3 cages this was 400 fish/m^3 (Coche 1977).

With protein-rich compounded feeds, good feeding efficiencies in tilapia cage culture are usually demonstrated by FCR's close to or lower than 2. FCR's between 1.0 and 2.0 are now attained by Campbell (pers. comm.) in the Ivory Coast with 25% protein sinking pellets and *S. niloticus*. With 36% protein floating pellets (about 10% moisture), FCR's below 1.0 have been attained with *S. aureus* in Puerto Rico (Jordan and Pagán 1973), probably in the presence also of additional natural food.

The presence of such additional food in the environment may, however, have a negative rather than a positive effect on the efficiency of utilization of the artificial feed. In the fertilized Alabama ponds (Table 12), 40% protein floating pellets fed at DFR 3% B to caged *S. aureus* gave a net FCR of 1.1 to 1.5 in the presence of moderate plankton blooms, but with dense blooms the net FCR rose to 3.3 to 6.3 (Armbrester 1972). As far as possible, one should, therefore, also take into account the natural productivity of the environment when fixing the DFR.

6. Examples of Intensive Culture

Some data for the intensive cage culture of *S. aureus* are summarized in Table 12. The fish densities used were usually high (above 400/m^3) and relatively small fish were harvested (130 g or less) because of the small sizes stocked (3 to 10 g) and/or the short culture period (2.5 to 3.0 months).

Growth rates and production were definitely higher in fertile fish ponds than in the rock quarry pond (Table 12D) although the feeding efficiency was best in the latter. It is worth noting that the FCR's were below 2.0, for all the examples given except in heavily fertilized ponds with dense algal blooms (Secchi disc depth visibility, 37 cm) and with 2.9 g fry as stocking material (Table 12A).

Table 13 summarizes data on the intensive cage culture of *S. niloticus*. The data for 20 m^3 cages (Campbell 1978b) are the first on intensive tilapia cage culture at the artisanal level. In all cases except the Philippine work (Guerrero 1979a, 1980a) sinking pellets were fed with a medium (20%) to high (40% or over) protein level mostly at DFR 4 to 6% B. The fish densities were generally low to medium: below 300 fish/m^3. In the Ivory Coast, a particular effort was made to produce commercial size fishes, averaging at least 200 g, which called for culture periods in the growout cages of 4 to 5 months. Some brief notes follow concerning each set of trials (Table 13).

Table 12. Summary of stocking, growth and harvesting data for the intensive cage culture of *Sarotherodon aureus*: using floating pellets unless otherwise stated.

Location	Feeding details	Stocking data Pm_i (g)	N/m^3	B_i (kg/m^3)	Growth data G (g/d)	MSG (%)	DRIB (%)	Harvesting data Pm_f (g)	B_f (kg/m^3)	MP (kg/m^3)	FCR	Culture period (mo)	References and remarks
A. Alabama, U.S.A.	40% protein DFR 3% B	2.9	600	1.7	0.56	579	4.65	39	41.0	15.7	1.1 to 1.5[c]	2.5[a]	Armbrester 1972: 0.25 m^3 cages (23 mm mesh) in fertilized ponds; mean values
	As above	2.9	600	1.7	0.76	790	5.13	73	56.4	21.9	3.3 to 6.3	2.5[b]	
B. Alabama, U.S.A.	40% protein DFR 3% B	10	500	5.0	1.36	408	2.99	128	64.7	20.6	1.3	2.9	Suwanasart 1972: 0.31 m^3 cages in fertile ponds; mean values; fed for 70 d only
		25	500	12.5	1.72	207	2.20	150	83.3	24.4	1.4		
	40% protein DFR 3% B	25	500	12.5	1.07	128	1.80	118	58.9	16.0	1.9	2.9	Suwanasart 1972: 0.12 m^3 cages in fertile ponds; mean values; fed for 70 d only; sinking pellets
C. Alabama, U.S.A.	40% protein DFR 3% B	13	857	11.2	1.03	238	1.67	174	149	26.5	1.8	5.2	Pagán 1973: 1 m^3 cages in fish ponds; an FCR of 1.2 was obtained for N286 fish/m^3
D. Puerto Rico	36% protein DFR 5% B	10	300	3.0	0.68	204	2.53	57	17.2	6.1	0.95	2.5	Jordan and Pagán 1973: 1 m^3 cages (mesh 8 mm) in a rock quarry pond
		10	400	4.0	0.55	165	2.28	48	19.4	6.6	0.90		
		10	500	5.0	0.52	156	2.21	46	23.1	7.8	0.91		

[a]in the presence of moderate plankton blooms
[b]in the presence of dense plankton blooms
[c]net values

Pm_i : mean weight at stocking
N : number of fish
B_i : mean biomass at stocking
G : average growth rate of individual fish during the culture period
MSG: average monthly specific growth rate

DRIB : daily rate of increase of the biomass calculated from $B_f = B_i (1 + i)^n$ where n is the culture period in days and i is $\frac{DRIB}{100}$
Pm_f : mean weight at harvest
B_f : mean biomass at harvest
MP : average monthly production ($B_f - B_i$) observed and recalculated on a 30-day basis
FCR : feed conversion ratio, on a net weight basis unless otherwise indicated

Table 13. Summary of stocking, growth and harvesting data for the intensive cage culture of *S. niloticus* including monosex male populations.

		Stocking data			Growth data				Harvesting data			Culture	
Country	Feeding details	Pm_i (g)	N/m^3	B_i (kg/m^3)	G (g/d)	MSG (%)	DRIB (%)	Pm_f (g)	B_f (kg/m^3)	MP (kg/m^3)	FCR	period (d)	References and remarks
Philippines	10% copra meal; 20% fishmeal and 70% rice bran as a mash: DFR 4% B	36	150	5.4	0.56	46	1.01	67	10.1	2.5	3.6	56	Guerrero 1980a: 1 m^3 cages (mesh 2.5 cm) in pond; mean results
	25% fishmeal and 75% rice bran as a mash: DFR 4% B	55	150	8.3	0.80	44	0.96	95	14.2	3.2	3.1	56	
Philippines	23% fish meal and 77% rice bran as moist pellets (24.2% dry weight of protein): DFR 5% B	20	250	5.0	1.33	200	2.72	100	25.0	10.0	2.5	60	Guerrero 1979a: 1 m^3 cages (mesh 1.9 cm) in pond
Ivory Coast	chicken feed as dry pellets (24.7% protein): DFR 4 to 6% B	16	268	4.3	1.05	197	1.52	175	41.9	7.6	2.8	151**	Coche 1977: 1 m^3 cages (mesh 2.5 cm) in the artificial Lake Kossou at 27 to 30°C; * DO < 5 mg/L for about 30 days; ** DO < 3 mg/L for about 20 days and < 5 mg/L for 2 to 2.5 mo
		22	218	4.8	1.20	164	1.29	207	34.6	5.8	3.4	154*	
		29	257	7.5	1.33	138	1.32	232	56.1	9.3	3.3	153**	
		29	349	10.1	1.18	122	1.27	197	60.9	10.8	3.2	142*	
		40	355	14.2	1.54	116	1.36	228	73.6	15.5	3.2	122*	
		40	488	20.5	1.39	104	1.37	168	71.9	16.7	3.0	92*	
		55	215	11.8	2.21	121	1.59	265	53.0	13.4	3.1	95	
Ivory Coast	see Table 9-B1 (20% protein): DFR 6 to 4% B	22	71	1.6	1.77	241	1.81	213	13.3	3.7	2.2	118**	Campbell 1978b: * 6 m^3 cages (mesh 2.5 to 1.4 cm); ** 20 m^3 cages (ibidem) in artificial Lake Kossou; sex ratio, average 84% males; DO >5 mg/L
		23	73	1.7	1.63	213	1.78	236	17.2	3.6	2.3	131**	
		33	185	6.1	2.33	212	1.97	308	49.9	9.8	2.4	108*	
	see Table 9-B2 (22% protein): DFR 6 to 4% B	32	177	5.6	1.98	186	1.64	278	42.1	8.8	2.1	124*	
		36	182	6.6	2.03	169	1.68	284	50.3	10.7	1.9	122*	
	see Table 9-B3 (20% protein): DFR 6 to 4% B	22	86	1.9	1.44	196	2.04	163	13.7	3.6	2.0	98**	
		23	74	1.7	1.22	159	2.08	118	8.5	2.6	2.0	78**	
		31	186	5.8	1.72	166	2.00	184	33.7	9.4	2.2	89*	
Central African Republic	see Table 9-C (41% protein): DFR 6% B	43	247	10.6	0.81	57	1.28	89	18.4	4.2	5.5	56	N'Zimasse 1979: 1 m^3 cages in fertilized pond; DO deficient (especially at night)

Table 13 (cont'd)

Country	Feeding details	Stocking data			Growth data			Harvesting data				Culture period (d)	References and remarks
		Pm_i (g)	N/m^3	B_i (kg/m^3)	G (g/d)	MSG (%)	DRIB (%)	Pm_f (g)	B_f (kg/m^3)	MP (kg/m^3)	FCR		
Belgium	commercial pellets (46% protein): DFR 7% B	55 54 59	600 400 100	33.2 21.7 5.9	1.0 1.2 1.6	56 67 83	1.73 1.69 2.00	86 91 107	55.5 36.0 10.7	18.3 14.2 4.8	3.3 2.8 2.2	30	Philippart et al. 1979: 0.5 m^3 cages in thermal effluent, average temperature 27.9°C
Ivory Coast	chicken feed pellets (24.7% protein): DFR 4 to 6% B	49	300	14.7	1.8	110	1.36	271	75.9	15.1	3.3	122	Coche 1977: 1 m^3 cages (mesh 2.5 cm) in lake; DO < 5 mg/L; monosex males
Central African Republic	see Table 9-C: DFR 6% B	82	122	10.0	1.25	46	0.96	187	22.5	4.5	5.6	85	N'Zimasse 1979: 1 m^3 cages in fertilized ponds; monosex males; low DO's
Belgium	commercial pellets (46% protein): DFR 7 reducing to 3% B	25 32 46 53 68	200 200 200 200 200	5.1 6.2 9.2 10.7 13.7	1.40 0.93 0.73 1.07 0.93	168 87 48 60 41	4.01 2.67 1.38 1.66 1.21	46 46 57 69 82	9.2 9.2 11.3 13.7 16.4	8.3 6.0 4.3 6.0 5.5	1.1** 1.4 1.7* 1.4* 2.0*	15	Philippart et al. 1979; 0.5 m^3 cages in thermal effluent, average temperature 25 to 27°C; monosex males; ** DFR = 7% B; * DFR = 3% B

Pm_i : mean weight at stocking
N : number of fish
B_i : mean biomass at stocking
G : average growth rate of individual fish during the culture period
MSG : average monthly specific growth rate
DRIB : daily rate of increase of the biomass calculated from $B_f - B_i (1 + i)^n$ where n is the culture period in days and i is $\frac{DRIB}{100}$
Pm_f : mean weight at harvest
B_f : mean biomass at harvest
MP : average monthly production: ($B_f - B_i$) observed and recalculated on a 30-day basis)
FCR : feed conversion ratio, on a net weight basis unless otherwise indicated
DFR : daily feeding rate as % B (biomass)
B : biomass
DO : dissolved oxygen

In the Philippines, using a relatively cheap feed (moist pellets at US$0.17/kg), Guerrero (1979a) appears confident of producing within 2 to 4 months 80 to 100 g marketable tilapias with FCR 2.5 after stocking at 5 to 20 g. An average production of 12 to 20 kg/m^3/month is attainable, giving an estimated monthly net income of US$6.75 to 10.80/1 m^3 cage.

Coche (1977) reports culture trials in the artificial Lake Kossou (Ivory Coast), from 1974 to 1975 beginning with 1 m^3 cages and pelleted chicken feed (24.7% protein). These demonstrated the excellent potential of *S. niloticus* intensive cage culture. Although reduced DO's were encountered for most of the culture periods, growth rates were usually about or above 1.2 g/day and the monthly production ranged from 9 to 15 kg/m^3, with fish densities and initial biomasses in the normally accepted range. Low feeding efficiencies were due mostly to the inadequate feed (which contained up to 30% undigested maize middlings) and to excessive DFR's. Campbell (1978b) used three improved feeds (Table 9—B1, B2, B3) and reduced DFR's in larger cages. His lower initial biomasses and densities (relative to cage size) resulted in better growth rates (1.4 to 2 g/day) but gave lower average productions with increasing cage size. Smaller cages (6 m^3) produced 9 to 11 kg/m^3/month and larger ones (20 m^3) 2.6 to 3.7 kg/m^3/month. Feeding efficiency was much better than Coche's (1977) results: FCR's, 1.9 to 2.2. It should be remembered, however, that Campbell used on average 84% male populations.

N'Zimasse (1979) used a heavily manured fish pond in the Central African Republic. The results are difficult to assess because of DO deficiencies, but the survival of *S. niloticus* was excellent. The growth rate, production and feeding efficiency were poor, especially for all-male populations.

Philippart et al. (1979) used increasing densities (up to 600 fish/m^3) and biomasses in pond-based cages in a thermal effluent. The feed was very rich in protein (46%). Neither growth rates nor FCR were as good as could have been expected in such small cages. The DFR was probably too high considering the size of the fish and the food quality. Good production was obtained with the two highest densities/biomasses, but at the cost of low feeding efficiencies.

Only limited data are available on the intensive cage culture of all-male *S. niloticus*. Coche (1977) suggested this as one way of increasing production: his fish, stocked at average weight 49 g, grew at an average rate of 1.8 g/day and gave a production of 15.1 kg/m^3/month on relatively poor quality feed. In the Central African Republic, N'Zimasse (1979) obtained poor results because of deficient environmental conditions and in Belgium (Philippart et al. 1979), the results for all-male culture were not as good as expected, although the feeding efficiencies were increased by adapting the DFR's to the fish sizes. Strangely enough, higher average production was experienced with smaller biomasses, which calls for some caution in interpreting these results. The relative brevity of the experiments might be partly responsible. It appears that further data are needed to assess the potential of monosex cage culture.

Table 14 summarizes the results of the intensive cage culture trials with

Table 14. Summary of feeding, stocking, growth and harvesting data for intensive cage culture of *Sarotherodon mossambicus*, *Tilapia guineensis* and mixed culture of tilapias with *Cyprinus carpio*.

Location	Species	Feeding details	Stocking data Pm_i (g)	N/m^3	B_i (kg/m^3)	Growth data G (g/d)	MSG (%)	DRIB (%)	Harvesting data Pm_f (g)	B_f (kg/m^3)	MP (kg/m^3)	FCR	Culture period (d)	References and remarks
Philippines	*S. mossambicus*	60% rice bran, 20% fish meal, 20% *Leucaena leucocephala* leaf meal: DFR, 10% B	10	50	0.5	0.65	195	2.15	68.4	3.4	0.97	4.0	90	Pantastico and Baldia 1979: cage mesh size, 5.4 mm; feed cost, US$0.09/kg.
Belgium	*T. guineensis*	Pellets (46% protein): DFR, 4.8% B	42	86	3.6	0.60	43	1.23	60	5.2	1.60	2.9	30	Philippart et al. 1979: 0.5 m^3 cages in thermal effluent; monosex males
		Soya pellets (30% protein): DFR, 4.8% B	42	86	3.6	0.40	26	0.82	53	4.6	0.96	5.0	30	
Taiwan	Hybrid (probably *S. mossambicus* x *S. niloticus*) with	Regular feeding	30	56	1.7	2.8	280	2.21	338	18.8	4.70	—	110	de la Cruz 1979; cage size 144 m^3; polyculture of two species.
	Cyprinus carpio	Regular feeding	150	21	3.2	16.8	338	2.36	2,000	41.7	10.50	—	110	

Pm_i : mean weight at stocking
N : number of fish
B_i : mean biomass at stocking
G : average growth rate of individual fish during the culture period
MSG : average monthly specific growth rate
DRIB : daily rate of increase of the biomass calculated from $B_f = B_i (1 + i)^n$ where n is the culture period in days and i is $\frac{DRIB}{100}$
Pm_f : mean weight at harvest
B_f : mean biomass at harvest
MP : average monthly production: $(B_f - B_i)$ observed and recalculated on a 30-day basis
FCR : feed conversion ratio, on a net weight basis unless otherwise indicated
DFR : daily feeding rate as % B (biomass)

results are also poor: the soya pellets did not improve on results obtained with fresh algal feeds (see F2 above) and animal protein pellets gave only slightly better results. This suggests that *S. mossambicus* and *T. guineensis* have low potential for cage culture.

7. Cage Culture of Mixed Species

In Indonesia, 10 to 25% *S. niloticus* are cultured with *Cyprinus carpio* in cages in a 20 ha lake at the Lido Station (Pedini pers. comm.). The cages are relatively large (9 m^2 x 1.25 m deep and 81 m^2 x 1.25 m deep) and a pelleted feed (32% protein) is given to the carps only at DFR 3% carp biomass. *S. niloticus* is here considered as a secondary crop. In Taiwan also an experimental cage (144 m^3) was stocked with tilapia (probably *S. niloticus* x *S. mossambicus* hybrids) and *C. carpio* (see Table 14). The growth of the tilapias was very good, even though far below that of common carp. The carp were so infested with parasites, however, that the local farmers now prefer tilapias for cage culture, even though their growth is slower and production smaller (de la Cruz 1979).

8. Brackishwater Culture

In the Ivory Coast *S. niloticus* is being cultured in cages placed in coastal lagoons where the salinity may reach 20‰. Some experiments are in progress, but until now results have been rather disappointing, the average growth rate varying from 0.20 to 1.17 g/day on a 15.5% protein feed (De Kimpe 1978). However, British Petroleum has established a commercial farm for intensive tilapia cage culture in Jacqueville (about 64 km west of Abidjan), including a hatchery/nursery and a feed mill. Both *S. niloticus* and *S. melanotheron (T. heudelotii)* are being tried in commercial size cages (27 and 54 m^3). The production potential of the farm is estimated at about 500 t/year (Campbell pers. comm.) but only preliminary data are available. Although *S. niloticus* has demonstrated an excellent growth rate (1.8 to 2.1 g/day), great mortalities have been experienced due to heavy parasitism. *S. melanotheron (T. heudelotii)*, a species endemic in the local lagoons, has also been tried with very little mortality but its real potential for cage culture has not yet been established.

Advantages, Constraints and Research Needs

1. Advantages

When compared with the more traditional methods of fish culture in ponds and tanks, cage culture presents some definite advantages (Coche 1978, 1979). In the particular case of tilapias, the major advantage is the possibility of controlling unwanted recruitment (Pagán 1975).

The other major advantages of cage culture include relatively high growth rates due to the continuous water exchanges, the limited space requirements, the production of large fish of a more uniform size than pond systems, and the possibility of greatly reducing production costs through a more precise adjustment of feeding rates.

2. CONSTRAINTS

Table 15 summarizes the major constraints to culturing tilapias in cages. The relative importance of these varies with location. The most important water quality parameter is the DO. Low DO's may appear once or twice a year as a periodical feature of the limnological cycle of the water body. At water temperatures from 26 to 30°C, special measures should be taken if the DO of surface water drops below 3 mg/L for several consecutive days, including lowered DFR or cessation of feeding and reduction in stocking density. In Lake Kossou, heavy *S. niloticus* mortalities have been recorded when DO's suddenly fell to 0.4 mg/L (Traore and Campbell 1976). Important mortalities in pond cages have also been observed for *S. aureus* following a DO reduction (Pagán 1973). This could be caused, in certain cases, by the cages and fish themselves, for example, in water bodies where the water circulation is limited, where the total cage volume has become proportion-

Table 15. Constraints to the cage culture of tilapias and suggested remedial action.

Constraint and target	Remedial action
A. Environmental	
1 Water quality	Modify the culture technique
2 Weather	Site selection/Anchoring system
3 Wild fish populations	Control wild fish populations
4 Aquatic predators	Control/Materials selection/Barrier
B. Inputs	
1 Equipment	Design/Construction/Durability/Cost
2 Fry/Fingerlings	Develop new methodology for mass production
3 Fish feeds	Formulation/Develop production methods
C. Fish Stocks	
1 Diseases/Parasites	Site selection/Control wild fish/Management
2 Security	Site selection/Full time watchmen
3 Marketing	Transport/Processing/Organization
D. Policy	
1 Research	Financial support/Training
2 Planning	Environmental potential/Inputs availability/Market potential
3 Development	Control/Extension service

ately too high, or where overfeeding takes place.

Regarding weather constraints, storm damage may occur if the farm site is not sufficiently protected. Such losses have occurred on Lake Ilopango in El Salvador (Hughes 1977) and in the Ebrie Lagoon in the Ivory Coast (Campbell pers. comm.).

Jordan and Pagán (1973) reported incursions and residence of wild *S. mossambicus* in cages for growing *S. aureus*. This greatly reduced feeding efficiency and production. The wild fish could also be vectors of parasites and diseases.

The shortage of tilapia seed is considered as one of the major present constraints to the development of the culture industry (Coche 1977, Guerrero 1979a, Anon. 1979a). A new methodology for the mass production of tilapia fry and fingerlings should be developed. Moreover, caged tilapias have practically no access to natural feed (apart from filter feeders in eutrophic water) and require essentially complete supplemental feed. The two main constraints here are the formulation of such feeds, e.g., in El Salvador (Hughes 1977) and Africa (ADCP 1980), and their availability and cost to the farmers.

3. Cage Design

The design and construction of adequate cages is still considered as a major problem in the Philippines (Anon. 1979a) and large cages used in El Salvador have proved difficult to harvest (Hughes 1977). In the Ivory Coast, the utilization of weak fibre netting material has resulted in great fish losses (Campbell 1978a). *S. niloticus* grazing on the settled algae on R250 tex nylon twine damaged it within 8 months sufficiently to necessitate a complete replacement by stronger netting (R470 tex). Following serious damage to fibre netting by crabs in brackishwater, imported plastic netting will replace this for future commercial production in the Ivory Coast (Campbell pers. comm.).

Damage to underwater structures by predators may also result in important losses, including the escape of the cultured fish. Such problems have been encountered in Indonesia with monitor lizards and turtles (Pedini pers. comm.) and in the Ivory Coast with iguana and large Nile perch *(Lates niloticus).* In brackish water, crabs have heavily damaged cage nets (Campbell pers. comm.). In such cases, either a predator net barrier should be used or the cage construction should be reinforced.

4. Disease Aspects of Cage Culture

Caged fish living in confined conditions are probably under greater stress than pond fish and are more susceptible to attack by parasites and diseases. Fortunately, the various tilapias cultured in cages have generally demonstrated high disease resistance. Only a few cases of health problems have been documented, mostly following increased stress due to elevation of salinity or fish biomass. In Puerto Rico, caged *S. aureus* reared in sea water (35‰ salinity) suffered chronic mortalities with open body lesions, fin rot and exophthalmus (Miller and Pagán 1973; Miller and Ballantine 1974). Pathogenic bacteria (*Aeromonas* and *Vibrio* spp.) were thought to be implicated. There is heavy pollution from raw sewage in the area.

In the Ivory Coast, a great decrease in *S. niloticus* production was observed in several cages where mycosis occurred (Coche 1977). Later, Campbell (pers. comm.) also observed cages with a high proportion of blind fish (*Diplostomum* ?). Susceptibility to all infections was higher in 1 m^3 cages when the biomass exceeded 73 kg/m^3. *S. niloticus* reared in shallow brackishwater (4 to 20‰ salinity) and relatively close to the lagoon sediments (0.5 m) suffered high parasitic burdens and mortalities (Campbell pers. comm.).

A high rate of mortality has been recorded in Alabama following outbreaks of bacterial diseases in pond cages with *S. aureus* (Pagán 1973). In Tanzania, a generalized *Saprolegnia* infection spread through a caged *T. zillii* population (Ibrahim et al. 1975).

5. Security

Poaching constitutes a major problem in cage culture. To ensure the security of commercial operations, Street (1978) has recommended in El Salvador the employment of full-time watchmen. Locating the cages close by the farmer's residence may also help solve the problem. In the Philippines, poaching is also mentioned among the main problems facing cage farmers (Guerrero 1979a).

6. Policy

Research support and rational planning of the development for tilapia cage culture are essential. Development should be supported by an adequate Extension Service, through which a continuous two-way flow of information can be established between farmers and researchers. Dramatic examples of the results of a lack of careful planning are available from the Philippines. In the 50 ha Lake Calibato, 170 fish cages containing about 10 million tilapias were present in 1978 (Anon. 1978). Within only a few years, the profit expected from tilapia cage culture fell by 50% because of such overcrowding. A similar situation has arisen in Lake Sampaloc (Sollows pers. comm.): see above. To bring such wild development under control, several planning steps are required: a survey of the lake's capacity for cage culture; setting guidelines for the siting/operation of the cages and licensing legitimate operators (Anon. 1978). A competent technical body is needed to oversee the development.

7. Research Needs

The research needs for tilapia cage culture in the Philippines have been recently reviewed (Anon. 1979a) and the following identified: the design and construction of the cages; the standardization of cage materials (which should ideally be sturdy, attract less fouling organisms and be nontoxic both to the fish and consumers) and the proper layout and positioning of the cages in water bodies.

Both in Latin America (ADCP 1978) and in Africa (ADCP 1980), tilapia cage culture has been selected as a priority culture system in the research programs of the future United Nations Development Programme (UNDP/ Food and Agriculture Organization of the United Nations (FAO) Aquaculture Regional Centres. The research needs identified include: design of cages suitable both from the functional and economical points of view; testing of construction materials; design of rafts, floats, anchors and other requirements for cage installations; feed formulation and preparation; feeding procedures; biology of the cultured stocks; economic evaluation and possible fish health hazards and their control.

Attention should also be given to the effects of cage culture on the environment. In particular, estimates of the optimum cage volume relative to that of the water body should be obtained, both for closed and open aquatic systems. These should become the basis for the rational future development of tilapia cage culture. Finally, a new methodology for the routine mass production of fry and fingerlings should be developed.

Prospects

1. Culture Potential of the Major Tilapia Species

Tables 16 and 17 summarize the culture potential in terms of production of the four main cultured species; *S. mossambicus* and *T. rendalli* have been both mostly raised in extensive and semi-intensive systems and *S. aureus* and *S. niloticus* in intensive systems. This probably reflects the geographical distribution of their cage culture (see Table 1) and makes it rather difficult to assess the real future potential of each of these four tilapias for cage culture on a worldwide basis. Further experiments will be required, especially for the semi-intensive cultivation of *S. aureus* and *S. niloticus* and the intensive cultivation of the other two species.

As the culture system is progressively intensified, growth and production increase, but, in the presence of moderate to dense algal blooms, relatively high values are attainable for *S. aureus* even in extensive systems. Indeed these results are even better than for the other tilapias in more intensive systems. Therefore, the extra cost of feeding should always be weighed against the production obtainable from natural foods for a given species and location. Such comparative data are lacking for *S. niloticus* which might also provide good results with algal blooms.

The current relative production potential for the four species may be summarized as: for extensive systems, *S. aureus* > *S. niloticus* > *S. mossambicus;* for semi-intensive systems, *T. rendalli* > *S. mossambicus* and for intensive systems, *S. aureus* > *S. niloticus* > *S. mossambicus. T. rendalli* fed with *Alocasia* leaves and wheat bran has shown a relatively good produc-

Table 16. Production performance of the four main species of tilapias used in cage culture systems: *Sarotherodon aureus; S. niloticus; S. mossambicus* **and** *Tilapia rendalli.* **Bracketed entries are representative of the range of values reported.**

	Stocking data			Growth data			Harvesting data		Culture	
	Pm_i (g)	N/m^3	B_i (kg/m^3)	G (g/d)	MSG (%)	DRIB (%)	B_f (kg/m^3)	MP (kg/m^3)	period (mo)	References and remarks
A. Initial stocking with small-size fingerlings										
S. aureus										
Extensive, with dense algal blooms	10	500	5.0	6.78	235	2.5	43.1	12.7	3.0	Suwanasart 1972
Intensive, 36% protein, floating pellets	10	500	5.0	0.5	156	2.2	23.1	7.8	2.5	Jordan and Pagán 1973
Intensive, 40% protein, floating pellets in fertile ponds	10	500	5.0	1.36	408	3.0	64.7	20.6	2.9	Suwanasart 1972
S. niloticus										
Semi-intensive, eutrophic lake + rice bran feeding	10	100	1.0	0.8	—	—	15	2.3	6.0	Anon. 1979a
Intensive, artificial lake, 25% protein feed	16	268	4.3	1.05	197	1.5	41.9	7.6	5.0	Coche 1977
S. mossambicus										
Extensive, eutrophic lake, experimental	10	50	0.5	0.36	110	1.3	1.3	0.8	3.0	Pantastico and Baldia 1979
Extensive, eutrophic lake, commercial	5 to 10	16 to 75	0.1 to 0.5	—	—	—	1.6 to 7.5	1.5 to 7	4 to 6	see Table 2
Semi-intensive eutrophic lake + 4% protein feed	10	50	0.5	0.5	157	1.9	2.8	0.8	3.0	Anon. 1979a
Intensive, eutrophic lake + 27% protein feed	10	50	0.5	0.65	195	2.2	3.4	1.0	3.0	Pantastico and Baldia 1979
T. rendalli										
Semi-intensive, fed *Alocasia* leaves	10	200	2.0	0.34	100	(1.0)	16.0	20	(7.2)	Popma 1978
B. Initial stocking with medium-size fingerlings										
S. aureus										
Extensive, with moderate algal blooms	25	500	12.5	0.72	86	1.7	54.8	14.1	3.0	Suwanasart 1972
Intensive, fertile ponds + 40% protein floating pellets	25	500	12.5	1.72	207	2.2	83.3	24.4	2.9	Suwanasart 1972
Intensive, fertile ponds + 40% protein sinking pellets	25	500	12.5	1.07	128	1.8	58.9	16.0	2.9	Suwanasart 1972

Table 16 (cont'd)

	Stocking data			Growth data			Harvesting data		Culture period	
	Pm_i (g)	N/m^3	B_i (kg/m^3)	G (g/d)	MSG (%)	DRIB (%)	B_f (kg/m^3)	MP (kg/m^3)	(mo)	References and remarks
S. niloticus										
Extensive, unfertilized pond, all males	25	200	5.1	0.60	68	(1.9)	6.8	3.5	(0.5)	Philippart et al. 1979
Intensive, 24% protein moist pellets	20	250	5.0	1.33	200	2.7	25.0	10.0	2.0	Guerrero 1979a
Intensive, artificial lake, 25% protein sinking pellets	22	218	4.8	1.20	164	1.3	34.6	5.8	5.1	Coche 1977
Intensive, artificial lake, 25% protein sinking pellets	29	257	7.5	1.33	138	1.3	56.1	9.3	5.1	Coche 1977
Intensive, artifical lake, 20% protein sinking pellets	31	186	5.8	1.72	166	2.0	33.7	9.4	3.0	Campbell 1978b
Intensive, unfertilized pond, 46% protein, all males	25	200	5.1	1.40	168	(4.0)	9.2	8.3	(0.5)	Philippart et al. 1979
Intensive, unfertilized pond, 46% protein, all males	32	200	6.2	0.93	87	(2.7)	9.2	6.0	(0.5)	Philippart et al. 1979
Intensive, artificial lake, 22% protein sinking pellets	32	177	5.6	1.98	186	1.6	42.1	8.8	4.	Campbell 1978b
T. rendalli										
Semi-intensive, fed *Alocasia* leaves	23	100	2.3	0.95	127	1.3	15.8	2.7	5	McLarney 1978
Semi-intensive, fed *Alocasia* leaves	27	225	6.0	0.50	56	1.0	18.8	3.2	4	Corredor 1978
Semi-intensive, fed *Alocasia* leaves + wheat bran	23	100	2.3	1.30	173	1.5	21.6	3.9	5	McLarney 1978

Pm_i : mean weight at stocking
N : number of fish
B_i : mean biomass at stocking
G : average growth rate of individual fish during the culture period
MSG : average monthly specific growth rate
DRIB : daily rate of increase of the biomass calculated from $B_f = B_i (1 + i)^n$ where n is the culture period in days and i is $\frac{DRIB}{100}$
B_f : mean biomass at harvest
MP : average monthly production: $(B_f - B_i)$ observed and recalculated on a 30-day basis

Table 17. Comparison of the production performances of *Sarotherodon aureus*, *S. niloticus*, *S. mossambicus* and *Tilapia rendalli* in three cage culture systems: extensive, semi-intensive and intensive. Bracketed entries are representative of the values reported. Data obtained from all authors cited in Table 16.

	Extensive culture				Semi-intensive culture				Intensive culture			
	G g/day	MSG (%)	MP (kg/m^3)	DRIB (%)	G (g/d)	MSG (%)	MP (kg/m^3)	DRIB (%)	G g/day	MSG (%)	MP (kg/m^3)	DRIB (%)
A. Initial stocking with small-size fingerlings (10 to 15 g)												
S. aureus	0.78	235	12.7	2.5	—	—	—	—	0.5 to 1.4	156 to 408	7.8 to 20.6	2.2 to 3.0
S. niloticus	—	—	—	—	0.8	—	2.3	—	1.05	197	7.6	1.5
S. mossambicus	0.36	110	0.8 to 4	> 1.3	0.5	157	0.8	1.9	0.65	195	1.0	2.2
T. rendalli	—	—	—	—	0.34	100	2.0	(1.0)	—	—	—	—
B. Initial stocking with medium-size fingerlings (20 to 30 g)												
S. aureus	0.72	86	14.1	1.7	—	—	—	—	1.1 to 1.7	128 to 207	16.0 to 24.4	1.8 to 2.2
S. niloticus	0.60	68	3.5	(1.9)	—	—	—	—	0.9 to 2.0	87 to 200	5.8 to 10	1.3 to 2.7
S. mossambicus	—	—	—	—	—	—	—	—	—	—	—	—
T. rendalli	—	—	—	—	0.5 to 1.3	56 to 173	2.7 to 3.9	1 to 1.5	—	—	—	—

G : average growth rate of individual fish during the culture period
MSG : average monthly specific growth rate
MP : average monthly production: ($B_f - B_i$) observed and recalculated on a 30-day basis
DRIB : daily rate of increase of the biomass calculated from $B_f = B_i (1 + i)^n$ where n is the culture period in days and i is $\frac{DRIB}{100}$

Table 18. A comparison of the production potentials of *Sarotherodon aureus, S. niloticus, Cyprinus carpio, Ictalurus punctatus* and *Salmo gairdneri* in intensive cage culture. Data averaged from various authors: see Table 16 for *Sarotherodon* spp. and Coche 1978, 1979 for other species.

	Growth (g/d)	Best average production ($kg/m^3/mo$)	Actual average production ($kg/m^3/mo$)	Estimated annual production potential (kg/m^3*)	Feeding	FCR
S. aureus	1.0 to 1.7	> 25 over 5 mo	10 to 20	200 to > 300	40% protein floating pellets	1.0 to 1.8
S. niloticus	1.0 to 2.3	> 15 over 4 mo	5 to 10	200 to > 300	25 to 30+% protein sinking pellets	1.9 to 2.2
C. carpio	3.3 to 7.7	> 35 over 3.5 mo	5 to 30	> 420	30 to 35+% protein	1.6 to 2.3
I. punctatus	1.5 to 2.2	25 over 8 mo	10 to 20	240	36% protein floating pellets	1.0 to 1.7
Salmo gairdneri	0.3 to 1.9	25 over 4 mo	1 to 10	300	40 to 46% protein	1.4 to 2.5

*mainly based on data from small cages (1 to 6 m^3); these values would be lower for larger cages.

tion potential, at low input cost. In intensive systems, both *S. aureus* and *S. niloticus* provide high production. It may be concluded that these tilapias have good production potentials, particularly in culture systems with low-energy inputs, where production costs may be kept relatively low. Even then, the average monthly production is above that experienced for super-intensive pond culture in Israel: estimated at about 0.5 kg/m^3/month, i.e., about 60 t/ha/year (Sarig and Arieli 1980).

2. Comparison with Other Cultured Species

The production potential of the *S. aureus* and *S. niloticus* in intensive systems is compared with that for *Cyprinus carpio, Ictalurus punctatus* and *Salmo gairdneri* in Table 18 (Coche 1978, 1979). The two tilapias can outperform *I. punctatus* and *Salmo gairdneri*, especially considering both the relative ease of tilapia culture and its efficient use of low-cost feed. *C. carpio* grows faster and has a definitely higher production potential, although this alone might not justify its preference as a cultured species over *S. aureus* or *S. niloticus*.

Discussion

MORIARTY: How much of a problem is there with fouling of the cages by filamentous algae and other organisms growing on the walls of the cages and restricting the water flowing through them?

COCHE: It depends on the environment where you are keeping your cages. In very rich environments, you can have problems, but in general tilapias help to clean the cages.

PULLIN: In the Philippines, tilapia actually eat the filamentous algae growing on the sides of the cages, so much so that it may be an important source of extra food for them.

COCHE: It can also be a problem for the cage netting. During this process, we have had tilapia damaging some of the netting.

NASH: Just a comment. I do not altogether accept your interpretation that as the cage size increases, the production tends to go down. I think that it does in regular, either hexagonal, square, or circular cages, but certainly in the early days of marine cage development in Scotland we found that the elongated cage opened up new dimensions of production for you, although structurally it was easier to make the square, hexagonal, or circular cage. Secondly, we cut down a great deal on food loss by having about two-thirds of the bottom being covered with some sort of plate which stopped the food going right through. The only problem was then that as we moved many of these cages up and down in the water, particularly if they were submerged cages, the phugoid motion often put the cages under tremendous stress and in fact cracked several of them. But certainly, if you have floating cages, I think that plates up to about 2/3 of the bottom structure are a help to reduce food loss.

HEPHER: Dr. Nash, could you explain the superiority of the elongated cage over the round one with respect to yield and production?

NASH: I think one is just more limited with a round or regular cage because it is largely the movement of the fish within the cage that maintains good circulation. With a regular

cage, you can get more dead pockets in the center as the fish can only move the water around so much. Also the cage area which can be made to face the water flow is limited. With an elongated cage, you can put the long side across the water flow and so maintain much better conditions, although putting it across a strong flow can give you additional structural problems.

ROBERTS: As Dr. Nash has indicated, we have had considerable experience of cage culture, particularly with salmonids in Scotland. Because of the very rapid development of freshwater cage culture in Scotland, we now have several 2 to 300 t units in relatively small lochs. We have a project at the Institute of Aquaculture measuring the environmental impact of cage culture in Scottish lochs. We are already finding that there is a very significant effect of having a 300 t unit in a loch. The effect seems principally to be not only in terms of the chemical water quality, but a more general effect on the fish fauna of the loch, for example, enlarged eel populations. Of course, the entrepreneur who has the fishing rights on the loch can say, "That's fine; I will put eels in the cages as well because the price of eels is so much higher than the price of trout!" But there are limits, certainly, as to safe stocking levels.

We have also studied samples from cage culture mortalities in the Philippines where the obvious cause was anoxia associated with a very highly eutrophic system. As biologists interested in cage culture we should hold back the entrepreneur as he will double and triple and quadruple his production without giving thought for tomorrow and tomorrow may be a very long day of misfortune if we are not careful.

One other thing which Dr. Coche raised in terms of cage culture was the possible differing risks of cage culture and other systems. We are not completely convinced that there is a significantly higher stress in cages compared with other high density culture systems. Stress of course would increase disease risks. But two particular groups of parasitic copepods, *Argulus* species in fresh water and salmon lice in salt water can hide out in the cage structures. They are not obligatorily living on the fish. As soon as you stress the fish, the adrenalin surge seems to give the parasites a message that they had better leave their hosts and they all disappear off the fish on to the cage-netting; so that if you pull fish out to treat them, you only get a small proportion of the parasites. When you put the fish back in, the parasites come back and reinfest the fish. Obviously, their hiding places are multiplied on fouled netting.

We have also experienced problems of wild fish infecting the cultured fish in the cages. In Scotland, the parasitic burden of the wild fish in the lakes with cages may increase to a significant extent and this annoys the anglers or capture fishermen. Part of the problem is that birds fishing in the cages facilitate the cycle of the fish parasites and spread them. Cages in our view have a very big future in aquaculture terms, but they have some specific disease problems, and must be monitored very carefully in terms of their environmental impact.

Diseases of Tilapias

R. J. ROBERTS AND C. SOMMERVILLE

Institute of Aquaculture
University of Stirling
Stirling, Scotland

ROBERTS, R.J. AND C. SOMMERVILLE. 1982. Diseases of tilapias, p. 247-263. *In* R.S.V. Pullin and R.H. Lowe-McConnell (eds.) The biology and culture of tilapias. ICLARM Conference Proceedings 7, 432 p. International Center for Living Aquatic Resources Management, Manila, Philippines.

Studies on intensively cultured and wild tilapias have shown that although they are more resistant to diseases than many other species, there is nevertheless a wide range of disease problems which can occur. To date no viral problems other than lymphocystis have been recorded. The bacterial pathogens include aeromonads, myxobacteria and particularly *Edwardsiella tarda*. A wide range of parasitic problems occur including particularly trichodinids, *Ichthyophthirius* and various intermediate stages of digenean flukes. Nutritional problems are a major difficulty in intensive culture, with aflatoxicosis a major cause of losses associated with poor quality storage of food ingredients.

Introduction

There are two principal reasons why diseases of the tilapias (genera *Tilapia* and *Sarotherodon*) have been less well studied than those of many other groups of cultured fishes. The first is that such fish are generally farmed in countries where diagnostic facilities are less than adequately developed, so that losses cannot be investigated properly; the second is that the culture of tilapias has only been intensified recently. At low stocking densities, environmental water quality is usually high and opportunities for infections to build up are thus limited. Equally, the observation of disease conditions is much more difficult at low densities in large water bodies than under the controlled conditions of the high density tank or cage systems.

Most of the early observations on diseases of tilapias have related to parasitic infections, often from wild fish and at low levels (Khalil 1971; Sarig 1971). These have generally shown no evidence of clinical effects on the fish and in most wild populations of tilapias it would seem that parasitism is a normal occurrence of little consequence. The present paper is restricted to a review of diseases of tilapias which occur in culture systems or those of wild fishes which have a clinical manifestation or potential significance for culture. Although the literature has been surveyed, and references are quoted where relevant, the majority of the information has been derived from the diagnostic files of the Institute of Aquaculture of the University of Stirling and from unpublished research by the authors and their students in the Institute.

Parasitic Diseases

Although a vast array of parasites has been recorded from tilapias and other cichlids most of them are from wild fish and no evidence of clinical effect is given. The number of reports from cultured fish is much smaller but clinical disease is much more frequently associated with infection in these circumstances. It is with clinical infection in cultured fish that the aquaculture industry is concerned, but in many cases such infections are derived from wild stock in the water supply and other hosts such as birds or invertebrates, or are introduced with fish brought in from wild sources for ongrowing or breeding.

1. Protozoan Parasites

a) Ichthyophthirius multifiliis. This parasite, which causes "Ich" or "white spot" is a relatively large ciliate which causes heavy losses in a wide range of cultured fish. It was first reported from tilapias by Paperna (1970) who recorded it in wild tilapias from east African rift valley species. It was believed to have been introduced to the area with mollies *(Mollienesia sphaenops)* brought in for mosquito control. Brock and Takata (1955) reported its introduction to the Hawaiian islands with imports of *S. mossambicus* from Singapore. Thus any area where it is not presently found would be advised to resist such importations very strenuously, unless in the form of young fish which have been comprehensively treated with formalin and securely quarantined before distribution. *Ichthyophthirius* has a closely defined optimal temperature range for growth of 20 to 24°C (Meyer 1970) and thus for tilapias it is unlikely that significant epizootics will occur in tropical areas unless the fish are highly stressed. It is in the subtropical areas when winter temperatures are both stressful to the tilapias and also pass through the optimum range for the parasite as they rise in spring that severe problems with this parasite will arise. It is a significant problem for fry since they are particularly susceptible to infection and lesions may be found in the nares, pharynx and gill as well as the skin. In intensive culture using recirculation systems, it is a particular problem because of the difficulty of treating the filter without destroying its bacterial flora. If the filter is isolated during treatment the parasite is able to lodge, and survive within it for a considerable period and thus infections can flare up when the circulation is restored. (Richards pers. comm.). Efforts to induce infections experimentally produce very variable results but natural infections can build up extremely rapidly, with the entire body surface of fry being covered with trophonts and tomites within 48 hr of their first being observed.

b) Trichodinids and Chilodonella spp. There are a number of ubiquitous protozoan parasites which commonly cause disease in cultured fish including

Trichodinella spp., *Tripartiella* spp., *Trichodina* spp. and *Chilodonella* spp. Only the latter pair has been reported as being of significance to cultured tilapias but it is likely that the others will also assume importance in the future. They all have a direct life cycle and reproduce by binary fission on the skin and gills. They are present on most fish in small numbers and characteristically cause problems in fish stressed by handling, poor feeding and especially where there has been a drop in temperature. As the temperature rises again, as with *Ichthyophthirius*, the parasites multiply on the stressed and debilitated fish.

Thus, they are a problem in Israel (Sarig 1971) and the Southern U.S.A. (Avault et al. 1968) where it is necessary to provide heated water facilities to overwinter fish and high fish density/low flow rates and minimal temperatures are the norm from economic necessity.

Sommerville (unpublished) has induced severe *Chilodonella* infection in *S. galilaeus* subjected to severe nutritional deficiency in a static water system. It was not possible to establish the infection by direct transfer to healthy, well-fed fish or to fish in clean water with good exchange. Fryer and Iles (1972) consider *Trichodina* to be a particular problem in mouthbrooders as these ciliates can invade the mouth and transmit the infection to fry. Sarig (1975) has also reported heavy infection from cultured tilapias in Ghana, stating that in many cases such infections were highly pathogenic and caused heavy losses, especially of small fish. Guerrero (pers. comm.) has also frequently implicated trichodinids in heavy, and often lethal infections of *S. mossambicus* fry undergoing hormone sex-reversal treatment.

c) Bodonid parasites. Ichthyobodo (=Costia) necatrix is a well recognized and highly pathogenic parasite of young, or severely stressed, salmonids (Robertson 1979). In tilapias, bodonids have been associated with disease in Alabama (Plumb, cited by Scott 1977) and in Israel (Sarig 1971). Sarig reported that they occurred infrequently in late autumn and winter and only on the gills of the affected fish, accompanied by *Trichodina* and *Glossatella*. In experimental populations of *S. mossambicus* held in recirculating systems at Stirling, Mohd-Shaharom (unpublished) has found mortalities associated with a bodinid parasite closely resembling *Ichthyobodo necatrix* in heavily stocked populations held at 26°C, which is a much higher temperature than previous reports.

d) Sporozoa. Myxosporidia occur commonly as cysts replete with spores in the tissues of most wild tilapias. Generally they are of the *Myxobolus/Myxosoma* group. Baker (1963), Fryer (1961a, 1961b) and Sommerville (unpublished) have all recorded myxosporidians from wild tilapias. They rarely show any evidence of significant pathological effect but in view of their life cycles, they are potentially highly significant for intensive earth pond culture, as is the case with *Myxobolus* and *Henneguya* infections in cultured carps in Bangladesh (Sommerville and Iqbal, unpublished). Earth pond culture greatly facilitates the parasite's life cycle since spores are released during decomposition of a fish carcass (or possibly from living fish) and for all known species require a period of potentiation in mud prior to the development of infectivity. These parasites are more likely to be a problem in countries where the height of the water table precludes seasonal drainage and desiccation of the ponds.

2. METAZOAN HELMINTH PARASITES

a) Monogenea. Monogenean parasites are found on both the gills and the skin of tilapias and a specific genus for the monogeneans of cichlid fish, *Cichlidogyrus*, was erected by Paperna in 1960. Paperna and Thurston (1968) found 16 different species from rift valley tilapias and Mohd-Shaharom and Sommerville (unpublished) have performed extensive studies on the incidence and pathogenesis of these monogeneans in cultured populations. *Cichlidogyrus sclerosus* was found to be the most widely spread, being commonly observed in fish from Southeast Asian origins such as Philippines (Duncan 1973) and Hong Kong, Singapore and Thailand (Sommerville, unpublished) as well as from Africa. Studies on the population dynamics and growth rate effects of *C. sclerosus* and *C. tubicirrus minutus* indicate that they are normally of low pathogenicity with little or no effect on growth rate. Pathological effects were quite obvious, however, in the form of focal hyperplasia of the branchial lamellae.

Gyrodactylus spp. are potentially of great significance to tilapias in culture. Clinical outbreaks of gyrodactyliasis have been recorded by Fryer and Iles (1972) in pond-reared tilapias in Uganda, where they were associated with corneal damage and by Sommerville and Haller (unpublished) from intensively cultured tilapias in Kenya. In the latter case mortalities appeared to be associated with handling and moving of fish from ponds to tanks. Mortalities always developed two weeks after transfer, tailing off after 5 to 6 weeks. This pattern was presumed to be related to the viviparous life cycle of the gyrodactylids. The increase in transmission opportunity was associated with the higher stocking density of a concrete tank system and the debility of badly handled fish.

b) Digenea. A great variety of digenean parasites are potentially capable of causing heavy losses in cultured tilapias. All, however, have a basically similar life cycle which involves three hosts: the first intermediate host, an aquatic or amphibious gastropod mollusc; the second intermediate host, a fish and the final host, a piscivorous vertebrate. The stage occurring in the fish is known as the metacercarial stage and is generally found encysted in the fish tissues. Virtually any part of the body can be involved but the metacercariae of many species of digeneans are site specific. The molluscan hosts are particularly common in water supply dams, inlet channels and drainage canals, where waste food and organic material accumulate. The fish may be infected on the farm or, as is frequently the case, in ponds or dams where fry are collected for stocking of rearing tanks. A wide range of metacercariae is observed in tilapias and because of the difficulty of identifying the larval form to species, they usually remain unspecified. Sommerville (unpublished) has investigated a wide range of digenean infections in intensively cultured tilapias and carried out experimental infections of final hosts to determine species. The principal groups of significance were clinostomes, neascids, haplorchids and diplostomulae and all were significant in different ways. Although the following are currently the only named digenean metacercariae to have been associated with disease in cultured tilapias, it seems likely that many others will be shown to be of significance as controlled intensive tilapia culture develops:

i) Clinostomes. (Family *Clinostomatidae*) The metacercariae of *Clinostomum* and *Euclinostomum* spp. are notable particularly for their large size—hence the common term "white grub" or "yellow grub" for the mature metacercariae *in situ*. In fry the growth of such a large parasite causes bulging and distortion of the body profile. As well as spoiling the appearance of the fish there is some evidence that in intensive culture, fish infected with clinostomes are also more susceptible to handling stresses (Balarin pers. comm.). The adult worms lie in the pharynx of a piscivorous bird and *Ardea goliath* has been recorded as being a particularly prevalent carrier of the infection.

ii) Neascus metacercariae. (Larval genus of Superfamily *Strigeoidea*, all belong to Family *Diplostomatidae*) A group of neascus metacercariae stimulate the accumulation of large numbers of melanocytes in the host capsule. Since they usually accumulate in the skin, the resulting black spots become very obvious even to the casual observer. However, although they have been reported from a wide range of tilapias from east and west Africa (Paperna and Thurston 1968; Sommerville unpublished;) they probably have only a limited effect on the fish host *per se* and their real significance is that they render fish unmarketable where infection is heavy.

iii) *Haplorchis* sp. (Family *Heterophyidae*) Metarcercariae of this genus are found in a wide range of freshwater food fishes including tilapias. They have a wide geographic distribution and have already been a cause for concern in the Philippines, Israel, Africa, India, Egypt, China and Japan (Sommerville in press).

Haplorchis pumilio was first recorded from tilapias by Witenberg in 1929 in Palestine. Like other species of this genus it has a wide distribution, summarized by Sommerville (in press) who has also recently found it in an intensive *S. spilurus* culture system in Kenya. Its cercariae were found in a high proportion of *Melanoides tuberculata:* the dominant mollusc in the tanks, drains, and ponds of the farm system. Experimental infections of tilapia fry showed that mortalities could result directly from penetration of the skin and muscle of fry by large numbers of cercariae and the migrations of metacercariae to accumulate at fin bases. It is assumed that the mortalities associated with mass penetration are related to loss of skin function, but other organs may be damaged as the parasites penetrate over the entire skin surface and migrate indiscriminately through tissues.

With less severe infections, experimental studies (Sommerville unpublished) have shown that the infected fish suffer no disadvantage when compared with uninfected fish as far as growth rate, condition or food conversion are concerned. However, the presence of this parasite in cultured food fish cannot be tolerated because of the risk it represents to human health when inadequately cooked fish are eaten. The main pathological effect when man is the final host is associated with the distribution of the eggs in the circulatory system, where they lodge in end-arteries, causing infarction. This can cause a variety of syndromes but the most significant are related to obstruction of myocardial, cranial or optic vessels. Since it is non-specific in its final host, piscivorous vertebrates other than birds, e.g., otters, dogs, cats, or even monitor lizards, can act as final hosts. Other larval heterophyids have been reported in tilapias in the Middle East, Paperna (1960) reporting *Heterophyes*

heterophyes as well as *Haplorchis pumilio* as common parasites of man and dogs in Egypt and Israel.

iv) Diplostomulae. (Larval genus of Family *Diplostomatidae*) The metacercariae of these strigeoids lodge in a variety of tissues within their second intermediate fish host, depending on the particular *Diplostomum* species involved. The most noticeable in clinical terms, however, are the metacercariae which are associated with blindness: the so-called eye flukes. There is a wide range of species of eye fluke, with a remarkable degree of site specificity so that some species localize in the lens, others in the retina and yet others in the aqueous or vitreous humour. Where only small numbers of unencysted metacercariae are involved, the sight is not significantly affected, but when large numbers of metacercariae locate in the eye they can produce total blindness (Plate 1). This causes loss of reflex pigmentation control and affected fish are thus more vulnerable to predation than would otherwise be the case, as well as being unable to find their food by visual means.

c) Cestodes, Nematodes and Acanthocephalans. Cestodes have not as yet been reported to cause any serious problems in cultured tilapias, although their presence in wild fish has been reported (Fryer and Iles 1972). Several nematode species have also been reported (Fryer and Iles 1972; Goldstein 1971) but apart from reports by Paperna (1964) and Scott (1977) of larval *Contracaecum* infection causing pathological effects, little is known of their significance. Scott found that in *S. alcalicus (S. grahami)* in Kenya up to 7 parasites might be found in the pericardial cavity (Plate 2) and that they had a significant effect on the growth of the fish. *Contracaecum* is a particularly unattractive parasite to the consumer as it can occur as large encysted worms throughout the muscle. Like some other anisakid parasites, it can also be zoonotic and it would seem important that it be excluded from cultured fish. Its life cycle involves ingestion of the egg by a free living crustacean which is ingested by the fish. When this then is eaten in turn by a piscivorous bird the life cycle is completed, and it would seem that it is only at this stage that the cycle can be readily broken. Since piscivorous birds are also responsible for many digenean infections of fish and also cause direct losses of cultured fish by predation, the netting or wiring of sites, tank coverings, and an active predator elimination policy, would seem to be highly desirable.

Plate 1. *Sarotherodon mossambicus* with bilateral cataract caused by a heavy diplostomulid metacercarial infection.

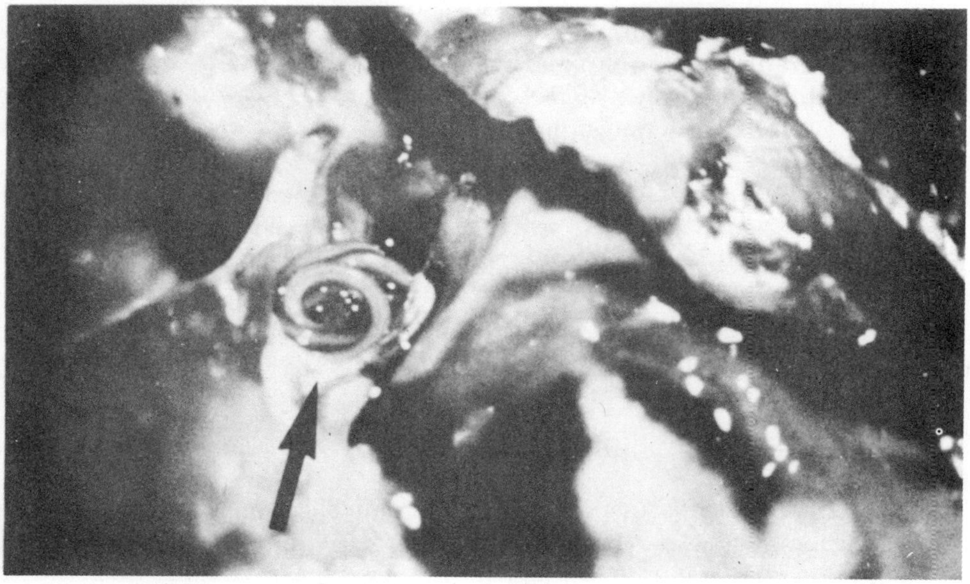

Plate 2. ***Contracaecum*** **within the pericardium of** ***Sarotherodon alcalicus (S. grahami).*** **Photograph courtesy of Mr. P.W. Scott.**

3. PARASITIC CRUSTACEANS

Numerous species of parasitic crustaceans have been reported from tilapias. Of these the most serious pathogens are the branchiuran *Argulus* spp. and the copepods *Ergasilus* spp. and *Lernaea* spp.. Sarig (1971) summarizes the reports of Fryer and Paperna and describes the problems experienced with infestations by these parasites. Fryer (1961b, 1965a, 1968) and Paperna (1969) found *Lernaea* to cause serious losses in ponds in Nigeria and Israel and in Malawi (Roberts unpublished), they were found to be a significant cause of mortality in *T. rendalli ('T. melanopleura')* culture in ponds. The parasites are deeply embedded in the dermis or musculature of the fish, and cause a severe necrotic ulcerative granulomatous response. The lesions render the fish unmarketable, and fish with even a single parasite feed poorly, and lose weight. *Lernaea tilapiae* is restricted to tilapias of Lake Malawi and has been described by Fryer and Iles (1972) as inhabiting the mouth and penetrating the tissues of the palate. In view of the finding of as many as 9 parasites in this site, all causing severe inflammatory lesions, they conclude that it must also interfere with feeding and mouthbrooding in addition to the more general effects of "anchor" penetration. The infective stage is the free swimming copepodid (the second larval instar) but a further four moults, accompanied by a final sexual differentiation and mating are necessary for the adult female to become pathogenic and insert her head into the skin where the feeding activity and growth of the "anchors" causes the severe reaction.

Argulus spp., the fish lice, are very mobile, unlike the sedentary *Lernaea*. Several species have been found on tilapias (Paperna 1964; Fryer 1968; Sarig 1971). Its life history is more simple than that of *Lernaea* and although it is an obligate parasite it can survive off the host for a limited period and since it lays its eggs on vegetation, it is a serious problem in standing waters. It can readily seek refuge around a pond if fish are removed for treatment. It can cause mortalities in its own right but usually it is responsible for facilitating secondary bacterial and fungal invasion by its feeding method, which causes open wounds.

Ergasilus spp. have been reported to cause epizootics in polyculture fish ponds in Israel (Sarig 1971) where carps, mullets, and tilapias are farmed together. Although the tilapias do become infected in such conditions, their parasite burdens are usually very low compared to those of the other two groups of fish species, possibly reflecting the higher resistance to skin penetration of the tilapias.

Bacterial Diseases

Tilapias are generally kept in water with a high organic load. This may be due either to deliberate fertilization to increase the production of food organisms in the water or to agricultural or sewage run-off. Such water conditions allow most of the recognized aquatic bacteria to occur at significant levels and under intensive culture conditions especially heavy mortality can result from infections with any of a wide range of recognized facultative pathogens.

Three clinical syndromes are associated with bacterial diseases of tilapias: a predominantly skin lesion syndrome associated with pathogenic myxobacteria; haemorrhagic septicaemia; and chronic granulomatosis associated with *Mycobacterium tuberculosis* infection.

1. Myxobacterial Infections

Myxobacteria, or 'slime' bacteria, are aquatic bacteria commonly found as commensals on fish skin. A few species are capable of becoming pathogenic under particular circumstances of environmental stress. The most common stressors inducing myxobacterial infections are high or low temperatures, but traumatic damage and low water quality, such as that caused by excessive silt or high ammonia levels associated with filtration failure in recirculation systems, can also induce the qualitative or quantitative changes in mucus secretion from the skin or gills which appear to trigger infection.

Flexibacter columnaris is the commonest myxobacterial pathogen in tilapia culture and is usually associated with high temperature stress. The lesions are opaque white, raised and whorl-like, or may take the form of a saddle-back of grey white epithelial necrosis over the dorsum of the back with a red haemorrhagic rim. These develop rapidly to form crateriform ulcers. The affected fish become very dark, slow moving and die quickly. Myxobacterial infection of the gills, particularly of fry is usually associated with heavy mucus production and the bacteria may be doing little more than obstructing gaseous exchange since treatment with surfactants, whether bacteriostatic or not, serves to remove both the mucus and the problem, provided environmental conditions are also improved. Both gill myxobacterial infection and dorsal fin-rot (also associated with myxobacteria) are usually associated with low water temperatures and Avault et al. (1968) stress the importance of maintaining overwintering temperatures of at least 14° C for this reason.

2. Haemorrhagic Septicaemia

Gram-negative bacterial septicaemias are the commonest bacterial causes of mortality in tilapia culture. Infection may arise from introduction of infected fish into a system but more often it is a function of the environmental conditions of that culture system.

Aeromonas hydrophila infection is the most common of the septicaemias and usually manifests itself by causing affected fish to darken, lose appetite, and cluster around exit screens. When examined they may have ulcers, or more frequently, areas of hyperaemia at the base of the pectoral and pelvic fins, and at the margins of the orbit (Haller 1974; Scott 1977). Internally the liver is usually pale and there may be focal haemorrhages over the visceral and peritoneal surfaces. Histologically there is haemopoietic necrosis and focal necrosis in the liver, heart or skeletal muscles, with accumulations of gram-negative bacteria ranged along strands of fibrin. The cellular inflammatory infiltrate is rarely marked but when it is present it generally comprises macrophages which often contain ingested melanosomes. Generally the condition is predisposed by handling trauma, poor nutrition, heavy parasitism or excessive fertilizing of ponds but sometimes severe outbreaks may occur without any obvious predisposing factor.

Occasionally pseudomonads, such as *Proteus* spp., β-haemolytic streptococci (Wu 1970) or *Edwardsiella tarda* (Roberts unpublished) may be associated with the condition. These are particularly common in ponds newly fertilized with human or animal faeces. Coliforms and *Salmonellae* of human health significance may also be ingested from such sources and there is therefore a risk of outbreaks of human gram-negative infections, although considering the extent to which night soil is used to fertilize fish ponds in Asia, their significance as a source of human infection would seem unlikely to be high.

Edwardsiella tarda infection provides a particularly intractable problem in intensive systems. Although clinically and histologically this condition is a typical bacterial septicaemia, losses are usually sporadic. It is extremely difficult to eradicate from the system and over a period of time such infections can be responsible for the loss of significant numbers of fish. Food medication with oxytetracycline or potentiated sulphonamides eliminates losses during the feeding period but they often recur once treatment has ceased. Ruthless culling of all poor growing, sluggish or darkened fish from the system and maintenance of best possible water quality and feeding standards are the most effective means of preventing losses.

Experimental infections with *Aeromonas salmonicida* (Almeida et al. 1968; Roberts unpublished) have shown that both the pigmented and the achromogenic strains of this common pathogen of temperate fishes can cause similar syndromes in tilapias and there seems no reason to suppose that further studies will not also reveal its presence in cultured tilapias as a significant pathogen.

3. TUBERCULOSIS

Infection with *Mycobacterium fortuitum*, commonly known as fish tuberculosis, is well known in all cichlids kept in aquaria (Nigrelli and Vogel 1963). In the wild it has been recorded only infrequently in tilapias, e.g., Roberts and Matthiesen (1979) recorded it in *S. andersonii* and *T. sparrmanii* from the Okavangu swamp in Botswana. In intensive culture it has only been recorded once in *S. niloticus* in intensive culture in Kenya (Roberts and Haller unpublished). Affected fish showed small focal granulomata in the liver, spleen and kidney. Mortalities were limited but again it seems likely that under certain circumstances, particularly if trash fish are fed, this condition could be a potential source of severe losses.

Mycotic Infections

Only two fungi are recognized as serious pathogens of tilapias although various others, such as demateacious moulds, are suspected of causing occasional mortalities. Phycomycetes of the genus *Saprolegnia* can be a cause of severe losses as with other cultured fish: again in association with traumatic damage from handling, from sexual aggression or at temperatures approaching the minimum range for the tilapias. Thus in Israel handling is avoided between October and May (Sarig 1971) and prophylactic spraying of ponds with malachite green is often carried out. *Saprolegnia* fungal growths are usually disposed as grey mats arranged in whorls over the surface of affected fish. Such infections are usually seen much more easily while the fish are in the water than when they are removed.

Infection with *Branchiomyces* spp. can be a significant cause of mortality when fish are reared in poor quality water with a very high level of decaying vegetable or other organic material. The fungus may invade the gill via the branchial vessels or the epithelium and cause massive destruction of the respiratory surfaces. Liming of the pond has been suggested as means of reducing losses but improvement of the water quality is a much more rational approach.

Viral Diseases

The only validated report of a viral disease of tilapias is the report by Paperna (1974) of lymphocystis infection of wild tilapias in east African lakes. This condition has caused severe problems in marine fish culture in Europe and in tropical aquaria and it represents a potential hazard for intensive culture systems. However, experimental vaccination studies have shown that susceptible marine fish can be readily protected (Roberts 1975 and unpublished) and it is likely that this would also apply to tilapias in culture.

Although only one virus infection has been recorded in tilapias it is inconceivable that a similar range of commensal and pathogenic viruses as found in the salmonids and other intensively cultured species is absent from the tilapias. It would seem to be only a matter of time before these are manifested in intensive culture given the rate of the development of the industry and the improvement of diagnostic facilities.

Toxic Conditions

Toxic conditions may arise from toxic substances in the diet or in the water. They range from high levels of metabolic wastes such as ammonia to gaseous supersaturation of the medium. The range of potential toxic agents is very wide but only small numbers have been definitely associated with losses.

1. Gaseous Supersaturation

In intensive culture systems, pumping or piping faults can result in supersaturation of the water with dissolved gases. When absorbed into the fish during respiration the change in partial pressure can result in the gas coming out of solution and blocking or rupturing blood capillaries. Bubbles of gas then accumulate at sites which vary depending on the age of the fish. In tilapia fry these supersaturation bubbles can form anywhere but are found particularly in the area of the yolk sac or, in older fish, in the gill and skin. In adults the scale and skin structure is such that normally the only clinical location of gas bubbles is on the gill although it is likely that small occlusive emboli are also distributed unnoticed, throughout the tissues, since affected stocks rarely grow particularly well thereafter.

2. Toxic Effects of Algae

Although some freshwater algae such as *Prymnesium parvum* are toxigenic *per se* the main deleterious effect of algae in tilapia culture is death from anoxia following an algal bloom. Generally the microorganisms responsible are of the genera *Microcystis*, *Anabaena*, *Oscillatoria* and *Spirulina*. According to Swingle (1967) the anoxia results from a sudden dominance of one particular species which multiplies phenomenally under favorable conditions to form a thick scum on the surface. This prevents light penetration, causes death of all submerged vegetation and restricts oxygenation to the upper 6 cm of the pond. Although fish can live in this upper layer any sudden drop of temperature, or other weather changes such as wind or rain, can induce overturn or mixing of the layers with resultant anaerobiosis of the entire pond and mass mortality of the fish. These usually die with the characteristic features of anoxia: a wide open mouth and dilated branchial chambers.

3. AFLATOXICOSIS

Aflatoxins are toxic compounds produced by the mould *Aspergillus flavus.* In salmonids they are now recognized as producing hepatomata at very low levels in supplementary feeds. In tilapia culture, storage of feed in poor conditions leads readily to growth of *Aspergillus.* Recent work has shown that tilapias receiving feeds with a high level of aflatoxins B_1 and B_2 show a haemorrhagic syndrome, characterized by severe haemorrhage into the branchial musculature below the dorsal commissure of the operculum, a wide range of internal haemorrhages, distinct depression of haemopoiesis and massive accumulation of haemosiderin in both the splenic and renal melanomacrophage centers (Haller and Roberts unpublished). Frank hepatic neoplasia is not seen but pre-neoplastic islands of well-defined hepatocytes, as described in rainbow trout *(Salmo gairdneri)* by Wales (1970) are seen. Given the problems of food storage in those humid tropical areas where intensive tilapia production has its greatest potential, this condition seems likely to be recorded with increasing frequency.

Neoplasia

Mawdeslay-Thomas (1972) suggested, somewhat flippantly, that visceral neoplasia of *Sarotherodon niloticus*, which he diagnosed from Egyptian tomb paintings, represented the earliest example of a documented fish disease. However there has been very little in the way of descriptions of neoplasia in tilapias since then. Haller and Roberts (1980) have described dual neoplasia in the form of a lymphocytic lymphoma and a renal tubular adenoma in the same specimen of *S. spilurus* but otherwise there have been no descriptions of neoplasia in the group. Again this is almost certainly a reflection of the lack of detailed observation rather than a true picture of incidence.

Anomalies and Deformities

Although the incidence of congenital anomalies in cultured tilapias would be expected to be similar to that in other genera, the following two specific conditions occur with a frequency which makes them a particular problem.

1. SPINAL DEFORMITIES

Spinal malformations are not uncommon in small numbers in any intensively reared species but in the tilapias a particular form of spinal deformity has been studied by Tave (1980 pers. comm.) and by Pullin and Roberts (unpublished). Although occasionally seen in *S. spilurus*, *S. mossambicus* and *S. niloticus*, the anomaly seems most frequent in the Auburn strain of *S. aureus.* Tave has defined two aberrant types, which he calls "saddle back" and "stump body" (Plate 3). "Saddle back" fish lack all or part of the dorsal fin and in extreme cases, some or all of the pectoral, pelvic and anal fins also. Such fish were, from his studies, also less resistant to diseases such as *Saprolegnia* fungus infection. In the "stump body" the body is compressed antero-posteriorly, with no two fish showing the same skeletal anomaly. The saddle back form has not been described elsewhere than in the Auburn fish and it is the stump body anomaly that is the one which is widely observed.

Plate 3a. Normal *Sarotherodon aureus* radiograph.

Plate 3b. Fish showing the saddle back spinal deformities (arrowed) and dorsal fin anomaly. (From material kindly supplied by Dr. D. Tave).

2. LARVAL ANOMALIES

In intensive culture, the establishment of hatchery technology, which allows detailed observations to be made on young fishes, has shown the prevalence of particular larval anomalies. Rothbard et al. (1980) have described the commonest of these in *Sarotherodon niloticus*, but others are also found.

Mortality Associated Specifically with High Density

It has long been a recognized phenomenon that in high density culture the growth rates of many tilapias tail off and there are unexplained mortalities. Such growth inhibition has been attributed to the presence of some inhibitory factor in the water. The phenomenon was first defined for *Tilapia* by Chen and Prowse (1964) and is discussed by Balarin and Hatton (1979) in relation to the "space factor" requirement defined by Swingle (1956) for goldfish *(Carassius auratus)* and other extensively farmed species. Recently some more specific information on the phenomenon (which is not universally recognized) has become available with the publication of preliminary findings by Henderson-Arzapalo et al. (1980) on a biochemical compound which they isolated from mucus and culture water of intensively cultured *S. mossambicus* which induced a syndrome suggestive of an acute anaphylactic reaction in *S. mossambicus*, *S. aureus*, *S. niloticus* and *T. zillii* but had no such effect on *Ictalurus punctatus*. This phenomenon shows a number of similarities with the findings of Scott (1977) who described a phenomenon which he called "shock syndrome" in *Sarotherodon spilurus spilurus* which he observed to go into an anaphylactoid state in high density stocking in Kenya (Plate 4).

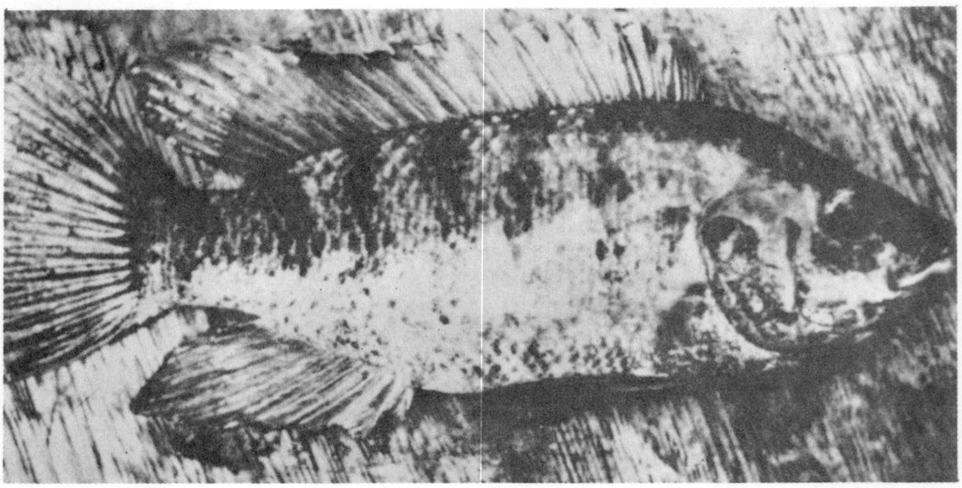

Plate 4. *Sarotherodon spilurus* in state of shock from high density intensive tank culture. Photograph courtesy of Mr. P.W. Scott.

Disease Prevention and Control

Since tilapia culture is still largely undeveloped despite the great potential of many of the species and their hybrids for culture in the developing countries, transglobal movements of fish are unfortunately still quite common. Indiscriminate transfers of young fish for on-growing or as broodstock hold dangers of many kinds. They have already led to problems from a fisheries point of view in countries such as Australia (Moriarty pers. comm.) but possibly even more significant, from the point of view of aquaculture and of native fish stocks, is the risk of the transfer of microbial pathogens from one area to another. Already two significant pathogens, *Ichthyophthirius multifiliis*, and the iridovirus of lymphocystis (see above), have been introduced in this way to stocks in Africa and Hawaii and, with increasing intensification of farm systems, may become constraints on production.

Thus, there is justification on both disease prevention and ecological grounds for national controls on stock transfers and importations. Where no such national regulations exist, the prudent importer will insist on fry importations being from a known and trusted source, will insist on heavy formalin and malachite green treatment on the farm prior to export and will perform similar prophylactic treatment within a secure and closed quarantine water system, (isolated and downstream from any production facility) within which such fish will be kept under close observation for at least a month, prior to release into a production system.

Ideally, on disease grounds a country, region or even an individual tilapia farm should be self-sufficient in fry production, but in view of the economic and management justifications, it seems likely that as in many other livestock production systems, the industry will stratify into high quality fry producers and fattening farms. The least that can be expected under these circumstances is that the aforementioned minimal quarantine and hygiene recommendations will be adopted as an insurance both for the individual and his national fisheries ecosystem.

The indiscriminate use of antibiotics in tilapia culture is a subject of some concern to medical and veterinary science. There is justification, on clinical grounds, for the use of oxytetracycline, or the potentiated sulphonamides, at *therapeutic* levels, in specific bacterial conditions. There is also some validity in using these drugs, again at therapeutic levels, in the specific traumatic situation involved in transporting large brood fish to overwintering facilities, in sub-tropical countries. The practice of indiscriminate so-called prophylactic use of antibiotics, however, is to be most severely deprecated and in view of its significance in human pathogen drug resistance induction, should in the authors' view be illegal in all countries. The use of the antibiotic chloramphenicol, which is the only reliable specific therapeutic for typhoid infections in man cannot be condoned at any time in fish culture.

Acknowledgments

The work on diseases of tilapias at the Institute of Aquaculture of the University of Stirling is part of a study of intensive tilapia culture techniques supported by the Overseas Development Administration. We are very grateful to our colleagues René Haller and John Balarin for their help with this program and Mr. I.H. MacRae, Dr. R.L. Oswald and Staff of the Institute for assistance with preparation of the manuscript. Dr. Ilan Paperna and Mr. Shmuel Sarig have also been very helpful to us in providing information from their long experience of tilapia culture in Israel.

Discussion

LOVSHIN: I don't think you mentioned anything about myxosporidian parasites. In cage culture in the Ivory Coast, they have just about had to stop using *Sarotherodon niloticus* because myxosporidians wipe them out. They are changing to a local species, *S. melanotheron (T. heudelotii)*.

ROBERTS: I did mention those in terms of pond culture, but I have only just learned of the example which you quote.

PULLIN: Dr. Roberts, can you say anything more on the reasons why cold temperatures always seem to bring disease problems, because this is not just so with tilapias, but for many different fish.

ROBERTS: Perhaps Dr. Avtalion could comment on this?

AVTALION: We have not performed a comprehensive study of this problem in tilapia, as we did in carp. We know that cold temperatures inhibit the immune response, namely, the ability to respond to different antigens. We found that the inhibitory threshold for antibody production was between 16 and 18°C in tilapia and 15°C in carp. At these low temperatures, both activities of bacteria and immune system are considerably inhibited. When spontaneous increase of environmental temperature occurs, the bacteria multiply faster than the immune system could react. This seems to be different when viral and parasitic diseases are considered.

ROBERTS: I think that it is temperature change that is a problem. If you have specific pathogen-free tilapia or carp or salmon or trout and if you can reduce the temperature very slowly, you can usually get away with it. Going down in temperature, each pathogen seems to have an optimum temperature range and you go through those ranges for particular strains or species of micro-organisms. At the same time your fish, particularly warmwater species like tilapias have to switch enzyme systems on and off at different temperatures and it would appear that for fishes in general the ability to do this is a little slower than the ability of micro-organisms to adjust and take advantage of the situation.

Fishes also have a very nice mechanism for repairing skin damage. If we damage our skin, then the cells round about proliferate and fill the hole. Fish have to use a different system since at low temperatures their rate of metabolism and therefore rate of cell multiplication is so low. In fish, the epidermis which is a thick layer, donates cells which migrate from a wide surrounding area to cover the lesion. This happens even at low temperatures within a few hours. In two or three hours, a reasonably small lesion will be completely covered by a layer one cell thick. This is probably important to both freshwater and marine fish, including carps. A problem remains, however, that if aeromonads and pseudomonads are deposited in the middle of a large lesion, then the epidermal cells come in a certain distance and then stop. Presumably, therefore, there is a secondary defense mechanism against such microorganisms for large lesions and for situations where fish are not able to make their cell migration system work fast enough (e.g. at the bottom of their temperature range). With large lesions, the whole skin gets very thin.

MIRES: What is your opinion of the use of antibiotics as prophylactics? Why do you say use these during fish transfers only?

ROBERTS: Well, antibiotics work and they are very important, but I am very fearful for the future of mankind if we carry on the way we are with the use of antibiotics. Antibiotics have revolutionized medicine, but aquaculturists, in attempting to revolutionize protein production, are going well on the way towards preventing us from keeping our ability to use some antibiotics, particularly in regions like Southeast Asia. Antibiotics are active against human bacterial pathogens as long as they are used at an adequate dosage. If they are used at lesser dosages, this selects for antibiotic-resistant strains. They can also pass on this resistance to completely different bacteria during reproduction. The most serious prospect is that aquatic bacteria in culture systems receiving low dosages of prophylactic will pass this resistance to human pathogens or animal bacterial pathogens. In manured farms, of course, we have very large amounts of human and animal enteric pathogens. In Southeast Asia, the development of drug-resistant strains of typhoid and cholera are serious possibilities.

MIRES: What about limited use of antibiotics during fish transfers, for example between tanks?

ROBERTS: Responsible culturists will of course use antibiotics sensibly and carefully in such situations, but in the developing countries, people don't understand the risks. It is very difficult to explain to a fish farmer in a developing country who has saved some fish with antibiotics why he should not continue low level treatments all the time. I can easily envisage the first chloramphenicol-resistant typhoid strain (and this is the only drug we have against typhoid) arising from its misuse in, for example, Southeast Asian catfish culture in ponds receiving human or animal wastes. Now, as far as using antibiotics during transfer, the first thing is, particularly for species such as silver carp which have very delicate skins and are big and very active fish, it is almost impossible to transport them without significantly damaging their skin. Also tilapias, particularly the large individuals with big strong spines can damage each other's skin and suffer spine breakage. It would, therefore, seem reasonable to treat such fish either by adding antibiotics to the transport water or as a pre-treatment by food incorporation to lessen infection of these lesions caused by transportation. I must stress, however, that this must be done with effective therapeutic levels of antibiotics, not lesser amounts. Again, I stress the dangers of the development of resistant strains of animal and human pathogens through ineffective dosage.

SESSION 4: CULTURE-RELATED TOPICS

Chairman's Overview

H. F. HENDERSON

The prospectus for this conference underlined the lack of an adequate research base as a major problem of the culture of tilapias, and suggested that lack of dialogue between fish biologists and culturists has contributed substantially to this problem. The latter may be open to dispute, but it is certainly an important task of this meeting to further such communication. The point about an inadequate research base, however, is undeniable and the reviews submitted provide ample evidence.

Without doubt, the major problem in tilapia culture is the reliable and economic production of suitable seed. Tilapias are too prolific in some culture operations and yet, as Mr. Mires points out, have too low a fecundity for easy mass rearing. There has been a great deal of practical research on ways of combating these two problems but much of this has been of a "bootstrap" variety. Without robust hypotheses about the sex-determining mechanisms in tilapias it has been very difficult to carry out effective practical research on efficient methods for the mass production of all-male hybrids. Without an understanding of the environmental and behavioral factors which trigger maturation, little success has been achieved in physiological suppression of reproduction in practical culture. Ignorance of the requirements of a complete diet for tilapias has made it necessary to revert to trial-and-error approaches in the adjustment of supplementary feeds appropriate to local conditions and using locally available feedstuffs. It makes a lot of difference in research whether one has to work from weak or from strong hypotheses.

Dr. Avtalion's paper suggests that we may now be close to understanding the mechanisms of sex-determination in tilapias. Regarding maturation, Professor Hyder's thesis that water dilution stimulates maturation and or spawning behavior, advanced quite a number of years ago, may not only be close to the mark but may have practical utility. Mr. Mires notes that Israeli culturists believe that periodic renewal of pond water restimulates spawning activity. There is also evidence, cited by Mires, that crowding, particularly of fry, suppresses reproductive activity.

Dr. Lovshin, in his review, also agrees that there has been a lack of basic science to underpin practical work on hybridization. He remarks that "fish culture researchers have generally solved their problems without explaining why the result was obtained." However in hybridization work he believes that if good results are to be obtained "basic research is needed by knowledgeable behavioralists and physiologists aimed at understanding the barriers to tilapia hybridization and then discovering methods to overcome these barriers."

There is another area where it seems to me that fish culturists have been slow to make use of available scientific knowledge. Genetic selection work cannot proceed efficiently without good knowledge of the pedigree of parental stocks, nor is it very effective if based on highly inbred stocks. In reviewing the various experiments cited in the papers on hybridization work in tilapias, I wondered how often contradictory reports on the effectiveness of selection for growth and other characters have been the result of very different levels of heterozygosity in the individuals used for the different experiments. It is very evident among those who have been working with experimental tilapia populations that (1) so-called pure strains have several times been derived from unrecognized accidental hybrids, and (2) many research stocks have passed through one or more "genetic bottlenecks" where the number of breeding pairs in the line has been reduced to well below twenty-five.

It would seem to me very valuable if, in the course of these discussions, we could identify other areas where "basic science barriers" stand in the way of efficient development of tilapia culture. I gather that Dr. Avtalion's work on electrophoretic markers was undertaken to determine the identity of farmed stocks. The work is clearly encouraging and the discovery of a male-specific protein is an exciting bonus.

I will turn now to some of the more specific aspects of fry and fingerling production, for example, the high cost of overwintering fry in temperate areas. Mr. Mires noted that *Sarotherodon* fry must be grown to at least 20 g to be successfully overwintered. Drs. Hepher and Pruginin remark that overwintering is economically feasible in Israel if all male fingerlings are selected for growout in the following summer: the market prefers large fish (400 g or more).

In Belgium, there is considerable interest in using warm water industrial effluents for tilapia culture. Dr. Coche tells me that in Belgium the restaurants want tilapias weighing at least 300 g and preferably about 350 g. There is possibly a similar market in the U.K. In many areas, however, thermal effluents may be more profitably utilized to produce hybrid male seed for growout elsewhere. Given that the cost of hybrid 50 g fingerlings is about 35% of all growout costs for large tilapias, as reported by Dr. Lovshin (quoting Tal and Ziv 1978), it would seem that the availability of inexpensive, heated water could make specialization in seed production quite profitable in some locations.

It was interesting to note, from Hepher and Pruginin's paper, that a late spring warmup could prove advantageous in the combined culture of mixed sex progeny (from natural spawnings) and hatchery-produced males, because of the suppression of spawning below 20°C. Where cold water is readily available to hold temperatures between 18° and 20° through the culture season, the advantages of the resulting depression of breeding might offset the reduced growth that would also occur.

Grading and sorting fry and fingerlings also seems to be an area where economies could be achieved by mechanization or simplification. Dr. Lovshin describes pen systems, as used by Pruginin in Uganda, and interconnected ponds, as used in Brazil, to allow the easy separation of hybrid fingerlings from their parents. The latter incorporated a set of screens between the ponds to sort out different sizes. Dr. Coche reports the separation of fingerlings by size to be an effective means of increasing the percentage of males in mixed sex culture in the Ivory Coast. Dr. Guerrero notes that Pruginin and Shell used a mechanical grader for this purpose.

The design of efficient structures for screening and sorting fish can be a complicated process. Solid gratings may be preferred to soft mesh as the geometry of flow can be quite important when they are operated in flowing water. Much more work of this sort seems to have been done in connection with the design of fish passes than by aquaculturists. The behavior of the species being sorted is also important. Tilapia fishermen tend to agree that adult fish, at least, are extremely wary of gill nets, but it is probably not known whether this is a question of retreat from a "new" object in the environment, or from some more specific stimulus.

In the control of reproduction of tilapias, it may be useful to make a distinction between the requirements for hybrid seed production systems, where almost complete control over wild spawning is needed, and the requirements of growouts where less control may be needed. Perhaps more work should be done on combining the various methods of inhibiting natural spawning and/or of increasing the male/female ratio of progeny. We should also determine whether the faster growth of males is genetically controlled.

Genetic Markers in *Sarotherodon* and Their Use for Sex and Species Identification

R. R. AVTALION

Department of Life Sciences
Bar-Ilan University
Ramat-Gan, Israel

AVTALION, R.R. 1982. Genetic markers in *Sarotherodon* and their use for sex and species identification, p. 269-277. *In* R.S.V. Pullin and R.H. Lowe-McConnell (eds.) The biology and culture of tilapias. ICLARM Conference Proceedings 7, 432 p. International Center for Living Aquatic Resources Management, Manila, Philippines.

Enzyme and other protein genetic markers in *Sarotherodon* and *Tilapia* are reviewed with special emphasis on their use for the identification of species and hybrids. Three different groups of markers: transferrins, esterases and male sex-protein (MSP) were found to be useful for the control of *Sarotherodon* parental breeding stocks used in Israel for the commercial production of F_1 all-male hybrids. Understandably, the tests for these markers were restricted to serum proteins, which could be collected from broodstock without damaging them. These serum markers were found to be species-specific, providing a tool for the identification of both parents and their hybrids, and therefore permitting the elimination of xenogeneic contaminants from the breeding stock.

The possible biological importance of MSP is discussed as a sex and species-specific marker and as a possible gene-product involved in sex regulation. Finally, an autosomal theory of sex determination in *Sarotherodon* is summarized and discussed.

Introduction

Sarotherodon species, the mouthbrooding tilapias, are currently being cultured in fish ponds in tropical and subtropical countries, and constitute an important source of animal protein in developing countries. A free-breeding culture of tilapias in a limited water space gives rise to an enormous quantity of small fish having no economical importance. Thus, the economic feasibility of *Sarotherodon* culture is essentially based on monosex culture of males. However, the manual separation of males from females is expensive and time consuming.

The culture of these fish made significant progress when it was shown that hybridization between some unrelated species of this genus gives rise to fertile F_1-hybrid broods, which present unusual sex ratios, including 100% males. This result was consistently obtained by different authors mainly in the crosses of *S. mossambicus* and *S. niloticus* ♀♀ with *S. hor-*

norum, *S. aureus*, *S. macrochir* and *S. jipe* ♂♂. Sex ratios (♀:♂) of 1:3 were consistently obtained in reciprocal crosses (Hickling 1960, 1963; Fishelson 1966a, 1966b; Chen 1969; Jalabert et al. 1971; Pruginin et al. 1975; Lovshin and Da Silva 1975; Haller pers. comm.). On the other hand, the back cross of the F_1 male hybrids and their female parents gives rise to equal numbers of males and females. In general, male hybrids give rise to progeny with a low percentage of males when crossed with their parents or grandparents. In practice, the infiltration of broodstocks by hybrids seems to be the reason for the decrease in the percentage of males in group spawning. Actually, for many reasons (e.g., human error, transfer of fry from one pond to another by predatory birds, occurrence of fry in the water used to fill the pond prior to stocking) it is quite impossible to avoid contamination of the parental species in large-scale commercial operations.

The main purpose of our work was to establish criteria, based on genetic, biochemical, electrophoretic and immunological markers, to identify parental species and their hybrids, in order to control broodstocks and to eliminate individuals presenting xenogeneic markers, and to carry out a comprehensive study on blood markers in tilapia.

Sampling for Serum Markers

For the commercial production of F_1-male hybrids in group spawning in Israel, *S. aureus* males are currently crossed with females of *S. niloticus* of different origins: *S. niloticus* (G) (originating from Ghana) and *S. niloticus* (V) (otherwise called *S. vulcani*, originating from Kenya). It is a matter of opinion whether to designate these as separate subspecies. In practice, it is difficult, if not impossible, to differentiate between the F_1 hybrid progeny of these crosses and their parents on the basis of external morphometric criteria. For this reason, a systematic check on the broodstocks is carried out using several blood protein markers. Once or twice a year, in various farms which produce F_1-male hybrids for commercial purposes, we test all the selected broodstocks (which are normally kept in aquaria) and take samples from *S. aureus* ♂♂ and *S. niloticus* ♀♀ destined for the mass production of F_1 male hybrids in ponds.

The test consists of withdrawing a small quantity of blood (0.1 to 0.4 ml) from the hemal arch of tagged fish, and testing their sera by polyacrylamide gel electrophoresis. Since the tested fish are destined for reproduction, the blood sampling is performed carefully to minimize injury. The mortality following sampling is normally very low, or zero, and even small fish of 8 to 10 g can be sampled.

The serum markers which were found significant for this purpose, and which are routinely tested using 6 and 7% polyacrylamide gels, are the transferrins, male sex-protein (MSP) and esterase isoenzymes. These markers, which exhibit sex and/or species-specific polymorphism, enable an easy identification of *S. aureus* and *S. niloticus* parents and their hybrids (Avtalion et al. 1975, 1976).

Serum Transferrins

Transferrins represent a major protein fraction in the serum of tilapias (4 to 8 mg/ml) and are involved with iron transport for the formation of hemoglobin. Tilapias exhibit marked polymorphism in their transferrin patterns (Chen and Tsuyuki 1970; Avtalion et al. 1976). This polymorphism seems to be controlled by different alleles, the genetics of which have still to be studied. The reading of transferrin patterns is difficult from 7% polyacrylamide gel, because their relative mobility (Rm 61 to 72.4%) is the same as that of the albumins and they are therefore often masked by that fraction. However, transferrin bands can be easily seen in 6% polyacrylamide gels, where the albumin spot (Rm 77 to 86%) runs faster than the transferrins (Rm 65 to 77%). At least 5 distinct transferrin bands in various combinations were identified in the above-mentioned species. They were numbered as bands 5 (Rm 78%) to 9 (Rm 64.6%) on the basis of their relative position in the electropherogram (Avtalion and Wojdani 1971). Plate 1 shows bands 6 to 9.

All the species present a slight intraspecific polymorphism. However, there are several interspecific differences which permit distinction between the different species (Avtalion et al. 1976). These interspecific differences were made more pronounced by eliminating from the parental stocks, individuals exhibiting overlapping transferrin markers (e.g., transferrin 9 in *S. aureus*) (Plate 2). After five years of selection, *S. aureus*, an endemic Israeli species, always presents transferrin 8 while the presence of transferrin 6 is still variable. On the other hand, *S. niloticus* from different origins always have the transferrins 7, 8 and 9: up to now it has not been

S. niloticus (Ghana)

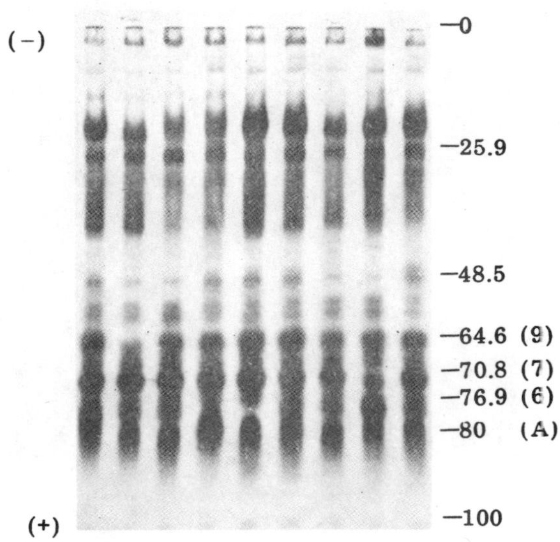

Plate 1. Electropherogram of *S. niloticus* (Ghana) in 6% polyacrylamide gel. Note the homogeneity of transferrins (7)-(9). (A) = Albumin; the figures are % relative mobilities.

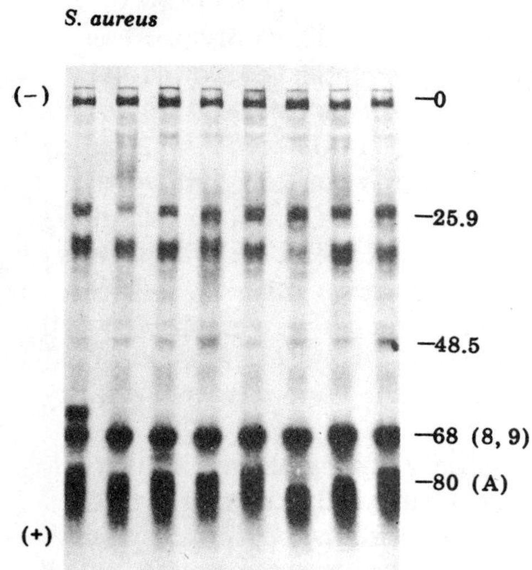

Plate 2. Electropherogram of *S. aureus* in 6% polyacrylamide gel. Individuals presenting xenogeneic bands (arrow) are normally eliminated from the breeding stocks. (A) = Albumin; (9) (8) are transferrins; the figures are % relative mobilities.

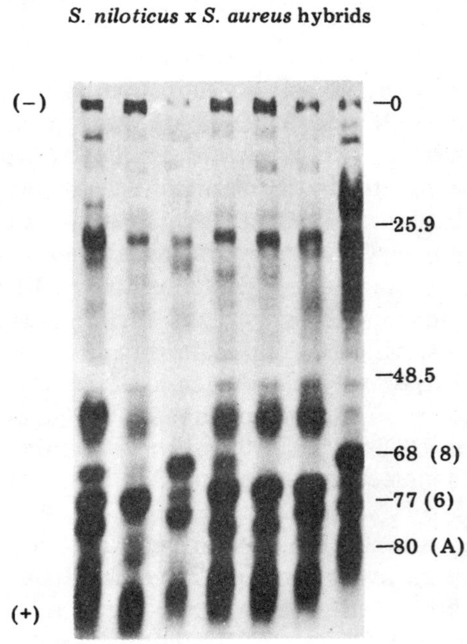

Plate 3. Electropherogram of *S. niloticus* ♀ x *S. aureus* ♂ hybrids in 6% polyacrylamide gel. All the possible transferrin combinations are found in F_1 hybrids. (A) = Albumin; (6) (8) are transferrins; the figures are % relative mobilities.

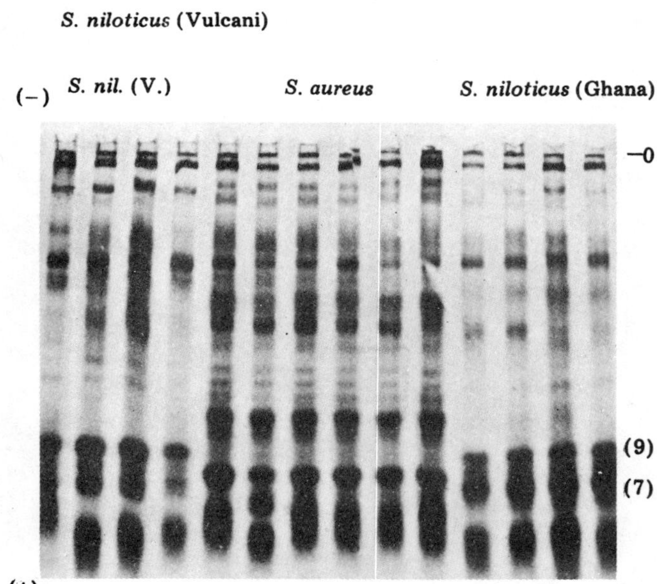

Plate 4. Differential identification of the selected monomorphic lines of *Sarotherodon* spp. From left to right: *S. niloticus* (Vulcani) where band 9 presents the higher density, *S. aureus*, where band 8 is constantly present and *S. niloticus* (Ghana) where band 7 is quantitatively more intense than the other transferrins.

possible to select for the elimination of band 8 in order to avoid any overlap with *S. aureus*. Bands 5 and 6 are sometimes present. The differential identification of these two species and their hybrids is based on the constant presence of bands 7 and 9 in all *S. niloticus* and their total absence in *S. aureus*. In the hybrids, however, all possible combinations occur (Plate 3). There are also some slight differences in this respect between *S. niloticus* (G) and *S. niloticus* (V). While both of them possess bands 7, 8 and 9, band 7 is quantitatively more intense in *S. niloticus* (G), whereas band 9 is more intense in *S. niloticus* (V) (Plate 4).

Enzymes

A study of some allozymic variations in *Tilapia* and *Sarotherodon* was carried out by Chen and Tsuyuki (1970). They studied serum esterase (SE), glucose-6-phosphate dehydrogenase (G6PD) in the liver and erythrocytes and lactate dehydrogenase (LDH) from serum and other tissues (muscle, eye, intestine, liver, ovary, heart, kidney and erythrocytes). Two species of mouthbrooder, *S. mossambicus*, *S. hornorum* and their hybrids, and two other species of substrate-spawners *(Tilapia zillii* and *T. rendalli*) were tested. No significant interspecific variations, in both SE and LDH were obtained within the *Sarotherodon* species, although enough variability was found to distinguish between *Tilapia* and *Sarotherodon*. On the other hand, the erythrocyte and liver G6PD showed significant interspecific polymorphism.

Isoenzymic variations in different cichlids from the sea of Galilee were investigated by Kornfield et al. (1979). Six different species were tested, three tilapias *(S. aureus*; *S. galilaeus* and *T. zillii)*, two *Tristamella* species *(Tr. sacra* and *Tr. simonis)* and a species of *Haplochromis (H. flaviijosephi)*. The interspecific similarities within these species were determined, mainly by investigating the interspecific variations in different enzyme systems (e.g., adenylate kinase, esterase, isocitrate dehydrogenase, lactate dehydrogenase, 6 phosphogluconate dehydrogenase). Twenty-one allozyme loci were resolved in all these species. Using Nei's coefficient (I_N) of genetic identity (Nei 1972), interspecific similarities, ranging from I_N = 0.95 within *Tristamella* and 0.92 within *Sarotherodon* to less than 0.25 for *Haplochromis*, were determined. A phenogram derived from these similarities shows that *Tristamella* is closely related to, but equidistant between, *Sarotherodon* and *Tilapia*.

Two serum enzyme systems (LDH and SE) are being studied in our laboratory in order to evaluate the interspecific variations in *Sarotherodon* spp. (Avtalion et al., in prep.). At least three different LDH bands could be shown, presenting an interspecific polymorphism, but no species-specific pattern could be shown. For this reason, LDH is not utilized in our systematic control test of broodstocks. The SE system shows no intraspecific polymorphism but *S. aureus* has a unique species-specific esterase band, located in the transferrin region of the electropherogram, having an Rm of 77.6%. The most interesting finding was that all the different *S. niloticus* tested, whether from local stocks (*S. niloticus* B, Ein Hamifrats Israel) or from different countries in Africa (Uganda, Ghana, Kenya), all possess a unique esterase band, electrophoretically distinct (Rm 74.3%) from that of *S. aureus.* All F_1 *S. niloticus* x *S. aureus* hybrids were found to possess both SE bands (Plate 5). Thus, the SE test was found very useful in controlling broodstocks. It is noteworthy that the same gels, after being stained for SE can undergo an additional staining for proteins (using Coomasie blue) or alternatively can serve for both LDH (upper part of the gel) and SE (lower part of the gel) tests.

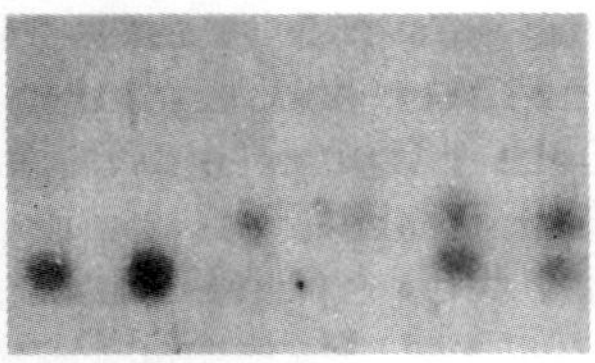

Plate 5. Electropherogram showing serum esterase patterns (E_1, E_2) of *Sarotherodon aureus* (SA), *S. niloticus* (SN) and their hybrids (HB). 6% polyacrylamide gel.

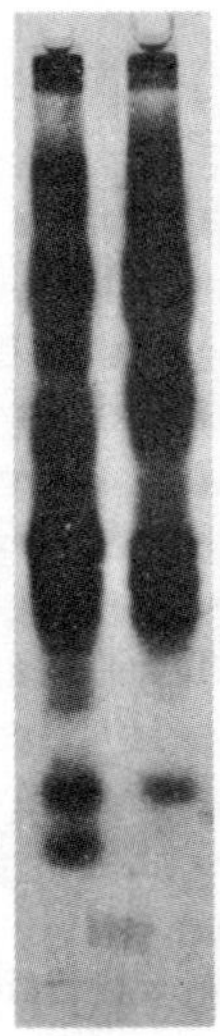

Plate 6. Electropherogram of *Sarotherodon aureus* serum in 7% polyacrylamide gel showing the male sex protein band (MSP) and the overlapping of transferrins (T) by the albumin band (A); m, male; f, female.

Male Sex-Protein (MSP)

Male-specific bands were found in 7% polyacrylamide gel electropherograms of *S. aureus*, one band (Plate 6); *S. niloticus*, 2 bands and *S. galilaeus*, 2 bands (Avtalion et al. 1975, 1976). The position of these bands in the electropherogram was found to be species-specific. This subject is now under study in our laboratory and some preliminary results have been obtained. This protein was purified using preparative electrophoresis and its molecular weight as determined by sodium dodecyl sulfate (SDS) electrophoresis is 40,100 Δ. Anti-MSP antibody raised in rabbits gave a single precipitin line when tested against MSP. However, this antibody was also reactive against female serum (Avtalion et al., in prep.). This result suggests that another serum protein might exhibit cross antigenicity with MSP, or that MSP exists in another form in females. At any rate, the antigenic determinants seem to be present in higher quantities in mature than in immature males. The *S. aureus* MSP was found to display cross antigenicity with *S. galilaeus* MSP. Some males were found to possess especially high quantities of this antigen, and we wonder whether these MSP-rich males are more capable than other males with less MSP, in their ability to give rise to a higher percentage of male offspring. This would be in agreement with our autosomal theory on sex determination in *Sarotherodon* (Hammerman and Avtalion 1979) which is summarized below.

Sex Determination

The autosomal theory suggests that the sex determination is by female and male gene products, which interact quantitatively (Avtalion and Hammerman 1978; Hammerman and Avtalion 1979). The crossing of males, which produce more male-sex regulating factors (SRF) with females, which produce less female-SRF, would provide a high percentage of male hybrids and *vice versa*. This balance theory is called "autosomal" because it is based on the assumption that the regulatory genes are located on autosomes (A,a) as well as on gonosomes (X, Y, W). This theory explains in a highly satisfactory manner the unusual sex ratios obtained by Chen (1969) who performed a comprehensive study in which hybrids of *S. mossambicus* and *S. hornorum* were extensively crossed between themselves, and were back-crossed with parents for up to four generations. This theory also explains most of the sex-ratio results obtained by Jalabert et al. (1971). It is noteworthy that a four-chromosome theory was originally suggested by Bellamy (1936) (XX-XY and WZ-ZZ) and modified by Gordon (1946, 1947) (3 sex chromosome theory, X, Y, Z) on the basis of their results on platyfish. This theory was reviewed by different authors (Kosswig and Oktay 1955; Anders and Anders 1963; and Kallman 1965) and was applied to *Sarotherodon* species by Hickling (1960), Chen (1969) and Jalabert et al. (1971) in order to explain the sex-ratio results they obtained in intercrossing these species. Since some of these sex-ratio results could not be explained on the basis of this theory, they all came to the conclusion that an autosomal influence on the sex determination process must be taken into account.

Assuming that autosomes indeed exert an influence on sex determination, the simplest system of sex-influencing chromosomes would consist of three gonosomes (X, W, Y) appearing as a complement of two, in any one of the possible combinations (XX, XY, WX, WW, WY or YY) and a combination of a pair of autosomes (AA, Aa or aa), all involved in primary sex determination. Within each pure species the pair of autosomes would be identical AA or aa. Thus, the complete set of chromosomes influencing sex determination in the pure species was suggested to be AAXX ♀ and AAXY ♂ in *S. niloticus* and *S. mossambicus* (homogametic female and heterogametic male), and aaWY ♀ and aaYY ♂ in *S. aureus* and *S. hornorum* (heterogametic female and homogametic male). The number of genotypes resulting from the combination of autosomes and gonosomes is 18. The sex of all the genotypes was determined on the basis of analysis of the sex ratio results obtained in Chen's crosses (Chen 1969).

The theoretical sex ratios which could be predicted for a cross between any male-female pair are presented in Table 1, where different theoretical sex ratios (♀:♂) of 0:1, 1:3, 3:5, 1:1, 9:7, 5:3, 3:1 and 1:0 can be observed. The sex ratios 3:1, 3:5 and 1:0 have not as yet been observed experimentally.

Table 1. Predicted sex-ratios (female:male) following crosses between males and females with different complements of autosomes (Aa) and gonosomes (WXY) influencing sex determination, AAYY = supermale; framed crosses generate all males (after Avtalion and Hammerman 1978).

		Males							
		AAYY	AaYY	AAWY	AAXY	aaYY	AaWY	AAWW	AaXY
Females	AAWX	0:1	0:1	1:3	1:1	0:1	3:5	1:1	1:1
	aaWY	0:1	1:3	1:3	1:3	1:1	1:1	1:1	1:1
	AaWW	0:1	1:3	1:3	1:1	1:1	1:1	1:1	5:3
	aaXY	0:1	1:3	1:3	1:3	1:1	1:1	1:1	1:1
	AAXX	0:1	0:1	1:1	1:1	0:1	1:1	1:0	1:1
	AaWX	0:1	1:3	3:5	1:1	1:1	9:7	3:1	5:3
	aaWW	0:1	1:1	1:1	1:1	1:0	3:1	1:0	3:1
	AaXX	0:1	1:3	1:1	1:1	1:1	5:3	1:0	5:3
	aaWX	0:1	1:1	1:1	1:1	1:0	3:1	1:0	3:1
	aaXX	0:1	1:1	1:1	1:1	1:0	3:1	1:0	3:1

AAYY = supermale; framed crosses generate all males.

Acknowledgments

This work was supported in part by a grant from the National Commission for Research and Development, Israel, and the GKSS & Geesthaght-Tesperhude, Germany. Thanks to Mr. J. Muravich for the excellent photographic work.

Discussion

LOVSHIN: I have an observation on your terms for degrees of maleness: males and supermales. With all-male hybrids in Brazil, we have observed that some fish sit on a nest just like females would and are courted by other males. We thought that we actually had some females and would get spawning, but we never did. There are obviously various degrees of maleness and femaleness, even in the all-male hybrid.

AVTALION: Yes. We have not of course studied this for all the crosses given in our autosomal theory because this would take a lot of time, but I agree that there is a balance between maleness and femaleness.

LOVSHIN: Regarding your theory of sex determination by sex chromosomes and autosomes, can you determine the genotype within a species by electrophoresis?

AVTALION: Not directly by this method. It is better done indirectly by performing crosses and analysis of the sex ratios of the progeny as I have indicated in my paper. Unfortunately, I do not have the facilities for this. Why don't you try out these crosses in America?

JALABERT: Does the quantity of male sex protein in the serum vary with time for a given male?

AVTALION: Yes, it can rise or decrease.

JALABERT: This could be related to sexual activity. Do you see any correlation between the levels of male sex protein and the intensity of sexual activity?

AVTALION: This is going to be tested by Mr. Mires and myself next year.

JALABERT: I think that this is very important.

ROBERTS: Our studies over the last two or three years on electrophoresis of muscle proteins are in agreement with Dr. Avtalion's. We have looked at possibly a wider range of natural populations and compared them with so called pure populations being used for culture. Many of these latter strains are in fact hybrids and we feel that this explains why many people in the past have not been able to repeat Chen and Prowse's work as they have not been using pure lines. We have not been able to make observations on male sex protein but we have found that in any hatchery situation where there is a limited number of females or a limited amount of nest-building area, certain males dominate the rest and adopt the courting habit. If the dominant male is removed from such a family, a new male becomes dominant to replace him. I also wonder about a correlation between male sex protein and dominance.

Tilapia Hybridization

L. L. LOVSHIN

International Center for Aquaculture
Auburn University
Auburn, Alabama 36844
U.S.A.

LOVSHIN, L.L. 1982. Tilapia hybridization, p. 279-308. *In* R.S.V. Pullin and R.H. Lowe-McConnell (eds.) The biology and culture of tilapias. ICLARM Conference Proceedings 7, 432 p. International Center for Living Aquatic Resources Management, Manila, Philippines.

Tilapia hybrids have demonstrated impressive growout potential using a wide variety of commercial and agricultural by-products as diets. The opportunity of stocking all-male tilapias without prior manual sex separation has tantalized fish culturists for many years. However, all-male tilapia hybrid culture has yet to be accomplished on a large commercial scale. Problems of maintaining pure genetic strains of broodstock and technical problems of producing commercial numbers of all-male tilapia hybrids have limited expansion. Research is needed to determine systems for mass producing all-male tilapia hybrid fingerlings.

When the technology for producing all-male tilapia hybrids on a commercial scale becomes available, culture will likely be limited to governments and wealthy farmers with money to build properly designed hatchery facilities and to hire trained biologists to manage the hatchery.

Introduction

Biologists working with tilapias have been experimenting for many years with methods to reduce or eliminate uncontrolled spawning in growout ponds. One method found to reduce tilapia reproduction is by hybridizing two species of tilapias resulting in offspring that are either all-male or contain a high percentage of males. This paper will attempt to determine the state of hybrid tilapia culture in the world today and to present the major strengths, limitations, and needs of this culture system. Unless otherwise noted, the female parent will appear first in all hybrid crosses.

Hickling (1960) was the first to report the possibility of producing all-male hybrids. He was attempting to produce a sterile hybrid with superior growth potential when he crossed *Sarotherodon mossambicus* with *Sarotherodon hornorum* (this fish was originally classified as *T. mossambica* (Zanzibar strain). To Hickling's surprise and satisfaction, crossing the female *S. mossambicus* with the male *S. hornorum* produced all-male hybrids. The hybrids were not "mules" or sterile but were fertile and capable of reproducing. These results stimulated other scientists to hybridize tilapias in the hope that other crosses would result in all-male offspring. In Uganda, Pruginin (1968) obtained 100% male hybrids by crossing *S. niloticus* with *S. hornorum*. The two crosses mentioned above have been repeated in many locations around the world with a high rate of success. Pruginin et al. (1975) reported the production of all-male offspring when *S. niloticus* was crossed with *S. aureus* in Israel. However, this cross did not consistently produce 100% male offspring. In west Africa, Lessent (1968) crossed *S. niloticus* with *S. macrochir* producing 100% male offspring. Unfortunately, the cross could not be regularly reproduced in ponds.

Pruginin et al. (1975) summarized other crosses that have produced 98 to 100% male offspring. Pruginin produced a high percentage of males by crossing *S. niloticus* with *S. variabilis*, *S. spilurus niger* with *S. hornorum*, *S. vulcani* with *S. hornorum*, and *S. vulcani* with *S. aureus*.

While the knowledge that all-male *Tilapia* hybrids could be produced has been available for 20 years, the commercial culture of all-male hybrids has remained very limited. In northeast Brazil, approximately 60 metric tons of the *S. niloticus* x *S. hornorum* hybrid was produced in 1979. An unknown quantity of the same hybrid is being produced in Panama. In Costa Rica, a small quantity of the *S. mossambicus* x *S. hornorum* hybrid is being raised.

Many other crosses have been reported resulting in skewed male sex ratios. Balarin and Hatton (1979) summarize the crosses performed and the percentages of male hybrids obtained. Of the many crosses tested, only two are being commercially used. Mires (1977) reports that the *S. niloticus* x *S. aureus* cross results in 80-90% males and is used commercially in Israel. Sarig (1979) writes that 38% or 2,230 t of all pond fish marketed in Israel were tilapias of which an unknown percentage were hybrids of *S. niloticus* x *S. aureus*. Chen (1976) reports that the *S. mossambicus* x *S. niloticus* cross is very popular in Taiwan and that fish farmers prefer this hybrid because of its superior growth. Kuo (1969) determined that 85% of the *S. mossambicus* x *S. niloticus* hybrids are males. Li (1978) states that over 22,000 metric tons of tilapias were raised in Taiwan in 1977. An unknown percentage of this production was the *S. mossambicus* x *S. niloticus* hybrid.

Genetic Determination and Selection

The method by which the sex of tilapias is determined is not well understood. Chen (1969) suggested that *S. mossambicus* has a homogametic female (XX) and a heterogametic male (XY) while the reverse is the case for *S. hornorum*, the male being homogametic (ZZ) and the female heterogametic (WZ). Chen demonstrated that all-male offspring could be obtained by crossing the female *S. mossambicus* (XX) with the male *S. hornorum* (ZZ).

S. mossambicus ♀ x *S. hornorum* ♂
XX ZZ

XZ
all-male hybrid

The reverse cross, female *S. hornorum* (WZ) with the male *S. mossambicus* (XY) resulted in three males for every female.

S. hornorum ♀ x *S. mossambicus* ♂			
WZ		XY	
WX	WY	XZ	YZ
♀	♂	♂	♂

It was necessary to postulate that the male-determining gene on the Z chromosome is more dominant than the female-determining gene on the X chromosome, while the male-determining gene on the Y chromosome is stronger than the female-determining gene on the W chromosome.

When the all-male hybrid (XZ) was backcrossed with female *S. mossambicus* (XX) or *S. hornorum* (WZ) the predicted 1:1 male to female ratio was obtained.

S. mossambicus ♀ x All-male hybrid			
XX		XZ	
XX	XX	XZ	XZ
♀	♀	♂	♂

S. hornorum ♀ x All-male hybrid			
WZ		XZ	
WX	WZ	XZ	ZZ
♀	♀	♂	♂

However, when further crosses were performed, the all-male hybrid (XZ) crossed with female hybrid (WX), the predicted Mendelian ratios did not appear. Chen suggested that differential mortality might have occurred: the female offspring having a higher rate of mortality than the males.

Differential mortality of female eggs or fry has been shown not to be the reason for the appearance of all-male offspring. Hickling (1960) first hypothesized that a lethal factor was killing the female eggs. However, a comparison of the *S. mossambicus* x *S. hornorum* all-male hybrid broods with pure strain *S. mossambicus* broods using females of equivalent sizes showed no difference in offspring numbers.

Lee (1979) demonstrated that fry numbers of *S. aureus*, *S. niloticus*, and *S. hornorum* hybrids were no different from fry numbers of the purebreds when these species were mated in aquaria.

Jalabert et al. (1971) carried out a similar series of genetic tests with the *S. niloticus* and *S. macrochir* hybrid crosses. Their findings and conclusions were similar to Chen's (1969). The ratios of female to male offspring in several crosses could not be explained by the four sex chromosome theory. Again, the simple genetic explanation of sex determination in *S. niloticus* with a homogametic female (XX) and heterogametic male (XY) and in *S. macrochir* with a heterogametic female (WZ) and homogametic male (ZZ) did not explain the results obtained in some crosses. Jalabert et al. (1971) hypothesized that autosomes were influencing sex determination in some hybrid crosses. Avtalion and Hammerman (1978) theorized that a pair of autosomes compliments the sex-determining chromosomes. Using the sex ratios obtained in Chen's crosses with *S. mossambicus* and *S. hornorum*, they were able to postulate the relative influence of the different chromosomes and the sex of all the genotypes. The sex of each genotype as well as the resulting theoretical sex ratios for all possible hybrid crosses were predicted. It remains to be seen if the autosomal theory of Avtalion and Hammerman holds true when the actual crosses are performed.

Commercial production of all-male hybrids has been difficult to maintain over a long period of time due to contamination of pure broodstock lines resulting in the appearance of varying proportions of females. At this time, there is no way to determine the genotype of tilapia broodstock routinely on a commercial scale. This problem has been especially troublesome in Israel with *S. niloticus* x *S. aureus* hybrids.

Avtalion et al. (1975) suggested electrophoretic comparisons of the blood proteins of broodstock to select pure genetic strains that will produce a high percentage of male offspring. However, this method has not been sufficiently developed to permit determination of the genotype and cannot be claimed as a complete method for ensuring all-male broods at this time.

Hulata et al. (in press) have proposed a system for determining genetically pure strains of *S. aureus* and *S. niloticus* so that all-male hybrids can be consistently produced. The system involves a series of aquarium spawnings of female *S. niloticus* and male *S. aureus* and their reciprocal cross. Those females and males of each species producing the correct male to female hybrid offspring ratios, 100% males for the *S. niloticus* x *S. aureus* cross and 3 to 1 males to females in the *S. aureus* x *S. niloticus* cross, will be mated with the selected individual of the opposite sex of the same species. The pure *S. aureus* male and *S. niloticus* female offspring from the selected matings will then be hybridized. This process will be continued until all-male offspring are consistently produced. The pure line broodstock of *S. aureus* and *S. niloticus* will then be isolated so that contamination will not occur again.

Fingerling Production: Installations and Systems

1. Artificial Environments

The maintenance of pure genetic lines to produce consistently all-male tilapia hybrids is difficult. Control of tilapia spawning and fry raising is often best accomplished in artificial environments: aquaria, concrete tanks, plastic pools, etc. Rothbard and Pruginin (1975) described a technique for induced spawning of *S. niloticus* and *S. aureus* to produce hybrids between the two species. The authors used aquaria that were 200 cm long, 50 cm wide, and 40 cm high in which one male *S. aureus* and 7 to 10 female *S. niloticus* were stocked. Immature fish were used and allowed to mature in the aquaria to form spawning families, which greatly reduced male aggression on the females. Spawning either took place naturally in the aquaria or the eggs were stripped from ripe females and fertilized with milt taken from the male. The eggs were incubated artificially and the young fry were raised in aquaria, concrete tanks, cages or ponds. Maintaining the water temperature between 25 and 29°C and daily artificial illumination of 12 hours permitted year-round spawning.

However, some hybrid crosses are very difficult to perform in aquaria because of male aggression, resulting in the death of the female. Lee (1979) eliminated female mortality in 100-l aquaria, stocked with 1 male and 3 females, by surgically removing the premaxilla of the male fish. The male fish continued his aggressive behavior but he was unable to damage the female without his premaxilla. This technique increased the number of successful aquarium spawnings. Lee also increased the frequency of spawning in female *S. niloticus* and *S. aureus* without decreasing egg production by removing the eggs from the mouths of the females and incubating them artificially. Spawning intervals of females from which eggs where removed were every 13 to 18 days as opposed to every 30 to 60 days for females that mouthbrooded eggs and young.

Lee produced tilapia hybrids by stocking 3 males and 9 females in 3-m diameter plastic pools of 3,500-l with or without a bottom substrate. Hulata et al. (1980) produced a number of tilapia hybrid crosses in 700-l plastic tanks stocked with 1 male and 5 to 10 females using the system described by Rothbard and Pruginin (1975). The *S. niloticus* x *S. hornorum* cross could not be successfully completed in these plastic tanks. However, when 3 males and 10 females were stocked in 4 m^2 concrete tanks, all-male offspring were obtained. I have also been unable to produce the *S. niloticus* x *S. hornorum* hybrid cross in 80-l aquaria or 1000-l cement-asbestos tanks. However, Lee (1979) produced the *S. niloticus* x *S. hornorum* hybrid in 100-l aquaria stocking one male and three females.

The use of artificial systems to produce tilapia hybrids permits control of the environment allowing year-round spawning in temperate climates. Taking the eggs from an incubating female increases the spawning frequency and thus, the number of hybrids that can be produced in a given period of time. Survival of eggs to fry is higher because of controlled conditions and better care. Finally, genetic control is much greater because of the reduced chance of contaminating pure genetic lines. However, some hybrid crosses are difficult to carry out in small, confined environments. There is no conclusive proof that an aquaria or tank-based hatchery will be able to produce commercial numbers of tilapia hybrids economically.

2. NATURAL ENVIRONMENTS

Tilapia hybrids have traditionally been produced in earthen ponds where the environment cannot be controlled and genetic lines are more difficult to maintain. Lovshin et al. (1977) and Lovshin and Da Silva (1975) describe several systems by which earthen ponds can be manipulated to produce and remove hybrid fingerlings and fry.

Lovshin (1980) describes a series of experiments performed in Brazil to determine the best method for producing all-male tilapia hybrid fingerlings by crossing *S. niloticus* x *S. hornorum*. One hundred and forty-two crosses were made over a 6-year period in 350-m^2 earthern ponds. A fingerling production system for use in Brazil was developed from these experiments and is detailed below (Figure 1). Plates 1-3 (see p. 308) illustrate some of the facilities and species in use in Brazil.

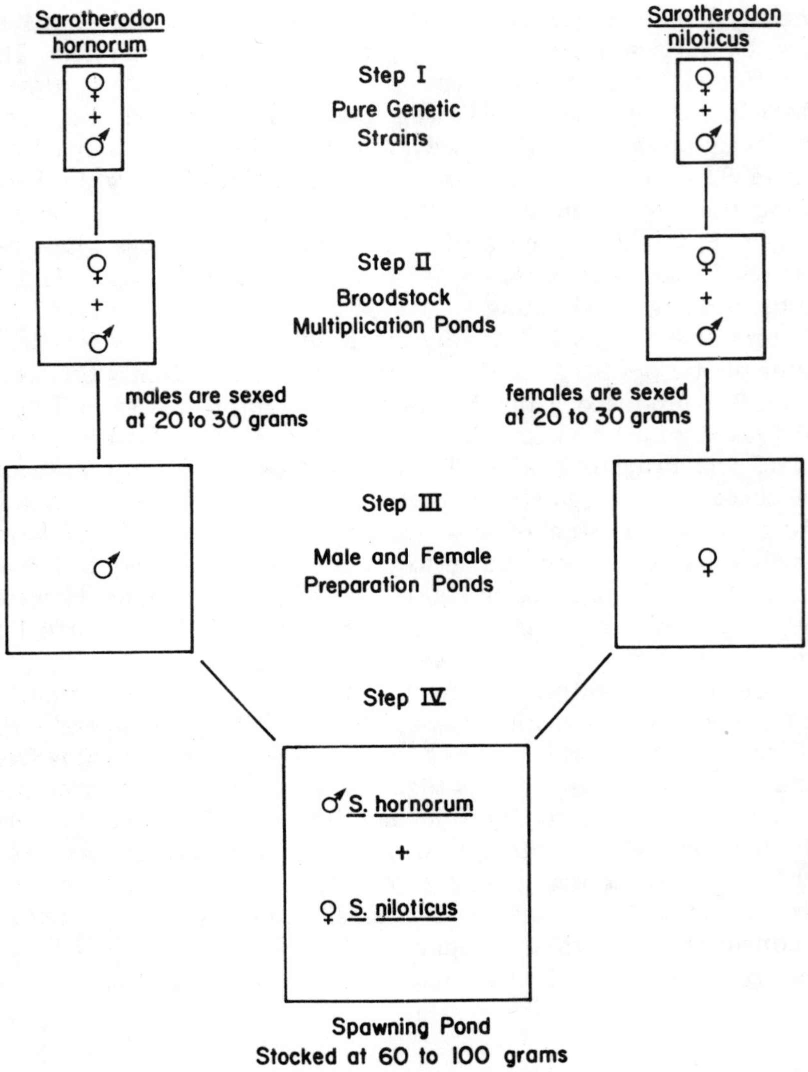

Figure 1. System used for producing all-male tilapia hybrids, *S. niloticus* x *S. hornorum* in Northeast Brazil (after Lovshin 1980).

Pure genetic strains that are known to give 100% male hybrid offspring should be carefully isolated from other tilapia species and hybrids so that contamination does not occur. This point cannot be overstressed, as a major drawback to producing tilapia hybrids is the failure to preserve pure genetic lines over long periods of time. If possible, pure strains should be held in tanks or small ponds isolated from the rest of the hatchery ponds and supplied with water free of natural tilapia populations. Well water would be the ideal water source but surface water can be used if care is taken. In Pentecoste, Brazil, pure stocks of *S. niloticus* and *S. hornorum* have been held for 7 years in 36-m^2 concrete tanks with earthen bottoms supplied with irrigation water. Water entering the tanks passes through several screen filters and a gravel filter before entering the tanks. The tanks are covered with nets to eliminate the entrance of predatory birds and animals. Thus, if at any time contamination of broodstocks is observed, the broodstock production ponds can be drained and restocked with pure strains taken from the concrete tanks.

Pure-line *S. niloticus* and *S. hornorum* fingerlings or adults are stocked into individual earthen spawning ponds to produce the number of broodstock needed to obtain commercial numbers of hybrid fingerlings. When pure *S. niloticus* and *S. hornorum* fingerlings reach 20 to 30 g and are still immature, they are sexed and the female *S. niloticus* and male *S. hornorum* are stocked into broodstock preparation ponds isolated from the opposite sex. If pond facilities are limited, fish can be stocked into cages to mature. Immature female *S. niloticus* and male *S. hornorum* should reach sexual maturity in 2 to 3 months: 60 to 100 g when stocked at 2 to 3/m^2 and fed 5% of their body weight daily. These fish should not be overstocked and should be well cared for so that numerous, healthy gametes are produced.

Mature male *S. hornorum* and female *S. niloticus* with swollen genital papillae are placed in the hybridization pond at the ratio of one male to one female, stocking one female/7 m^2 of pond surface area. The water depth of the spawning ponds should be less than 1 m. The broodstock are fed agricultural by-products (bran and oilseed cake) at 5% of their body weight daily during the hybridization period. After 2.5 months, the spawning ponds are drained and the hybrid fingerlings transferred to a nursery pond where they are stocked at up to 10/m^2 and fed agricultural by-products to attain further growth. Lovshin and Da Silva (1975) recommended that spawning ponds be drained after 3 months to avoid backcrossing with the female *S. niloticus*, but now a 2.5 month spawning period is recommended as mature hybrids have been found before 3 months. When transferring hybrid fingerlings to nursery ponds, a number (25 to 50) of large, sexable fingerlings should be checked to determine whether all-male hybrids have been produced.

Upon draining a hybrid spawning pond female hybrids are sometimes found or an extremely large number of small fry, 1-2 cm in length, that are too small to sex at harvest but contain a high percentage of females when sexed at a later date. The presence of female hybrids is almost always the result of backcrossing between a mature all-male hybrid and female *S. niloticus*: human error. After a 2.5 month spawning period, the pond has to be completely dried to eliminate any small hybrids that may remain. If the spawning pond cannot be dried, the pond should be carefully poisoned.

Every little puddle is a potential hiding place and small fry and fingerlings can live for weeks in a footprint filled with water. Partial pond filling before poisoning facilitates the elimination of fish in many small holes. Screens filtering pond water supplies should be checked daily and repaired when needed. Predatory birds should also be controlled.

Broodstock "families" that have produced sufficient numbers of hybrid fingerlings are immediately transferred to a freshly prepared spawning pond where a new 2.5 month cycle is begun. Broodstock "families" that have not given adequate fingerling numbers are eliminated and new broodstock are used. Broodstock "families" that produce low numbers of hybrid fingerlings in the first spawning cycle will usually produce poorly on subsequent spawning cycles.

Preliminary studies, (Lovshin 1980) have shown that broodstock used for more than one spawning season should be replaced because of declining fingerling production after 3 to 4 hybridization cycles or when the broodstocks are 14 to 17 months old: assuming they were stocked when sexually mature at 5 to 6 months old. The average number of hybrid fingerlings produced in a 350-m^2 pond when 50 female *S. niloticus* are stocked with 50 male *S. hornorum* is approximately 2,700 in 2.5 months.

Berrios-Hernandez (1979) demonstrated that cannibalism by fingerlings on newly hatched fry greatly reduced fingerling production of *S. aureus* in 7-m^2 plastic pools. Partial harvesting of the fry at weekly intervals increased fingerling production 35 times compared with total fingerling harvest after a 4 month spawning period. Initial studies done in Pentecoste, Brazil have not been able to demonstrate an increase in hybrid fingerling production with partial harvesting at 10 day intervals. While cannibalism by fingerlings is an important factor in fry survival in confined environments, it is not known what influence cannibalism has in a larger spawning pond environment. When hybrid spawning ponds are drained, an attempt should be made to grade and stock hybrid fingerlings and fry separately to minimize cannibalism in nursery ponds.

Pretto-Malca (1979) describes a system, similar to that used in Brazil, to produce *S. niloticus* x *S. hornorum* hybrids in Panama. The principal difference is that the all-male hybrid fry are partially harvested at 5 day intervals with a fine mesh seine and transferred to concrete tanks for further growth. The hybrid fry and small fingerlings are graded by size for stocking into separate tanks to reduce cannibalism. The fry are fed zooplankton in the concrete tanks.

Mires (1977) and Morissens (1977) outlined the procedure used in Israel to produce hybrid fingerlings of female *S. niloticus* and male *S. aureus* (Figure 2). The hybrid fingerling production system used in Israel is more complicated and time-consuming than the hybrid fingerling system used in Brazil and Panama because of the temperate climate, lack of genetic purity of the broodstock, and the need to hand-sex the less than 100 percent male hybrid fingerlings. Pure genetic strains of *S. niloticus* and *S. aureus* are selected electrophoretically and the broodstock placed in aquaria to spawn. The pure *S. niloticus* and *S. aureus* fry produced are either raised in the laboratory until reaching 50 g or raised in the laboratory up to 2 to 3 g when they are transferred to earthen ponds for growth to 50 g. After reaching

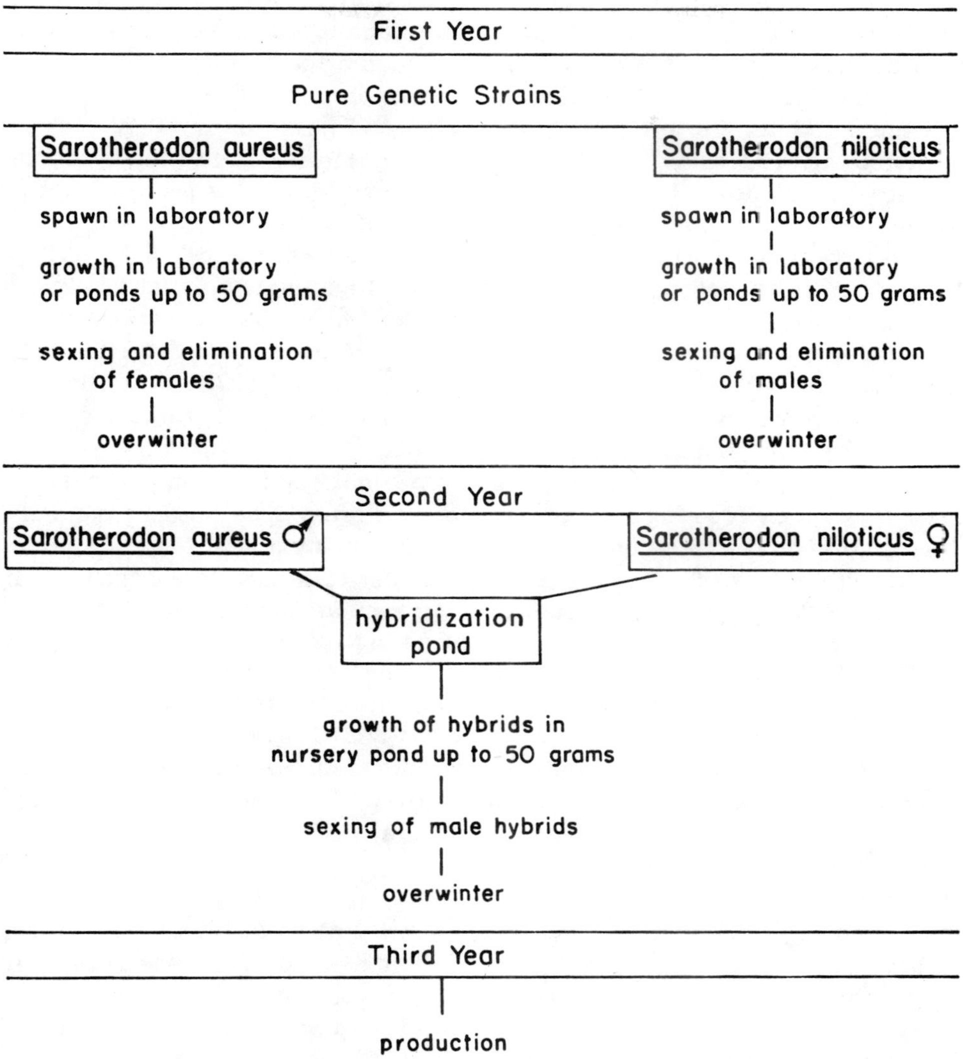

Figure 2. System used for producing tilapia hybrids, *Sarotherodon niloticus* x *S. aureus* in Israel (after Morissens 1977).

50 g, the fish are sexed and only the female *S. niloticus* and male *S. aureus* are kept. Often the female *S. niloticus* and male *S. aureus* are tagged so that they can be identified from wild, unselected tilapia. Selected female *S. niloticus* and male *S. aureus* broodstock are overwintered in deep earthen ponds, small ponds or concrete tanks covered with polyethylene sheets or supplied with heated water. In the spring the selected broodstock are stocked into earthern ponds of approximately 1 ha with a water depth of about 50 cm. The stocking density is 150 to 200 *S. aureus* males with 500 *S. niloticus* females/ha. Mires (pers. comm.) reports that the fish farming kibbutz where he works has doubled hybrid fingerling production by increasing the number of females stocked to 1,000/ha and altering the ratio of females to males to 1:1.

The first harvest of hybrid fingerlings is 1.0 to 1.5 months after stocking the broodstock and is continued at approximately 3 week intervals. The hybrid fingerlings, 1 to 2 g, are harvested with a seine in the feeding areas where the fingerlings concentrate. No mention is made of the length of time the tilapia adults are left in the spawning pond before removal to prevent backcrossing. The small hybrid fingerlings are transferred to nursery ponds and stocked at 30,000/ha. The nursery ponds also serve as growout ponds for carp *(Cyprinus carpio)* adults and fingerlings.

The hybrids are grown until they reach a sexable size of 50 to 100 g. In most cases the percentage of males is not high enough to permit direct stocking into growout ponds so the hybrid fingerlings have to be sexed to eliminate the females. Hybrid males must be overwintered for stocking the following summer in growout ponds. Tal and Ziv (1978) state that the production cost of 50 g hybrid fingerlings in Israel is very expensive: as much as 35% of the market price of the tilapia.

Pruginin (1967) describes a system used in Uganda to produce 100% male hybrids by crossing *S. niloticus* x *S. hornorum*. The need to regularly drain hybrid spawning ponds was a handicap in many areas of Uganda. Also, the handling of small fingerlings during transfer to nursery ponds caused 25 to 30% mortality. To reduce the need for regular draining and transfer of small hybrids to nursery ponds, a pen spawning system was tested. In a 0.2 ha earthen pond, pens measuring 8.8 x 6.0 m were placed in shallow water. The pens were made of welded steel fencing with a 2.5 to 3.5 cm mesh. Each pen was stocked with 8 female *S. niloticus* and 6 male *S. hornorum*. The small hybrid fry were able to pass through the mesh into the spawning pond where they would grow to 30 g before removal. The larger hybrid fingerlings were unable to reenter the spawning pens at maturity so the chances of backcrossing were reduced.

Lovshin and Da Silva (1975) divided a 350-m^2 spawning pond in the shallow end with a fence to form a 100-m^2 enclosure to facilitate the removal of broodstock after the spawning period. The mesh of the fence permitted the hybrid fingerlings to pass out of the spawning area containing the parents. This allowed easy removal of the adults when the spawning pond was lowered. This system worked fine in most cases, but, for each 3.5 month fingerling production period (consisting of 2.5 months of spawning and 1 month of additional growing time) 1 month of spawning time was lost while the fry were growing.

Lovshin and Da Silva (1975) describe a spawning-nursery pond designed to continually produce all-male tilapia hybrids and eliminate the need to handle hybrid fry. Figure 3 shows a modified design that is being built in Brazilian government hatcheries to produce tilapia hybrids. Mature adults are stocked in the 400-m^2 upper spawning pond. When the spawning period has terminated, the dam boards dividing the spawning and nursery units are removed and the water from the spawning pond drains by gravity into the nursery pond carrying hybrid fingerlings and fry. The broodstock are retained in the spawning pond by a screen. Large fingerlings descending from the spawning to nursery pond can be collected with net graders and stocked immediately. Small fingerlings and fry needing more growth before handling pass into the nursery pond. The spawning pond can then be pre-

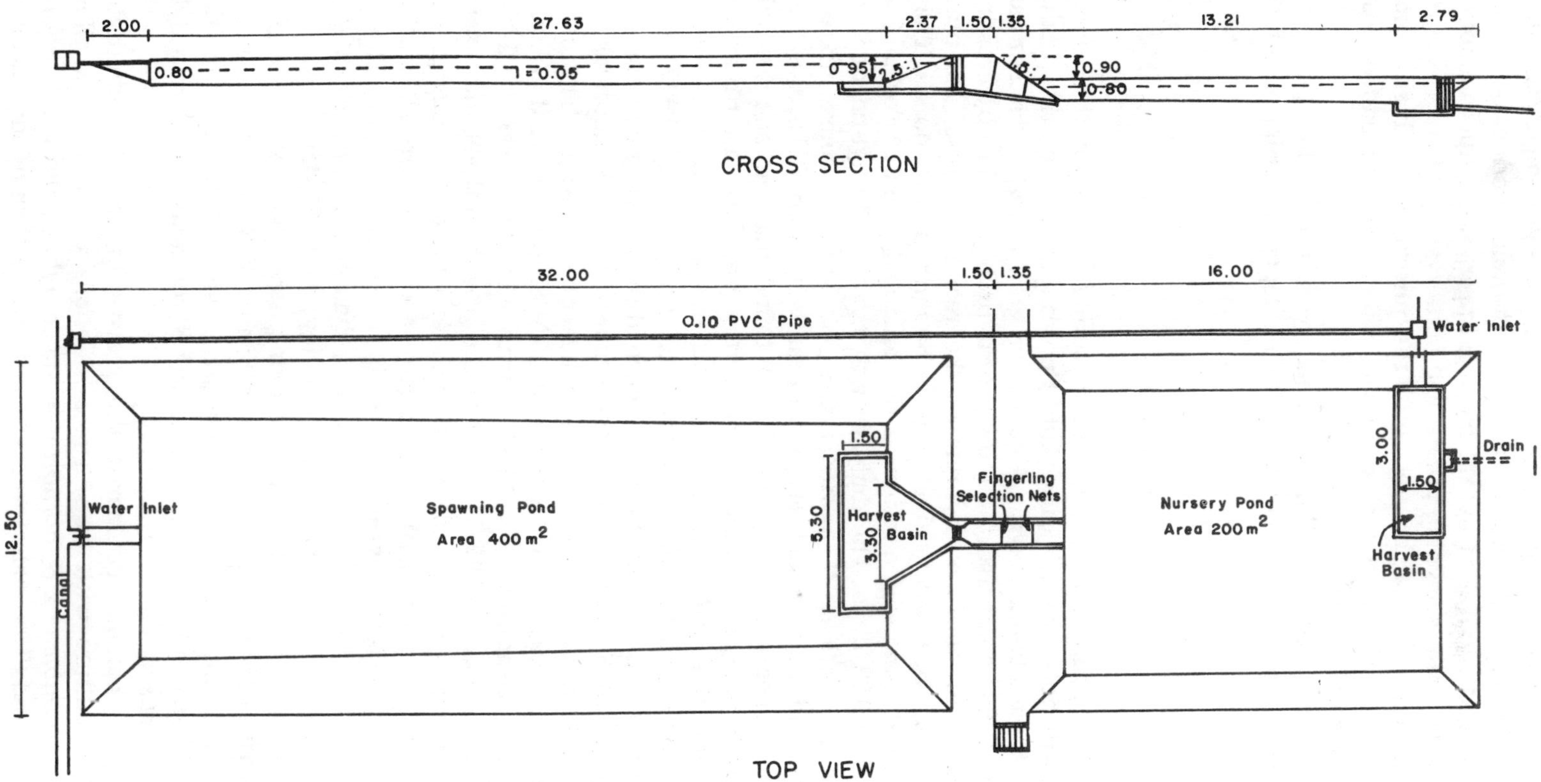

Figure 3. Design of tilapia hybrid spawning-nursery pond used in Northeast Brazil (all measurements in meters).

pared for a new spawning period while the fry are growing in the nursery pond below. These ponds should allow for continual production of tilapia hybrids, increased survival of hybrid fry and fingerlings, reduced amount of labor needed to drain spawning ponds, and conservation of water (as water from the spawning pond can be used to fill the nursery pond). It still remains to be determined if the above advantages will offset the added cost of building these spawning-nursery ponds.

Consistent production of all-male tilapia hybrids in earthen ponds takes more dedication and care than any other tilapia fingerling production system.

Growth

One of the apparent advantages of culturing male hybrids is their increased growth potential. Hickling (1968) notes that male hybrids produced by crossing *S. mossambicus* with *S. hornorum* grew faster than either parent stock. Kuo (1969) compared the growth of hybrids produced by crossing *S. mossambicus* x *S. niloticus*, hybrids of the reciprocal cross, and pure *S. mossambicus* and *S. niloticus* fingerlings in 600-m^2 earthen ponds. The fish were stocked at $1/m^2$, fed a ration of rice bran and peanut cake, and were harvested after 122 days. The daily growth rates of the hybrid of *S. mossambicus* x *S. niloticus*, *S. niloticus* x *S. mossambicus*, pure mixed sex *S. niloticus*, and pure mixed sex *S. mossambicus* were 1.16, 0.85, 0.74, and 0.59 g, respectively. Pruginin (1967) found that the all-male hybrid produced by crossing *S. niloticus* x *S. hornorum* grew 30% faster than mixed sex fingerlings of *S. niloticus* and 40% faster than mixed sex fingerlings of *S. hornorum* over a 126 day growing period. Lovshin et al. (1977) demonstrated that there was no statistical difference in the growth rate between the all-male hybrid of *S. niloticus* x *S. hornorum* and male *S. niloticus* grown separately and in polyculture with supplementary feeding. The all-male hybrid grew an average of 1.6 g/day and male *S. niloticus*, 1.3 g/day. The all-male hybrid demonstrated an 18% growth advantage over the 180-day culture period. Dunseth (1977) compared the growth of the *S. niloticus* x *S. hornorum* all-male hybrid, male *S. niloticus* and male *S. aureus* grown in polyculture with channel catfish, grass carp, and silver carp. No statistical difference was detected between the growth of the 3 male tilapias. However, the tilapia hybrid grew 7% faster (2.8 g/day) than male *S. aureus* (2.6 g/day) and 14% faster than male *S. niloticus* (2.4 g/day) over 160 days.

Pruginin et al. (1975) were unable to demonstrate a difference in growth rate between the hybrids of *S. niloticus* x *S. aureus*, *S. vulcani (S. niloticus)* x *S. aureus* and their parents. Yashouv and Halevy (1967) did not find a significant difference in growth between the *S. niloticus* x *S. aureus* hybrid and pure *S. aureus*.

Hulata (pers. comm.) compared the growth of three tilapia hybrids grown in polyculture with carp. The *S. niloticus* x *S. hornorum* all-male hybrid, *S. mossambicus* x *S. hornorum* all-male hybrid, and the *S. niloticus* x *S. aureus* hybrid (70% males) were stocked into earthen ponds with mirror carp (var. of *Cyprinus carpio*) and cultured for 105 days. The *S. niloticus* x *S. hornorum* hybrid grew the fastest (1.8 g/day) the *S. mossambicus* x *S. hornorum* hybrid at 1.4 g/day and the *S. niloticus* x *S. aureus* hybrid at 1.3 g/day. The results are preliminary as carp mortalities in several ponds may have biased the data. Pruginin (1967) found no significant differences between the growth of all-male hybrids *S. niloticus* x *S. hornorum* and *S. niloticus* x *S. aureus*.

Production

1. MONOCULTURE

Pruginin (1968), working with the all-male hybrid of *S. niloticus* x *S. hornorum* in Uganda, reported yields of 800 kg/ha/year when 1,500 hybrids/ha were stocked. When hybrids were stocked at 8,000/ha the rate of growth up to 50 g did not differ from that of fingerlings stocked at lower rates. After reaching 50 g, hybrid fingerlings previously maintained at high densities were then transferred to growing ponds at a density of 1,000 to 1,500/ha. Under these conditions, the individual weight gain of hybrids was 1.5 to 3.0 g/day to reach 200 to 450 g after 100 to 150 days. It was not stated whether supplementary feeds or fertilizers were used in these trials. Hickling (1962) reported that all-male hybrids produced at Malacca, Malaysia, reached weights of about 550 g in 6 months when stocked at 1,000/ha, giving a total production of 552 kg/ha/6 months from production ponds receiving 1,000 kg/ha of limestone and 22.4 kg of triple superphosphate (TSP)/ha.

Pond experiments in Ivory Coast by Lazard (1973) resulted in the production of 1.396 kg/ha/year of all-male hybrids *(S. niloticus* x *S. hornorum)* using TSP applied at the rate of 13.5 kg/ha every 2 weeks. The fish were stocked at 10,000/ha at an average weight of 2 g and after 180 days reached an average weight of 98 g.

Lovshin et al. (1977) also studied the growth and production of the same all-male hybrid in 355-m^2 earthen ponds in Brazil. Water entering the ponds had a pH of 7.8 to 8.3 and total alkalinity of 140 to 150 mg/l. An early experiment tested three treatments replicated three times each at two levels of stocking, 5,600 and 8,960 fish/ha. The treatments were a control, culture with cattle manure, and culture with supplemental feed (50% castor bean meal and 50% wheat bran) fed at 3% of the fish biomass per day for 6 days/week. Ponds receiving cattle manure were fertilized once a week with 840 kg/ha. Hybrids were stocked at an average weight of 7 g to determine growth during a 253-day period. The average total productions for ponds stocked at 5,600/ha were 300, 804 and 980 kg/ha for the control ponds, manured ponds, and fed ponds, respectively. The corresponding figures for ponds stocked at 8,960/ha were 277, 1,016 and 1,778 kg/ha. The fish grew best in the fed ponds at 8,960/ha: average individual weight, 229 g; feed conversion ratio (FCR), 2.7. The differences in production between the two levels of stocking and between treatments at each level of stocking were significant ($p < 0.05$).

A second experiment was performed using different culture methods (organic fertilizer, inorganic fertilizer and organic fertilizer plus feeding, all ponds stocked with 8,960 fish/ha averaging 21 g over 356 days. Ponds in one treatment received 1,400 kg/ha/week of cattle manure, while ponds in a second treatment were given TSP and ammonium sulfate (AS), each at a rate of 29 kg/ha every 2 weeks. Ponds in the third treatment were fertilized with 1,400 kg/ha/week of cattle manure and also received a ration of 50% castor bean meal and 50% wheat bran fed at 3% of the fish biomass per day, 6 days a week. All treatments were replicated twice. The average total fish productions were: organic fertilizer, 1,341 kg/ha; inorganic fertilizer, 1,856 kg/ha, and organic fertilizer plus supplemental feed, 4,883 kg/ha. The corresponding average final weights of fish were 154 g, 215 g and 565 g. Total production with feeding plus fertilizer exceeded that with inorganic and organic fertilizer by 163% and 264%, respectively. These differences were significant ($p < 0.05$).

Augusto and Melo (cited in Lovshin 1980) tested the effects of AS (20% N), TSP (46% P_2O_5), AS + TSP and no fertilizer (control) on the growth of all-male hybrids *(S. niloticus* x *S. hornorum)* in 355-m^2 ponds for 190 days. The AS and TSP treatments were replicated three times while the AS + TSP treatment and the control were replicated twice. Ponds in the AS and TSP treatments received equal amounts of N and P_2O_5 (280 g per application). Ponds in the AS + TSP treatment received half this amount (140 g per application) so that the total amount of N plus P_2O_5 applied was equivalent to 280 g. Ponds fertilized with AS or TSP received 1.42 kg/pond application (40 kg/ha) and 0.61 kg/pond application (17 kg/ha), respectively. Ponds receiving both fertilizers had 0.76 kg/pond application of AS (20 kg/ha) and 0.31 kg/pond application of TSP (8.7 kg/ha). Ponds were fertilized every 15 days by placing the fertilizer on a platform located 20 cm below the water surface. All fertilized ponds received an initial application of fertilizer 15 days before fish were stocked. Ponds were stocked with fish averaging 20 g at 10,000/ha. The average productions of the control, TSP, AS + TSP and AS treatments were 350, 832, 977 and 1,016 kg/ha, respectively. Fertilizers increased the total production two to three times that of the control. TSP alone was not as effective as AS alone or the combined treatment.

Paiva et al. (cited in Lovshin 1980) tested the growth of the same all-male hybrid fed three types of agricultural by-products found in northeast Brazil. Nine 355-m^2 earthen ponds were stocked with 400 fish (i.e., 11,250/ha), average weight 15 g. Castor bean meal (30% protein), babacu cake (a palm nut with 21% protein), and cottonseed cake (21.5% protein) were fed daily at 3% of the average biomass of fish in each treatment 6 days a week. Feeding rates were recalculated monthly based on seine samples. Treatments were replicated three times in a random block design. The culture period was 238 days. The average total fish productions for castor bean meal, babacu cake, and cottonseed cake were 3,444, 2,605, and 2,264 kg/ha, respectively. These differences were significant ($p < 0.05$). Fish fed castor bean meal were heavier at harvest (323 g) than fish fed babacu (245 g) or cottonseed cake (239 g), but there was no significant difference between fish fed babacu and cottonseed cake. FCR's for castor bean meal, babacu cake, and cottonseed cake were 2.4, 2.6, and 2.7, respectively.

Brazilian biologists (cited in Lovshin 1977) investigated the growth and production of the same all-male hybrid at higher stocking levels over a 367-day culture period in 350-m^2 earthen ponds, with no replication (Table 1). The fish were fed a ration of 50% babacu cake and 50% cottonseed cake at 5% of pond fish biomass/day, 5 days a week. Feeding rates were recalculated each month based on pond seine samples. Each pond was fertilized with a total of 19 kg (540 kg/ha) of both TSP and AS applied at 2-week intervals over the first 7 months of the experiment. Productions were the highest yet obtained in research with hybrid tilapia in Brazil: very impressive considering that only vegetable materials were fed to the fish. Increasing the stocking rates had a positive effect on production without greatly reducing growth. Feed conversion was good up to 6 months, average at 10 months and poor at 12 months. From the tenth month until harvest, FCR's increased sharply because of extremely poor water quality and high standing crops. The maximum feeding rate was 20 kg/pond (564 kg/ha) in the pond stocked at the rate of 31,000 fish/ha.

Da Silva and Lovshin (cited in Lovshin 1977) tested the culture of the same all-male hybrid in conjunction with the fattening of pigs in Pentecoste, Brazil. Three ponds of 1,000-m^2 were stocked with 25 g hybrid fingerlings at the rate of 8,000/ha. One pig sty was constructed on the margin of each pond and 7 pigs (70/ha of pond water) averaging 17 kg were placed in each sty. The pigs were not allowed to enter the ponds. The sties were cleaned daily and all waste products washed into the ponds. The pigs were fed a daily ration of 5% of their body weight consisting of 35% manioc, 20% wheat bran, 15% corn, 15% babacu cake, and 20% grass. The ration contained 10% protein. After 189 days, 1,490 kg/ha of fish were harvested averaging 205 g. The pigs averaged 60 kg each. The FCR for the pigs was 7.1 and 5.9 for combined pigs and fish production.

Da Silva et al. (unpublished data) carried out a second experiment to study the performance of the all-male hybrid with a waste loading equivalent of 120 pigs/ha of pond water. Hybrids averaging 29 g were stocked in three 1,000-m^2 ponds at the rate of 10,000/ha and received only the organic wastes washed daily into the ponds from adjacent pig sties. The pigs were fed a daily ration of 5% of their body weight consisting of 24% babacu cake, 45% corn, 5% meat meal, 25% elephant grass, and salt. After 180 days, 2,800 kg/ha of fish were harvested averaging 272 g. FCR for the pigs was 7.6 and 5.5 for combined pig and fish production.

Da Silva et al. (cited in Lovshin 1977) carried out a third experiment with six pigs per sty (60/ha of pond water), fed a daily ration of corn, babacu cake, meat meal, and grass (i.e. 11.6% total protein) at 5% of their body weight. Three 1,000-m^2 ponds were stocked with 10,000 fish/ha averaging 31 g. The fish were fed cottonseed cake at 2% of their body weight/day, 6 days a week. Wastes from the pig sties were washed into the ponds daily. After 193 days, 3,043 kg/ha of fish were harvested averaging 304 g. The FCR for pigs was 6.4 and the FCR for cottonseed cake to fish was 1.7.

Da Silva et al. (cited in Lovshin 1977) stocked three 1,000-m^2 ponds with 25 g hybrid fingerlings at 8,000/ha. The ponds were fertilized with 50 kg/wk (500 kg/ha/wk) of chicken manure taken from a commercial

Table 1. Summary of the results of culture experiments in Brazil in which all-male tilapia hybrids (*Sarotherodon niloticus* ♀ x *S. hornorum* ♂) were stocked at different levels in 350-m^2 earth ponds and grown for 367 days. The figures are from single ponds. Source: various Brazilian biologists cited in Lovshin (1977).

	Stocking level (fish/ha)									
	13,000	15,000	17,000	19,000	21,000	23,000	25,000	27,000	29,000	31,000
Fish stocked/pond	455	525	595	665	735	805	875	945	1,015	1,085
Average initial weight (g)	25	27	26	38	22	22	23	23	25	25
Average final weight (g)	452	466	495	456	492	384	444	416	353	410
Production/pond (kg)	197	245	279	303	354	309[a]	371	375	356	419
Production/ha (kg)	5,629	7,000	7,971	8,657	10,114	8,828	10,599	10,714	10,171	11,971
Survival (%)	96	100	95	100	98	100	96	95	99	94
Feed conversion ratio at 6 months	3.3	2.8	2.7	3.2	2.4	2.8	3.3	3.0	3.0	2.7
Feed conversion ratio at 10 months	4.5	4.6	4.0	4.5	4.4	4.1	3.8	4.2	4.3	3.7
Feed conversion ratio at 12 months	6.0	6.3	6.2	6.2	6.2	6.7	5.9	6.2	6.3	6.2
Ration fed/pond (kg)	1,124	1,465	1,648	1,732	2,093	1,947	2,074	2,168	2,067	2,413

[a]Included are 62 kg of fish that died due to poor water quality 2 days before harvest.

Table 2. Summary of results obtained with the all-male tilapia hybrid (*Sarotherodon niloticus* ♀ x *S. hornorum* ♂) cultured with organic and inorganic fertilizers and rations in the northeast of Brazil.

Stocking rate/ha	Fertilizer	Fish ration	Average stocking wt (g)	Average final wt (g)	Net yield kg/ha	Conversion ratio	Growth g/day	Net yield kg/ha/day	Culture period (days)
5,600	control	—	7	58	288		0.2	1.1	253
5,600	Cow manure 840 kg/ha/wk	—	7	166	264	36	0.6	3.0	253
5,600		50% wheat bran, 50% castor bean meal, 3% body weight/d	7	185	941	2.7	0.7	3.7	253
8,960	— control	—	8	36	179		0.1	0.7	253
8,960	Cow manure 840 kg/ha/wk	—	7	148	927	30	0.6	3.7	253
8,960	—	50% wheat bran, 50% castor bean meal, 3% body weight/d	7	229	1,680	2.7	0.9	6.6	253
8,960	Cow manure 1,400 kg/ha/wk	—	22	154	1,159	50	0.4	3.3	356
8,960	TSP[1]-29 kg/ha/15 d AS[2]-29 kg/ha/15 d	—	22	215	1,660		0.7	4.7	356
8,960	Cow manure, 1,400 kg/ha/wk	50% wheat bran, 50% castor bean meal, 3% body wt/d	20	565	4,760	3.6	1.6	13.4	356
10,000	control	—	20	43	151		0.1	0.8	192
10,000	TSP-17 kg/ha/ 15 d	—	19	101	644		0.4	3.4	192
10,000	AS-40 kg/ha/15 d	—	19	118	829		0.5	4.3	192
10,000	TSP-8.7 kg/ha/15 d	—	20	122	778		0.5	4.0	192
10,000	TSP-56 kg/ha/15 d AS-56 kg/ha/15 d	Rice bran, 3% body wt/d	60	340	1,648	2.8	1.6	14.7	180
11,250	—	Castor bean meal, 3% body wt/d	15	323	3,276	2.4	1.3	13.8	238
11,250	—	Palm nut cake, 3% body wt/d	14	245	2,447	2.6	1.0	10.3	238
11,250	—	Cottonseed cake, 3% body wt/d	15	239	2,095	2.7	0.9	8.8	238
15,000	TSP-540 kg/ha/yr AS-540/kg/ha/yr	50% palm nut cake	27	466	6,578	6.3	1.2	17.9	367
21,000	TSP-540 kg/ha/yr AS-540/kg/ha/yr	50% palm nut cake, 50% cottonseed cake, 5% body wt/d	22	492	9,627	6.2	1.3	26.2	367
25,000	TSP-540 kg/ha/yr AS-540/kg/ha/yr	50% palm nut cake, 50% cottonseed cake, 5% body wt/d	23	444	10,000	5.9	1.1	27.2	367
31,000	TSP-540 kg/ha/yr AS-540/kg/ha/yr	50% palm nut cake, 50% cottonseed cake, 5% body wt/d	25	410	11,169	6.2	1.0	30.4	367
8,000	70 pigs/ha of water		25	205	1,290		1.0	6.8	189
10,000	120 pigs/ha of water		29	272	2,510		1.4	13.9	180
10,000	60 pigs/ha of water	cottonseed cake, 2% fish body wt/d	31	304	2,733	1.7	1.4	14.2	193
8,000	Chicken manure, 500 kg/ha/wk		25	186	1,150	10	0.9	6.1	189

[1]TSP = Triple Superphosphate
[2]AS = Ammonium Sulfate

chicken farm. The chicken manure was 79% organic matter. After 189 days, an average of 1,350 kg/ha of fish were harvested, averaging 186 g. The FCR for chicken manure to fish was 10. A summary of research results performed on the all-male tilapia hybrid in northeast Brazil is found in Table 2.

Collis and Smitherman (1978) raised all-male *S. niloticus* x *S. hornorum* hybrids in 400-m^2 ponds in Auburn, Alabama. Fish averaging 29 g were stocked at 10,000/ha. Three ponds were each fertilized with the equivalent of 28,381 kg/ha of fresh cattle manure (5,392 kg of dry matter) divided into daily applications over a 103-day culture period. Fish in three other ponds were fed a commercial catfish diet (36% protein) at 3% of their body weight per day. The average net productions in the manured and fed ponds were 1,646 and 2,663 kg/ha, respectively. The corresponding average fish weights at harvest were 200 g and 318 g. The FCR's for manure and catfish diet on a dry matter basis were 3.3 and 1.3.

Lovshin et al. (unpublished data) studied the effect of introducing *S. niloticus* females and the resulting recruitment on the growth and production of all-male *S. niloticus* x *S. hornorum* hybrids. Nine 350-m^2 earthen ponds were stocked with 10,000 all-male hybrids/ha averaging 11 g. Three ponds were then stocked with an additional 2.5% female *S. niloticus* and 3 other ponds with an additional 5% female *S. niloticus*. The female *S. niloticus* averaged 16 g. The fish were fed agricultural by-products at 5% of the body weight of the hybrids per day, 6 days a week over a 273-day culture period. Seine samples were taken monthly to determine the growth of the hybrids and recalculate feeding rates. The average weights of the hybrids from the 100% male, 2.5% female, and 5% female treatments were 481 g, 218 g, and 285 g, respectively. The corresponding average productions of all-male hybrids were 4,286, 1,809 kg/ha and 2,186 kg/ha. *S. niloticus* recruits increased the total fish production in the 2.5% female and 5% female treatments to 4,309 and 5,451 kg/ha, respectively. The growth of all-male hybrids was equal in the three treatments until the fifth month when the 100% male treatment showed a distinct advantage. It was possible to produce a 200-g fish in 6 months in the all-male treatment, but 7 months were required in the two treatments with females. While the 100% male treatment showed a significant difference ($P < 0.05$) in hybrid growth and production compared with the treatments containing females, harvestable hybrids were still obtained in the treatments with tilapia recruitment.

Greenfield et al. (1977) demonstrated the profitability of raising all-male hybrids using agricultural by-products as feeds in Northeast Brazil.

2. Polyculture

In Israel, Tal and Ziv (1978) state that the majority of *S. niloticus* x *S. aureus* hybrids are raised in polyculture with common and silver carp. The tilapia assume the role of a pond cleaner, maintaining the water in good culture condition by consuming organic material and waste feeds that would otherwise decompose and pollute the pond environment. In most cases, the tilapia are able to increase total fish production without significantly reducing growth or production of the other species. The tilapia hybrids are normally 30 to 50% of the total fish population and their yield is 20 to 30% of the total.

In Brazil, Lovshin et al. (1977) tested the culture of the *S. niloticus* x *S. hornorum* all-male hybrid with mirror carp *(Cyprinus carpio)* to determine whether the addition of mirror carp would increase total fish production. Mirror carp raised alone were stocked at 2,240/ha, tilapia hybrids in monoculture at 8,960/ha and tilapia hybrids and mirror carp combined at 8,960/ha and 1,400/ha, respectively. All ponds received applications of cattle manure at 1,400 kg/ha/wk for 5 months after which applications were suspended. Rice polishings were fed throughout the 245-day culture period at 3% of the body weight of fish/day, 6 days a week. At harvest the average productions were: carp alone, 812 kg/ha; tilapia hybrids alone, 3,993 kg/ha and combined culture, 3,567 kg/ha. There were no significant differences in total production ($P < 0.05$) between treatments with the tilapia hybrid alone and the combined hybrid-carp cultures. However, these two treatments both produced significantly more harvestable fish than with carp alone. Moreover, the combined culture averaged 105 kg/pond of tilapia hybrids and carp raised on 295 kg of feed, while 108 kg of tilapia hybrids stocked alone required 441 kg of feed. Thus, less feed was required to raise an equal weight of hybrids and carps than was needed to raise hybrids alone. The average weight of tilapia hybrids raised with carp was 285 g, while that of the hybrids cultured alone was 353 g.

Da Silva et al. (1978) studied the influence of the all-male tilapia *(S. niloticus* x *S. hornorum)* in combined culture with tambaqui *(Colossoma macropomum)* a fruit-eating characid native to the Amazon River. Six 355-m^2 ponds were stocked with 25 g tambaqui at the rate of 5,000/ha. Three of the ponds were stocked with an additional 5,000 all-male tilapia hybrids per hectare, with an average weight of 18 g. The fish were fed daily on a pelleted chicken ration (17% protein) at 3% of the average body weight of tambaqui only in the afternoon, 6 days a week for 365 days. Feeding rates were recalculated monthly based on seine samples. The monoculture of tambaqui produced an average of 6,683 kg/ha with fish averaging 1.50 kg. The FCR was 2.8. The combined culture produced an average of 5,640 kg/ha of tambaqui averaging 1.20 kg and 3,299 kg/ha of tilapia hybrids averaging 0.75 kg. The FCR for the combined culture was 1.8. The addition of tilapia hybrids increased fish production by a total of 2,256 kg/ha compared to the monoculture of tambaqui without increasing the quantity of feed or the worsening the FCR.

To further test the influence of the all-male tilapia hybrid on tambaqui culture, Da Silva et al. (unpublished data) stocked the equivalent of 10,000 tambaqui per hectare together with 3,000, 4,000, and 5,000 all-male *(S. niloticus* x *S. hornorum)* hybrids/ha in 355-m^2 ponds. Each tilapia hybrid stocking rate was replicated three times. The average initial weights of the tambaqui and tilapia hybrids were 39 g and 13 g, respectively. The fish were fed a pelleted chicken ration (17% protein) at 3% of the average body weight of tambaqui in each treatment/day, 6 days a week for 360 days. Feeding rates were recalculated monthly based on seine samples.

The average tambaqui productions were 7,453, 7,201 and 7,779 kg/ha, with average weights of 760, 785 and 770 g for the 3,000, 4,000, and

5,000 all-male tilapia treatments, respectively. The corresponding tilapia productions were 2,224, 3,045, and 3,327 kg/ha with average weights of 770, 725 and 702 g and the total average productions were 9,677, 10,246 and 11,106 kg/ha. The FCR's for combined tambaqui and tilapia production were 3.4, 3.2 and 2.9. Increasing the stocking rates of the tilapia hybrid therefore increased the total fish production per pond without significantly affecting the growth or production of tambaqui. A summary of research results performed with tambaqui and the all-male tilapia hybrid can be found in Table 3.

Miscellaneous Aspects

1. Cold Tolerance

Chervinski and Lahav (1976) demonstrated that *S. aureus* native to Israel was more cold tolerant than *S. niloticus* introduced from Africa. The hybrid produced by crossing *S. niloticus* x *S. aureus* was as cold tolerant as *S. aureus. S. aureus* and the *S. niloticus* x *S. aureus* hybrid began dying at 9°C while *S. niloticus* died at 11°C.

Lee (1979) tested the cold tolerance of *S. aureus, S. niloticus, S. hornorum*, and their hybrids. All fish tested were acclimated to 21°C and the temperature was decreased about 0.8°C/hr until 50% mortality was recorded. For the purebred fish, *S. aureus* had the lowest thermal tolerance limit (6.7°C) and *S. hornorum* the worst (10°C). The cold tolerance limits of hybrids of *S. aureus* x *S. niloticus* were not significantly different ($P > 0.05$) from the pure *S. aureus* but they were significantly more cold tolerant than pure *S. niloticus* and the *S. niloticus* x *S. hornorum* hybrid. These results suggest that cold tolerance is a specific attribute of *S. aureus* and is probably transmitted to its hybrids.

2. Catchability

Lovshin et al. (1977) mentioned that the *S. niloticus* x *S. hornorum* all-male hybrid is easy to seine like its male parent *S. hornorum*, whereas *S. niloticus* is very difficult to seine and lies on its side in the bottom mud as the seine passes over.

Dunseth (1977) tested the catchability of male *S. aureus*, male *S. niloticus*, and the *S. niloticus* x *S. hornorum* all-male hybrid in replicated seine trials in 400 m^2 ponds. An average of 2% of the populations of male *S. niloticus* and *S. aureus* could be caught in the first seine haul. However, 50% of the all-male hybrid population was caught in the first seine haul.

Discussion

1. Fingerling Production

Fingerling production is a major constraint to commercial culture of tilapia hybrids, especially where 90 to 100% males are required. Maintaining pure genetic strains that give a high percentage of male hybrids takes great care and dedication on the part of hatchery workers. Large earthen ponds are not ideal units for holding pure genetic lines. Small, controllable earthen ponds, fiberglass, or concrete tanks are preferred.

A small number of pure broodstock couples can produce enough fingerlings to initiate even a large earthen pond tilapia hybrid fingerling production operation. If or when the stocks become contaminated, they can be eliminated and pure genetic lines reintroduced. Facilities for producing hybrids should be laid out so that ponds containing pure tilapia broodstock are isolated from each other as well as from hybrid spawning ponds. Hybrid nursery ponds should also be isolated from hybrid spawning ponds. In essence, at least three groups of fish are involved in any hybrid hatchery operation: the two parental stocks and the progeny. Care should be taken to eliminate any contact between groups except in the spawning ponds.

It is my opinion that the primary obstacle to producing commercial numbers of all-male tilapia hybrids is the low number of progeny produced per spawning. In Brazil, an average of 2,763 all-male hybrids were produced in 2.5 to 3.0 months when 50 female *S. niloticus* were stocked with 50 male *S. hornorum* in a 350-m^2 pond. It seems reasonable to assume that the average number of eggs produced per female was about 400. Thus, if survival of eggs to fingerlings was 100%, about 7 female *S. niloticus* spawned i.e. 14% of the females stocked. Even if the survival of eggs to fingerlings was not 100%, it seems unlikely that more than 20% of the females spawned. This, moreover, assumes that a female spawned only once per spawning period. If some of the females were able to spawn more than once then the percentage of spawning females would have been even less.

It appears at first glance that the number of hybrid fingerlings produced is small because the females are dying during some stage of development: hence the all-male survival. However, Hickling (1960) and Lee (1979) proved this to be untrue; females that spawn produce normal numbers of hybrid offspring. Lovshin (unpublished data) demonstrates that there is a relation between the number of females stocked in a 350-m^2 spawning pond and the number of hybrid fingerlings produced per female (Table 4). These data were collected over a 6-year period and do not consider the size of the female *S. niloticus* stocked, the variation in ratio of *S. hornorum* to *S. niloticus* stocked, or length of the spawning period (which varied from 71 to 111 days). The fewer the female *S. niloticus* per unit area of spawning pond the greater the number of hybrid fingerlings produced, calculated on a per female basis. Thus, it appears that the percentage of females that actually spawn increases with a decrease in the stocking density of females. It is not known why this happens. Increasing the stocking density of females increased the number of hybrids produced up to a value of 1 ♀/7 m^2, but further increases up to 1 ♀/4.7 m^2 and 1 ♀/3.5 m^2 failed to give hybrid increases. There appears to be some behavioral or chemical factor present that reduces the compatibility of female *S. niloticus* and male *S. hornorum* so that at high densities of female and/or male broodstock the percentage of females that will spawn is greatly reduced. The role of the male *S. hornorum* and the impact that surplus males or *S. niloticus* females may have on male territorial and sexual behavior must be considered. The densities of males given in

Table 3. Summary of results obtained culturing tambaqui, *Colossoma macropomum*, in monoculture and polyculture with the all-male tilapia (T.) hybrid (female *Sarotherodon niloticus* x male *S. hornorum*).

Species	Stocking rate/ha	Ration	Average stocking wt (g)	Average final wt (g)	Net yield kg/ha	Conversion ratio	Growth g/day	Net yield kg/ha/day	Culture period days
Tambaqui	5,000	Pelleted chicken ration (17% protein) 3% of body wt/d	25	1,496	6,558	2.8	4.0	18.0	365
Tambaqui	5,000	Pelleted chicken ration (17% protein) 3% of body wt/d of tambaqui only	25	1,189	5,515	2.8	3.2	15.1	365
T. hybrid	5,000		18	748	3,209		2.0	8.8	365
combined	10,000				8,724	1.8		23.9	
Tambaqui	10,000	Pelleted chicken ration (17% protein) 3% of body wt/d of tambaqui only	39	760	7,063	4.4	2.0	19.6	360
T. hybrid	3,000		15	770	2,179		2.1	6.1	360
combined	13,000				9,242	3.4		25.7	
Tambaqui	10,000	Pelleted chicken ration (17% protein) 3% of body wt/d of tambaqui only	36	785	6,841	4.7	2.1	19.0	360
T. hybrid	4,000		10	725	3,005		2.0	8.3	360
combined	14,000				9,846	3.2		27.3	
Tambaqui	10,000	Pelleted chicken ration (17% protein) 3% of body wt/d of tambaqui only	42	770	7,359	4.1	2.0	20.4	360
T. hybrid	5,000		12	702	3,267		1.9	9.1	360
combined	15,000				10,626	2.9		29.5	

Table 4. The effect of female *Sarotherodon niloticus* stocking density on the production of all-male hybrid fingerlings *(S. niloticus* x *S. hornorum)* in 350-m^2 ponds in Brazil over periods ranging from 71 to 111 days (after Lovshin 1980).

	Number of female *S. niloticus* in the spawning pond, i.e.,/350 m^2						
	6	10	15	25	50	75	100
No. of replicates	5	9	6	27	61	17	9
No. of male *S. hornorum*/pond	3	5	10	5-50	10	15	20
Average no. of fingerlings produced/pond	2,117	2,350	1,643	1,475	2,763	2,167	1,502
Range of the average no. of fingerlings produced	580 to 3,619	93 to 3,639	821 to 2,439	0 to 7,149	0 to 8,443	473 to 5,295	607 to 2,526
Average no. of fingerlings per female	353	235	110	59	55	29	15
Estimated no. of females spawning[a]/pond	5	6	4	4	7	5	4
Estimated % of females spawning	83	60	27	16	14	7	4

[a]Assuming 400 eggs per spawn, 100% survival of eggs to fingerlings, and females spawned only once in each spawning period.

Table 2 do not appear excessive for 350-m^2 ponds. Note should also be taken of the large variation in hybrid fingerlings produced when the same density of females is used (Table 4), especially the zero figures.

To what extent reduced fingerling production exists in other hybrid crosses is not known. Tal and Ziv (1978) state that the production of *S. niloticus* x *S. aureus* hybrids is much lower than from single-species crosses. Lessent (1968) also had problems obtaining all-male hybrid fingerlings by crossing *S. macrochir* x *S. niloticus* in ponds. However, Hickling (1960) stated that he had little problem producing *S. mossambicus* x *S. hornorum* all-male hybrid progeny in ponds.

I have observed many cases of the all-male hybrid of *S. niloticus* x *S. hornorum* backcrossing with the female *S. niloticus*. In a pond containing female *S. niloticus*, male *S. hornorum*, and the all-male hybrid, the fingerlings produced are fathered almost exclusively by the all-male hybrid. The sexual behavior of the all-male hybrid appears to present fewer compatibility barriers with the female *S. niloticus* than the male *S. hornorum*.

Whatever the causes, the low numbers of fingerlings produced and the level of technology needed for pure-strain maintenance place restrictions on the commercial culture of the *S. niloticus* x *S. hornorum* hybrid. It is my opinion, based on present knowledge, that the culture of all-male hybrid tilapias is not to be recommended for the majority of farmers in tropical developing countries where the financial resources to construct properly designed hatcheries and hire qualified hatchery staff are lacking. Contamination of pure genetic strains of tilapia broodstock and backcrossing in spawning ponds are likely to occur. The governments in most tropical developing countries do not have the money to build and staff hybrid tilapia hatcheries to produce and distribute the fingerlings needed to culture this fish on a wide scale. In most cases, for tilapia culture to have a wide economic and nutritional impact in a country, the private sector has to become involved in fingerling production.

It is my opinion that some system of culturing a pure tilapia species is better suited to most developing countries because the production of pure species fingerlings can be accomplished with few problems by most farmers.

2. Growout

If the low production of fingerlings is the major disadvantage in culturing all-male tilapia hybrids then the growout of hybrids to marketable size is the major advantage. Stocking all-male tilapia without having to hand-sex is very advantageous. The hybrids grow rapidly and uniformly when fed a wide range of commercial and agricultural by-product diets. The addition of chemical or organic fertilizers to pond water further increases production and improves food conversion efficiency. Most tilapia hybrids are resistant to disease and to low levels of dissolved oxygen. This allows high rates of feeding and fertilizing resulting in elevated productions.

The culture of tilapia hybrids that are less than 100% male can also be very profitable depending on the percentage of males and the size of the fish to be marketed. Fish of about 200 g can be raised in 6 to 7 months with feeding or addition of organic fertilizer even if a small percentage of females are present. If larger fish (⩾ 400 g) are needed, then additional hand-sexing or a predatory fish may have to be stocked with the less than 100% male hybrids to control recruitment and allow a longer growing period.

3. THE EVALUATION OF TILAPIA HYBRIDS

The evaluation of hybrids for culture performance is not easy because few comparative data on fingerling production and growout are available. Only two crosses, *S. niloticus* x *S. hornorum* and *S. mossambicus* x *S. hornorum* are known to consistently produce all-male offspring. The male *S. hornorum* appears to be the most dependable male parent of 100% male tilapia offspring. Several other crosses will, however, produce more than 75% males consistently.

The maintenance of pure genetic lines is easier for some hybrid crosses than for others. The *S. niloticus* x *S. hornorum* hybrid, when 10 to 15 cm long, is relatively easy to distinguish from either parent on physical appearance alone by a trained biologist. Differences in physical appearance allow hatchery workers to easily determine the presence of contaminated broodstocks or mature hybrids in spawning ponds. The hybrids produced by crossing *S. mossambicus* x *S. hornorum* and *S. niloticus* x *S. aureus* are difficult to distinguish from their parents even by the trained eye. The chances of pure broodstock becoming contaminated with these hybrids are therefore increased. However, it is known that female *S. niloticus* crossed with male *S. hornorum* produces a low number of hybrid offspring. *S. hornorum* and *S. mossambicus* are close relatives as are *S. niloticus* and *S. aureus*. These pairs of species have similar reproductive behavior and coloration which may result in increased spawning and hybrid fingerling production, than, for example, *S. hornorum* with *S. niloticus*.

In countries where winter temperatures are low enough to threaten the survival of tilapias, hybrids with *S. aureus* as a parent should be considered because of their superior cold tolerance.

Future Research Needs

Fingerling production is the area in most need of research. If all-male hybrid tilapia culture is going to have an increased impact, then improved methods of consistently and economically producing all-male hybrid fingerlings on a commercial scale will have to be worked out.

A further pressing need is for information concerning the genetics, reproductive behavior and physiology of tilapias in relation to hybridization. What exactly is the genetic mechanism that produces all-male progeny from some crosses? Why are reduced numbers of fingerlings produced in some hybrid crosses when compared with pure species fingerling production? What are the visual and/or chemical cues which determine spawning compatibility between species? What other factors may be influencing the frequency with which two tilapia species hybridize? Does close taxonomic proximity of two species to be hybridized mean similar spawning behavior resulting in increased spawns and hybrid fingerlings?

Fish culture researchers have traditionally tried to answer these questions with practical research that often solved the problem without explaining why the result was obtained. Seven years of such work with the *S. niloticus* x *S. hornorum* hybrid has not provided the required increase in hybrid fingerling numbers on which to base large commercial cultures. I believe that the answer, if one exists, is to be found in basic research by knowledgeable behavioralists and physiologists aimed at understanding the barriers to tilapia hybridization and then discovering methods to overcome them.

Further investigation is needed to determine the best installations for producing tilapia hybrids. What are the comparative advantages of producing hybrids in small units, where a high degree of control is possible, over a well-designed pond hatchery? Which system is more economical and technically feasible?

Other areas still in need of further research are the optimum stocking densities and sex ratios of broodstock, the length of time for which they can be used before replacement, and the influence of broodstock nutrition on fecundity and fingerling survival.

Other tilapia species should be hybridized to discover new crosses that will produce 100% male offspring. The discovery of an all-male herbivorous hybrid would be very valuable.

Little comparative data is available on the growth and production of hybrids. Studies are needed to determine which hybrid crosses respond best to feeding and which to fertilizing so that intelligent choices of hybrids can be made.

The vast majority of cultured tilapia hybrid crosses give less than 100% male progeny. These fish are often marketed at a small size (Taiwan) or the males are hand-sexed so that larger fish can be raised (Israel). The commercial culture of all-male hybrids presents problems that have still to be researched. The consistent production of 100% all-male hybrid fingerlings in sufficient numbers to stock a large growout operation has not been worked out.

The level of technology needed to raise all-male hybrids is beyond the reach of most farmers in tropical developing countries and great care should be exercised in introducing all-male hybrid culture into countries with no fish culture tradition. Culture of pure species appears to be the logical first step in introducing tilapia culture to such countries. However, in countries with a tradition of fish culture, where trained biologists are available and money exists to build hybrid fingerling hatcheries, culture of all-male tilapia hybrids may hold great potential if researchers are able to improve fingerling production techniques.

Acknowledgments

I would like to thank the National Department of Works Against the Droughts (DNOCS) and the Centro de Pesquisas Ictiologicas for their support and patience over the 6.5 years I spent in Fortaleza, Brazil, performing fish culture research. I want to specifically thank my Brazilian colleagues in fish culture, Amaury B. da Silva, Antonio Carneiro-Sobrinho, Fernando R. de Melo, F. Ari Pinheiro, and J. Anderson Fernandes for their collaboration and friendship. Without the efforts of the biologists mentioned above much of the research reported in this paper could not have been accomplished. Questions pertaining to unpublished data should be directed to one of the above biologists at the following address: Centro de Pesquisas Ictiologicas, Av. Bezerra de Menezes, 2045, Fortaleza, Ceara, Brazil, 60.000.

Discussion

BOWEN: Can you tell us if you have any idea of the fate of the agricultural products or by-products which you used as feed? Did the fish consume these readily?

LOVSHIN: There is no doubt about it. They consumed them immediately. You could see the feeding activity and when you examined the guts they would be crammed with the materials. Now, as far as digestibility is concerned, that is another matter. They can probably digest some of them—at least a certain percentage of the available nutrients—but a lot of their growth comes from the incorporation of nutrients into bacterial detrital feeds. We found that to really get the best growth in the pond culture of tilapia hybrids, we had to reach a certain level of pond fertility.

BOWEN: Is there any correlation between that level of fertility and feeding habits? Do the feeding habits change at that level of fertility?

LOVSHIN: That is a good question. I don't know. All I can say is that we always used extra supplemental food until pond fertility reached a certain point, and lowered the feeding rate thereafter because we knew that the fish were getting a lot of natural feed from the pond.

BOWEN: Secondary plant substances have been mentioned as obstacles to the utilization of some agricultural by-products. Did you ever note anything that could be suspected as a toxic effect from any of these substances which you used?

LOVSHIN: No. We used a castor bean meal which can be toxic, but ours had been detoxified by roasting. It was a good product for us because the animal husbandry people, pig and cattle raisers, do not like to use castor bean meal because it gives their animals intestinal problems. It does not kill them, but it irritates the intestines and causes diarrhea. We could get this particular material very cheaply. In my paper, there is a comparison between castor bean meal, cottonseed cake and palm nut cake, and the castor bean meal is by far and away the best because it has the highest protein level. We used the whole, ground castor bean seed which has an outer shell which is very hard and indigestible. In an aquarium, the bottom would be literally covered with these hulls, but we never had any toxicity problems.

PULLIN: You have already touched on the problem of compatibility between crosses of different species. What do you think about the technique of premaxilla clipping to reduce male aggression in confined environments?

LOVSHIN: In a confined environment, male aggression can be a problem, but we have never had any such problem in a pond environment which allows the females to escape from aggression. We never had any mortality from male aggression in the ponds we used, which were about 350 m^2 but when we brought fish into more confined environments, 3 to 4 m^2, the problem did occur. During the time I was in Brazil, I did not know about Lee's technique of cutting off the premaxilla, but at Auburn University, it works fine and we have been using it quite a bit. The male fish is still aggressive. It still chases the female but, because the premaxilla has been removed, he cannot wound her as easily. Sometimes he will still run her around so much that I think she could die of exhaustion, but he cannot break the skin.

JALABERT: What evidence do you have for the purity of the males which you used?

LOVSHIN: The only evidence I have is that we consistently got 100% all-male progeny. We have been doing this in Brazil now for seven or eight years.

JALABERT: In group spawning or single pair spawning?

LOVSHIN: Group spawning in Brazil, but 100% males have been obtained from single pair spawning using stocks which we sent to Israel. Also, our stocks have produced 100% males in Panama, in the United States and in Puerto Rico so I am pretty certain of the purity of the strains.

HENDERSON: You mentioned the problems of the reduced fry production with the hybrid crosses. Do you think this is a problem of reduced fecundity or incompatibility?

LOVSHIN: I don't think it is basically fecundity, that is an egg production problem. The number of eggs produced by actual spawning females is not abnormally low. I am convinced that the problem is compatibility. There are certain females that will spawn readily and repeatedly both in aquaria and tanks in normal pure species crosses. Now the question is, can you select for this, i.e., take those females out, mate them with a normal male, take the offspring, test them, select good females again, etc. Can we select genetically for good spawning in pure species crosses and for compatibility in hybrid crosses? These are questions for the geneticists.

GUERRERO: Is there any evidence of hybrid vigor when you compare the growth performance of hybrids with pure species progeny?

LOVSHIN: A good question and the answer is not clear from the literature because most hybridization has been to produce all-males from pure stock male and female. Comparisons between the growth of an all-male population and mixed sex or female populations are complicated because we know that males grow faster than females—both hybrid and pure species males. Where we compared all-male hybrids with the *males* of the other species, we found that the differences in growth were not very great. In Brazil, we found a difference, although I am talking here of a 10% to 15% advantage for the all-male hybrid at best and sometimes this was not statistically significant. In Israel, we found several cases where there was no difference at all.

ROBERTS: At the Institute of Aquaculture in Stirling, we have been producing very pure lines using normal and sex-reversed (hormone-treated) sibling crosses. We found a very significant growth promotion effect from the anabolic steroid treatment used for sex reversal. More important, by crossing pure lines, we can produce hybrid all-male progeny which, after 6 months or so, are very much larger than the pure lines. The performance of these hybrids cannot be due to any residual effects of anabolic steroids.

HENDERSON: The question is, is the growth improvement which you observe for the hybrids comparable to that for the sex-reversal treatment alone?

GUERRERO: We found significant differences in growth of normal males and those which are sex-reversed females. The latter remain slower growers compared with the normal males and with all-male hybrids.

JALABERT: Dr. Lovshin, in your maintenance of pure strains, have you seen any signs of adverse affects of inbreeding?

LOVSHIN: It is difficult to say. We suspect a reduction in growth rate but we have to check this. The *Sarotherodon niloticus* and *aureus* used at Auburn University, elsewhere in the USA and some of the Latin American work all stem from an initial introduction to the USA of about 25 individuals of each species from the Ivory Coast. Over the years, we kept about 25 or 30 broodstock of each species and we let them reproduce. Every six months to one year, we would clean everything out and replace the broodstock with 25 or 30 younger fish.

JALABERT: This is not a large enough population.

LOVSHIN: I agree with you.

HENDERSON: Incidentally, there was a consultation meeting in Rome in June 1980 sponsored by UNEP on the conservation of genetic stocks of fish. One thing that did come out of this was a recommendation on the minimum size of the stocks which you need keep to avoid the problems of inbreeding. I can provide an overview of this meeting to anyone who is interested.

EDITORS: In Panama, the cross *S. niloticus* ♀ x *S. hornorum* ♂ male has now been abandoned for all-male hybrid production in favor of the *S. mossambicus* ♀ x *S. hornorum* ♂ cross. This gives very high fingerling production and 100% males. Fry are reared in earthen ponds and then grown to fingerling size in open concrete ponds as in the Ivory Coast (see Coche, this volume). This information was received from R. Pretto Malca.

Plate 1. Phase I of the Rodolpho von Ihering Fish Culture Research Center, Pentecost, Brazil. (Reprinted by kind permission of the International Center for Aquaculture, Auburn University.)

Plate 2. *Sarotherodon hornorum* male, all-male hybrid *(S. niloticus* x *S. hornorum)* and *S. niloticus* female (top to bottom).

Plate 3. A 36 m^2 concrete-sided, earthern-bottomed tank used to maintain pure strains of tilapia for hybridization in Northeast Brazil. When in use, the tank has a filtered water supply and is covered with a net to avoid contamination from wild stocks. (Reprinted by kind permission of the International Center for Aquaculture, Auburn University.)

Control of Tilapia Reproduction

R. D. GUERRERO III

College of Inland Fisheries
Central Luzon State University
Muñoz, Nueva Ecija, Philippines

Present address:
Aquatic Biosystems
Bay, Laguna
Philippines

GUERRERO, R.D. III. 1982. Control of tilapia reproduction, p. 309-316. *In* R.S.V. Pullin and R.H. Lowe-McConnell (eds.) The biology and culture of tilapias. ICLARM Conference Proceedings 7, 432 p. International Center for Living Aquatic Resources Management, Manila, Philippines.

Uncontrolled reproduction gives excessive recruitment and low yields of harvestable-size tilapias from culture ponds. Several effective methods have been developed and applied for controlling this in commercial farms including monosex culture, the use of predators and cage culture. Each method, however, has its advantages and disadvantages. Other methods that have been tested are stock manipulation, the use of irradiation and chemosterilants, reproduction inhibitors such as methallibure, salinity, light and temperature. Further research is needed to determine the practicality of these methods for commercial culture.

Introduction

Tilapias are an important foodfish in many tropical and subtropical countries. More than 20 species of tilapia have been cultured in developing countries where animal protein is lacking. Tilapias are considered suitable for culture because of their high tolerance to adverse environmental conditions, their relatively fast growth and the ease with which they can be bred.

The main drawback to the worldwide culture of tilapias is their excessive recruitment in ponds which results in low yields of harvestable-size fish. Where the acceptable market size is 150 g or more, this becomes a critical problem.

In general, tilapias have high breeding rates. Their fecundity ranges from a few hundred eggs to several thousands per spawning. Under pond conditions, precocious breeding and stunting have been reported (Fryer and Iles 1972; Cridland 1962) which assist excessive recruitment.

The various methods for controlling the reproduction and recruitment of tilapias are reviewed in this paper with a brief discussion on the advantages and disadvantages of each method. To be practical, a method must be easy to apply, effective and economical.

Monosex Culture

1. Rationale

In the tilapias, the male in general grows faster than the female (van Someren and Whitehead 1960a, 1960b; Fryer and Iles 1972; Guerrero and Guerrero 1975; Anon. 1979b). This appears to be genetically controlled. Monosex male culture therefore gives faster growth and eliminates reproduction. Monosex tilapias are obtained by manual or mechanical (grader) separation of the sexes or by production of monosex broods through hybridization or sex reversal.

2. Manual Sexing or Grading

Manual sexing of tilapia has been suggested and tested by several workers (Hickling 1963; Meschkat 1967; Shell 1967; Guerrero and Guerrero 1975). The sexes are distinguished by examination of the urinogenital papillae. Two orifices are present in the female papilla and one in the male (Vaas and Hofstede 1952). Although manual sexing is laborious and requires some skill, it is applied commercially in Israel. Sexing of 50 g or larger fish is easily done. One man can segregate about 2,000 male tilapias in a working day (Lovshin and Da Silva 1975). In many countries where the method has been introduced, however, it has failed. The major disadvantages of the method are human error in sexing and the wastage of females. Bardach et al. (1972) reported that hand sexing is 80 to 90% accurate. The use of mechanical graders for separating larger sized males was tried by Pruginin and Shell (1962) and Bard et al. (1976). The accuracy of this method, however, has been questioned by Balarin and Hatton (1979).

3. Monosex Hybrids

The production of all-male progeny from *Sarotherodon mossambicus* (female) x *S. hornorum* (male) hybridization was first described by Hickling (1960). Five other crosses have been reported to produce all-male F_1 hybrids: *S. niloticus* x *S. hornorum* (Pruginin 1967), *S. niloticus* x *S. macrochir* (Jalabert et al. 1971), *S. niloticus* x *S. aureus* (Fishelson 1962) and *S. niloticus* x *S. variabilis* and *S. spilurus niger* x *S. hornorum* (Pruginin 1967). Other crosses that have yielded 93 to 98% males are male *S. niloticus* x *S. leucostictus* (Pruginin 1965) and *S. niloticus* x *S. spilurus niger* (Pruginin 1965).

Several theories have been proposed to explain these ratios. Hickling (1960) and Chen (1969) suggested that the sex-determining mechanisms XX female-XY male and WZ female-ZZ male are both found in tilapias. However, not all the sex ratios of tilapia hybrid progeny can be explained by these. Hammerman and Avtalion (1979) have presented a model which takes into account the possible sex-determining effects of autosomes as well as sex chromosomes.

Culture of tilapia hybrids has resulted in high yields of harvestable-size fish (Hickling 1960; Pruginin 1967; Lovshin et al. 1977; Avault and Shell 1967). Although hybrid vigor has been reported by Hickling (1960) and Avault and Shell (1967), the findings of Pruginin (1967) and Lovshin et al. (1977) showed otherwise. One other advantage of using crosses which give monosex hybrid progeny is that the wastage of females from manual sexing is eliminated.

The main disadvantages of such crosses are: difficulty in maintaining pure parental stocks that consistently produce 100% male offspring (Pruginin et al. 1975), poor spawning success (Lee 1979) and incompatibility of breeders resulting in low fertility (Lovshin et al. 1977).

4. SEX-REVERSAL

The use of steroid hormones for production of monosex broods has proved effective in many tilapias. Artificial sex-reversal of genotypic females of five species has been achieved using methyltestosterone and ethynyl-testosterone (Table 1) and feminization of genotypic males has been induced with estrogens: ethynylestradiol, estrone and diethylstilbestrol (Table 2). Treatment is *per os* for periods ranging from 15 to 60 days in tanks or aquaria. Survival of fry under hormone treatment does not differ significantly from controls indicating no differential mortality (Guerrero 1974). Effectiveness of the sex reversal treatment depends on the kind and dosage of steroid used, the method of administration, time and duration of treatment, and on the species (Yamamoto 1969). Shelton et al. (1978) recommend oral treatment of 9 to 11 mm *S. aureus* fry at a density of 2,600/m^2 or less with ethynyltestosterone dosage of 60 mg/kg of feed for 6 weeks at 25 to 29°C for production of all-male broods.

Sex-reversal using androgens on a commercial scale has been shown to be feasible by Guerrero (1979b) and Koplin et al. (1977). The disadvantages of the method are the need for holding facilities for treatment of large numbers of fry (Mires 1977) and the sophisticated skill needed for its application.

Hopkins (1979) produced all-male progeny of *S. aureus* with sex-reversed females (genotypic males) treated with ethynylestradiol. He also produced all-female fry with sex-reversed males of *S. niloticus* treated with androgens. The low percentage of spawns with monosex fish, however, points out that further research is needed to establish the practicality of the method.

Table 1. The treatment of tilapia fry with androgens to produce all-male broods (M = methyltestosterone; E = ethynyltestosterone).

Species	Compound	Treatment: Level of food incorporation (mg/kg)	Duration (d)	% Males	References
Sarotherodon mossambicus	M	10-40	60	95-100	Clemens and Inslee (1968)
	E	50	40	100	Guerrero (1979b)
S. niloticus	M	40	60	100	Jalabert et al. (1974)
	M	15-50	42	96-98	Guerrero and Abella (1976)
	M	30-60	25-29	99-100	Tayamen and Shelton (1978)
	E	30-60	25-29	98-100	Tayaman and Shelton (1978)
S. aureus	M	30	18	98	Guerrero (1975)
	E	30-60	18	98-100	Guerrero (1975)
	E	30	22	90-100	Sanico (1975)

Table 2. The treatment of tilapia fry with estrogens to produce all-female broods.

Species	Compound	Treatment: Level of food incorporation (mg/kg)	Duration (d)	% Females	References
Sarotherodon mossambicus	ethynylestradiol	50	19	100	Nakamura and Takahashi (1973)
	estrone	200	56	99	Guerrero and Guerrero (1976)
S. niloticus	diethylstilbestrol	100	25	91	Tayamen and Shelton (1978)
S. aureus	ethynylestradiol*	100	42	90	Hopkins et al. (1979) Jensen and Shelton (1979)

*With 100 mg/kg methallibure

The Use of Predators to Control Recruitment

Effective predators on young tilapias include many piscivorous fishes, such as *Elops hawaiiensis*, *Megalops cyprinoides*, *Micropterus salmoides*, *Ophicephalus striatus*, *Cichla ocellaris*, *Lates niloticus*, *Clarias lazera*, *Hemichromis fasciatus* and *Cichlasoma manguense*. In most cases, predator-prey stocking ratios have been determined (Table 3). Where effective predators are used, high yields of harvestable-size tilapia are reported (Swingle 1960; Lovshin 1975; Dunseth and Bayne 1978). Total production of tilapias, however, is significantly reduced as the recruits are eaten (Maar et al. 1966; Lovshin 1975). Moreover, difficulty in obtaining stocks of the desirable predator has limited application of this population control method (Balarin and Hatton 1979).

Table 3. Fish predators used for the effective control of recruitment of tilapias.

Prey	Predator	Stocking Ratio (Predator:Prey)	References
Sarotherodon mossambicus	*Elops hawaiiensis*	1:10 and 1:20	Fortes (1979)
	Megalops cyprinoides	1:10	Fortes (1979)
	Micropterus salmoides		Swingle (1960)
S. niloticus	*Cichla ocellaris*	1:15	Lovshin (1975)
	Clarias lazera	1:10	Bard et al. (1976)
	Lates niloticus	1:20-1:84	Planquette (1974)
S. shiranus	*Clarias* sp.	1:10-1:20	Meecham (1975)
T. rendalli			
S. aureus	*Cichlasoma managuense*	1:4-1:8	Dunseth and Bayne (1978)
Not specified	*Hemichromis fasciatus*	1:48	Bardach et al. (1972)

The Use of Stock Manipulation Methods

Reproduction of tilapias appears to be inhibited by high stocking densities (Allison et al. 1979). Swingle (1960) reported that increasing stocking rates of fingerlings decreased rates of reproduction. He suggested the presence of a repressive factor affecting reproduction. In *S. mossambicus*, a substance believed to be present in the mucus has been found to cause an autoallergic response at high densities (Henderson-Arzapalo et al. 1980).

Culture of tilapia in cages at high densities has limited reproduction (Pagan 1975; Coche 1979; Guerrero 1980a). Pagan (1969) suggested that failure of tilapia to reproduce in cages is due to an alteration of reproductive behavior that prevents fertilization or that the eggs pass through the cage.

The Use of High Stocking Densities

The continuous harvesting of tilapias from ponds to reduce their population has been proposed by Hickling (1960). A similar method suggested is harvesting before the fish become sexually mature (Swingle 1960). No data are available to evaluate such methods for tilapias but Payne (1970) reported that regular seining of ponds with *S. esculentus* and *T. zillii* reduced the fry population. Mortalities were believed to have occurred from physical damage or deoxygenation.

The Use of Irradiation, Chemosterilants and Reproduction Inhibitors

Al-Daham (1970) observed a decrease in the gonadosomatic index and growth rate of *S. aureus* fry exposed to high doses of ^{60}Co gamma ray irradiation. Nelson et al. (1976, as cited by Balarin and Hatton 1979) found no obvious effects on the germ cells of 7 to 8 week old *T. zillii* fry treated with ^{60}Co gamma radiation for 35 days.

Destruction of the gonads of *S. aureus* fry was induced by Eckstein and Spira (1965) using estrogens at concentrations of 50 and 100 μg per liter of aquarium water for a period of 3 to 4 weeks. Al-Daham (1970) inhibited brood production in tilapias with the chemosterilants metepa and tetramine administered at concentrations of 20 ppm for 2 and 3 months and 0.8 ppm for 2 and 3 months, respectively. Sterile male tilapia were produced with the treatment of ^{90}Sr at 10^{-10} and 10^{-6} Ci/liter (Voronina 1974, as cited by Balarin and Hatton 1979).

Using methallibure, a compound which blocks synthesis or release of pituitary gonadotropins, Dadzie (1974) suppressed gonadal development in *S. aureus.* The treatment enhanced the growth of female fish and delayed spawning. A major constraint in the future development of methallibure for controlling reproduction is the discovery of its teratogenic effect in swine (see discussion in Balarin and Hatton 1979).

The reproduction of tilapias also appears to be affected by salinity. Chimits (1955) reported that *S. mossambicus* did not reproduce in salinities above 30‰. Similarly, Chervinski and Yashouv (1971) found that *S. aureus* did not reproduce in saltwater ponds with salinities of 36.6 to 44.6‰. *S. niloticus* fry were not found in brackishwater ponds in salinities of 15 to 30% (Dureza, pers. comm.). Ang (pers. comm.) observed that no hatching of *S. niloticus* eggs occurred in aquaria in salinities of 18‰ and higher.

Light and temperature strongly influence the spawning of tilapias. Cridland (1962) showed in laboratory experiments that sexual maturity of *T. zillii* was delayed by "strong periodic illumination for 12-hr periods" and by low temperatures. The sexual precocity of tilapias in ponds and swamps has been related to light (Chimits 1955; Lowe (McConnell) 1958). Mires (1974) indicated that temperature influenced the sex ratio of *S. aureus*: a higher percentage of females was associated with low temperatures.

Conclusion

It is evident that three methods can be used for the control of tilapia reproduction: monosex culture, the use of predators and stocking at high densities. Each method, however, has disadvantages.

The use of stock manipulation techniques, irradiation, chemosterilants and reproduction inhibitors have shown promise in limiting tilapia reproduction on an experimental scale, but further research is needed to determine the practicality of these methods for commercial use.

Discussion

HENDERSON: Dr. Guerrero, do you have any idea of the comparative cost of sex reversal by hormone treatment compared with that of manual sexing?

GUERRERO: For manual sexing in the Philippines, we pay twenty centavos (US$0.03) per fish and a worker can sex 2,000 fish per day. The daily wage is up to US$3/day, on average about $2.60. Sexable size is taken as 10 g or above. Hormone sex-reversal is of course labor saving, but you can see that our labor costs are low. Moreover, we have to use alcohol for dissolving the hormone for food incorporation. For every kilogram of feed, with hormone we use one liter of alcohol. This is quite expensive unless you can recycle the alcohol. We consider that it costs about 20 Philippine pesos (about US$3) to treat about 1 million fry. These costs would be only for feed and labor, ignoring the depreciation cost of facilities. If we sell these sex-reversed fingerlings, we get about 150 Philippine pesos (which is about US$20) per thousand so we still make a good profit. Sex-reversal techniques are fine if you have practical experience, but in general, I would not recommend them in a developing country like ours because of the relatively high capital requirements and the need for skilled workers. There are no commercial hatcheries using sex-reversal in the Philippines at present.

ROBERTS: We use similar methods to yours but with a 35-day treatment period and 30 ppm dose regime. Going back to the anabolic steroid growth promotion effects, we find that this anabolic effect is so significant and gives the fry such a head start that it is worth doing quite irrespective of any sex-reversal objectives. Recently, we had indications that sex-reversed females grew better than normal females and normal males. We are also studying the residence time of steroids in our fish after cessation of treatment using radioactively-labelled compounds. They are generally undetectable after 5 days. This is important as many of the developing countries are borrowing their food and drug legislation direct from the USA and we have to show that this short-term fry treatment is very different from caponisation in the poultry industry which uses long-lasting steroid implants. We have had our best sex-reversal results with *S. mossambicus* and *S. spilurus* and our worst with *T. zillii*.

It may be of interest that the black coloration of species like *S. mossambicus* does not produce sales resistance in Africa—unlike in Southeast Asia.

HEPHER: I wonder whether space is a severe limiting factor on the whole sex reversal operation. If you talk in terms of raising 2,000 fry/m^2 then you will need 15 of such tanks to serve 1 ha of monoculture pond with a normal stocking ratio of about 30,000/ha. Therefore, think how many tanks you would need for a whole farm. Won't this be a bottleneck?

GUERRERO: Our systems are as yet experimental but the production per unit area could probably be improved by using raceways with a flow-through system.

HEPHER: The second bottleneck which I can envisage is obtaining sufficient numbers of fry at exactly the right age for sex-reversal treatment.

GUERRERO: Using our net enclosure (hapa) system, we can produce 10,000 fry/ 500 m^2/day.

MORIARTY: Has anyone tried treating with hormones in the tank water? I believe that aquarists have done this with other species.

GUERRERO: Some people have tried this with tilapias but it has not been effective. Of course hormone injections are effective—more so than food incorporation but this is not practical for sex-reversing fry.

MORIARTY: What about a large quantity of hormone in the water?

GUERRERO: The hormones are not water soluble.

CHERVINSKI: I think I am right in saying that *S. aureus* has never been sex-reversed by hormone treatments.

GUERRERO: It has been done successfully for *S. niloticus, aureus* and *mossambicus.*

PHILIPPART: Regarding the use of flow-through systems, we succeeded last year in producing sex-reversed *S. niloticus* in such systems. We had to regulate food input very carefully. We treated the fish for 29 to 30 days.

A Study of the Problems of the Mass Production of Hybrid Tilapia Fry

D. MIRES

Chief Extension Worker in Fish Culture
Ministry of Agriculture
Kibbutz Ein Hamifrats D.N. Ashrat. 25210.
Israel

MIRES, D. 1982. A study of the problems of the mass production of hybrid tilapia fry, p. 317-329. *In* R.S.V. Pullin and R.H. Lowe-McConnell (eds.) The biology and culture of tilapias. ICLARM Conference Proceedings 7, 432 p. International Center for Living Aquatic Resources Management, Manila, Philippines.

Comparative data on the spawning capacity and spawning frequency of female tilapias demonstrate successful spawning in intraspecific and interspecific crosses. There is, however, great variability in the fecundity of individual females. Spawning incompatibility exists between tilapia species, the degree of incompatibility varying with the species combination. The influence of the sex ratio of the parental stock and other factors which reduce male aggressiveness are discussed. There are three main influences on the production of tilapia fry: genetic, behavioral and environmental.

Introduction

For the last ten years or so, fish farmers in various parts of the world have searched for efficient ways of rearing all-male broods of tilapias (Hickling 1960; Fishelson 1962; Pruginin 1967; Chen 1969; Mires 1973; Lovshin and Da Silva 1975; Mires 1977), yet the problem of mass-rearing these fish remains. After many years of hard work, some of the Israeli fish hatcheries have succeeded in making a major breakthrough in this field (Mires 1977) but this young industry still has to overcome the problem of the low fry production from interspecific spawns. This paper discusses various aspects of this problem.

Possible Reasons for the Low Production of Tilapia Fry in Spawning Ponds

There are quite a few possible reasons for the low production of tilapia fry in commercial spawning ponds. The main ones are probably: too low a density of broodstock; incompatibility in interspecific crosses; inappropriate broodstock sex ratios; inadequate spawning techniques and high fry mortality. More information is needed on each of these possibilities.

The production of tilapia fry in spawning ponds depends on three main factors: the spawning frequency of each female in the parental stock, the number of eggs produced in each spawn, and the survival of the fry. These aspects will be considered for intraspecific and interspecific crosses.

Egg and Fry Production: General Considerations

It is important to define the term 'fecundity'. Lowe (McConnell) (1955b) defined fecundity as "the number of young produced during the life time of an individual". This definition is perfectly acceptable for natural conditions, but it is very hard to use in culture or in experimental conditions. In this paper it is proposed to change the time limit from "life time" to 12 months starting from the first spawning.

It has been stressed that the number of eggs in each spawn depends on the length and/or weight of the spawner (Fryer and Iles 1972) but Peters (1959) and Lee (1979) found different results. Lee (1979) found that for some individual spawners there was a trend of increasing clutch size through successive spawns, but this was not uniform (see also Tables 1 and 3). Babiker and Ibrahim (1979) showed correlations between the length or the weight of *S. niloticus* and the number of oocytes in the ovaries, but there are doubts as to whether these correlations can forecast the actual quantities of eggs or fry obtainable from a given female during a given period under various conditions.

Tables 1, 2, 3, and 5 show a big variability not only in the spawning frequency between spawners of different species but also in the total number of eggs produced by the females within the groups. Moreover Lee (1979) showed a 20 to 30% difference in the number of eggs produced within female groups of two different species, stocking three females per aquarium.

Peters (1959) showed an exact correlation between the weight of the clutches and the body weight of the females; he also stressed that "fish either produce relatively few large eggs or relatively many small eggs". It seems possible therefore that on an annual basis small but more fecund females can and will probably produce more eggs than big but poor spawners.

Spawning procedures and conditions can also influence the production of fry in the spawning ponds: for example the rate of water changes. There is no substantial data on the influence of this factor on fry yields from spawning ponds, but for aquaria most of the hatchery operators in Israel believe firmly that periodical water change will usually cause a renewal of spawning activity. Commercial fry producers also believe that the change of water in spawning ponds has a beneficial effect on spawning frequency. This is probably due to the flushing out of metabolites. It seems that overcrowding in ponds, tanks, etc., may induce earlier maturity than in nature and a larger production of smaller sized eggs (Fryer and Iles 1972).

Feeding conditions also have an influence on egg and fry production (Miranova 1977). For example, *S. mossambicus* females spawned more frequently when they were underfed, and the total amount of eggs was higher than when they were fed abundantly. This response to unfavorable conditions has probably permitted many tilapia species to adapt to harsh and changing environment during their evolution (Fryer and Iles 1972). Lately, the Ein Hamifrats fish farmers (unpublished) have observed that when a spawning pond becomes overcrowded with fry, the spawning activities of parent fish cease completely. In these conditions, there are no mouthbrooding females and the few nests present seem to be unattended.

Intraspecific Spawning

A lot of work has been published on the reproductive behavior of tilapia spp., but very little has been said about the individual spawning potential of females. The only possible way to study this closely is under aquarium conditions, where the females are kept in separate aquaria or in separate compartments of the same aquarium, because in spawning groups a hierarchy may become established in aquaria and cause some dominant females to spawn more than others (Fishelson 1966a; Mires 1973; Rothbard 1979).

During the first working period in the Ein Hamifrats hatcheries, some valuable data were gathered on the individual spawning potential of *S. niloticus* B and *S. vulcani* (Mires 1977). A few females weighing 400 to 500 g were introduced into separate compartments in aquaria and kept at a constant temperature of 25 to 28°C year-round, fed with 25% protein pellets. The day on which each female was ready to spawn was recognized by the form of the genital papilla being much larger and more erect than usual. Such a female was transferred to an aquarium where one male of the same species was held. If the female was ready to spawn, the male would immediately start courting her; but if not he would start chasing her around and eventually bite her severely. In such a case the female was removed and placed back in her compartment.

After spawning, the female was separated from the male by a screen, and was left to mouthbrood her eggs for a few days. The larvae were then removed from her mouth, and cared for in Zuger bottles for the rest of their larval development. When a female spawned without the presence of a male the spawn was recorded, but unfortunately the eggs were not counted. The results for one group of *S. vulcani* and another group of *S. niloticus* B are given in Tables 1 and 2. Although the females were kept in separate compartments of the same aquarium, it is possible that slight environmental differences existed (corners versus center). Their spawning frequency was very variable and they did not spawn year-round despite the controlled temperature. In both groups the maximum number of individual spawns was 7: represented in both cases by only one female. It is interesting that the proportions of prolific and poor spawners were similar for both *S. niloticus* and *S. vulcani.* The same phenomenon exists with interspecific spawns.

McBay (1961) gathered data on the spawning of what was then called *T. nilotica,* later re-identified as *S. aureus.* The fish were paired in 40-liter aquaria and the egg or fry production was recorded and counted. Out of 17 females, one spawned 6 times, two spawned 5 times, one spawned 4, and all the rest spawned three times or less.

Table 1. The spawning frequency and egg production of isolated individual *Sarotherodon vulcani* females (400-500 g each) studied from February 1, 1973 to January 1, 1974 (after Mires 1977 and Mires unpublished data).

Female no.		Date of spawning							Total egg count	Average eggs/ spawn*	No. of spawns	Average interval (days)
	Date	24/3-	24/4-	15/5-	3/6-	24/6-	26/7-	5/9-				
1	No. eggs	1,600	**	**	**	**	**	**	1,600	1,600	7	
	Interval (days)	31	21	19	21	32	40					27
	Date	10/3-	8/4-	27/5-	24/6-	3/7-	5/8-					
2	No. eggs	1,400	**	1,500	1,200	**	1,000		5,100	1,275	6	
	Interval (days)	39	47	28	9	33						31
	Date	25/3-	4/4-	9/5-	7/8-							
3	No. eggs	1,000	**	1,150	**				2,150	1,075	4	
	Interval (days)	10	35	90								45
	Date	27/3-	28/4-	18/7-								
4	No. eggs	**	**	**					—	—	3	
	Interval (days)	32	81									56
	Date	25/3-	21/9-									
5	No. eggs	**	**						—	—	2	
	Interval (days)	150										150
	Date	14/3-										
6	No. eggs	**							—	—	1	
	Interval (days)											

*Based on the number of eggs in fertile spawns only.
**Infertile spawns: eggs not counted.

Unfortunately, some of the females died during the experiment and the times of death were not recorded. It is therefore impossible to know whether the spawning of some of the females ceased because of their death or because of other reasons. It is, however, interesting to observe that here again, although an optimal temperature was maintained throughout the experiment, the spawning frequency of the females differed.

In the Kibbutz Ein Hamifrats fish hatcheries, where several families of *S. niloticus* and *S. aureus* are kept in 500-liter aquaria, individual females of both species have never spawned more than six or seven times in one year and the average spawning frequency of any given group of females in a family has not exceeded four. Moreover the period over which spawning occurs is restricted to five or six consecutive months even when the temperature is optimal year-round.

In other intraspecific spawns in aquaria (Lee 1979) very big differences were observed in the spawning frequency and fecundity of various females.

Out of ten *S. aureus* females, one spawned 8 times, one 7 times, two 6 times, one 5 times, four 4 times and one 3 times. Out of nine *S. niloticus* Ivory Coast one spawned 5 times, five 4 times and three 3 times. Yashouv (1958) stated that in Israeli fish ponds, two-year old *S. aureus* usually spawn three or four times from May to October.

Table 2. The spawning frequency and egg production of isolated individual *Sarotherodon niloticus* B (400-500 g each) studied from February 1, 1973 to January 1, 1974 (after Mires 1977 and Mires unpublished data).

Female no.		Date of spawning							Total egg count	Average eggs/ spawn*	No. of spawns	Average interval (days)
1	Date	7/2-	4/3-	3/5-	5/6-	6/7-	30/7-	5/9-				
	No. eggs	200	1,600	**	**	**	1,000	**	2,800	933	7	
	Interval (days)	25	57	33	31	24	35					34
2	Date	26/2-	24/4-	2/5-	29/5-	23/7-	20/8-					
	No. eggs	900	**	**	**	1,300	**		2,200	1,100	6	
	Interval (days)	57	8	27	56	28					35	
3	Date	20/5-	9/6-	13/7-	4/8-							
	No. eggs	**	**	**	1,000				1,000	1,000	4	
	Interval (days)	15	34	21								23
4	Date	4/3-	29/3-	4/5-								
	No. eggs	**	**	700					700	700	3	
	Interval (days)	25	36									30
5	Date	24/5-	18/6-	2/9-								
	No. eggs	1,090	1,600	**					2,690	1,345	3	
	Interval (days)	25	76									50
6	Date	21/6-	26/7-									
	No. eggs	**	**								2	
	Interval (days)	35										35

*Based on the number of eggs in fertile spawns only.
**Infertile spawns: eggs not counted.

Interspecific Spawning

1. Aquaria and Tanks

The successful crossing of two different tilapia species in aquaria has always been a serious challenge to researchers and hatchery operators. It was made easier with the help of induced spawning techniques (Rothbard and Pruginin 1975) but it seems improbable that this can be used widely. Recently interspecific spawns have been obtained in tanks (Hulata et al. 1980). Hulata and co-workers have been studying interspecific spawns in 600-liter tanks since 1978 and have kindly permitted the use of some of these data (Tables 3, 4 and 5). As in Tables 1 and 2, there is an enormous variability in fecundity between females of the same group, even with constant environmental conditions year-round.

Table 3. The spawning frequency and egg production of *Sarotherodon niloticus* Ghana females in interspecific spawning crosses with *S. aureus* males weighing 50 g each over a 12-month period (Hulata et al. 1980 and Hulata unpublished data).

Female nc		Date of spawning								Total egg count	Average eggs/ spawn*	No. of spawns	Average interval (days)
1	Date	6/4-	7/5-	21/6-	8/7-	23/8-	14/9-	9/10-	26/10-				
	No. eggs	82	41	**	230	171	298	220	399	1,441	205	8	
	Interval (days)	31	45	17	46	22	25	17					29
2	Date	10/6-	8/7-	12/8-	6/9-	2/10-	26/10-						
	No. eggs	274	375	360	467	158	172			1,806	301	6	
	Interval (days)	28	35	24	26	24							27
3	Date	18/7-	15/8-	6/9-	2/10-	22/10-	26/11-						
	No. eggs	283	55	**	247	368	272			1,225	245	6	
	Interval (days)	27	22	28	20	35							26
4	Date	8/7-	15/8-	6/9-	2/10-	22/10-	26/11-						
	No. eggs	251	298	451	214	449	383			2,046	341	6	
	Interval (days)	38	21	27	20	34							28
5	Date	5/8-	17/9-	6/11-									
	No. eggs	254	390	284						928	309	3	
	Interval (days)	42	49										45
6	Date	26/8-	24/9-										
	No. eggs	414	200							614	307	2	
	Interval (days)	26											26
7	Date	14/9-											
	No. eggs	210								210	210	1	
	Interval (days)	—											—
8	Date												
	No. eggs									0	0	0	
	Interval (days)												
9	Date												
	No. eggs									0	0	0	
	Interval (days)												

*This figure represents only the average amount of eggs in fertile spawns.
**Infertile spawns: eggs not counted.

2. PONDS

For interspecific spawning in ponds, it is practically impossible to obtain any information on the fecundity of individual spawners and only the total fecundity of the broodstock can be studied. Crosses between 3 groups of nine *S. niloticus* Ivory Coast females and three *S. hornorum* males (mean individual weight of both sexes, 45 g) in triplicate experimental ponds yielded 223, 280, and 404 fry/pond for a 72-day period (Lovshin and Da Silva 1975). The same authors reported no significant difference in hybrid fry production from two spawning groups: *S. niloticus* Ivory Coast females x *S. hornorum* males (10 ♀♀:5 ♂♂ and 6 ♀♀:3 ♂♂). There was a high variability in fry production within the replicates of the same group: 93 to 2,981 for the 10:5 group and 580 to 3,619 for the 6:3 group. The spawning period in both cases was 72 days. Two factors could have caused this big variability: some of the females could have spawned more often than others (thus causing a temporary higher yield) and some of the spawning groups may have had a greater number of more fecund females than others.

3. The Effect of the Sex Ratio of the Parental Stock

In a field experiment (Mires, unpublished) done in two spawning ponds (0.4 and 1.0 ha) at the Kibbutz Ein Hamifrats, two different parental sex ratios were tested over a whole spawning season: *S. niloticus* B females x *S. aureus* males, 1:1 and 3:1. The basic assumption was that in this kind of spawn, the females do not have a strong urge to spawn with males of a different species and therefore a stronger "male pressure" could improve the situation. This assumption was based on the fact that in aquarium conditions it often happens that two or more females are ready to spawn at the same time, while there is only one male available. In such cases it often happens that the eggs of one of these females are not fertilized. The 1:1 ratio gave higher fry production than the 3:1 (800 fry per female in the 3:1 ratio compared to 1,100 in the 1:1 ratio).

The beneficial effect of the "male pressure" ratio has again been shown recently (Lovshin 1980) when for *S. niloticus* Ivory Coast females x *S. hornorum* males, a 1:2 sex ratio gave significantly higher fry production than 2:1 or 1:1.

4. Differences in Spawning Behavior; Incompatibility Problems

Differences in spawning behavior have been described for various species of tilapias (Fryer and Iles 1972; Lee 1979). The incompatibility between the species used in interspecific crosses have usually been studied in aquaria or in tanks, but not in nature.

Tables 3, 4 and 5 show that *S. niloticus* Ghana females do not have any incompatibility problems in crosses with *S. aureus* males. In the last experiment (Table 3), the average spawning frequency was just as high as in individual intraspecific spawns (Tables 1 and 2). However, *S. aureus* and *S. niloticus* Ivory Coast females did encounter serious problems while spawning with *S. niloticus* Ivory Coast and *S. aureus* males respectively, and therefore, the number of spawns obtained from these crosses was very low.

In interspecific crosses in tanks between *S. niloticus* Ivory Coast females and *S. hornorum* males, some of these incompatibility problems can be overcome by surgical removal of the male premaxilla (Lee 1979). However, male aggression is only one of the problems. Other factors that may cause low fry productivity in interspecific crosses include differences in courting behavior, in mating color display and in the form of the nest. All these problems and others may exist separately or simultaneously and are very hard to overcome in ponds.

Table 4. The spawning frequency and egg production of *Sarotherodon aureus* females in interspecific spawning crosses with *S. niloticus* Ivory Coast males 50 g each over a 12-month period (after Hulata et al. 1980 and Hulata unpublished data).

Female no.		Date of spawning					Total egg count	Average eggs/ spawn*	No. of spawns	Average interval (days)
1	Date	16/6-		4/7-		16/8-				
	No. eggs	**		348		88	348	348	3	
	Interval (days)		18		43					30
2	Date	7/8-								
	No. eggs	**							1	
	Interval (days)									
3	Date	5/7-								
	No. eggs	**							1	
	Interval (days)									
4, 5, 6,	Date									
	No. eggs								0	
	Interval (days)									

*Based on the number of eggs in fertile spawns only.
**Infertile spawns: eggs not counted.

Table 5. The spawning frequency and egg production of *Sarotherodon niloticus* Ivory Coast females in interspecific spawning crosses with *S. aureus* males 50 g each over a 12-month period (after Hulata et al. 1980 and Hulata unpublished data).

Female no.		Date of spawning							Total egg count	Average eggs/ spawn*	No. of spawns	Average interval (days)
1	Date	27/7-		19/8-		7/9-		5/10-				
	No. eggs	341		273		460		464	1,538	384	4	
	Interval (days)		23		18		18					19
2	Date	3/2-										
	No. eggs	136								136	136	1
	Interval (days)											
3, 4, 5, 6,	Date											
	No. eggs										0	
	Interval (days)											

*Based on the number of eggs in fertile spawns only.
**Infertile spawns: eggs not counted.

Table 6. Fry production (*Sarotherodon niloticus* ♀ x *S. aureus* ♂ hybrids) from 3 spawning ponds at Ein Hamifrats fish farm over the four month summer spawning season, harvested with a 5 mm mesh net.

Pond no.	1	2	3
Area (ha)	1.5	1.3	0.7
Parental stock			
Females	1,400	1,057	579
Males	900	1,084	300
Removal of fry			
End of July	650,000	600,000	160,000
End of August	230,000	192,000	220,000
Total	880,000	792,000	380,000
No. of fry/♀	628	749	656

The Mass Production of Hybrid Tilapia Fry

Five or six million hybrid tilapia fry per year are now being cultured commercially within the Israeli fish farming system, produced by a few farmers who have acquired a lot of recent experience. However, only a part of these produces pure-strain tilapias. It seems, therefore, that the future of the Israeli tilapia culture industry lies in the hands of a few specialist hatchery operators.

The problem of low fry production per spawner has been solved by holding more parental stocks. Low productivity of spawners is, however, only one of the problems that have an influence on the mass production of fry. In countries with a warm temperate climate, like Israel, overwintering of the fry is another problem. In Israel today, this has been recognized as a serious limiting factor, and many farmers are experimenting with various facilities which will help overcome it.

Tilapia fry under 20 grams are hardly able to survive the Israeli winter temperatures. To overcome this problem, young fry must be introduced into the nursing ponds early enough to reach or exceed this weight before the winter. This takes about 60 days and the growing season ends at the beginning of November. Therefore, the last date at which it is still worthwhile to start nursing is September. This restricts the usable spawning period to four months only: from May to the end of August.

In addition to these restrictions, the Israeli hybrid fry producers prefer to harvest with nets of at least 5 mm mesh, because smaller mesh nets collect a lot of mud and cause big fry losses. At the end of the spawning season, therefore, hundreds of thousands of the small fry are left unharvested and are not recorded as part of the total production of the spawning ponds. During the four-month spawning period, not more than two or three spawns can be expected from each female. Therefore a 400 g *S. niloticus* female cannot produce more than about 2,500 fry per season. In practice, the best yields obtained from females of this size in the Kibbutz Ein Hamifrats fish ponds, using a 1:1 sex ratio and counting all small fry, never exceeded 1,500 fry/female. The normal average production, using a 5 mm mesh net, is usually less than 1,000 fry/female (Table 6). Therefore, we could only expect to increase production by about 100% from existing facilities, even if all the problems cited above were solved.

Discussion

With this information we can attempt a new synthesis of the interrelationships of the various factors influencing intraspecific and interspecific spawning in tilapias and point out the factors that will have to be studied in order to overcome the problem of low fry production. The various factors fall into three main categories: genetical, behavioral and environmental.

1. Genetic Improvement

There is a very big phenotypical individual variability in the fecundity of the various tilapias. This variability exists within individuals of the same species as well as between the different species. Therefore, females whether from the same or from different species, will not necessarily produce the same amount of eggs even when crosses occur under identical conditions. This big variability has probably been the cause of some of the discrepancies between results of previous experiments. It seems that the smaller the number of spawners in a spawning pond, the bigger the chances will be of obtaining inconsistent results. If more fecund genotypes can be selected for cultured strains then higher yields of fry can be expected from both intraspecific and interspecific crosses.

2. Behavioral Factors

The different behavioral patterns that exist between the various species cannot be altered, although techniques like the removal of the male premaxilla may reduce behavioral incompatibility in some cases. Other solutions to this problem may be altering the conventional parental sex ratios, especially by using a higher male to female ratio, or using species with a closer behavioral pattern.

3. Environmental Factors

In normal environmental conditions female tilapias of any species will always spawn according to their natural individual spawning capacity, but the number of fertilized eggs and the production of fry seem to be mainly determined by behavioral factors. The improvement of environmental factors will most probably have a beneficial effect on fry production from any kind of cross.

In the immediate future it seems that the mass production of hybrid tilapia fry will still depend on the number of spawners used for any given cross.

Acknowledgments

I would like to thank Dr. G. Hulata and the genetic team in Dor station, for allowing me to use some of their data. I would also thank Dr. Hulata, Dr. Ernesto and Mr. Aldridge for reading the manuscript and for their remarks.

Discussion

NOAKES: In your Tables 1 to 3 giving the spawning performance of females, are these individual females in isolated aquaria or are they all in the same tank?

MIRES: Tables 1 and 2 refer to individual females in individual cells.

NOAKES: Then these were in separate water bodies with no contact between them?

MIRES: No, they were separated by partitions but in a common water body. There would be the possibility of chemical contact.

NOAKES: The reason for my question is that the apparent pattern suggested to me some kind of hierarchical arrangement, or some kind of inhibition of spawning of some females by others. Whatever the reason, the pattern of spawning by females is quite repeatable.

MIRES: I would say that there was no hierarchical arrangement. A hierarchy would probably show in the amount of eggs spawned, whereas here, successful spawning females always showed full spawning aggression and spawned normally.

NOAKES: It just occurred to me that the other females may sense chemically that a given female has spawned close by and may be inhibited by this.

MIRES: If this is so, I have no knowledge of it. There was of course a water change in this system.

PHILIPPART: Have you found differences in the number of spawn obtained from different-sized females?

MIRES: The data we are discussing here were from comparisons with females of equal size. They show a big variability in spawning frequency. I can say, however, that within a spawning family in aquaria in my hatchery it always happens that a few of the females are doing all the work and the remainder are not doing very much.

CHERVINSKI: Do you think that by draining your spawning ponds you are getting higher fry production than by not draining?

MIRES: I believe definitely yes, and I would like to suggest why. We know that there are definite differences in fecundity between species. Also *within* a species there are differing individuals with higher or lower fecundity. Pond drainage and water change will therefore improve production both from inter- and intraspecific spawning.

CHERVINSKI: Perhaps by doing this we are getting rid of some waterborne chemical factor produced during reproduction which at high fry densities inhibits further reproduction?

MIRES: I don't believe so.

LOVSHIN: What is your source of water?

MIRES: In the hatchery, it is the city water supply; in outdoor ponds it is a mixture of irrigation water, well water and water reclaimed from sewage effluent.

LOVSHIN: I was just thinking that even if you do keep wild tilapia out, they could be a source of chemical factors in the water, but I do not really think this is important. What is very significant in my view is that by draining the pond and removing the small fry you are greatly reducing the scope for cannibalism and this is probably the basis of your improved production. Partial selective harvesting or draining on a one- or two-month basis is always a good thing. I would not discount the possibility of a chemical stimulatory effect of water change on spawning or vice versa on an inhibitory effect by chemical factors at very high fish densities, but I have experienced continuation of spawning at very high densities (see Henderson-Arzapalo et al. 1980—Editors).

ROBERTS: Much of our hatchery work is done in Edinburgh prison by long-term prisoners. They take great pride in keeping the fish and preparing high protein diets, such as earthworms, for them. We maintain there pure lines of *S. niloticus*, *S. mossambicus* and *S. spilurus.* We can get regular spawning virtually every month (say 28 to 35 days) from given broodstock of all these species over an 18-month period. After this, they are less effective as broodstock and we choose to replace them although they are still fecund. These are fish which receive very careful attention and the best possible diet in small aquaria. I realize that this would not be economical on a commercial scale and that our situation is very different from Mr. Mires' hatchery. Our results suggest, however, that you can smooth out the variability in spawning by optimizing all conditions.

MIRES: Do your results apply to all your fish on an individual basis?

ROBERTS: To virtually every fish.

MIRES: Are they in separate tanks?

ROBERTS: They are kept in tanks in groups of 3 or 4 females per male.

HENDERSON: This suggests that we know very little about optimum husbandry methods in commercial hatcheries.

ROBERTS: I think that the main factor with our fish is their good husbandry and diet, including live food.

LOVSHIN: Regarding the age of broodstock and the length of time for which you can use them, we found in Brazil that after 1 year of spawning activity, i.e., after 3 or 4 spawnings (after which time our fish would be one to one-and-a-half years old) the number of fry produced per unit body weight would fall by about 50%. We therefore began to replace such fish, which were about 300 g in weight, with smaller fish, say 60 g, and we found that such continuous replacement gave better production than carrying on with the big fish. I cannot altogether explain this as of course larger fish produce larger total numbers of fry per spawning than small fish. I suspect that it is a question of spawning frequency. The smaller fish are more active and frequent spawners.

NASH: Does the fecundity decrease in the larger older spawners, or does the viability of the eggs decrease?

LOVSHIN: I have no way of telling from pond work. We should expect larger numbers of eggs from larger fish.

MIRES: I have observed in aquarium spawnings that the viability of eggs is just as good from 6-year old fish as from younger broodstock.

LOWE-McCONNELL: In the wild, with batch-spawning species such as *S. esculentus* and also *S. leucostictus* in Lake Victoria, the number of eggs per batch gets lower with successive batches in females of comparable size. Robin Welcomme has demonstrated this.

MIRES: I know this, but in aquaria we have not observed a similar situation. On the contrary, the number per batch sometimes goes up.

General Discussion on the Biology and Culture of Tilapias

Compiled by R.S.V. Pullin

Abstract

A report is given on a wide ranging discussion on the biology and culture of tilapias. The topics covered include speciation, species for aquaculture, establishment and conservation of pure strains, the concepts of stunting and a switch between somatic growth and gametogenesis/spawning, the physiology of digestion in tilapias, detritus and detritivory, feed formulation, food presentation, broodstock nutrition, broodstock management, mass seed production, recruitment control, integrated farming and wastewater reuse, and diseases of tilapias.

Classification and Speciation

Dr. Trewavas began by outlining the classification and interrelationships of the tilapias. It is generally accepted that the mouthbrooding species arose from ancestral substrate-spawning *Tilapia* and from a narrow group of these. Prof. Peters thinks that this may have happened several times, including the comparatively recent evolution of the species of the Cameroon crater-lake, Barombi Mbo. For this reason he thinks it inappropriate to unite those of more ancient origin with the more recent ones in one genus separate from *Tilapia*. He would retain the generic name *Tilapia* for all of them. But the mouthbrooders are alike, not only in their reproductive arrangements but also in their feeding habits and structures, and the *Sarotherodons* in Barombi Mbo are in Dr. Trewavas' opinion derived from species that were already mouthbrooders and microphagous feeders. This is not to say that there are no mouthbrooders derived independently from *Tilapia*, but these have not developed the same feeding structures as *Sarotherodon*.

There is more than one way of expressing such opinions in the nomenclature, and it is proposed to adopt the name *Sarotherodon* for all the mouthbrooding tilapias (but see Preface and Addendum, p. 11-12). There is no doubt that the east and central mouthbrooding species belong to a branch or branches that have long been separated from the west African. This may be expressed by regarding them as separate subgenera, a rank that may be disregarded by those not specialists in taxonomy.

Methods of speciation were discussed and the main point emerging was that speciation among the tilapias has been almost exclusively by specialization following geographical separation, i.e., allopatric. Among the notable exceptions where sympatric speciation is thought to have occurred, that in Lake Barombi Mbo is the most interesting. This 2.5 km diameter lake has 11 species of endemic cichlids, including four which are arguably

Sarotherodon spp., although their reproductive behavior has not been fully described. Dr. Trewavas speculated that the present situation had probably arisen from more than one source species, and included some fascinating scenarios requiring successive introductions of local riverine species with extreme spates and specialization/radiation/changing lake conditions in the intervals between spates. The Malawi species flock is another interesting example.

The difficulty of detecting modes of speciation was also apparent from studies on Lake Jipe in which it was not clear whether the present lacustrine species had arisen sympatrically from incursions of *S. pagani* from the Pagani River or whether *S. jipe* was itself originally present in the river. Also in west Africa, the lower Nile and Israel, *S. niloticus* and *S. aureus*—which are so alike that some museum workers have confused them in the past—are sometimes found together. Whether this is a secondary coming-together or a result of some sympatric speciation is difficult to say. These species do have slightly different ecological niches, for example in feeding habits, and sexual dichromatism and dimorphism are more marked in *aureus* than in *niloticus*.

Allopatric speciation has been the general rule for tilapias. For example, *S. mossambicus*, which occurs in the lower Zambezi and Limpopo Rivers, *S. spilurus* in Kenya (this species is so similar to *mossambicus* that they are difficult to separate in preserved material), *S. urolepis* in the Rufigi River and *S. hornorum* in Zanzibar, are all part of the same group. They are of similar appearance with a long snout and long jaws especially in breeding males. *S. hornorum* is particularly easy to distinguish on characters such as fin-ray counts.

The species living in the east African alkaline lakes, such as Lake Magadi, were also discussed. These species, e.g., *S.a. alcalicus* share a common problem with tilapias inhabiting river systems that are prone to severe drying, namely, a very restricted environment. In the alkaline lakes, the waters habitable by fish are restricted to those after the hot spring waters (about 42° C) have cooled slightly and before their salt content becomes too elevated by evaporation. Such species have many things in common. They generally have low number of gill rakers, vertebrae and fin-rays; they are very aggressive and active despite their high population density and restricted space, and they breed at an early age and small size. In Lake Natron, very small *S.a. alcalicus* males set up their breeding pits adjacent to those of mature males but in Lake Magadi small *S.a. grahami* interfere with larger spawning fish. Such populations show in natural waters the early breeding characteristics that tilapia culturists would like to avoid. They merit much fuller study. It would be a mistake, however, to call these wild fish 'stunted' as they are often in good condition.

The total number of species of *Tilapia* and *Sarotherodon* is a matter of opinion. Dr. Trewavas estimated that there are approximately 30 *Tilapia* and 46 *Sarotherodon* species. She pointed out that Thys van der Audenaerde regards as species populations of *S. melanotheron* that she considers subspecies. Dr. Trewavas has also recognized 7 subspecies of *S. niloticus*, some of which have yet to be named.

It is obvious that transplantation of species into natural waters, dams and aquaculture systems has further complicated an already complex situation.

Plate 1. Typical *S. mossambicus* from lower/middle Zambezi. Photo by M.S. Caulton.

Plate 2. *S. mossambicus* introduced in the Far East.

In some natural waters which have received multiple introductions, e.g., Lake Naivasha, the tilapia populations have changed year by year through hybridization and interspecific competition. The classic definition of a species, based on reproductive isolation, is obviously of limited use for tilapias, especially in aquaculture where so many different hybrids have been produced. There is a need to distinguish between postulational and operational definitions of species. For aquaculturists, an operational definition is required, i.e., a species must be recognizable either from morphology or biochemical tests.

There was a brief discussion on whether species developed through allopatric speciation would be incompatible with respect to reproduction if brought together. The consensus of opinion was that tilapias usually hybridize relatively easily. Differences in reproductive behavior are probably the most important factor where interbreeding is difficult. It is possible that bringing species together in an artificial lake or culture environment may facilitate hybridizations which would not occur in a natural environment.

Several speakers commented on the change in appearance of transplanted species. To illustrate this, Dr. Caulton showed a picture of typical *S. mossambicus* from the lower/middle Zambezi (Plate 1). These large deep-bodied fish are much larger at first maturity (around 500 g for breeding males) than is commonly seen in introduced populations in the Far East (Plate 2).

Species for Aquaculture

The species currently being used for aquaculture are a good representative group of the tilapias. The use of additional tilapia species for aquaculture is controversial. It could be justified where, for example, a previously unoccupied niche in a polyculture system could be filled or where additional species could outperform existing cultured species in growth and/or reproductive performance. Additional species would also open up new possibilities for hybridization work, but the maintenance of genetically pure lines is not an easy task. Hybridization work has concentrated largely on the production of all-male progeny rather than on demonstration of hybrid vigor, production of new hybrids for brackishwater culture, etc.

There are some other cases in which new species could prove useful. For example, Prof. Roberts stated that *S. spilurus spilurus* performs very well in brackish and seawater culture unlike *S. spilurus niger.* Small decorative species, such as *S. alcalicus grahami*, could have a future in the aquarium trade, although sex-reversal or other sterility-inducing techniques are needed to make this a continuing commercial proposition. Attempts to sex-reverse *S. alcalicus* fry have so far been unsuccessful and it is possible that such small, early maturing tilapias have fry whose sex is determined before release from the parental mouth and are therefore not easily treated with sex-reversing steroids.

T. rendalli merits more attention in culture where vegetation is available for feeding. Although *S. hornorum* x *S. mossambicus* hybrids produce all-male progeny, *S. spilurus* x *S. mossambicus* crosses do not. This is probably indicative of a closer relationship between the latter two species.

Establishment and Conservation of Pure Strains

Very few pure strains of tilapias are available to culturists. Only where stocks are continuously tested by electrophoretic markers and selected can genetic purity be ensured, e.g., the *S. aureus* stocks at Mr. Mires' hatchery at Kibbutz Ein Hamifrats, Israel, tested by Prof. Avtalion. Elsewhere, culturists are relying on wild stocks, often contaminated by hybridization with introduced species, or on introductions made many years ago from very restricted gene pools, e.g., about 30 *S. aureus* individuals sent from Israel to Auburn University, Alabama, have been the ancestors of widespread introductions for culture in Southeast Asia and Latin America, where it is not surprising that they are showing signs of inbreeding depression.

There is a clear need for collections of known strains of tilapias for research and culture work. The establishment of such collections was discussed and the following points were made:

1. Collections would have to be very carefully maintained by painstaking workers in high-security premises and replicated at several different locations.
2. Tilapia broodstock collections are probably not a function for regional aquaculture centers in tropical developing countries where such work would be difficult to carry out and would normally take lower priority than production-orientated research.
3. The location of some collections in colder countries could prove advantageous as escapees would not survive to cause environmental problems.
4. The facilities needed to accommodate large numbers of broodstock of different species and strains would be very costly.
5. Tilapia researchers have much to learn from domestic animal and plant breeders in planning such collections.

It was agreed that the methodology for maintaining pure strains needed further development. Cryopreservation of semen could prove a useful future technique for storing male genetic material. At present, the collections of tilapias maintained in Israel, the United Kingdom (Stirling University), the Philippines (Central Luzon State University) and the U.S.A. (Auburn University) were recognized as the most significant. In the absence of funding or plans for larger collections, it was stressed that natural populations of tilapias in African waters should be carefully protected. Transplantations should be avoided wherever possible to assist in the conservation of genetic material for the future. For example, the natural populations of *S. hornorum* are probably uncontaminated at present. There is a clear need for cataloguing critical sites, such as the small lakes in the Cameroons and possibly Lake Bosumtwi. The International Union for the Conservation of Nature (IUCN) is aware of the vulnerable nature of these populations.

Three main objectives were identified concerning the genetic improvement of cultured tilapias: 1. The guaranteed supply of high quality pure strain or hybrid fry for farmers; it was recognized that these would be expensive and that restriction of distribution to monosex progeny may be desirable. 2. Conservation of wild tilapia stocks. 3. The absolute necessity for *all* researchers to work with known material.

Prof. Roberts gave a very useful example of the confusion which can arise if the last of these objectives is not fulfilled. At Stirling University, experimental infections of *S. mossambicus* with a monogenean (*Cichlidogyrus* sp.) would take only in a strain known as 'B' from a type location in Africa. *S. mossambicus* strain 'A' which had come via Singapore and was slightly contaminated was resistant to infection. A parasitologist working with 'A' only would have drawn the erroneous conclusion that this particular *Cichlidogyrus* sp. was not a parasite of *S. mossambicus*.

The Concept of 'Stunting' and a 'Switch' Between Somatic Growth and Gametogenesis/Spawning

(The concept of a simple 'switch' is perhaps false, as somatic growth demonstrably does not cease during gametogenesis. The greatest growth check occurs from the onset of spawning through to mouthbrooding—Editors.)

The phenomenon of stunting was discussed in relation to the concept of a switch from somatic growth to gametogenesis leading to first maturity. The phenomenon of maturation at a small size is common to both guarders and mouthbrooders and to riverine and endemic lacustrine species. For example, *S. esculentus* normally take about three years to mature as 20-cm fish in Lake Victoria, but will mature after only five months as 10-cm fish in aquaria. This ability to shift the timing of the 'switch' has great adaptive value particularly in lakes which dry up periodically, e.g., Lake Rukwa and Lake Chilwa, where fish migrate to inflowing streams as the lakes dry, breed at a small size, and then repopulate the lakes when they fill again.

Dr. Lowe-McConnell's data for *S. niloticus* suggest that breeding size is low if condition factor is low in the population. This suggests that the quantity and/or quality of available food may be a factor. Other possible factors include water chemistry, temperature, salinity, dissolved oxygen and parasite burdens.

There are further clues in results of field biology studies. For example, the *S. niloticus* population in Lake George remained unfished from the 1920's to 1952 because of tsetse fly infestation of the area; their minimum breeding size was about 28 cm. With increasing fishing pressure from 1952 to 1972, the minimum breeding size fell to about 18 cm. Field studies have shown that the majority of the fish have always concentrated around the lake margins. Beyond 100 m offshore the bottom becomes a loose flocculent ooze which is unusable for tilapia nests. This suggests that there is a severe limitation on breeding sites and that the removal of large fish by the fishermen has allowed smaller individuals to come in and occupy these sites. The mesh size of fishing (gill) nets has been reduced from 15 cm to about 10 cm. In the early days of this fishery (1950's), very few females smaller than 28 cm had ripe ovaries, and therefore it was gametogenesis, not just spawning, that was restricted to larger individuals.

In Lake Sibaya, it was stated that only the large males can find breeding sites when the lake is at a low level: the smaller individuals take up sites when the water rises. In Lake Valencia (Venezuela), again the cichlid population is found only around the periphery where they form nesting arenas. Diving observations indicate that much wider areas offshore have suitable substrates, but perhaps behavioral factors are involved in the selection of inshore sites.

In Lake Kariba (total area 5,000 km^2), fish have experienced the change from a riverine to a lacustrine environment, and for *S. mortimeri* and possibly some *S. mossambicus* this has involved a change from a situation with relatively scarce to one with abundant breeding sites. In the fast-flowing Zambezi, fish were able to breed only in bays and backwaters; now it has about 2,000 km^2 of lake waters of appropriate depth. With this change, the minimum breeding size has increased from 3-400 g to 600 g. Middle Zambezi fishes in the Kariba area are now called *S. mortimeri* although some *S. mossambicus* may have been moved around and hybridization may have occurred.

S. a. grahami introduced to Lake Nakuru has changed the whole ecology of the lake. For example, the population of birds has increased from a few casual visitors to vast numbers of residents feeding on the fish. *S. alcalicus* has thrived in this lake and appears to breed at a larger size than in its more restrictive alkaline lake habitats, although the breeders are probably still very young. This is perhaps analogous to the situation with milkfish *(Chanos chanos)* in hypersaline lagoons on Christmas Island, which mature late and at a small stunted size compared to open sea fish. *S. a. grahami*, however, may grow quickly in restricted, high temperature alkaline lakes and breed when small and *young*. The fast growth could explain its low vertebral counts. These questions will not be resolved until more data are collected on growth rates for 'stunted' and other populations in restricted and open habitats.

The above examples were discussed at length but there was no general agreement on their implications for culturists. The fact that culturists have frequently observed mouthbrooding species spawning in cages suggests that the hypothesis concerning shortage of breeding sites in lakes may be incorrect. A cage does, however, provide a solid substrate, with visual cues. In the Lake George example, it was felt that there was insufficient knowledge of the simple effects of fishing selection and the genetic make up of the population: apparently, a few 40 cm fish are still catchable far offshore.

There were also conflicting reports on the effects of environmental factors on maturation and spawning. For example, nest building is rarely seen below 60 to 70 cm depth in Israeli tilapia culture but *S. mossambicus* breeds normally at 2 m in Lake Sibaya. Indeed, the preferred depth for mating pits seems to be a specific character, c.f. the Malawi species flock. It was suggested that culturists could test the Lake George hypothesis in ponds by adding large fish to outcompete the cultured crop for breeding sites. The idea here is that the large fish would inhibit spawning by the cultured crop which would therefore keep growing. This idea was not received very enthusiastically as it would still mean accepting recruitment from successful large breeders. An alternative suggestion was the development of weak or non-functional males which would fail to fertilize eggs and cause these to be rejected by the females thereby reducing the females' non-feeding period.

Many plants and animals adjust their life-history strategy to maximize the intrinsic growth rate of the population ('r'). The 'decision' when to produce gametes and breed involves a tradeoff with respect to any individual's contribution to the population. By delaying breeding, the individual increases its size and energy stores for the future whereas by breeding early, it reduces its exposure to mortality risks and shortens the generation time.

This theoretical framework can be fitted to all the tilapia examples given. For example, in Lake George, there is a classical response of earlier breeding with exploitation (fishing mortality). In Lake Sibaya, *S. mossambicus*, when they reach about 8 cm, are exposed to heavy predation from birds and *Clarias* if they remain in the rich feeding grounds inshore and here they breed at a small size. The theoretical framework, however, does not tell us the mechanism for the switch to breeding. The mechanism is probably some sort of energy constraint rather than a simple behavioral or ecological factor such as competition for breeding sites or mates.

The theoretical framework does, however, tell us where to look for the mechanism or at least for ways to manipulate it without having to unravel all the detailed behavioral, endocrinological and other physiological interactions. The basic theories of population dynamics and natural selection tell us to look for correlations between environmental changes and reproductive output.

Dr. Jalabert proposed a model of a balance beam between somatic growth and gonad development with genetic and environmental factors controlling its swing. It was generally agreed that the difficulty, expense and complexity of endocrinological and other physiological research made it unlikely that more research in these fields would bring rapid payoffs for aquaculture in the form of techniques to manipulate the switch. The best approaches were identified as behavioral studies, environmental manipulation and work on the genetics of maturation age and size. It was felt, however, that some useful techniques for culturists could emerge from attempts to identify and synthesize pheromones or other waterborne factors. There was further discussion of the existing experimental evidence and the best possible clues for environmental studies.

The question of available oxygen was debated and it was pointed out that 1 kg of small-sized tilapias (say individual weight about 10 g) require about twice as much oxygen as 1 kg of, say, 200 g fish. It is possible, therefore, that in a pond situation with densely crowded small fish, the available oxygen for digestive/oxidative processes becomes so reduced that metabolic pathways are affected. Perhaps, as a result of this, nutrients begin to be channelled into gonadal products. Also, as the gill surface area of fish limits their ability to take up oxygen, the relative growth of this to the overall size of the fish could be involved in the 'switch' in environments with variable oxygen availability (see Pauly 1981—Editors).

There are two factors that could encourage a fish to continue to channel all its assimilated nutrients into somatic growth rather than gonadal products, unlimited food supply and a low risk of mortality. In culture, the questions of food supply and quality are relatively easy to investigate. The risk or 'expectation' of mortality is more difficult, however, and can only be expressed in vague terms as stress factors, e.g., oxygen stress, salinity stress, thermal stress, density stress.

The possibility of genetic determination of age at first maturity was then discussed. It was stated that if age at first maturity is genetically controlled, then the problem for a culturist is how to get the maximum growth within this period. Evidence was cited against genetic determination of first spawning age in the tilapias. The example of *S. esculentus* breeding at two to three years of age in natural waters and five months in captivity was repeated, but the prospect of genetic selection for later spawning was not dismissed.

The consensus was that gross environmental changes are the overriding factors determining age/size at spawning. The examples of *S. mossambicus* breeding at 6 g after five months in aquarium conditions and the fished population of *S. niloticus* in Lake George now spawning at 18 cm (estimated age 1 yr) as opposed to 28 cm (estimated age 1.5 yr or older) were repeated. Dr. Caulton also cited some personal observations on *S. mossambicus* which bred two or three times over a two-year period in aquaria and were about 50 g in weight. When transferred to a more spacious environment—in this case a *Macrobrachium* farm raceway—the fish stopped further breeding until they reached about 250 g and were 3 yr old. This suggests that the flexibility of timing maturation and spawning extends beyond the age/size at first maturity and effects spawning frequency as well. In Lake Sibaya, *S. mossambicus* reaches first maturity in one or two years.

The experiences of experimental culturists in Israel, the Philippines and the U.K. indicate that the age at first maturity for a given species is relatively constant, i.e., to within a matter of a few weeks at a given location, but that great variability occurs between different populations at different experimental stations. *S. niloticus* and *S. aureus* are later spawners than *S. mossambicus*.

In the Philippines, the age for first spawning of *S. mossambicus* is about two months and for *S. niloticus* about three months in pond systems. Stunted fish of 5 to 10 g spawn after two to three months as well as larger fish. It has also been observed that *S. andersonii* in Zambia, living at relatively low temperatures, breed fairly late (at about 15 months in ponds) and at a large size. In natural waters here they also take 3 years or more to reach first maturity. When this species was transferred to Tanzania, maturation was noted as early as two to five months at a very small size.

In Israel, it was of note that the necessity to overwinter fish gave very little scope for trying to alter the age at first maturity. The greatest potential appears to be in tropical locations where 'young of the year' are cultured in shorter production cycles.

It was also noted that while manipulation of stocking density, salinity, temperature and other environmental factors may be biologically feasible as a means to influence gametogenesis and spawning, any method chosen for commercial application must be economically feasible also. For example, it may be possible to delay or suppress reproduction by lowering the temperature to 18 to 20°C, but this would be expensive and growth rates would be severely reduced.

The effects of light and daylength were discussed and it was pointed out that a given species usually spawns at a fixed time within a 24-hour cycle in aquarium conditions. There could be scope for further investigation in this, especially as gametogenesis in trout has been suppressed experimentally by continuous illumination. The effect of temperature may be different for fish of different ages or sizes and more important than the effects of feed quality and quantity. If so, then *S. mossambicus*, first spawning at four months in a relatively cool location, could spawn at, say, two months at a hotter location, irrespective of the feeding regimes.

Nutrition

1. The Physiology of Digestion in Tilapias

Dr. Moriarty summarized the digestive physiology of microphagous tilapias, i.e., those feeding on phytoplankton, blue-green algae (cyanobacteria) and detritus (including bacterial proteins). Previous studies have shown that although the stomach in these fish is structured such that some food can pass from esophagus to intestine through the anterior end only, it is a true stomach with a pyloric sphincter and a very important cycle of acid secretion.

The sequence of events observed in large individuals having a relatively large stomach is as follows. In the early morning (6:00 A.M.), when feeding commences, the stomach is collapsed (contracted) and secretes little or no acid. For the first few hours of feeding, therefore, some of the food passes through the anterior end of the stomach straight from esophagus to intestine. As feeding increases, however, the stomach becomes progressively distended and takes in food. It also begins to secrete acid. By 10:00 A.M., acid secretion is maximal and over the next 1 or 2 hours the pH in the ventral portion of the stomach falls below 1.8, at which point acid lysis of bacterial and algal cell walls commences. Over the 6-hour feeding period from 10:00 A.M. to 4:00 P.M. ingested food can either pass into this very acidic portion of the stomach for complete lysis or pass through only the acidic portions (pH $>$ 2.0) or bypass the stomach more or less completely. After feeding stops, the rate of movement of food through the stomach slows down, thus allowing all the food material to be exposed to a low pH. Complete lysis and digestion of all food occurs, whereas bacterial or blue-green algal cells that pass rapidly through the stomach and thus are not exposed to a high acid concentration are incompletely digested. The assimilation efficiency may be as high as 80% for blue-green algae ingested during the period of peak acidity, but it is near zero at the commencement of feeding and variable at other times. The mean assimilation efficiency over the whole feeding period is probably 40 to 50%. For small tilapias and for *Haplochromis nigripinnis*, which has a similar digestive physiology, the overall assimilation efficiency is higher, probably because the acid concentration is high throughout their small stomachs.

Fish culturists were advised to take note of the very slow passage of food through the tilapia intestine during non-feeding periods (4:00 P.M. to 6:00 A.M.) in systems where natural or *ad libitum* feeding is available. The assimilation efficiency is very high during non-feeding periods. This suggests that the most efficient utilization of food in cultured fish could probably be obtained by a series of feeding/non-feeding cycles, i.e., a feeding period sufficient to distend the stomach and stimulate maximum acidity followed by a non-feeding period to allow maximum lysis of the ingested food and slow passage through the intestine. With controlled feeding, perhaps two or three such cycles could be applied each day. Further work is needed to determine the effect of rate of movement through the intestine on assimilation rate and efficiency, and thus whether such feeding cycles are advantageous.

Dr. Bowen's work on *S. mossambicus* in Lake Valencia was discussed. It was noted that high-quality proteinaceous material is usually completely assimilated within the first quarter of the intestine whereas the peptidic fractions of detrital agregates (see below) are digested and assimilated all along the intestine. It was agreed that although such tilapias do produce a pepsin (pepsinogen has been demonstrated) there is probably no proteolytic digestion in the stomach. The stomach function in *S. niloticus* has become specialized and restricted to acid lysis: proteolytic digestion is confined to the intestine.

Dr. Bowen pointed out that *S. mossambicus* differs from the above pattern (which was largely taken from studies on *S. niloticus* in Lake George, feeding on lake plankton) in that it can fill its stomach within 10 minutes to one hour, depending on the size of the fish. Observations on trawled specimens of *S. mossambicus* (size range 4 to 25 cm) from Lake Sibaya and Lake Valencia show that the degree of stomach acidity depends on stomach fullness. The pH of an empty (resting) stomach is from 4 to 7 whereas that of a full stomach is around 1.5. Therefore, with *S. mossambicus*, there is no long lag period to reach maximum efficiency. The first batch of diatoms ingested can still pass through the gut without lysis, but the very rapid stomach filling and acid secretion mean that the system can work at close to 100% assimilation efficiency thereafter.

There are perhaps some lessons here for culturists. Phytoplankton feeding *S. niloticus* in ponds would largely resemble the Lake George situation and operate well below maximum assimilation efficiency. Any species that can fill its stomach rapidly, however, such as fish receiving supplementary feed or *S. mossambicus* ingesting large quantities of detritus will be much more efficient. Of course economics will dictate what is possible in a commercial situation: phytoplankton is cheaper than supplemental feed!

There was further discussion on the functional anatomy of the tilapia gut. *S. mossambicus* was said to have a relatively muscular stomach which maintains its shape and size when empty better than *S. niloticus* in which the stomach shrinks to a very small size. Dr. Moriarty showed, however, that the stomach of *S. niloticus* at maximum distension holds about 50% of the total ingested material: the other 50% is spread out along the very long intestine. In *S. mossambicus*, the distended stomach holds only about 10% of the total ingested food, the remainder being in the intestine.

Dr. Caulton described the digestive physiology of *T. rendalli* for comparison. The most important aspect of this is the rasping of plant material by the pharyngeal teeth. Within one hour from the commencement of feeding the entire gut of *T. rendalli* can be found packed with plant material of which virtually none is digested, despite pharyngeal teeth disruption. At this time, the stomach pH is about 4. This suggests that *T. rendalli* wastes a lot of food, but it should be noted that not only does it normally have an overabundant food supply but also it can consume very large quantities very quickly, e.g., 3 g in 10 minutes even for very small fish. After about one hour of feeding, the stomach pH falls to about 1.4, lysis begins and assimilation efficiency increases. The overall assimilation efficiency is probably around 50%.

The stomach residence time for food is very important in *T. rendalli* not only because gastric acid assists lysis of the plant cells, but also because denaturation of the protein by acid seems to be a necessary pretreatment for its digestion by enzymes in the intestine. It is well known that trypsins, which are present in tilapias, act better on previously denatured proteins so the acids here could be assisting proteolytic digestion. This is possibly analogous to the action of renin coagulating milk protein in the mammalian gut before digestion.

2. Detritus and Detritivory

Detritus is a very complex mixture of living and non-living components. It is erroneous to regard the food value of detritus (including plant wastes, such as straw and grass clippings) as merely the production of microbial protein built from the waste substrate. Non-living organic matter plays the principal role in nutrition of many detritivores.

The methodology available to sort out the various components of detritus was discussed. Light microscopy is useful to give rough estimates of bacterial numbers and sizes, especially with epifluorescence techniques. Bacteria and blue-green algae can be easily counted when they are made to fluoresce against the non-living matrix. Electron microscopy (EM) has revealed the complexity of detritus. Bacterial slime layers and capsules are visible using EM and include proteinaceous material (stainable with osmium tetroxide), polysaccharides (stainable with Ruthenium Red) and lipids. These are all presumably of some nutritional value to detritivores. Bacterial populations can also be estimated by determining the muramic acid content of detritus. However, even simple analysis into living and non-living material is difficult. ATP measurements will give some idea of this but say nothing of the types of organisms present. It is well known that plant material decomposing in aquatic ecosystems first suffers a drop in N-content for a few days and then the N-content increases. This has always been assumed to be entirely due to the obvious colonization of the material by microorganisms. As the material ages, however, some of its N-content derives from chemical processes, e.g., precipitation/complexing and is found as refractory nitrogenous compounds in the detrital aggregate.

Dr. Bowen's work in Lake Valencia shows that well-established detritus contains very significant amounts of these nitrogenous compounds in addition to microbial protein. Dr. Moriarty questioned the methodology used to estimate these compounds and also Dr. Bowen's conclusion that they were not protein. The extraction methods used to estimate these compounds—6N HCl for 24 hr at 100°C—would in fact hydrolyze all peptidic material to its constituent amino acids. It was also considered significant that while these compounds in Dr. Bowen's work were not extracted by 0.1 N KOH and were therefore not considered proteins, this method does not always give complete protein extraction. Their exact nature must await further study with improved methodology. Dr. Bowen accepted that methodology needed improvement but emphasized that even with the existing crude methodology, his work had demonstrated that microbial protein could account for only a portion of the N-content of detritus. The food value of the refractory nitrogenous compound requires further study.

Detritus and detritivory are very important in pond aquaculture. It was suggested that agricultural wastes could stimulate microbial production in ponds by providing large surface areas for bacteria. The use of animal manures to give high concentrations of suspended colloidal particles for bacterial colonization was considered very important as well as the development of benthic detrital aggregates. This discussion centered on Israeli experience with cow and chicken manure in which the main thesis was that as yields from anaerobic bacterial production are only about 10% of those from aerobes, a system of suspended colloidal material for aerobic bacteria to work on is the best. Frequency of manuring is very critical in the maintenance of sufficient dissolved oxygen for aerobic bacteria and for normal fish growth. Pond aeration can help here. Daily manuring is recommended and rates as high as 180 kg dry matter/ha/day have been used. Therefore, pond management techniques are already being used to maximize bacterial production. Tilapias are the most responsive of all fish to such techniques.

Dr. Lovshin recalled that pretreatment of fish ponds with manure was a long established technique designed to build up plankton populations. It was recognized that this technique must also encourage detrital production as well and that perhaps the recommended time lags between manuring and stocking should be re-examined. This requires a full investigation, i.e., not only how to pretreat to get maximum bacterial and detrital production but also how to continue treating to maintain these through growout. The food value of such sources was also debated. Obviously, before culturists are encouraged to develop detrital systems to feed their fish, it must be demonstrated that these will supply all the required dietary components especially essential amino acids, essential fatty acids, vitamins, etc. The consensus of opinion was, however, that high bacterial diversity and the chemical complexity of detritus made it unlikely that deficiencies in essential nutrients would arise in well-managed detrital-based culture: for example, in manured ponds. Experience with marine fish larval rearing has shown that some long-chain polyunsaturated fatty acids present only in some phytoplankton species are essential for growth and survival, but this situation was not considered comparable to detritivorous adult fish feeding in a pond where the diversity of food items is very high.

Dr. Bowen pointed out that detritivorous fish are very selective feeders in natural waters and suggested that a painstaking approach to defining the essential dietary requirements of tilapias (as had been done for example for salmonids with test diets deficient in individual amino acids and vitamins) was probably not required in this context. Why not just characterize the physical and chemical nature of their selected types of detritus? This approach would help to determine whether it is the microbial biomass or the other components of detritus that are most important in detritivore nutrition in different situations. It was stressed that the nutritional requirements and feeding habits of juvenile fish are often different to those of adults. For example, the juveniles of the microphagous tilapias, up to about 5 to 10 cm, often feed on copepods. As they switch to microphagy, mucous glands develop at the back of the buccal cavity and pharynx. The mucus captures microorganisms for passage to the gut. This switch from zooplankton feeding to microphagy has been observed in culture and in natural waters, e.g., Lake George, and indicates a high requirement for animal protein in juvenile life. The changes in digestive physiology associated with this change in diet have yet to be investigated.

3. Feed Formulation

Prof. Roberts stated that the optimal protein levels, protein/energy ratios and essential amino acid requirements for *S. aureus* and *S. mossambicus* had recently been determined by work at Stirling University for the formulation of complete, pelleted diets. Information was also being gathered on the nutritional value of locally available agricultural products in the tropics. These studies show essentially that these tilapias need the same essential amino acids as other fish but that their requirements for individual amino acids are different. In some control feeding trials using high protein commercial trout diets, tilapias showed a hyperproteinosis syndrome: loss of orientation, swollen bodies and death. The histopathology of this condition was acute circulatory collapse and histamine release from mast cells. Conversely, Dr. Lovshin reported that trout chow with 40% protein is fed routinely to tilapias in ponds and cages with very satisfactory results and no deleterious effects.

Tilapias can accept very high levels of carbohydrates in supplemental feeds. For example, Dr. Coche described work with *S. niloticus* in the Ivory Coast in which diets with up to 45% carbohydrate were accepted.

The use of growth promoters was discussed. It was generally agreed that anabolic steroids could not be incorporated into feeds on a large scale because of cost. Prof. Roberts repeated, however, that sex-reversed fry gain a useful growth boost over non-treated fish from the steroids used: the treatment lasts for 35 days. This effect, and the steroid residues, disappear within 5 days of cessation of treatment when the fry are transferred to growout tanks. The use of virginiamycin (e.g., Biazon) was also discussed. This has been used as a growth promotor in grouper culture in Malaysia, but not apparently with tilapias. Despite its name, it is not an antibiotic.

4. Food Presentation

Dr. Coche noted that some pond culture work, for example, Miller's work with tilapias in the Central African Republic, has shown that powdered supplemental feed could give yields as high as pelleted feeds. This suggests that the high cost of pelleting can be avoided. Dr. Lovshin supported this observation from studies in experimental ponds at Auburn University, Alabama, in which meal diets gave comparable yields to pelleted diets. It was agreed, however, that pelleted feeds were desirable for cage culture.

Dr. Coche reported that very good results have also been obtained in the Ivory Coast with mixed feeds presented as a "mash" shaped into large balls, placed in the middle of cage covering nets and lowered underwater. For mariculture, unpelleted wet feeds are generally used.

The problem of food wastage in cage culture was discussed. Although a variety of chemical binders is available for pellets, these can cause problems, for example, the kidney disorders observed in experimental marine flatfish culture in the U.K. Prof. Roberts mentioned that freshwater-moistened diets are preferable for tilapias cultured in seawater as these supply dietary water and reduce osmotic work below that required for salt regulation following seawater drinking. Dr. Caulton commented that in the tropical developing countries, 2% dried green banana powder is the best and cheapest binder available and is used in commercial feeds for *Macrobrachium* culture. However, other materials such as boiled plantains, potatoes, manioc or any other starch source can be used as cheap binders at 10 to 20% of the diet. They would be cheaper and more available than 2% dried green banana powder in many developing countries.

The extent of food wastage from commercial tilapia cages does not appear to have been quantified. All commercial growers use sinking as opposed to the more expensive floating pellets. The ability to manufacture floating pellets may be absent in many of the developing countries. Food wastage seems to depend on the degree of powdering of the pellets during storage and handling (this powder is easily washed from the cage), the pattern of water currents and the rate of feeding. Water currents are particularly important. Dr. Nash pointed out that elongated cages which set parallel to the current can be useful in maximizing food retention if suitable end baffles can be fitted: there are, however, structural limitations to this in strong tidal streams.

It was agreed that tilapias in cages can be very voracious feeders similar to salmonids. Feeding at a very slow continuous rate seems, however, to be very advantageous. Dr. Coche reported that Campbell in the Ivory Coast has used a very slow feeder in which a rope hanging down into the water below the hopper is moved by gentle water currents to allow a trickle of pellets to come out. This simple device could have advantages over more sophisticated automatic and demand feeders. Dr. Philippart stated that in experimental tank trials in Belgium, tilapias in excess of 200 g fed five or six times/day grew better than those fed more frequently, e.g., twice/hour. It was agreed that juvenile fish require more frequent feeding than larger fish: usually about 10 feeds/day.

5. Broodstock Nutrition

It was agreed that there is very little information available on the nutritional requirements of tilapia broodstock. There is, for example, no information on the effects of nutrition on fecundity in tilapias apart from general observations that well-fed fish have higher counts of oocytes at all stages than poorly-fed fish. Prof. Roberts reported work at Stirling University which indicates that the dietary protein requirements of broodfish are significantly lower than those of immature fish in the most active phase of somatic growth. The *quality* of protein required for broodstock diets is, however, very high. Earthworms are an excellent protein source. For a number of species, it has been shown that broodstock can be spawned at approximately monthly intervals for as much as 18 months given good nutrition. Also, for *S. mossambicus* in tanks, all the animal protein, for example fish meal, in the diet can be replaced with high quality algal protein. For spawning ponds, however, broodstock pelleted diets should contain animal protein.

Dr. Guerrero reported that a commercial animal feed company in the Philippines is developing a diet for *S. niloticus* broodstock. The starting point for this is a salmonid-type diet with high energy content and 25% protein, at least half of which is animal protein, largely from fish meal. The natural foods present in spawning ponds are a very important part of broodstock nutrition.

6. Summary on Nutrition

To summarize on nutrition, it is clear that vastly different approaches are needed to optimize the nutrition of tilapias in extensive and intensive culture systems. The feeding niches in extensive systems need fuller definition, particularly the microbial ecology of detrital aggregates, while for intensive culture, least cost formulation of complete diets is still awaited for most culture situations.

Seed Supply

1. Broodstock Management

In spawning tanks, the territorial aggression of males is the major factor limiting stocking density. In Israel, however, as Mr. Mires pointed out, mixed sex broodstock are overwintered in extremely dense conditions (50 kg fish/m^3 of water) at 15 to 17°C. This does not affect their reproductive performance when put out to ponds. Presumably, the low winter temperatures suppress aggressive behavior.

Dr. Lovshin mentioned the possible advantages of sexing broodstock while they are still immature and then keeping the sexes in separate ponds until they are ready to spawn, as is done for carp breeding. This method gives a high degree of control for handling pure stocks and for hybridization work.

The best sex ratios to use for high broodstock performance were then discussed. Experimental work in the Philippines has shown that a ratio of one male to one female gives better results than one male to two or three females for *S. niloticus* and *S. aureus* broodstock in plastic pools and aquaria. This result is, however, in conflict with some other work, for example Campbell working in the Ivory Coast has found that high female:male ratios of 5 or even 7 or more to one has given very intensive reproduction of *S. niloticus* in shallow raceway-type ponds. The best sex ratio varies with specific breeding behavior. For example, in *S. melanotheron (T. heudelotii)* in the Ivory Coast, in which the male is the mouthbrooding parent, it is 1:1 in ponds.

In contrast to this, Dr. Lovshin's work in Brazil and the concept of 'male pressure' suggests that even higher numbers of males may be advantageous. Dr. Lovshin pointed out, however, that there is a difference between intraspecific or pure strain crosses, in which sex ratios of about one male to three females seemed to be generally accepted, and the interspecific hybrid crosses where he had harvested more fingerlings from ponds with 'male pressure': two males to one female. This may be due simply to the constant availability of males to court any ripe female, but more work on sex ratios in ponds and restricted environments is urgently needed.

It is not clear whether chemical stimuli play an important part in the reproductive behavior of tilapias. In salmonids, males use chemical recognition to select females that are ready to spawn. If such chemical stimuli are also produced by tilapias, they could be major factors in mate selection and in reducing male aggression against ripe females. Dr. Noakes pointed out that there were fundamental differences between the salmonids and the lek breeding tilapias and similar groups, like the centrarchids. In the salmonids, the female excavates the nest and the male undoubtedly recognizes and chooses his mate partly by chemical recognition. In the lek breeding tilapias and centrarchids, however, it appears to be the female that makes the choice, and of course, the male that excavates the breeding pit. For the centrarchids, it has been shown that a male will attempt to court any fish or object that comes into his breeding territory. If this also applies to the tilapias, it argues against chemical stimuli having importance over behavioral cues. Dr. Lowe-McConnell added the observation that *S. karomo* in the field wander through the breeding arenas and appear to inspect not only the males but also their breeding pits before selecting a mate. This was observed in very clear water.

The problem of male aggression appears to be very different in restricted environments, such as tanks and aquaria, from the situation in ponds. In the former, males can kill females or cause ovarian regression by repeated attacks, whereas in ponds, the females have space for escape and are rarely adversely affected. Mr. Mires described aquarium observations in which ripe females were keen to enter the males' breeding territories and the males pushed and encouraged them to enter. If this also occurs in ponds with a preponderance of males, it means that ripe females are continually bombarded by courting males. This could have a stimulatory effect and help to account for the good results of hybrid crosses with 'male pressure'. The question of breeding coloration was raised but it was felt that this had little relevance in ponds, where visibility was often very low.

2. Mass Seed Production

The discussion centered first on the possible alternatives to spawning ponds for mass fry and fingerling production. The example of mass fry production using cages (hapas) was described by Dr. Guerrero. Prof. Roberts commented that a commercial farm in Kenya was producing large numbers of pure strain, hybrid and sex-reversed tilapia fry for sale within Kenya, Nigeria, other African countries and the Middle East. Details of the system used were not available.

Dr. Philippart mentioned that an experimental system of 4-m^2 fiberglass tanks had also been developed using recirculated heated effluents in Belgium. The system, with 15 to 16 spawners/tank produces about 500 fry/tank/day for short periods (5 to 7 days) of intense reproductive activity and about 150 fry/tank/day on average over longer periods (50 days). Small females of around 100 g perform better than larger fish. The development of intensive recirculatory raceway systems for fry production was discussed but it was felt that these would be costly both in construction and energy requirements.

It was agreed that most of the large-scale commercial production of tilapia fry was still done in spawning ponds. Mr. Mires stated that in Israel, producers have changed from small to large spawning ponds because the demands of buyers were constantly changing with respect to size and numbers of fish. Large spawning ponds become essentially early nursery ponds as well. Fry or fingerlings can be harvested and graded to meet demands with less labor and management than from numerous small ponds. At the Kibbutz Ein Hamifrats hatchery, the use of large spawning ponds has reduced the monthly labor requirement for fry/fingerling harvest to four hours work by a team of five to six people.

The development of tilapia hatcheries was felt to be the key to expansion of the culture industry. A small number of large hatcheries with well-trained personnel was considered to be a better prospect than large numbers of hatcheries scattered throughout the rural areas of developing countries. This would allow controlled production of good strains, hybrids, etc. It was recognized, however, that distribution difficulties could arise in island systems like the Philippines and in areas where roads and communication systems are very poor and fuel prices high. In these cases, local hatcheries would be essential. Tilapias are among the hardiest of fish and survive transportation very well. Prof. Roberts commented that tilapia fry brought overland from Israel to Scotland in fruit lorries suffered no mortality despite exposure to cold temperatures.

Several participants expressed the view that private sector hatcheries afforded the best hope for a rapid and efficient increase in tilapia seed production. However, the economic viability of tilapia seed production has been tested in very few countries, such as Israel and Taiwan. Government hatcheries or government subsidies to private sector producers may be needed in the future.

Culture Techniques and Systems

1. Recruitment Control

The discussion concentrated on methods of recruitment control other than monosex culture of sex-reversed or hybrid progeny, which it was felt were adequately discussed elsewhere. Attention was drawn to the high success (84%) of male fry production by size selection alone in the Ivory Coast. There was also a brief discussion on the destruction of urinogenital papillae. This has been accomplished at Auburn University, Alabama, by hot wire cautery, but the papillae regenerate and the fish spawn normally.

The main method discussed was the use of predators to remove unwanted recruits. It was recognized that although tilapia recruits have different feeding preferences to their parents, they do compete for food and oxygen in culture ponds.

Dr. Pullin commented that *Channa striata* was being studied at Central Luzon State University in the Philippines to control tilapia recruitment in an *S. niloticus* (85%):*Cyprinus carpio* (15%) polyculture system. Very low numbers of this predator (20 to 30 stocked as fingerlings) could obliterate tilapia recruitment in 0.1 ha ponds containing 850 or 1,700 *S. niloticus* stocked at average weight 3 g and grown for 90 days, during which first maturity is reached and spawning occurs. If the culture period is extended to 180 days, however, the predators are unable to cope with the increased level of recruitment and up to 200 to 300 kg of tilapia recruits/pond can be present at harvest. Therefore, the size and fecundity of the breeding tilapias and the length of the culture period must be considered, not just numbers of fish. *Channa striata* is a very useful controlling predator and has a high value as a by-catch. It is used in commercial culture of tilapias in Taiwan—largely in monosex culture where it mops up any recruits resulting from inaccurate sexing or poor quality hybrid progeny.

Dr. Coche pointed out that smaller predators than *Channa* should also be considered. Although of negligible value as a by-catch, they have the advantage of always selecting the smallest recruits as prey and there is, therefore, no danger of them eating the cultured crop as well as unwanted recruits. He also reported work using *Clarias lazera* in the Cameroons. This species ceased to be an effective controlling predator when the water became very turbid and it was also useless in systems where high protein supplemental food was given to the cultured tilapias: it preferred to eat the supplemental food.

Dr. Lovshin reported that in his experience predator control was the most reliable method of controlling recruitment. He also voiced the opinion that there was a very great potential for managed fish production from the thousands of livestock watering/irrigation ponds in tropical developing countries by using predator-prey systems. Guidelines for stocking and harvesting would have to be worked out as for the bass-bluegill systems described by Swingle for the southern U.S.A.

2. Integrated Farming and Wastewater Reuse

It was agreed that tilapias were ideal species for integrated agriculture-aquaculture farming systems, particularly in tropical developing countries. It was recognized that the public health aspects of such systems needed fuller investigation.

The use of wastewater (sewage) to fertilize tilapia ponds was considered a very controversial topic. Dr. Hepher described Israeli work in which human bacterial pathogens had been shown to penetrate the muscle and internal organs of fish above certain threshold levels of bacterial populations in the water. The thresholds vary with species: silver carp are the most susceptible to contamination, common carp are intermediate and tilapias the most resistant. Tilapias have particularly high thresholds for *Salmonella* and *Shigella* penetration. The thresholds also vary greatly with stress. They become lowered in stressful conditions such as low dissolved oxygen and high ammonia. Fish which have become contaminated with pathogens in muscle and internal organs take a very long time to depurate in clean water: usually over one month. Fish which have taken pathogenic bacteria into their guts only can be depurated in 2 to 3 days.

Mr. Mires voiced great concern at the use of wastewater in aquaculture and drew attention to the adverse publicity that it could give to fish culture in general. It was his opinion that public attitudes would be strongly against this method of growing fish even if the health hazards could be controlled. It was pointed out, however, that fertilization of fish culture systems with human wastes has a long history in Asia and that human waste recycling through aquaculture should not be abandoned because of public attitudes in the developed countries.

The buildup of heavy metals in manured ponds was also discussed. Dr. Hepher stated that in Israel, both common carp and tilapias remained largely unaffected by heavy metal buildup. The levels of heavy metals in the pond sediments increase with heavy manuring but the fish are present in the ponds for such a short time that heavy metal levels in their tissues always remain well below World Health Organization recommended safe limits. In particular, common carp take up virtually no heavy metals at all, presumably because they lack the highly acidic stomach of the tilapias.

It was agreed that much more information was required on the public health aspects of the use of organic manures and wastewater in aquaculture.

Diseases of Tilapias

Dr. Roberts summarized the present importance of diseases in the commercial culture of tilapias. There are very few serious disease problems and these are localized in nature, for example, infestation with *Lernaea* in a few Southeast Asian and Indian locations. The myxobacterial diseases and bacterial septicaemias are generally diseases of bad husbandry and can be avoided but such disease problems will probably increase as tilapia culture expands. Vaccines may be available in the future for septicaemias.

Although only one virus has been isolated from tilapias at present and produces no serious pathological effects, more viruses are bound to occur in intensively cultured fish. There are also risks for the future with respect to public health. The misuse of antibiotics in fish ponds receiving human wastes could create antibiotic-resistant strains of human pathogens. Also, a parasite such as *Haplorchis* could ruin a culture industry if it can be shown that infection is caused by farmed fish.

The transportation of broodstock and fry has attendant risks of disease transfer. If a large fry supply industry develops, then control measures will have to be undertaken. These must be realistic, however, not like the extreme quarantine measures recently applied to the introduction of coho salmon to the U.K. for experimental work, which included complete confinement in recirculation systems for long periods and sterilization of effluents. A realistic series of measures is needed, including routine monitoring of broodstock health, disinfection of fry with formalin and malachite green both before despatch and on receipt, isolation from natural waters and other stocks on receipt (quarantine) and the destruction, preferably by burning, of all packing materials. All inspection, certification and licensing should be matters for official government scientists or other trained personnel. Stirling University has been involved in training fish disease experts from Southeast Asia and Africa and in liaison work with Thailand where the Government is developing a routine disease diagnostic system for the catfish industry.

Dr. Pullin asked whether the possibility of shipping disinfected tilapia eggs had been investigated as it was impossible to give whole fish a clean bill of health. The shipment of disinfected eyed ova is routine in the salmonid culture industry and greatly reduces the risks of transfer of all diseases apart from intraovarian viruses. It was agreed that tilapia eggs can be incubated in water provided that they are kept moving by slow aeration or agitation and that transportation of disinfected eggs in self-contained units could be viewed as a future method.

Dr. Lovshin drew attention to the difficulties of treating tilapia diseases, particularly in the rural areas of developing countries where even basic chemicals such as formalin and potassium permanganate are either unavailable or very costly. Prof. Roberts stated that the best treatments in any disease situation were usually improvement in water quality and reduction in stocking density. It was agreed that formalin is the most useful general chemical for disease treatment, but that salt can also prove useful against ectoparasites. Details of treatment methods applicable to tilapia culture in rural areas have still to be worked out.

Consensus Statement and Research Requirements

A. Consensus Statement

The conference participants agreed on the following:

1. The Value of Fundamental Research

Fundamental research on the taxonomy, genetics, physiology and ecology of tilapias has produced a large amount of information of direct benefit to the culture industry and merits increased and sustained financial support.

2. Standardization

The scientific nomenclature used for the tilapias remains a matter of controversy among taxonomic experts. For the sake of uniformity, the genera *Sarotherodon* for the mouthbrooders and *Tilapia* for the substrate-spawners are accepted for the published proceedings of this conference.

To avoid confusion, all descriptions of hybrid crosses should be given with the female parent first. The various laboratories establishing collections of so-called pure strains of tilapias should collaborate to standardize the nomenclature. The geographical origin and history of transplantation of strains should be documented and information on electrophoretic and other genetic markers exchanged on a regular basis. The present situation is very confused and should not be allowed to worsen.

Researchers should use fish of known origin and history and the fullest possible information on these should be given in the methodology of all published works. Ideally, all fundamental research should use pure strains of fish but this must be seen as a long-term objective, particularly in the developing countries.

3. Conservation of Genetic Material

Collections of pure strains of tilapias should be established both to improve the genotypes of cultured stocks and to supply standard material for research. It should be recognized that some strains and hybrids developed by commercial operators could be the subject of industrial patents. Collections should be replicated at several sites and should maintain sufficient numbers of broodstock to avoid inbreeding depression, except where inbred lines are developed intentionally.

Information should be collected on the sites at which pure wild stocks of tilapias can still be found. Aquatic reserves should be established at critical locations to conserve these stocks and any rare species, especially to protect them from any contamination by fish introductions.

4. Fish Introductions and Transfers

The continuing widespread introductions and transfers of tilapias are a cause for concern, and there is a clear need for the involvement of competent technical bodies to advise on and control these in the future. The possible adverse effects of introductions are ecological damage, contamination or elimination of endemic wild stocks and transfer of pathogens.

Introductions and transfers of tilapias will, however, be essential for the future development of the culture industry, especially as new strains and hybrids are developed. The risks of pathogen transfer can be minimized by moving only early life history stages (which are less prone to carry pathogens than broodstock) from reputable suppliers, by enforcing medication and inspection of fish and the destruction of packing materials on arrival.

5. Health Aspects

The use and abuse of antibiotics in fish culture requires urgent control *and legislation*. Control measures to contain the spread of fish diseases should also be considered by appropriate authorities, including the right of officials to inspect fish and farms for the presence of diseases, to restrict fish sales and movements as and when necessary, and to require farmers by law to notify the appropriate authorities of outbreaks of designated serious diseases.

6. Information Resources

Recent bibliographies and reviews have collated much of the large volume of literature on the biology and culture of tilapias but have missed a considerable amount of so-called "grey" literature, particularly reports and documents with a limited circulation from Africa, Asia and Latin America and material published in local languages.

There will be a continuing need for information collection and dissemination (for example, as special bibliographies) as the literature on tilapias continues to grow. To facilitate this and to lessen the volume of grey literature, published material should, wherever possible, be in a form for direct input into abstracting services and computerized data bases (i.e., having an abstract in English, ISSN number, key words etc., as appropriate. The following could have roles to play in the future collection and dissemination of information: the regional aquaculture centers of FAO, the International

Collection "Cichlid Papers" Reference Service (Parkstrasse 15, D-5176 Inden-Lucherberg, Federal Republic of Germany) and ICLARM.

B. Research Requirements

Table 1 summarizes the research priorities identified by the Conference. The division into near-term and sustained work is arbitrary: all are worthy of sustained research support. Support for those in the left-hand column could result in rapid payoffs for the culture industry.

The following notes amplify some of the topics:

1. Genetics

Applied research on the applied genetics of tilapias can have rapid payoffs for the culture industry and merits sustained support as the industry expands. It should be recognized, however, that all work on genetic improvement is high-risk (high-investment) research.

Although some of the tilapias currently available have good culture characteristics, there is much room for improvement by selection of strains for fast growth, higher fecundity and later maturation. The screening of new species and hybrids for culture in freshwater could be beneficial; it is required urgently for brackish and seawater culture where there are few culturable finfish species available of which the life cycles have been closed in captivity. Future studies on hybridization should, therefore, include the development of hybrids which perform well in saline waters. Hybridization work should also include studies on hybrid vigor, and should not be restricted to the search for crosses resulting in all-male progeny. For the developing countries, however, the improvement of cultured strains should have priority over hybridization studies as the continuous development of hybrids requires elaborate facilities for the isolation and the characterization of parents.

The elucidation of sex-determination mechanisms in the tilapias is a pressing need in order to explain the sex-ratios of progeny from the various hybrid combinations either in current use or for future development. The greatest benefit to the culture industry from this would be the reliable production of 100% all-male progeny.

2. Reproduction

The design of systems for mass fry and fingerling production is the most important single requirement for the culture industry. The private sector is expected to develop such systems rapidly if given the necessary biological data and basic guidelines from researchers. This requires technological as well as practical extension work by leading aquaculture research centers.

The other important research areas related to reproduction can be summarized as behavioral studies. For example, broodstock performance is likely to be controlled by behavioral factors. Compatibility in hybrid crosses is one area in which near-term research and application of the existing published information on the reproductive behavior of tilapias would be useful.

Table 1. Summary of research priorities on the biology and culture of tilapias. A broken line between the columns indicates scope for both near-term and sustained work on the adjacent topic(s).

Near-term research priorities for both public-funded institutions and the private sector.	Priority research areas for sustained work, which would benefit from institutional collaboration.
1. Genetics	
Screening of new species and hybrids for freshwater and saltwater culture	
Selection for fast growth, higher fecundity and later maturation	
Elucidation of sex determination mechanisms	
2. Reproduction	
Design of mass fry and fingerling production systems	Variability in reproductive performance
Behavioral studies relevant to broodstock performance	Suppression of gametogenesis by environmental, behavioral or physiological manipulation
Chemical communication in reproductive behavior	
Cryopreservation of gametes	
Recruitment control by predators	
3. Growth and Nutrition	
Nutritional requirements of young, growing and mature fish	Feeding niche dynamics in polyculture
Incorporation of local materials in formulation of supplementary feeds	Metabolic pathways in relation to growth and gametogenesis
Feeding behavior in culture systems	
Digestive physiology in relation to feeding practices in culture	
4. Facilities and Equipment	
Development of integrated farming systems	
Pond and cage design	
5. Environment	
Pen, cage and pond effluent studies	
6. Fish Health	
Diseases and pathology	
Acclimation and stress reactions in culture systems	

Variability in reproductive performance has many causes, including possibly a genetic basis. Attempts should be made to explain the large differences in time to first maturation, fecundity, spawning frequency, etc. observed by different workers within a single species under different conditions. This work overlaps the narrower aim of suppressing gametogenesis by environmental, behavioral or physiological manipulation.

The importance of communication via dissolved organic compounds (pheromones) should also be investigated. If chemical communication is significant in the reproductive behavior of the tilapias, then the culturists could conceivably develop techniques either to encourage or suppress spawning by chemical means.

Cryopreservation of gametes is a useful technique for storing genetic material cheaply. Experience with other fish indicates that only spermatozoa are amenable to cryopreservation. Techniques for tilapia semen should be developed.

Control of tilapia recruitment, especially in pond culture, remains a difficult problem. Recruitment control by predators offers a viable solution for culture in both developed and developing countries. Research is needed to identify suitable predator species and to study predator-prey relationships to develop management techniques.

3. Growth and Nutrition

The dietary requirements of the important cultured species and hybrids must be defined so that supplemental feeds can be formulated on a sound technological basis. Because supplemental feeds are a major cost item in intensive or semi-intensive culture, studies on the physiology of digestion and assimilation in relation to feeding rate and frequency are important. The use of locally available dietary components can reduce feeding costs, but must be nutritionally adequate. More attention should be paid to *published information* on the feeding behavior and digestive physiology of wild fish and more research performed on these topics using fish in culture systems both with and without supplemental feeding. The study of feeding niches in polyculture systems is also important as there is evidence that some tilapias can cross from niche to niche which could reduce the number of species for a specific system and also maximize production.

4. Facilities and Equipment

The design of integrated agriculture-aquaculture farming systems is a high priority area for further research. While such developments must consider the public health aspects of producing human food from agricultural wastes, it is important that any apparent health hazards are assessed against those that exist anyway in normal agricultural and aquacultural practice. For example, by using animal manures to fertilize ponds, are more health hazards created than those already present in intensive animal production systems or in culture pond water?

5. Environment

Methodologies and standards are needed to assess environmental impact of effluents from aquaculture systems. This applies to pens and cages in enclosed or semi-enclosed bodies of water and to ponds, whether as periodic run-off or when draining to harvest (particularly in fertilized systems).

6. Fish Health

Sustained research on the parasites and diseases of tilapias and their pathology is essential as the culture industry continues to expand.

Acclimatization and stress reactions of fish in culture systems are poorly understood and have wide implications for growth, survival, reproductive performance and susceptibility to pathogens. Studies in both these areas are needed.

References

ADCP. 1978. Aquaculture Development and Coordination Programme, Outline of an applied research and experimental development programme for the Latin American Regional Centre for Aquaculture. ADCP/REP/78/5. 26 p. UNDP/FAO, Rome.

ADCP. 1980. Aquaculture Development and Coordination Programme, Outline research programme for the African Regional Aquaculture Centre, Port Harcourt, Rivers State, Nigeria. ADCP/REP/80/12. 36 p. UNDP/FAO, Rome.

Abdel-Malek, S.A. 1972. Food and feeding habits of some Egyptian fishes in Lake Quarun. I. *Tilapia zillii* (Gerv.) B. according to different length groups. Bull. Inst. Oceanogr. Fish. Cairo 2: 204-213.

Al-Daham, N.K. 1970. The use of chemosterilants, sex hormones, radiation and hybridization for controlling reproduction in *Tilapia* species. Auburn University, Auburn, Alabama. 145 p. Ph.D. dissertation.

Alkholy, A.A. and S.A. Abdel-Malek. 1972. Food and feeding habits of some Egyptian fishes in lake Quarun. Part I. *Tilapia zillii* (Gerv.) A. according to different localities. Bull. Inst. Oceanogr. Fish. Cairo 2: 185-201.

Allanson, B.R. and R.G. Noble. 1964. The tolerance of *Tilapia mossambica* (Peters) to high temperature. Trans. Amer. Fish. Soc. 93: 323-332.

Allanson, B.R., A. Bok and N.I. van Wyk. 1971. The influences of exposure to low temperature on *Tilapia mossambica* Peters (Cichlidae). II. Changes in serum osmolarity, sodium and chloride concentrations. J. Fish Biol. 3: 181-185.

Allison, R., R.O. Smitherman and J. Cabrero. 1979. Effects of high density culture and form of feed on reproduction and yield of *Tilapia aurea*, p. 168-170. *In* T.V.R. Pillay and W.A. Dill (eds.) Advances in aquaculture. Fishing News Books Ltd., Farnham, Surrey, England.

Almeida, L.J., E.J. Da Silva and Y.M. Freitas. 1968. Microorganisms from some tropical fish diseases. J. Fish. Res. Board Can. 25: 197-201.

Anders, A. and F. Anders. 1963. Genetisch bedingte XX- und XY- YY- beim wilden *Platypoecilus maculatus* aus Mexiko. Z. Vererbungslehre 94: 1-18.

Anon. 1978. Lake Tilapia threatened by too many cages. Fish Farming Int. 5(3): 48.

Anon. 1979a. Available technology gaps in tilapia farming in the Philippines. Asian Aquaculture 2(6): 3, 6-7.

Anon. 1979b. Research highlights. ICA (Int. Cent. Aquaculture, Auburn Univ.) Commun. 2(4): 1.

Anon. 1979c. Tilapia production at the Freshwater Fisheries Station. Asian Aquaculture 2(7): 6.

Armbrester, W. 1972. The growth of caged *Tilapia aurea* (Steind.) in fertile farm ponds. Proc. Annu. Conf. Southeast Assoc. Game Fish Comm. 25: 446-451.

Aronson, L.R. 1945. Influence of the stimuli provided by the male Cichlid fish *Tilapia macrocephala* on the spawning frequency of the female. Physiol. Zool. 18: 403-415.

Aronson, L.R. 1949. An analysis of reproductive behavior in the mouthbreeding cichlid fish *Tilapia macrocephala* (Bleeker). Zoologica (N.Y.) 34: 133-158.

Aronson, L.R. 1951. Factors influencing the spawning frequency in the female cichlid fish *Tilapia macrocephala* (Bleeker). Trans. N.Y. Acad. Sci. 2: 33-42. *Also in* Amer. Mus. Novit. 1484: 1-26.

Aronson, L.R. 1957. Reproductive and parental behaviour, p. 271-304. *In* M.E. Brown (ed.) The physiology of fishes. Vol. 2. Academic Press, New York.

Arrignon, J. 1969. L'élevage de *Tilapia mossambica* comme animal de laboratoire. Verh. Int. Ver. Theor. Angew. Limnol. 17: 650-661.

Aston, R.J. and D.J. Brown. 1978. Heated effluents and fish culture, p. 39-60. *In* C.M.R. Pastakia (ed.) Conference on fish farming and wastes. Institute of Fisheries Management and Society of Chemical Industry, London.

Atz, J.M. 1954. The peregrinating Tilapia. Anim. Kingdom 57(5): 148-155.

Avault, J.W. and E.W. Shell. 1967. Preliminary studies with the hybrid of tilapia *Tilapia nilotica* x *Tilapia mossambica*. FAO Fish. Rep. 44: 237-242.

Avault, J.W., E.W. Shell and R.O. Smitherman. 1968. Procedures for overwintering tilapia. FAO Fish. Rep. 44: 343-345.

Avtalion, R.R. and A. Wojdani. 1971. Electrophoresis and immunoelectrophoresis of sera from some known F_1 hybrids of *Tilapia*. Bamidgeh 23: 117-124.

Avtalion, R.R., Y. Pruginin and S. Rothbard. 1975. Determination of allogeneic and xenogeneic markers in the genus *Tilapia*. I. Identification of sex and hybrids in Tilapia by electrophoretic analysis of serum proteins. Bamidgeh 27: 8-13.

Avtalion, R.R., M. Dyczymier, A. Wojdani and Y. Pruginin. 1976. Determination of allogeneic and xenogeneic markers in the genus of *Tilapia*. II. Identification of *T. aurea*, *T. vulcani* and *T. nilotica* by electrophoretic analysis of their serum proteins. Aquaculture 7: 255-265.

Avtalion, R.R. and I.S. Hammerman. 1978. Sex determination of *Sarotherodon (Tilapia)*. I. Introduction to a theory of autosomal influences. Bamidgeh 30: 110-115.

Axelrod, H.R. and L. Burgess. 1973. Breeding aquarium fishes. Book 3. T.F.H. Publ., Neptune City, N.J.

Babiker, M.M. and H. Ibrahim. 1979. Studies on the biology of reproduction in the cichlid *Tilapia nilotica* (L.): gonadal maturation and fecundity. J. Fish Biol. 14: 437-448.

Badenhuizen, T.R. 1967. Temperatures selected by *Tilapia mossambica* (Peters) in a test tank with a horizontal temperature gradient. Hydrobiologia 30: 541-554.

Baerends, G.R. and J.M. Baerends-Van Roon. 1950. An introduction to the study of ethology of cichlid fishes. Behaviour, Suppl. 1: 1-242.

Baerends, G.R. and G.T. Blockzijul. 1963. Gedanken über das Entsehen von Formdivergenzen zwischen homologe Signalhandlugen verwandter Arten. Z. Tierpsychol. 20: 517-528.

Bailey, R.G. and P. Denny. 1978. The 1974 biological survey of Nyumba ya Mungu reservoir, Tanzania. Concluding remarks. Biol. J. Linn. Soc. 10: 151-157.

Bailey, R.G., S. Churchfield, T. Petr and R. Pimm. 1978. The ecology of fishes in Nyumba ya Mungu reservoir, Tanzania. Biol. J. Linn. Soc. 10: 109-137.

Baker, H.G. and G.L. Stebbins. 1965. The genetics of colonizing species. Academic Press, New York.

Baker, J.R. 1963. Three new species of *Myxosoma* (Protozoa: *Myxosporidia*) from East African freshwater fish. Parasitology 53: 289-292.

Balarin, J.D. and J.P. Hatton. 1979. Tilapia. A guide to their biology and culture in Africa. Unit of Aquatic Pathobiology, University of Stirling, Scotland.

Balon, E.K. 1959a. Die Entwicklung der Texas-Cichlide (*Herichthys cyanoguttatus* Baird et Girard) nach dem Schlüpfen. Zool. Anz. 162: 339-355.

Balon, E.K. 1959b. Die Beschuppungsentwicklung der Texas-Cichlide (*Herichthys cyanoguttatus* Baird et Girard). Zool. Anz. 163: 82-89.

Balon, E.K. 1960. Über die Entwicklungstufen des Lebens der Fische und ihre Terminologie. Z. Wiss. Zool. Abt. A 164: 294-314.

Balon, E.K. 1964. Verzeichnis und ökologische Characteristik der Fische der Donau. Hydrobiologia 24: 441-451.

Balon, E.K. 1971. The intervals of early fish development and their terminology (a review and proposals). Vest. Cesk. Spol. Zool. 35: 1-8. (In Czech)

Balon, E.K. 1974. Fish production of a tropical ecosystem, p. 249-573. *In* E.K. Balon and A.G. Coche (eds.) Lake Kariba: a man-made tropical ecosystem in Central Africa. Monographiae Biologicae 24. Dr. W. Junk bv Publ., The Hague.

Balon, E.K. 1975a. Ecological guilds of fishes: a short summary of the concept and its application. Verh. Int. Ver. Theor. Angew. Limnol. 19: 4230-4239.

Balon, E.K. 1975b. Reproductive guilds of fishes: a proposal and definition. J. Fish. Res. Board Can. 32: 821-864.

Balon, E.K. 1975c. Terminology of intervals in fish development. J. Fish. Res. Board Can. 32: 1663-1670.

Balon, E.K. 1976. A reply to Dr. Richards' comments. J. Fish. Res. Board Can. 33: 1254-1256.

Balon, E.K. 1977. Early ontogeny of *Labeotropheus* Ahl, 1927 (Mbuna, Cichlidae, Lake Malawi), with a discussion on advanced protective styles in fish reproduction and development. Environ. Biol. Fishes 2: 147-176.

Balon, E.K. 1978. Reproductive guilds and the ultimate structure of fish taxocenes: amended contribution to the discussion presented at the mini-symposium. Environ. Biol. Fishes 3: 149-152.

Balon, E.K. 1979a. The theory of saltation and its application in the ontogeny of fishes: steps and thresholds. Environ. Biol. Fishes 4: 97-101.

Balon, E.K. 1979b. The juvenilization process in phylogeny and the altricial to precocial forms in the ontogeny of fishes. Environ. Biol. Fishes 4: 193-198.

Balon, E.K. 1980. Charrs: salmonid fishes of the genus *Salvelinus*, perspectives in vertebrate science 1. Dr. W. Junk bv Publ., The Hague.

Balon, E.K. 1981a. About processes which cause the evolution of guilds and species. Environ. Biol. Fishes 6: 129-138.

Balon, E.K. 1981b. Reproductive guilds of fishes: amendments and an evolutionary proposal. Environ. Biol. Fishes 6. (In press)

Balon, E.K. 1981c. Saltatory processes and altricial to precocial forms in the ontogeny of fishes. Amer. Zool. 21 (In press)

Balon, E.K. and A.G. Coche, editors. 1974. Lake Kariba: a man-made tropical ecosystem in Central Africa, Pt. II. Monographiae Biologicae 24. Dr. W. Junk bv Publ., The Hague.

Balon, E.K., W.T. Momot and H.A. Regier. 1977. Reproductive guilds of percids: results of the paleogeographical history and ecological succession. J. Fish. Res. Board Can. 34: 1910-1921.

Banks, J.W., M.J. Holden and R.H. Lowe-McConnell. 1965. Fishery report, p. 21-42. *In* E. White (ed.) First Scientific Report Kainji Biological Research Team. Cyclostyled, University of Liverpool.

Bard, J., P. de Kimpe, J. Lazard, J. Lemasson and P. Lessent. 1976. Handbook of tropical fish culture. Centre Technique Forestier Tropical, Nogent-sur-Marne, France.

Bardach, J.E., J.H. Ryther and W.O. McLarney. 1972. Aquaculture: the farming and husbandry of freshwater and marine organisms. Wiley-Interscience, New York.

Barlow, G.W. 1974. Contrasts in social behavior between Central American cichlid fishes and coral-reef surgeon fishes. Amer. Zool. 14: 9-34.

Barlow, G.W. 1978. Introduction, p. 1-8. *In* Scientific American reader in ichthyology. Freeman, San Francisco.

Barlow, G.W. and R.F. Green. 1970. The problems of appeasement and of sexual roles in the courtship behavior of the blackchin mouthbreeder, *Tilapia melanotheron* (Pisces: Cichlidae). Behaviour 36: 84-115.

Baron, J.C. 1974. Sur la présence dans la baie de Gorée de *Tilapia guineensis* (Bleeker 1962). Bull. Inst. Fondam. Afr. Noire, Ser. A 36(4): 934-936.

Barr, W.A. 1968. Patterns of ovarian activity, p. 164-238. *In* E.J.W. Barrington and C. Barker Jorgensen (eds.) Perspectives in endocrinology. Academic Press, New York.

Barrington, E.J.W. 1957. The alimentary canal and digestion, p. 109-161. *In* M.E. Brown (ed.) The physiology of fishes. Vol. 1. Academic Press, New York.

Bauer, J. 1968. Vergleichende Untersuchungen zum Kontaktverhalten verschiedener Arten der Gattung *Tilapia* (Cichlidae, Pisces) und ihrer Bastarde. Z. Tierpsychol. 25: 22-70.

Baylis, J.R. 1978. Paternal behaviour in fishes: a question of investment, timing or rate? Nature (London) 276: 738.

Baylis, J.R. 1981. The evolution of parental care in fishes, with reference to Darwin's rule of male sexual selection. Environ. Biol. Fishes 6: 223-251.

Bayoumi, A.R. 1969. Notes on the occurrence of *Tilapia zillii* (Pisces) in Suez Bay. Mar. Biol. (Berl.) 4: 255-256.

Beadle, L.C. 1974. The inland waters of tropical Africa. An introduction to tropical limnology. Longman, London.

Beamish, F.W.H. 1970. Influence of temperature and salinity acclimation on temperature preferenda of the euryhaline fish *T. nilotica*. J. Fish. Res. Board Can. 27(7): 1209-1214.

Beamish, F.W.H. 1974. Apparent specific dynamic action of large-mouth bass, *Micropterus salmoides*. J. Fish. Res. Board Can. 31: 1763-1769.

Bell-Cross, G. 1968. The distribution of fishes in central Africa. Fish. Res. Bull. (Zambia) (1965-66) 4: 3-20.

Bell-Cross, G. 1972. Fish fauna of the Zambezi River system. Arnoldia (Rhodesia) 5(29): 1-19.

Bell-Cross, G. 1973. The fish fauna of the Buzi River System in Rhodesia and Mocambique. Arnoldia (Rhodesia) 6(8): 1-14.

Bell-Cross, G. 1974. A fisheries survey of the Upper Zambezi river system. Occas. Pap. Natl. Mus. Rhod. B5(5): 279-338.

Bell-Cross, G. 1976. The fishes of Rhodesia. National Museum, Salisbury, Rhodesia.

Bellamy, A.W. 1936. Inter-specific hybrids in *Platypoecilus*: one species ZZ-ZW, the other XY-XX. Proc. Natl. Acad. Sci. USA 22: 531-535.

Ben-Tuvia, A. 1959. The biology of the cichlid fishes of Lakes Tiberias and Huleh. Bull. Res. Counc. Isr., Sect. B: Zool. 8B: 152-188. *Also in* Bull. Sea Fish. Res. Stn., Isr. 27: 153-188. 1960.

Ben-Tuvia, A. 1978. Fishes, p. 407-430. *In* C. Serruya (ed.) Lake Kinneret: (Lake Tiberias, Sea of Galilee) Monographiae Biologicae 32. Dr. W. Junk bv Publ., The Hague.

Bergstrand, E. and A.J. Cordone. 1970. Exploratory bottom trawling in Lake Victoria. Rep. East Afr. Freshwater Fish. Res. Org. 1969: 42-52.

Bern, H.A. 1975. Prolactin and osmoregulation. Amer. Zool. 15: 937-948.

Berns, S., E.H. Chave and H.M. Peters. 1978. On the biology of *Tilapia squamipinnis* (Günther) from Lake Malawi (Teleostei: Cichlidae). Arch. Hydrobiol. 84 : 218-246.

Berrios-Hernandez, J.M.B. 1979. Evaluation of management and stocking techniques for maximum production of *Tilapia aurea* fry and fingerlings in plastic pools. Auburn University, Auburn, Alabama. 73 p. M.S. thesis.

Billard, R. 1978. Testicular feed-back on the hypothalamo-pituitary axis in rainbow trout (*Salmo gairdneri* R.) Ann. Biol. Anim. Biochim. Biophys. 18: 813-818.

Billard, R. and R.E. Peter. 1977. Gonadotropin release after implantation of anti-estrogens in the pituitary and hypothalamus of goldfish, *Carassius auratus*. Gen. Comp. Endocrinol. 32: 213-220.

Billard, R., A. Solari and A.M. Escaffre. 1974. Méthode d'analyse quantitative de la spermatogénèse des poissons téléostéens. Ann. Biol. Anim. Biochim. Biophys. 14: 87-104.

Billard, R., E. Burzawa-Gerard and B. Breton. 1970. Régénération de la spermatogenese du Cyprin hypophysectomisé *Carassius auratus* L. par un facteur gonadotrope hautement purifié de Carpe. C.R. Acad. Sci. Paris, D 271: 1896-1899.

Billard, R., B. Breton, A. Fostier, B. Jalabert and C. Weil. 1978. Endocrine control of the teleost reproduction cycle and its relation to external factors: Salmonid and Cyprinid models, p. 37-48. *In* P.J. Gaillard and H.H. Boer (eds.) Comparative endocrinology. Elsevier-North Holland Publ. Co., New York.

Bishai, H.M. 1962. Experiments on the suitability of broad beans (*Vicia faba* L.) as a diet for fish. Hydrobiologia 20(1): 31-39.

Bishai, H.M. 1965. Resistance of *Tilapia nilotica*. 1. To high temperatures. Hydrobiologia 25: 433-488.

Blache, J., M.F. Miton, A. Stauch, A. Iltis and G. Loubens. 1964. Les poissons du Tchad et du bassin adjacent du Mayo Kebbi. O.R.S.T.O.M., Paris.

Blake, B.F. 1977. Lake Kainji, Nigeria: a summary of the changes within fish population since the impoundment of the Niger in 1968. Hydrobiologia 53(2): 131-137.

Blüm, V. 1973. Experimente mit teleosteen Prolaktin. Zool. Jahrb. Abt. Allg. Zool. Physiol. Tiere 77: 335-347.

Blüm, V. 1976. Die Rolle des Prolaktins bei der Cichlidenbrutpflege. Forschr. Zool. 22: 310-333.

Blüm, V. and K. Fiedler. 1964. Der Einfluss von Prolactin auf das Brutpflegeverhalten von *Symphysodon aequifasciata axelrodi L.P.* Schultz (Cichlidae, Teleostei). Naturwissenschaften 51: 149-150.

Blüm, V. and K. Fiedler. 1965. Hormonal control of reproductive behaviour in some Cichlid fish. Gen. Comp. Endocrinol. 5: 186-196.

Blüm, V. and K.M. Weber. 1968. The influence of prolactin on the activity of steroid-3β-ol dehydrogenase in the ovaries of the Cichlid fish *Aequidens pulcher*. Experientia 24: 1259-1260.

Blumer, L.S. 1979. Male parental care in the bony fishes. Q. Rev. Biol. 54: 149-161.

Bond, C.E. 1979. Biology of fishes. W.B. Saunders Co., Philadelphia.

Borgstrom, G. 1978. The contribution of freshwater fish to human food, p. 469-491. *In* S.D. Gerking (ed.) Ecology of freshwater fish production. Blackwell Scientific Publ., Oxford.

Bourn, D.M. 1974. The feeding of three commercially important fish species in Lake Chilwa, Malawi. Afr. J. Trop. Hydrobiol. Fish. 3(2): 135-145.

Bowen, S.H. 1976a. Feeding ecology of the cichlid fish *Sarotherodon mossambicus* in Lake Sibaya, KwaZulu. Rhodes University, Grahamstown. 135 p. Ph.D. dissertation.

Bowen, S.H. 1976b. Mechanism for digestion of detrital bacteria by the cichlid fish *Sarotherodon mossambicus* (Peters). Nature (London) 260: 137-138.

Bowen, S.H. 1978. Chromic acid in assimilation studies—a caution. Trans. Amer. Fish. Soc. 107(5): 755-756.

Bowen, S.H. 1979. A nutritional constraint in detritivory by fishes: the stunted population of *Sarotherodon mossambicus* in Lake Sibaya, South Africa. Ecol. Monogr. 49(1): 17-31.

Bowen, S.H. 1980a. Determinants of the chemical composition of periphytic detrital aggregate in a tropical lake (Lake Valencia, Venezuela). Arch. Hydrobiol. 87(2): 166-177.

Bowen, S.H. 1980b. Detrital nonprotein amino acids are the key to rapid growth of tilapia in Lake Valencia, Venezuela. Science. 207: 1216-1218.

Bowen, S.H. 1981. Digestion and assimilation of periphytic detrital aggregate by *Tilapia mossambica*. Trans. Amer. Fish. Soc. 110: 241-247.

Bowmaker, A.P., P.B.N. Jackson and R.A. Jubb. 1978. Freshwater fishes, p. 1181-1230. *In* M.J.A. Werger and A.C. van Bruggen (eds.) Biogeography and ecology of southern Africa. Dr. W. Junk bv Publ., The Hague.

Bowman, D. 1977. Comparacion entre *Tilapia aurea* Steindachner, y *Tilapia mossambica* Peters en estanques de El Salvador. FAO Fish. Rep. 159 (Vol. 1): 78-90.

Boyd, C.E. and C.P. Goodyear. 1971. Nutritive quality of food in ecological systems. Arch. Hydrobiol. 69(2): 265-270.

Breder, C.M. Jr. and D.E. Rosen. 1966. Modes of reproduction in fishes. Natural History Press, Garden City, New Jersey.

Breton, B. and C. Weil. 1973. Effects du LH-FSH-RH synthétique et d'extraits hypothalamiques de Carpe sur la secrétion d'hormone gonadotrope *in vivo* chez la Carpe *Cyprinus carpio*. C.R. Acad. Sci. Paris, D. 277: 2061-2064.

Breton, B., B. Jalabert, R. Billard and C. Weil. 1971. Stimulation *in vitro* de la libération d'hormone gonadotrope par un facteur hypothalamique chez la Carpe (*Cyprinus carpio* L.) C.R. Acad. Sci. Paris, D 273: 2591-2595.

Breton, B., C. Weil., B. Jalabert and R. Billard. 1972. Activité reciproque des facteurs hypothalamiques de béliers et de poissons téléostéen sur la secrétion *in vitro* des hormones gonadotropes c-HG et LH respectivement par des hypophyses de Carpe et de bélier. C.R. Acad. Sci. Paris, D 274: 2530-2533.

Breton, B., B. Jalabert and R. Billard. 1973. Pituitary and plasma gonadotropin levels and spermatogenesis in the goldfish, *Carassius auratus*, after methallibure treatment. J. Endocrinol. 59: 415-420.

Breton, B., B. Jalabert and C. Weil. 1975a. Caractérisation partielle d'un facteur hypothalamique de libération des hormones gonadotropes chez la Carpe (*Cyprinus carpio* L.), étude *in vitro*. Gen. Comp. Endocrinol. 25: 405-415.

Breton, B., B. Jalabert and A. Fostier. 1975b. Induction de décharges gonadotropes hypophysaires chez la Carpe (*Cyprinus carpio* L.) à l'aide du citrate de cis-clomiphéne. Gen. Comp. Endocrinol. 25: 400-404.

Breton, B., B. Jalabert and P. Reinaud. 1976. Purification of gonadotropin from rainbow trout (*Salmo gairdneri* Richardson) pituitary glands. Ann. Biol. Anim. Biochim. Biophys. 16: 25-36.

Breton, B., A. Fostier, B. Jalabert and C. Weil. 1980. Apport des connaisances fondamentales au contrôle du cycle reproducteur des poissons d'étang: limites et perspectives. *In* R. Billard (ed.) La pisciculture en étang. INRA, Paris. (In press)

Brock, V.E. and M. Takata. 1955. Contributions to the problems of bait capture and mortality together with the experiments in the use of *Tilapia* as live bait. Rep. Ind. Res. Adv. Counc. Hawaii 49: 38-40.

Brown, B.M. 1975. Fishes, p. 199-229. *In* B.A. Whitton (ed.) River ecology, studies in ecology, Vol. 2. Blackwell Sci. Publ., Oxford.

Brown, J.H. 1971. The desert pupfish. Sci. Amer. 225: 104-110.

Bruton, M.N. 1979. The fishes of Lake Sibaya, p. 162-245. *In* B.R. Allanson (ed.) Lake Sibaya. Monographiae Biologicae 36. Dr. W. Junk bv Publ., The Hague.

Bruton, M.N. and B.R. Allanson. 1974. The growth of *Tilapia mossambica* Peters (Pisces: Cichlidae) in L. Sibaya, S. Africa. J. Fish Biol. 6: 701-715.

Bruton, M.N. and R.E. Boltt. 1975. Aspects of the biology of *Tilapia mossambica* Peters (Pisces: Cichlidae) in a natural freshwater lake (Lake Sibaya, South Africa). J. Fish Biol. 7: 423-446.

Buddington, R.K. 1979. Digestion of an aquatic macrophyte by *Tilapia zillii*. J. Fish Biol. 15(3): 449-456.

Buddington, R.K. 1980. Hydrolysis-resistant organic matter as a reference for measurement of fish digestive efficiency. Trans. Amer. Fish. Soc. 109: 653-656.

Burgis, M.J. and I.G. Dunn. 1978. Production in three contrasting ecosystems, p. 137-158. *In* S.D. Gerking (ed.) Ecology of freshwater fish production. Blackwell Sci. Publ., Oxford.

Burgis, M.J. 1978. Case studies of lake ecosystem at different latitudes: the tropics. The L. George ecosystems. Verh. Int. Ver. Theor. Angew. Limnol. 20(2): 1139-1152.

Burgis, M.J., J.P.E. Darlington, I.G. Dunn, G.G. Ganf, J.J. Gwahaba and L.M. McGowan. 1973. The biomass and distribution of organisms in Lake George, Uganda. Proc. R. Soc. London Ser. B 184: 271-298.

Burns, R.G. and R.R. Stickney. 1980. Growth of *Tilapia aurea* in ponds receiving poultry wastes. Aquaculture 20: 117-121.

Burzawa-Gerard, E. 1971. Purification d'une hormone gonadotrope hypophysaire de poisson téléostéen, la Carpe (*Cyprinus carpio* L.). Biochimie 53: 545-552.

Callard, I.P., G.V. Callard, V. Lance, J.L. Bolaffi and J.S. Rosset. 1978. Testicular regulation in nonmammalian vertebrates. Biol. Reprod. 18: 16-43.

Campbell, D. 1978a. La technologie de construction des cages d'élevage de *Tilapia nilotica* (L.) dans le Lac de Kossou, Côte d'Ivoire. Rapp. Tech. 45. 31 p. Authorité Aménagement Vallée du Bandama, Centre Dével. Pêches Lac Kossou.

Campbell, D. 1978b. Formulation des aliments destinés à l'élevage de *Tilapia nilotica* (L.) en cages dans le Lac de Kossou. Côtes d'Ivoire. Rapp. Tech. 46. 26 p. Authorité Aménagement Vallée du Bandama, Centre Dével. Pêches Lac Kossou.

Capart, A. 1955. L'échosondage dans les lacs du Congo Belge. Bull. Agric. Congo 46(5): 1075-1104.

Carey, T.G. 1965. Breeding behaviour of *Tilapia macrochir*. Fish. Res. Bull. (Zambia) (1962/63): 12-13.

Castenholz, R.W. and C.E. Wickstrom. 1975. Thermal streams, p. 264-285. *In* B.A. Whitton (ed.) River ecology. Blackwell Scientific Publ., Oxford.

Caulton, M.S. 1975a. Diurnal movement and temperature selection by juvenile and sub-adult *Tilapia rendalli* Boulenger. Trans. Rhod. Sci. Assoc. 56: 51-56.

Caulton, M.S. 1975b. The ability of the cichlid fishes *Tilapia rendalli* Boulenger, *Tilapia sparrmanii* A. Smith and *Hemihaplochromis* (= *Pseudocrenilabrus*) *philander* (M. Weber) to enter deep water. J. Fish Biol. 7: 513-517.

Caulton, M.S, 1976. The importance of pre-digestive food preparation to *Tilapia rendalli* Boulenger when feeding on aquatic macrophytes. Trans. Rhod. Sci. Assoc. 57: 22-28.

Caulton, M.S. 1977a. The effect of temperature on routine metabolism in *Tilapia rendalli* Boulenger. J. Fish Biol. 11: 549-553.

Caulton, M.S. 1977b. A quantitative assessment of the daily ingestion of *Panicum repens* L. by *Tilapia rendalli* Boulenger (Cichlidae) in Lake Kariba. Trans. Rhod. Sci. Assoc. 58(6): 38-42.

Caulton, M.S. 1978a. Tissue depletion and energy utilization during routine metabolism by sub-adult *Tilapia rendalli* Boulenger. J. Fish Biol. 13: 1-6.

Caulton, M.S. 1978b. The importance of habitat temperature for growth in the tropical cichlid *Tilapia rendalli* Boulenger. J. Fish Biol. 13: 99-122.

Caulton, M.S. 1978c. The effect of temperature and mass on routine metabolism in *Sarotherodon (Tilapia) mossambicus* (Peters). J. Fish Biol. 13: 195-201.

Caulton, M.S. 1979. The biology and farming of Tilapia in Southern Africa. Backhouse, Pietermaritzburg.

Caulton, M.S. and B.J. Hill. 1973. The ability of *Tilapia mossambica* to enter deep water. J. Fish Biol. 5: 783-788.

Caulton, M.S. and B.J. Hill. 1975. The effect of temperature on the ability of *Tilapia mossambica* Peters to enter deep water. J. Fish Biol. 7: 221-226.

Caulton, M.S. and E. Burzell. 1977. The relationship between changes in condition and body composition in young *Tilapia rendalli* Boulenger. J. Fish Biol. 13: 1-6.

Chapman, D.W., W.H. Miller, R.G. Dudley and R.J. Scully. 1971. Ecology of fishes in the Kafue River. FI:SF/ZAM 11 Technical Report 2. 66 p. + 50 figs. FAO, Rome. (Prepared from Univ. Idaho, Moscow, Idaho, U.S.A.)

Chen, F.Y. 1969. Preliminary studies on the sex-determining mechanism of *Tilapia mossambica* Peters and *T. hornorum* Trewavas. Verh. Int. Ver. Theor. Angew. Limnol. 17: 719-724.

Chen, F.Y. and G.A. Prowse. 1964. The effect of living space on the growth rate of fish. Ichthyologica 3: 11-20.

Chen, F.Y. and H. Tsuyuki. 1970. Zone electrophoretic studies on the plasma proteins of *Tilapia mossambica* and *T. hornorum* and their F_1 hybrids, *T. zillii* and *T. melanopleura*. J. Fish. Res. Board Can. 27: 2167-2177.

Chen, T.P. 1976. Aquaculture practices in Taiwan. Fishing News Books Ltd., Farnham, Surrey, England.

Chervinski, J. 1961a. Laboratory experiments on the growth of *Tilapia nilotica* in various saline concentrations. Bamidgeh 13(1): 8-13.

Chervinski, J. 1961b. On the spawning of *Tilapia nilotica* in brackish water during experiments in concrete tanks. Bamidgeh 13(1): 30.

Chervinski, J. 1961c. Study of the growth of *Tilapia galilaea* (Artedi) in various saline conditions. Bamidgeh 13: 71-74.

Chervinski, J. 1966. Growth of *Tilapia aurea* in brackish water ponds. Bamidgeh 18: 81-83.

Chervinski, J. 1971. Sexual dimorphism in *T. aurea* and *T. zillii* from Dor and from L. Tiberias. Bamidgeh 23(1): 56-59.

Chervinski, J. 1972. Occurrence of *Tilapia zillii* (Gervais) (Pisces, Cichlidae) in the Bardawil lagoon in northern Sinai. Bamidgeh 24: 49-50.

Chervinski, J. 1974. Sea bass, *Dicentrarchus labrax* Linne (Pisces, Serranidae) a "policefish" in freshwater ponds and its adaptability to saline conditions. Bamidgeh 26: 110-113.

Chervinski, J. 1975. Sea bass, *Dicentrarchus labrax* (Linne) and *D. punctatus* (Bloch) (Pisces, Serranidae), a control fish in freshwater. Aquaculture 6: 249-256.

Chervinski, J. and A. Yashouv. 1971. Preliminary experiments on the growth of *Tilapia aurea* (Steindachner) (Pisces, Cichlidae) in sea water ponds. Bamidgeh 23(4): 125-129.

Chervinski, J. and E. Hering. 1973. *Tilapia zillii* (Gervais) (Pisces, Cichlidae) and its adaptability to various saline conditions. Aquaculture 2(1): 23-29.

Chervinski, J. and M. Lahav. 1976. The effect of exposure to low temperature on fingerlings of local tilapia *(Tilapia aurea)* (Steindachner) and imported tilapia *(Tilapia vulcani)* (Trewavas) and *Tilapia nilotica* (Linne) in Israel. Bamidgeh 28: 25-29.

Chervinski, J. and M. Zorn. 1974. Note on the growth of *Tilapia aurea* (Steindachner) and *Tilapia zillii* (Gervais) in seawater ponds. Aquaculture 4: 249-255.

Chien, A.K. 1973. Reproductive behaviour of the angelfish *(Pterophyllum scalare)* (Pisces: Cichlidae). II. Influence of male stimuli upon the spawning rate of females. Anim. Behav. 21: 457-463.

Chimits, P. 1955. Tilapia and its culture: a preliminary bibliography. FAO Fish. Bull. 8: 1-33.

Chimits, P. 1957. The tilapias and their culture. A second review and bibliography. FAO Fish. Bull. 10(1): 1-24.

Chua, T.E. and S.K. Teng. 1977. Floating fishpens for rearing fishes in coastal waters, reservoirs and mining pools in Malaysia. Fish. Bull. 20. 36 p. Min. Agric. (Malaysia), Kuala Lumpur.

Clemens, H.P. and T. Inslee. 1968. The production of unisexual broods by *Tilapia mossambica* sex-reversed with methyltestosterone. Trans. Amer. Fish. Soc. 97: 18-21.

Coche, A.G. 1977. Premiers résultats de l'élevage en cages de *Tilapia nilotica* (L.) dans le Lac Kossou, Côte d'Ivoire. Aquaculture 10: 109-140.

Coche, A.G. 1978. Revue des pratiques d'élevage des poissons en cages dans les eaux continentales. Aquaculture 13: 157-189.

Coche, A.G. 1979. A review of cage fish culture and its application in Africa, p. 428-441. *In* T.V.R. Pillay and W.A. Dill (eds.) Advances in aquaculture. Fishing News Books Ltd., Farnham, Surrey, England.

Coe, M.J. 1966. The biology of *Tilapia grahami* Boulenger in Lake Magadi, Kenya. Acta Trop. 23: 146-177.

Coe, M.J. 1967. Local migration of *Tilapia grahami* Boulenger in Lake Magadi, Kenya in response to diurnal temperature changes in shallow water. East Afr. Wildl. J. 5: 171-174.

Coe, M.J. 1969. Observation on *Tilapia alcalica* Hilgendorf an endemic cichlid fish from Lake Natron, Tanzania. Rev. Zool. Bot. Afr. 80: 1-14.

Collis, W.J. and R.O. Smitherman. 1978. Production of Tilapia hybrids with cattle manure or a commercial diet, p. 43-54. *In* R.O. Smitherman, W.L. Shelton and J.H. Grover (eds.) Culture of exotic fishes symposium proceedings. Fish Culture Section, American Fisheries Society, Auburn, Alabama.

Colombo, L., H.A. Bern, J. Pieprzyk and D.W. Johnson. 1973. Biosynthesis of 11 deoxy-costicosteroids by teleost ovaries and discussion of their possible role in oocyte maturation and ovulation. Gen. Comp. Endocrinol. 21: 168-178.

Conover, R.J. 1966. Assimilation of organic matter by zooplankton. Limnol. Oceanogr. 11: 338-345.

Corbet, S.A., J. Green, J. Griffith and E. Betney. 1973. Ecological studies on crater lakes in West Cameroon: Lakes Koto and Mboandong. J. Zool.170(3): 309-324.

Corredor, G.G. 1978. Ensayo comparativo sobre producción de *Tilapia rendalli* Boulenger a densidad alta, libres y en jaulas, alimentada con hojas de bore *(Alocasia macrorhiza)*. Centro Piscicolo Experimental, Inf. Técn. 2: 79-84. Univ. Caldas, Manizales, Colombia.

Courtenay, W.R. and C.R. Robins. 1973. Exotic aquatic organisms in Florida with emphasis on fishes: a review and recommendations. Trans. Amer. Fish. Soc. 102(1): 1-13.

Courtenay, W.R. and D.A. Hensley. 1979. Range expansion in southern Florida of the introduced spotted tilapia, with comments on its environmental impress. Environ. Conserv. 6: 149-151.

Cowey, C.B. and J.R. Sargent. 1972. Fish nutrition. Adv. Mar. Biol. 10: 383-494.

Cridland, C.C. 1961. Laboratory experiments on the growth of *Tilapia* spp. The reproduction of *Tilapia esculenta* under artificial conditions. Hydrobiologia 18: 177-184.

Cridland, C.C. 1962. Laboratory experiments on the growth of *Tilapia* spp. The effect of light and temperature on the growth of *T. zillii* in aquaria. Hydrobiologia 20: 155-166.

Da Silva, A.B., A. Carneiro-Sobrinho, F.R. de Melo and L.L. Lovshin. 1978. The intensive mono and polyculture of Tambaqui, *Colossoma macropomum* and Pirapitinga, *Colossoma bidens,* with the all-male Tilapia hybrid, female *Sarotherodon niloticus* and male *Sarotherodon hornorum.* Paper presented at the Second Symposium of the Latin American Aquaculture Association, 13-17 November 1978, Mexico City, Mexico. (In Portuguese with English abstract)

Dadzie, S. 1969. Spermatogenesis and the stages of maturation in the male Cichlid fish *Tilapia mossambica.* J. Zool. 159: 399-403.

Dadzie, S. 1974. A preliminary report on the use of Methallibure in *Tilapia* culture. Afr. J. Trop. Hydrobiol. Fish. 4: 127-141.

Dadzie, S. 1980. Recent changes in the fishery of a new tropical man-made lake, Lake Kamburu, Kenya. J. Fish Biol. 16: 361-367.

Dadzie, S. and M. Hyder. 1976. Compensatory hypertrophy of the remaining ovary and the effects of methallibure in the unilaterally ovariectomized *Tilapia aurea.* Gen. Comp. Endocrinol. 29: 433-440.

Daget, J. 1952. Observations sur la ponte de *Tilapia zillii* (Gervais), poissons de la famille des Cichlidae. Ann. Mag. Nat. Hist. 5(12): 309-310.

Daget, J. 1954. Les poissons du Niger Supérieur. Mém. Inst. Fr. Afr. Noire 36: 1-391.

Daget, J. 1956. Mémoires sur la biologie des poissons du Niger-Moyen. II. Recherches sur *Tilapia zillii* (Gerv.). Bull. Inst. Fr. Afr. Noire 18, Ser. A: 165-223.

Daget, J. 1957. Données récentes sur la biologie des poissons dans le delta central du Niger. Hydrobiologia 9: 321-347.

Daget, J. 1968. The distribution of freshwater fishes in western Africa, p. 7. *In* French Institute of Black Africa (ed.) International Atlas of West Africa, 1. Zoogeography. Int. Publ. Serv., N.Y.

Davis, G.E. and G.E. Warren. 1971. Estimation of food consumption, p. 227-248. *In* W.E. Ricker (ed.) Methods for assessment of fish production in freshwater. IBP Handbook No. 3. Blackwell Scientific Publ., Oxford.

Dawkins, R. 1976. The selfish gene. Oxford University Press, New York.

Dawkins, R. 1978. Replicator selection and the extended phenotype. Z. Tierpsychol. 47: 61-76.

Dawkins, R. and T.R. Carlisle. 1976. Parental investment, mate desertion and a fallacy. Nature (London) 262: 131-133.

De Bont, A.F. 1950. La reproduction en étangs des *Tilapia melanopleura* (Dum.) et *macrochir* (Blgr.) C.R. Conf. Piscicole Anglo-Belge, Elizabethsville 1949. Communications (Brussels) 8: 303-312.

De Kimpe, P. 1971. Feeds for tilapia and catfish. FAO Aquaculture Bull. 3(4): 4.

De Kimpe, P. 1978. Première expérience de pisciculture en enclos flottant en lagunes en Côtes d'Ivoire et perspectives de développement de cette technique en Afrique tropicale. Bois For. Trop. 180: 59-64.

De la Cruz, C.R. 1979. Observations on freshwater aquaculture and extension in Malaysia, Thailand, Japan and Taiwan. ICLARM (Int. Cent. Living Aquat. Resources Manage., Manila) Newsl. 2(1): 6-8.

De Silva, S.S. and J. Chandrasoma. 1980. Reproductive biology of *Sarotherodon mossambicus*, and introduced species, in an ancient man-made lake in Sri Lanka. Environ. Biol. Fishes 5: 253-259.

De Vlaming, V.L. 1974. Environmental and endocrine control of teleost reproduction, p. 13-83. *In* C.B. Schreck (ed.) Control of sex in fishes. Virginia Polytechnic Institute and State University. VPI-SG; 74-01. Virginia Polytechnic Institute and State University, Blacksburg.

Denny, P., D.W. Bowker and R.G. Bailey. 1978. The importance of littoral epiphyton as food for commercial fish in the recent African man-made lake, Nyumba ya Mungu reservoir, Tanzania. Biol. J. Linn. Soc. 10(1): 139-150.

Denzer, H.W. 1968. Studies on the physiology of young *Tilapia*. FAO Fish. Rep. 44(4): 356-366.

Dixon, J.E.W.W. and M.J. Blom. 1974. Some aquatic vertebrates from the Namib Desert, South West Africa. Madoqua, ser. 2 Sci. Pap. Namib Desert Res. Stn. 3: 69-73.

Dodd, J.M. 1975. The hormones of sex and reproduction and their effects in fish and lower chordates: twenty years on. Amer. Zool. 15: 137-171.

Donaldson, E.M., F. Yamazaki, H.M. Dye and W.W. Philleo. 1972. Preparation of gonadotropin from salmon *(Oncorhynchus tshawytscha)* pituitary glands. Gen. Comp. Endocrinol. 18: 469-481.

Donnelly, B.G. 1969. A preliminary survey of *Tilapia* nurseries on Lake Kariba during 1967/68. Hydrobiologia 34: 195-207.

Dubois, J.T. 1959. Note sur la chimie des eaux du lac Tumba. Acad. R. Sci. Outre-Mer (Brussels) Bull. Seances 5: 1321-1324.

Dudley, R.G. 1974. Growth of *Tilapia* of the Kafue floodplain Zambia: predicted effects of the Kafue Gorge Dam. Trans. Amer. Fish. Soc. 103(2): 281-291.

Dudley, R.G. 1979. Changes in growth and size distribution of *Sarotherodon macrochir* and *Sarotherodon andersoni* from the Kafue floodplain, Zambia, since construction of the Kafue Gorge Dam. J. Fish Biol. 14: 205-223.

Dudley, R.G. and R.J. Scully. 1980. Changes in experimental gillnet catches from the Kafue floodplain, Zambia, since construction of the Kafue Gorge Dam. J. Fish Biol. 16: 521-537.

Duerre, D.C. 1969. Report to the Government of Zambia on fishery development in the Central Barotse floodplain (Second phase). FAO/UNDP (Technical Assistance) Reports on Fisheries TA Reports 2638. 80 p. FAO, Rome.

Duncan, B.L. 1973. *Cichlidogyrus sclerosus* Paperna and Thurston from cultured *Tilapia mossambica*. Kalikasan: Philipp. J. Biol. 2: 154-158.

Dunn, I.G. 1972. The commercial fishery of L. George, Uganda (E. Africa). Afr. J. Trop. Hydrobiol. Fish. 4(1): 109-120.

Dunseth, D.R. 1977. Polyculture of channel catfish, *Ictalurus punctatus*, Silver carp, *Hypophthalmichthys molitrix*, and three all-male tilapias, *Sarotherodon* spp. Auburn University, Auburn, Alabama. 59 p. Ph.D. dissertation.

Dunseth, D.R. and D.R. Bayne. 1978. Recruitment control and production of *Tilapia aurea* (Steindachner) with the predator, *Cichlasoma managuense* (Günther). Aquaculture 14: 383-390.

Dussart, J. 1963. Contribution à l'étude de l'adaptation des Tilapia (Pisces, Cichlidae) à la vie en milieu mal oxygéné. Hydrobiologia 21: 328-341.

Eccles, D.H. 1975. Fishes of the African Great Lakes as candidates for introduction into large tropical impoundments. J. Fish Biol. 7: 401-405.

Eckstein, B. 1970. Metabolic pathways of steroid biosynthesis in ovarian tissue of a teleost, *Tilapia aurea*. Gen. Comp. Endocrinol. 14: 303-312.

Eckstein, B. and Y. Katz. 1971. Steroidogenesis in post and pre-spawned ovaries of a Cichlid fish, *Tilapia aurea*. Comp. Biochem. Physiol. A 38: 329-338.

Eckstein, B. and M. Spira. 1965. Effect of sex hormone on gonadal differentiation in a cichlid, *Tilapia aurea*. Biol Bull. (Woods Hole, Mass.) 129: 482-489.

Edwards, R.R.C., D.M. Finlayson and J.H. Steele. 1972. An experimental study of the oxygen consumption, growth and metabolism of the cod (*Gadus morhua* L.) J. Exp. Mar. Biol. Ecol. 8: 299-309.

Elder, H.Y., D.J. Garrod and P.J.P. Whitehead. 1971. Natural hybrids of the African cichlid fishes *Tilapia spilurus nigra* and *T. leucosticta*: a case of hybrid introgression. Biol. J. Linn. Soc. 3: 103-146.

El-Zarka, S. 1956. Breeding behaviour of the Egyptian cichlid fish, *Tilapia zillii*. Copeia 2: 113-123.

El-Zarka, S., A.H. Shaheen and A.A. Aleem. 1970a. Reproduction of *Tilapia nilotica* L. Bull. Inst. Oceanogr. Fish. Cairo 1: 193-204.

El-Zarka, S., A.M. Maghraby and K. Abdel-Hamid. 1970b. Studies on the distribution, growth and abundance of migrating fry and juveniles of mullet in a brackish coastal lake (Edku) in the United Arab Republic. Stud. Rev. Gen. Fish. Counc. Mediterr. 46. 19 p. FAO, Rome.

Elliott, J.M. and W. Davison. 1975. Energy equivalents of oxygen consumption in animal energetics. Oecologia (Berl.) 19: 195-201.

Emlen, J.M. 1973. Ecology: an evolutionary approach. Addison-Wesley, New York.

FAO. 1976. FAO's current field projects in aquaculture. Lake Kossou Fishery Development Project, Ivory Coast. FAO Aquaculture Bull. 8(1): 22-23.

FAO. 1977a. From research institutions. Cage culture of tilapia. FAO Aquaculture Bull. 8(2): 8.

FAO. 1977b. Aquaculture development. Ivory Coast. FAO Aquaculture Bull. 8(2): 21.

FAO. 1977c. Actas del simposio sobre acuicultura en America Latina, Montevideo, Uruguay, 26 Noviembre a 2 de Diciembre, 1974. FAO Inf. Pesca 159. 3 volumes. FAO, Rome.

FAO. 1978. 1977 Yearbook of fishery statistics. Catches and landings. Vol. 44. 343 p. FAO, Rome.

Fagade, S.O. 1971. The food and feeding habits of *Tilapia* species in the Lagos Lagoon. J. Fish Biol. 3: 151-156.

Fagade, S.O. 1974. Age determination in *T. melanotheron* (Rüppell) in the Lagos Lagoon, Lagos, Nigeria, with a discussion of the environmental and physiological influence basis of growth markings in the tropics, p. 71-77. *In* T.B. Bagenal (ed.) The ageing of fishes. Unwin Brothers, Surrey, England.

Farmer, S.W. and H. Papkoff. 1977. A teleost *(Tilapia mossambica)* gonadotropin that resembles luteinizing hormone. Life Sci. 20: 1227-1232.

Farmer, S.W., H. Papkoff, T.A. Bewley, T. Hayashida, R.S. Nishioka, H.A. Bern and C.H. Li. 1977. Isolation and properties of teleost prolactin. Gen. Comp. Endocrinol. 31: 60-71.

Fish, G.R. 1951. Digestion in *Tilapia esculenta*. Nature (London) 167: 900-901.

Fish, G.R. 1955. The food of *Tilapia* in East Africa. Uganda J. 19: 85-89.

Fish, G.R. 1956. Some aspects of the respiration of six species of fish from Uganda. J. Exp. Biol. 33: 186-195.

Fish, G.R. 1960. The comparative activity of some digestive enzymes in the alimentary canal of *Tilapia* and perch. Hydrobiologia 15: 161-178.

Fishelson, L. 1962. Hybrids of two species of the genus *Tilapia* (Cichlidae, Teleostei). Fishermen's Bull. Haifa 4: 14-19. (In Hebrew with English abstract)

Fishelson, L. 1966a. Cichlidae of the genus *Tilapia* in Israel. Bamidgeh 18: 67-80.

Fishelson, L. 1966b. Untersuchungen zur vergleichenden Entwickslungsgeschichte der Gattung *Tilapia* (Cichlidae, Teleostei). Zool. Jahrb. Abt. Anat. Ontog. Tiere 83: 571-656.

Fishelson, L. and W. Heinrich. 1963. Some observation on the mouthbreeding *T. galilaea* L. (Pisces: Cichlidae). Ann. Mag. Nat. Hist. 6(13): 507-508.

Fishelson, L. and D. Popper. 1968. Experiments on rearing fish in salt water near the Dead Sea, Israel. FAO Fish. Rep. 44(5): 244-245.

Fontaine, M. 1976. Hormones and the control of reproduction in aquaculture. J. Fish. Res. Board Can. 33: 922-939.

Fortes, R.D. 1979. Evaluation of tenpounder and tarpon as predators on Java tilapia in brackishwater ponds. Auburn University, Auburn, Alabama. 98 p. Ph.D. dissertation.

Fry, F.E.J. 1971. The effect of environmental factors on the physiology of fish, p. 1-98. *In* W.S. Hoar and D.J. Randall (eds.) Fish physiology, Vol. VI. Academic Press, New York.

Fryer, G. 1961a. Observations on the biology of the Cichlid fish *Tilapia variabilis* Boulenger in the northern waters of Lake Victoria (East Africa). Rev. Zool. Bot. Afr. 64: 1-33.

Fryer, G. 1961b. The parasitic Copepoda and Branchiura of Lake Victoria and the Victoria Nile. Proc. Zool. Soc. London 137: 41-60.

Fryer, G. 1965a. A new isopod of the genus *Lironeca*: parasitic on a cichlid fish of Lake Tanganyika. Rev. Zool Bot. Afr. 71: 376-384.

Fryer, G. 1965b. Predation and its effects on migration and speciation in African fishes: a comment. Proc. Zool. Soc. London 144: 301-322.

Fryer, G. 1968. The parasitic crustacea of African freshwater fishes: their biology and distribution. J. Zool. 156: 45-95.

Fryer, G. and T.D. Iles. 1969. Alternative routes to evolutionary success as exhibited by African cichlid fishes of the genus *Tilapia* and the species flocks of the Great Lakes. Evolution 23: 359-369.

Fryer, G. and T.D. Iles. 1972. The cichlid fishes of the great lakes of Africa: their biology and evolution. T.F.H. Publ., Neptune City, New Jersey. Also published by Oliver and Boyd, Edinburgh.

Fukusho, K. 1968. The specific difference of temperature response among cichlid fishes genus *Tilapia*. Bull. Jap. Soc. Sci. Fish. 34(2): 103-111.

Fukusho, K. 1969. The specific difference of salinity tolerance among cichlid fishes genus Tilapia and histological comparison of their kidneys. Bull. Jap. Soc. Sci. Fish. 35(2): 148-155.

Furse, M.T., R.G. Kirk, P.R. Morgan and D. Tweddle. 1979. Fishes: distribution and biology in relation to changes, p. 175-208. *In* M. Kalk, A.J. MacLachlan and C. Howard-Williams (eds.) Lake Chilwa: studies of changes in a tropical ecosystem. Monographiae Biologicae 35. Dr. W. Junk bv Publ., The Hague.

Furukawa, A. and H. Tsukahara. 1966. On the acid digestion method for the determination of chronic oxide as an index substance in the study of digestibility in fish. Bull. Jap. Soc. Sci. Fish. 32: 502-506.

Gaigher, I.G. 1973. The habitat preferences of fishes from the Limpopo river system, Transvaal and Mozambique. Koedoe 16: 103-116.

Garcia, R.P. 1979. Cage culture: a promising alternative. Fish. Today (Philippines) 2(3): 32-35.

Garrod, D.J. 1959. The growth of *Tilapia esculenta* Graham in Lake Victoria. Hydrobiologia 12: 268-298.

Gee, J.M. 1968. A comparison of certain aspects of the biology of *Lates niloticus* (L.) in endemic and introduced environment in E. Africa, p. 251-259. *In* L. Obeng (ed.) Man-made lakes: the Accra symposium. Ghana Universities Press, Accra.

Gee, J.M. 1969. Trawling results within northern waters of Lake Victoria. Rep. East Afr. Fish. Res. Org. 1968: 15-27.

George, T.T. 1975. Observations on the growth of *Tilapia nilotica* L. in tropical freshwater fish ponds treated with different fertilizers. Paper presented at the FAO/CIFA Symposium on Aquaculture in Africa, 30 September-2 October 1975. Accra, Ghana. CIFA/75/SE 11. 16 p. FAO, Rome.

George, T.T. 1976. Introduction and transplantation of cultivable species into Africa. FAO/CIFA Tech. Pap. 4 (Suppl. 1): 407-432.

Gerking, S.D. 1959. The restricted movements of fish populations. Biol. Rev. 34: 221-242.

Ghosh, K.K. 1974. On the choice of growth curve for Indian major carps—von Bertalanffy or Gompertz. Indian Soc. Agric. Stat. 26: 57-70.

Gleastine, B.W. 1974. A study of the cichlid *Tilapia aurea* (Steindachner) in a thermally modified Texas reservoir. Texas A & M University, Texas. 258 p. M.S. thesis.

Glucksman, J., G. West and T.M. Berra. 1976. The introduced fishes of Papua New Guinea with special reference to *Tilapia mossambica*. Biol. Conserv. 9(1): 37-44.

Godinez, G. and F. Jose. 1976a. Cultivo de *Tilapia aurea* en jaulas experimentales de un m^2 en lagos de El Salvador. Serv. Recursos Pesq., DGRNR, MAG El Salvador.

Godinez, G. and F. Jose. 1976b. Cultivo de *Tilapia aurea* en jaulas comerciales de 32 m^2 en el Lago Ilopango. Serv. Recursos Pesq. DGRNR, MAG El Salvador.

Goldstein, R.J. 1971. Diseases of aquarium fishes. T.F.H. Publ., Neptune City, New Jersey.

Gophen, M. 1980. Food sources, feeding behaviour and growth rates of *Sarotherodon galilaeum* (Linnaeus) fingerlings. Aquaculture 20(2): 101-115.

Gordon, M. 1946. Interchanging genetic mechanisms for sex determination in fishes under domestication. J. Hered. 37: 307-320.

Gordon, M. 1947. Genetics of *Platypoecilus maculatus*. IV. The sex determining mechanism in two wild populations of the Mexican platyfish. Genetics 32: 8-17.

Gosse, J.P. 1963. Le milieu aquatique et l'écologie des poissons dans la région de Yangambi. Ann. Mus. R. Afr. Cent. (Sér. 8 Sci. Zool.) 116: 113-249, + 1-10.

Gould, S.J. 1977. Ontogeny and phylogeny. Harvard University Press, Cambridge, Mass.

Gould, S.J. 1979. On the importance of heterochrony for evolutionary biology. Syst. Zool. 28: 224-226.

Green, J., S.A. Corbet and E. Betney. 1973. Ecological studies on crater lakes in West Cameroon. The blood of endemic cichlids in Barombi Mbo in relation to stratification and their feeding habits. J. Zool. 170: 299-308.

Green, J., S.A. Corbet, E. Watts and O.B. Lan. 1978. Ecological studies on Indonesian lakes. The montane lakes of Bali. J. Zool. 186: 15-38.

Green, R.F. 1980. A note on K-selection. Amer. Nat. 116: 291-296.

Greenfield, J.E., E.R. Lira and J.W. Jensen. 1977. Economic evaluation of *Tilapia* hybrid culture in northeast Brazil. FAO Fish. Rep. 159 (vol. 1): 328-340. (In Spanish with English abstract)

Greenwood, P.H. 1965. On the cichlid fishes of Lake Nabugabo, Uganda. Bull. Br. Mus. (Nat. Hist.) Zool. 12: 315-357.

Greenwood, P.H. 1966. The fishes of Uganda. The Uganda Society, Kampala.

Greenwood, P.H. 1974. The cichlid fishes of Lake Victoria, East Africa: the biology and evolution of a species flock. Bull. Br. Mus. (Nat. Hist.) Zool. Suppl. 6: 1-34.

Greenwood, P.H. 1976. Fish fauna of the Nile, p. 127-141. *In* J. Rzoska (ed.) The Nile, biology of an ancient river. Monographiae Biologicae 29. Dr. W. Junk bv Publ., The Hague.

Guerrero, R.D. 1974. The use of synthetic androgens for the production of monosex male *Tilapia aurea* (Steindachner). Auburn University, Auburn, Alabama. 112 p. Ph.D. dissertation.

Guerrero, R.D. 1975. Cage culture of male and female *Tilapia mossambica* with and without supplementary feeding in a fertilized pond. CLSU (Cent. Luzon State Univ., Philipp.) Sci. J. 9(2): 18-20.

Guerrero, R.D. 1977. Production of tilapia fry in floating net enclosures. FAO Aquaculture Bull. 8(3-4): 4.

Guerrero, R.D. 1979a. Cage culture of Tilapia in the Philippines. Asian Aquaculture 2(11): 6.

Guerrero, R.D. 1979b. Culture of male *Tilapia mossambica* produced through artificial sex-reversal, p. 166-168. *In* T.V.R. Pillay and W.A. Dill (eds.) Advances in aquaculture. Fishing News Books Ltd., Farnham, Surrey, England.

Guerrero, R.D. 1980a. Studies on the feeding of *Tilapia nilotica* in floating cages. Aquaculture 20: 169-175.

Guerrero, R.D. 1980b. Development prospects and problems of *Tilapia* cage culture in the Philippines. International Foundation for Science and Centre de Recherches Océanographiques Meeting on Aquaculture in Africa (1979), Provisional Report 7: 293-298. IFS, Stockholm.

Guerrero, R.D. and L.A. Guerrero. 1975. Monosex culture of male and female *T. mossambica* in ponds at three stocking rates. Kalikasan: Philipp. J. Biol. 4: 129-134.

Guerrero, R.D. and L.A. Guerrero. 1976. Use of estrone for the production of monosex female *Tilapia mossambica*. Inland Fisheries Project Philippines. Tech. Rep. No. 9: 119-122. Freshwater Aquaculture Center, Central Luzon State University, Muñoz, Nueva Ecija, Philippines.

Guerrero, R.D. and T.A. Abella. 1976. Induced sex-reversal of *Tilapia nilotica* with methyltestosterone. Fish. Res. J. Philipp. 1: 46-49.

Günther, A. 1889. On some fishes for the Kiliman-Njaro district. Proc. Zool. Soc. London 1889: 70-72.

Gwahaba, J.J. 1973. Effects of fishing on the *Tilapia nilotica* (L.) population in L. George, Uganda, over the past twenty years. East Afr. Wildl. J. 11: 317-328.

Gwahaba, J.J. 1975. The distribution, population density and biomass of fish in an equatorial lake, Lake George, Uganda. Proc. R. Soc. London Ser. B 190: 393-414.

Gwahaba, J.J. 1978. The biology of cichlid fishes (Teleostei) in an equatorial lake (Lake George, Uganda). Arch. Hydrobiol. 83: 538-551.

Habel, M.L. 1975. Overwintering of the cichlid *Tilapia aurea* produces fourteen tons of harvestable fish in a south Alabama bass-Bluegill public fishing lake. Prog. Fish-Cult. 37(1): 31-32.

Halevy, A. 1979. Observations on polyculture of fish under standard farm pond conditions at the Fish and Aquaculture Research Station, Dor, Israel, during the years 1972-1977. Bamidgeh 31: 96-104.

Haller, R.D. 1974. Rehabilitation of a limestone quarry. Report of an environmental experiment. Bamburi Portland Cement Co. Ltd., Mombasa, Kenya.

Haller, R.D. and R.J. Roberts. 1980. Dual neoplasia in a specimen of *Sarotherodon spilurus spilurus* (Gunther) (= *Tilapia spilurus*). J. Fish. Dis. 3: 63-67.

Hamada, A. and T. Ida. 1973. Studies on specific dynamic action in fishes. I. Various conditions affecting the value of measurement. Bull. Jap. Soc. Sci. Fish. 39: 1231-1235.

Hammerman, I.S. and R.R. Avtalion. 1979. Sex determination in *Sarotherodon (Tilapia)*. II. The sex ratio as a tool for the determination of genotype-A model of autosomal influence. Theor. Appl. Genet. 55: 177-187.

Hanon, L. 1975. Adaptations morphologiques et comportementales à l'incubation buccales chez les poissons cichlides: oeufs et alevins. Ann. Soc. R. Zool. Belg. 105: 169-192.

Harbot, B.J. 1975. Preliminary observations on the feeding of *Tilapia nilotica* in L. Rudolf. Afr. J. Trop. Hydrobiol. Fish. 4(1): 27-37.

Harding, D. 1966. Lake Kariba. The hydrology and the development of fisheries, p. 7-20. *In* R.H. Lowe-McConnell (ed.) Man-made lakes. Academic Press, London.

Harper, A.E. 1967. Effects of dietary protein content and amino acid pattern on food intake and preference, sec. 6, vol. 1, article 29, p. 399-410. *In* C.F. Code and W. Heidel (eds.) Handbook of physiology. American Physiology Society, Washington, D.C.

Hauser, W.J. 1975. An unusually fast growth rate for *Tilapia zillii*. Calif. Fish Game 61: 54-56.

Hauser, W.J. 1977. Temperature requirements of *Tilapia zillii*. Calif. Fish Game 63: 228-233.

Heinrich, W. 1967. Untersuchungen zum Sexualverhalten in der Gattung *Tilapia* (Cichlidae, Teleostei) und bei Artbastarden. Z. Tierpsychol. 24: 684-754.

Henderson-Arzapalo, A., R.R. Stickney and D.H. Lewis. 1980. Immune hypersensitivity in intensively cultured *Tilapia* species. Trans. Amer. Fish. Soc. 109: 244-247.

Hepher, B. 1978. Ecological aspects of warm water fishpond management, p. 447-468. *In* S.D. Gerking (ed.) Ecology of freshwater fish production. Blackwell Scientific Publ., Oxford.

Hickling, C.F. 1960. The Malacca *Tilapia* hybrids. J. Genet. 57: 1-10.

Hickling, C.F. 1962. Fish culture. Faber and Faber, London.

Hickling, C.F. 1963. The cultivation of Tilapia. Sci. Amer. 208: 143-152.

Hickling, C.F. 1966. On the feeding process in the white amur, *Ctenopharyngodon idella.* J. Zool. 148: 408-419.

Hickling, C.F. 1968. Fish hybridization. FAO Fish. Rep. 44: 1-11.

Hildemann, W.H. 1959. A Cichlid fish, *Symphysodon discus,* with unique nurture habits. Amer. Nat. 93: 27-34.

Hirschfield, M.F. 1980. An experimental analysis of reproductive effort and cost in the Japanese medaka, *Oryzias latipes.* Ecology 61: 282-292.

Hoar, W.S. 1969. Reproduction, p. 1-72. *In* W.S. Hoar and D.J. Randall (eds.) Fish physiology. Vol. III. Academic Press, New York.

Hoar, W.S. and Y. Nagahama. 1978. The cellular sources of sex steroids in teleost gonads. Ann. Biol. Anim. Biochim. Biophys. 18: 893-898.

Hodgkiss, I.J. and H.S.H. Man. 1977a. Stock density and mortality assessment of *Sarotherodon mossambicus* (Cichlidae) in Plover Cove Reservoir, Hong Kong. Environ. Biol. Fishes 1: 171-180.

Hodgkiss, I.J. and H.S.H. Man. 1977b. Age composition, growth and body condition of the introduced *Sarotherodon mossambicus* (Cichlidae) in Plover Cove Reservoir, Hong Kong. Environ. Biol. Fishes 2: 35-44.

Hodgkiss, I.J. and H.S.H. Man. 1978. Reproductive biology of *Sarotherodon mossambicus* (Cichlidae) in Plover Cove Reservoir, Hong Kong. Environ. Biol. Fishes 3: 287-292.

Hofstede, A.E. 1957. Report to the government of Israel on inland fisheries. FAO/UNDP (Technical Assistance) Reports on Fisheries TA Reports 327. 56 p. FAO, Rome.

Hogg, R.G. 1976. Established exotic cichlid fishes in Dade County, Florida. Fla. Sci. 39(2): 97-103.

Holden, M.J. 1963. The population of fish in dry season pools of the R. Sokoto. Fish. Publ. Colon. Off. H.M.S.O., London 19: 1-58.

Hopkins, K.D. 1979. Production of monosex tilapia fry by breeding sex reversed fish. Auburn University, Auburn, Alabama. 44 p. Ph.D. dissertation.

Hopkins, K.D., W.L. Shelton and C.R. Engle. 1979. Estrogen sex reversal of *Tilapia aurea.* Aquaculture 18: 263-268.

Hopson, A.J. 1972. A study of the Nile Perch in L. Chad. Overseas Res. Publ. No. 19. 23 p. H.M.S.O., London.

Hora, S.L. and T.V.R. Pillay. 1962. Handbook of fish culture in the Indo-Pacific region. FAO Fish. Biol. Tech. Pap. 14. 204 p.

Horn, H.S. 1978. Optimal tactics of reproduction and life-history, p. 411-429. *In* J.R. Krebs and N.B. Davies (eds.) Behavioural ecology. An evolutionary approach. Blackwell Scientific Publ., Oxford.

Huet, M. 1959. Profiles and biology of Western European streams as related to fish management. Trans. Amer. Fish. Soc. 88: 155-163.

Huet, M. 1970. Traité de pisciculture. Fourth ed. Editions Ch. de Wyngaert, Brussels.

Huet, M. 1971. Textbook of fish culture: breeding and cultivation of fish. Fishing News Books Ltd., West Byfleet, Surrey, England. Translated from French by H. Kahn.

Hughes, D.G. 1977. Progress report on fisheries development in El Salvador, August 1974-May 1976. Res. Dev. Ser. 15. 16 p. Int. Cent. Aquaculture, Auburn Univ., Auburn, Alabama.

Hulata, G., S. Rothbard and G.W. Wohlfarth. 1980. Genetic approach to the production of all-male progeny of *Tilapia.* Paper presented at the German-Israeli symposium on aquaculture, March 10-11, 1980. Shefayim, Israel.

Hyder, M. 1969. Gonadal development and reproductive activity of the Cichlid fish *Tilapia leucosticta* (Trewavas) in an equatorial lake. Nature (London) 224: 1112.

Hyder, M. 1970a. Gonadal and reproductive patterns in *Tilapia leucosticta* (Teleostei:

Cichlidae) in an equatorial lake: Lake Naivasha (Kenya). J. Zool. 162: 179-195.

Hyder, M. 1970b. Histological studies on the testes of pond specimens of *Tilapia nigra* (Gunther) (Pisces: Cichlidae) and their implications on the pituitary-testis relationship. Gen. Comp. Endocrinol. 14: 198-211.

Hyder, M. 1972. Endocrine regulation of reproduction in *Tilapia*. Gen. Comp. Endocrinol. Suppl. 3: 729-740.

Hyder, M. and M.A. Kirschner. 1969. Detection and estimation of testosterone in the testes of *Tilapia leucosticta* (Pisces, Cichlidae). J. Endocrinol. 44: 281-282.

Hyder, M., A.V. Shah and M.A. Kirschner. 1970. Effect of chorionic gonadotropin in testicular histology and testosterone production in *Tilapia leucosticta* (Teleostei: Cichlidae). Endocrinology 97: 819-922.

Hyder, M., A.V. Shah and A. Stockell-Hartree. 1979. Methallibure studies on Tilapia. III. Effects of Tilapia partially purified gonadotropic fractions on the testes of methallibure treated *Sarotherodon spilurus* (= *Tilapia nigra*). Gen. Comp. Endocrinol. 39: 475-480.

Hyder, M., A.V. Shah, C.M. Campbell and S. Dadzie. 1974. Methallibure studies on Tilapia. II. Effect of Tilapia pituitary homogenate (TPH), human chorionic gonadotropin (HCG) and testosterone propionate (TP) on the testes of methallibure-treated *Tilapia nigra*. Gen. Comp. Endocrinol. 23: 245-255.

Hynes, H.B.N. 1970. The ecology of running waters. Liverpool University Press, Liverpool

Ibrahim, K.H., T. Nozawa and R. Lema. 1975. Preliminary observations of cage culture of *Tilapia esculenta* Graham and *Tilapia zillii* (Gervais) in Lake Victoria, at the Freshwater Fisheries Institute, Nyegezi, Tanzania. Afr. J. Trop. Hydrobiol. Fish. 4(1): 121-125.

Idler, D.R. and T. Bun Ng. 1979. Studies on two types of gonadotropins from both salmon and carp pituitaries. Gen. Comp. Endocrinol. 38: 421-440.

Iles, T.D. 1971. Ecological aspects of growth in African cichlid fish. J. Cons., Cons. Int. Explor. Mer. 33: 362-384.

Iles, T.D. 1973. Dwarfing or stunting in the genus *Tilapia* (Cichlidae) a possibly unique recruitment mechanism, p. 247-254. *In* B.B. Parish (ed.) Fish stocks and recruitment. Rapp. P.-V. Réun. Cons. Int. Explor. Mer, Charlottenlund.

Iles, T.D. and M.J. Holden. 1969. Bi-parental and mouthbrooding in *Tilapia galilaea* (Pisces, Cichlidae). J. Zool. 158: 327-333.

Ilg, L. 1952. Über larvale Haftorgane bei Teleostern. Zool Jahrb. Abt. Anat. Ontog. Tiere. 72: 577-600.

Ita, E.O. 1976. Observations on the present status and problems of inland fish culture in some northern states of Nigeria and preliminary results of cage culture experiments in Kainji Lake, Nigeria. FAO/CIFA Tech. Pap. 4 (Suppl. 1): 241-247.

Ita, E.O. 1978. An analysis of fish distribution in Kainji Lake, Nigeria. Hydrobiologia 58(3): 233-244.

JFRO (Joint Fisheries Research Organization, Northern Rhodesia and Nyasaland). Annual Reports. Government Printer, Lusaka (replaced by Zambia Fisheries Research Bulletin from 1965).

Jackson, P.B.N. and K.H. Rogers. 1976. Cabora Bassa fish populations before and during the first filling phase. Zool. Afr. 11(2): 373-397.

Jalabert, B. 1976. *In vitro* oocyte maturation and ovulation in rainbow trout *(Salmo gairdneri)*, northern pike *(Esox lucius)*, and goldfish *(Carassius auratus)*. J. Fish. Res. Board Can. 33: 974-988.

Jalabert, B., B. Breton and C. Bry. 1980. Evolution de la gonadotropine plasmatique t-GtH synchronisation des ovulations par injection de 17α-hydroxy 20β dihydroprogestérone chez la Truite arc-en-ciel (*Salmo gairdneri* R.). C.R. Acad. Sci. Paris, Ser. D. (In press)

Jalabert, B., B. Breton and R. Billard. 1974. Dosage biologique des hormones gonadotropes des poissons par le test de maturation *in vitro* des ovocytes de Truite. Ann. Biol. Anim. Biochim. Biophys. 14: 217-228.

Jalabert, B., J. Moreau, P. Planquette and R. Billard. 1974. Determinisme de sexe chez *Tilapia macrochir* et *Tilapia nilotica*. Action de la methyltestosterone dans l'alimentation des alevins sur la differentiation sexuelle; proportion des sexes dans la descendance des males "inversés". Ann. Biol. Anim. Biochim. Biophys. 14: 729-739.

Jalabert, B., P. Kammacher and P. Lessent. 1971. Sex determination in *Tilapia macrochir* x *Tilapia nilotica* hybrids. Investigations on sex ratios in first generation x parent crossings. Ann. Biol. Anim. Biochim. Biophys. 11: 155-165. (In French with English summary)

Jensen, G.L. and W.L. Shelton. 1979. Effects of estrogens on *Tilapia aurea*: implications for production of monosex genetic male tilapia. Aquaculture 16: 233-242.

Jensen, K.W. 1957. Determination of age and growth of *T. nilotica, T. galilaea, T. zillii* and *Lates niloticus* by means of their scales. K. Nor. Vidensk. Selsk, Forh. 30: 150-157.

Job, S.V. 1969a. The respiratory metabolism of *Tilapia mossambica* (Teleostei). I. The effect of size, temperature and salinity. Mar. Biol. (Berl.) 2(2): 121-126.

Job, S.V. 1969b. The respiratory metabolism of *Tilapia mossambica* (Teleostei). II. The effect of size, temperature, salinity, and partial pressure of oxygen. Mar. Biol. (Berl.) 3: 222-226.

Johnels, A.G. 1954. Notes on fishes from the Gambia River. Ark. Zool. 6(2): 327-411.

Johnson, R.P. 1974. Synopsis of biological data on *Sarotherodon galilaeus*. FAO Fish. Synop. No. 90. 42 p.

Jones, A.J. 1972. The early development of substrate-brooding cichlids (Teleostei: Cichlidae) with a discussion of a new system of staging. J. Morphol. 136: 255-272.

Jordan, D.T. and F.A. Pagán. 1973. Developments in cage culture of *Tilapia aurea* in a rock quarry pond in Puerto Rico. Paper presented at the 10th Annual Meeting of the Association of the Islands Marine Laboratories of the Caribbean, 4-7 October 1973. San Juan, Puerto Rico.

Jubb, R.A. 1974. The distribution of *Tilapia mossambica* Peters, 1852 and *Tilapia mortimeri* Trewavas, 1966 in Rhodesian waters. Arnoldia (Rhodesia) 6(25): 1-14.

Jubb, R.A. and P. Skelton. 1974. Note on *Sarotherodon placida* (Trewavas 1941). Arnoldia (Rhodesia) 7(4): 1-4.

Jubb, R.A. and I.G. Gaigher. 1971. Check list of the fishes of Botswana. Arnoldia (Rhodesia) 5(7): 1-22.

Jubb, R.A. and H.M. Jubb. 1967. Freshwater fishes of southern Africa. A.A. Balkema, Cape Town.

Jubb, R.A. and F.L. Farquharson. 1965. The freshwater fishes of the Orange River drainage basin. S. Afr. J. Sci. 61: 118-125.

Kalman, K.D. 1965. Genetic and geography of sex determination in the poeciliid fish, *Xiphophorus maculatus.* Zoologica (N.Y.) 50: 151-190.

Kamal Pasha, S.M. 1964. The anatomy and histology of the alimentary canal of a herbivorous fish *Tilapia mossambica* Peters. Proc. Indian Acad. Sci., Sect. B 59-60: 340-349.

Kapetsky, J.M. 1974. The Kafue River floodplain: an example of pre-impoundment potential for fish production, p. 497-523. *In* E.K. Balon and A.G. Coche (eds.) Lake Kariba: a man-made tropical ecosystem in Central Africa. Monographiae Biologicae 24. Dr. W. Junk bv Publ., The Hague.

Katz, Y. and B. Eckstein. 1974. Changes in steroid concentration in blood of female *Tilapia aurea* (Teleostei, Cichlidae) during initiation of spawning. Endocrinology 95: 963-967.

Katz, Y., B. Eckstein, R. Ikan and R. Gottlieb. 1971. Estrone and estradiol-17β in the ovaries of *Tilapia aurea* (Teleostei, Cichlidae). Comp. Biochem. Physiol. 40B: 1005-1009.

Keenleyside, M.H.A. 1979. Diversity and adaptation in fish behaviour. Springer-Verlag, Berlin.

Kelly, H.D. 1956. Preliminary studies on *Tilapia mossambica* Peters relative to experimental pond culture. Proc. Annu. Conf. Southeast Assoc. Game Fish. Comm. 10: 139-149.

Kenmuir, D.H.S. 1973. Observations on a breeding pair of *Tilapia rendalli* (Boulenger 1896) in an experimental tank at Lake Kariba Fisheries Research Institute. Hydrobiologia 43(3-4): 365-370.

Khalil, L.F.T. 1971. Check list of the helminth parasites of African freshwater fishes. Technical Communication No. 42. 80 p. Commonwealth Institute of Helminthology, St. Albans.

Khoo, H.K. and E.S.P. Tan. 1980. Review of rice-fish culture in Southeast Asia, p. 1-14. *In* R.S.V. Pullin and Z.H. Shehadeh (eds.) Integrated agriculture-aquaculture farming systems. ICLARM Conference Proceedings 4. International Center for Living Aquatic Resources Management, Manila and the Southeast Asian Regional Center for Graduate Study and Research in Agriculture (College), Los Baños, Laguna, Philippines.

King, J.A. and R.P. Millar. 1979. Heterogeneity of vertebrate luteinizing hormone-releasing hormone. Science 206: 67-69.

Kirilenko, H.S., G.B. Mel'nikov, L.R. Grinberg and N.I. Nazimirova. 1975. The digestibility of the crude protein and neutral fat of some microalgae by *Tilapia mossambica*. J. Ichthyol. 15: 151-155.

Kirk, R.G. 1972. A review of the recent developments in tilapia culture with special reference to fish farming in the heated effluents of power stations. Aquaculture 1: 45-60.

Koestler, A. 1967. The ghost in the machine. Hutchinson, London.

Koplin, S.J., J.G. Woiwode and S.E. Jenkins. 1977. The feasibility of commercial production of sex-reversed *Sarotherodon mossambicus* utilizing flow-through tanks at three hatcheries. Unpublished Report, U.S. Peace Corps Volunteers. 6 p.

Kornfield, I.L., U. Ritte, C. Richler and J. Wahrman. 1979. Biochemical and cytological differentiation among cichlid fishes of the sea of Galilee. Evolution 33: 1-14.

Korycka, A. and B. Zdanowski. 1980. Some aspects of the effect of cage fish culture on lakes, with special reference to heated lakes. Paper presented at the EIFAC symposium on new developments in the utilization of heated effluents and of recirculation systems for intensive aquaculture, 28-30 May 1980. Stavanger, Norway. 10 p.

Kosswig, C. and M. Oktay. 1955. Die Geschlechtsbestimmung bei den Xiphophorini (Neue Tatsachen und neue Deutungen). Istanbul Univ. Fen Fak. Hidrobiol. Arastirma Enst. Yayin. B2: 133-156.

Krebs, C.J. 1978. Ecology: the experimental analysis of distribution and abundance. 2nd edition. Harper and Row, New York.

Kryzhanovsky, S.G. 1956. Development of clupeid fishes. Tr. Inst. Morfol. Zhibotn. Severtsova 17: 1-255. (In Russian)

Kuo, H. 1969. Notes on hybridization of tilapia. Jt. Comm. Rural Reconstr. (Chinese-American) Fish. Ser. 8: 116-117.

Kuroda, T. 1977. Japanese aquaculture with thermal water from power plants—present conditions and problems, p. 6. *In* Atoms in Japan. Japanese Atomic Industrial Forum, Inc., Tokyo.

Kutty, M.N. 1972. Respiratory quotient and ammonia excretion in *Tilapia mossambica*. Mar. Biol. (Berl.) 16(2): 126-133.

Lagler, K.F., J.E. Bardach, R.R. Millet and D.R.M. Passino. 1977. Ichthyology. Wiley, New York.

Lamarque, P. and Cl. Closset. 1975. Etude des conditions de pêche à l'électricité dans les eaux tropicales. II. Etudes conduites au Malawi (1974). Bull. Cent. Etud. Rech. Sci. Biarritz 10(4): 658-684.

Lamarque, P., Y. Therézien and N. Charlon. 1975. Etude des conditions de pêche à l'électricité dans les eaux tropicales. I. Etudes conduites à Madagascar et en Zambie (1972). Bull. Cent. Etud. Rech. Sci. Biarritz 10(3): 403-554; 10(4): 575-665.

Landau, R. 1979. Growth and population studies on *Tilapia galilaea* in Lake Kinneret. Freshwater Biol. 9: 23-32.

Langford, F.H., F.J. Hare and R.D. Casaway. 1978. Status and harvest of introduced *Tilapia aurea* in Florida Lakes, p. 102-108. *In* R.O. Smitherman, W.L. Shelton, and J.H. Grover (eds.) Culture of exotic fishes symposium proceedings. Fish Culture Section, American Fisheries Society, Auburn, Alabama.

Lanzing, W.J.R. 1974. Sound production in the cichlid *Tilapia mossambica* Peters. J. Fish. Biol. 6: 341-347.

Lanzing, W.J.R. 1978. Effect of methallibure on gonad development and carotenoid content of the fins of *Sarotherodon mossambicus (Tilapia mossambica)*. J. Fish. Biol. 12: 181-185.

Lanzing, W.J.R. and C.C. Bower. 1974. Development of colour patterns in relation to behaviour in *Tilapia mossambica* (Peters). J. Fish Biol. 6: 29-41.

Lanzing, W.J.R. and D.R. Higginbotham. 1976. Scanning microscopy of pharyngeal and oral teeth of the teleost *Tilapia mossambica*. Hydrobiologia 48(2): 137-139.

Latif, A.F.A. 1976. Fishes and fisheries of Lake Nasser, p. 299-307. *In* J. Rzoska (ed.) The Nile: biology of an ancient river. Monographiae Biologicae 29. Dr. W. Junk bv Publ., The Hague.

Lauenstein, P.C. 1978. Intensive culture of Tilapia with geothermally heated water, p. 82-85. *In* R.O. Smitherman, W.L. Shelton and J.H. Grover (eds.) Culture of exotic fishes symposium proceedings. Fish Culture Section, American Fisheries Society, Auburn, Alabama.

Lauzanne, L. 1972. Régimes alimentaires des principales espèces de poissons de l'archipel oriental du lac Tchad. Verh. Int. Ver. Theor. Angew. Limnol. 18: 636-646.

Lauzanne, L. 1978a. Croissance de *S. galilaeus* dans le lac Tchad. Cybium 3: 5-14.

Lauzanne, L. 1978b. Étude quantitative de l'alimentation de *Sarotherodon galilaeus* (Pisces, Cichlidae) du lac Tchad. Cah. O.R.S.T.O.M. Sér. Hydrobiol. 12: 71-81.

Lauzanne, L. and A. Iltis. 1975. La sélection de la nourriture chez *T. galilaea* (Pisces: Cichlidae) du lac Tchad. Cah. O.R.S.T.O.M. Sér. Hydrobiol. 9(3): 193-199.

Lazard, J. 1973. Experiments with mineral fertilizer (phosphate). Cent. Tech. For. Trop. Nogent-sur-Marne, France, Notes Doc. June: 1-38. (In French)

Le, L.Q., G.D. Nguyen and V.Q. Nguyen. 1961. First statement on the influence of temperature on *Tilapia mossambica* Peters, acclimatized in Vietnam from 1951. Zh. Obshch. Biol. 22(6): 444-451. (In Russian with English summary)

Le Cren, E.D. 1951. The length-weight relationship and seasonal cycle in gonad weight and condition in perch *(Perca fluviatilis)*. J. Anim. Ecol. 20: 201-219.

Le Cren, E.D. and R.H. Lowe-McConnell. 1980. The functioning of freshwater ecosystems. IBP Synthesis, Vol. 22. Cambridge Univ. Press, Cambridge and New York.

Le Roux, P.J. 1956. Feeding habits of young of four species of *Tilapia*. S. Afr. J. Sci. 53(2): 33-37.

Lee, J.C. 1979. Reproduction and hybridization of three cichlid fishes, *Tilapia aurea* (Steindachner), *T. hornorum* (Trewavas) and *T. nilotica* (Linnaeus) in aquaria and in plastic pools. Auburn University, Auburn, Alabama. 84 p. Ph.D. dissertation.

Lelek, A. 1968. The vertical distribution of fishes in the Ebo Stream and notes to the fish occurrence in the Lake Bosumtwi, Ashanti, Ghana. Zool. Listy 17: 245-252.

Lema, R.B.G. and K.H. Ibrahim. 1975. Observations on the introduction of *Tilapia andersonii* (Castelnau) into Tanzania from Zambia. FAO/CIFA Tech. Pap. 4 (Suppl. 1): 433-439.

Lessent, P. 1968. Hybridization of the genus *Tilapia* at the Fish Culture Research Station at Bouake, Ivory Coast. FAO Fish. Rep. 44(4): 148-159. (In French with English abstract)

Lewis, D.S.C. 1974. The effects of the formation of L. Kainji (Nigeria) upon the indigenous fish population. Hydrobiologia 45: 281-301.

Lewontin, R.C. 1965. Selection for colonizing ability, p. 77-94. *In* H.G. Baker and G.L. Stebbins (eds.) The genetics of colonizing species. Academic Press, New York.

Li, L.K., Z.T. Nguyen and W.K. Nguyen. 1961. The first results of investigation of the effect of temperature on the fish *Tilapia mossambica* Peters acclimatized in Vietnam since 1961. Zh. Obshch. Biol. 22(6): 444-451.

Li, Y-P. 1978. Aquaculture in Taiwan. Chinese-American Joint Commission on Rural Reconstruction, Fisheries Series, Taipei. 10 p.

Liao, P.B. and R.D. Mayo. 1974. Intensified fish culture combining water reconditioning with pollution abatement. Aquaculture 3: 61-85.

Liem, K.F. 1973. Evolutionary strategies and morphological innovations: cichlid pharyngeal jaws. Syst. Zool. 22: 425-441.

Livni, N. 1971. Ovarian histochemistry of the fishes *Cyprinus carpio*, *Mugil capito* and *Tilapia aurea* (Teleostei). Histochem. J. 3: 405-414.

Loiselle, P.V. 1977. Colonial breeding by an African substratum spawning cichlid fish, *Tilapia zillii* (Gervais). Biol. Behav. Ser. 2(2): 129-142.

Loiselle, P.V. and G.W. Barlow. 1978. Do fishes lek like birds?, p. 31-76. *In* E. Reese and F.J. Lighter (eds.) Contrasts in behavior. McGraw-Hill, New York.

Lombard, G.L. 1959. A preliminary guide to fish farming in the Transvaal. Fauna Flora

(Transvaal) 10: 17-69. (In Dutch and English)

Lotan, R. 1960. Adaptability of *Tilapia nilotica* to various saline conditions. Bamidgeh 12: 96-100.

Lovshin, L.L. 1975. Progress report on fisheries development in northeast Brazil. Res. Dev. Ser. No. 9. 11 p. Int. Cent. Aquaculture, Auburn Univ., Auburn, Alabama.

Lovshin, L.L. 1977. Progress report on fisheries development in northeast Brazil. Res. Dev. Ser. No. 14. 11 p. Int. Cent. Aquaculture, Auburn Univ., Auburn, Alabama.

Lovshin, L.L. 1980. Progress report on fisheries development in northeast Brazil. Res. Dev. Ser. No. 26. 15 p. Int. Cent. Aquaculture, Auburn Univ., Auburn, Alabama.

Lovshin, L.L. and A.B. da Silva. 1975. Culture of monosex and hybrid tilapia. FAO/CIFA Tech. Pap. 4 (Suppl. 1): 548-564.

Lovshin, L.L., A.B. da Silva and J.A. Fernândes. 1977. The intensive culture of all male hybrid of *Tilapia hornorum* (male) x *T. nilotica* (female) in Northeast Brazil. FAO Fish. Rep. 159(1): 162-179. (In Spanish with English abstract)

Løvtrup, S. 1974. Epigenetics—a treatise on theoretical biology. Wiley, New York.

Lowe, R.H. 1952. Report on the *Tilapia* and other fish and fisheries of Lake Nyasa, 1945-47. Fish. Publ. Colon. Off., H.M.S.O., London 1(2): 1-26.

Lowe, R.H. 1953. Notes on the ecology and evolution of Nyasa fishes of the genus *Tilapia*, with a description of *T. saka* Lowe. Proc. Zool. Soc. London 122: 1035-1041.

Lowe (McConnell), R.H. 1955a. New species of *Tilapia* from Lake Jipe and the Pangani River, E. Africa. Bull. Br. Mus. (Nat. Hist.) Zool. 2(12): 349-368.

Lowe (McConnell), R.H. 1955b. The fecundity of *Tilapia* species. East Afr. Agric. J. 21: 45-52.

Lowe (McConnell), R.H. 1956a. The breeding behaviour of *Tilapia* species in natural waters: observations on *T. karomo* Poll and *T. variabilis* Blgr. Behaviour 9: 140-163.

Lowe (McConnell), R.H. 1956b. Observations on the biology of *Tilapia* in L. Victoria, E. Africa. East Afr. Fish. Res. Org. Suppl. Publ. No. 1: 1-72.

Lowe (McConnell), R.H. 1957. Observations on the diagnosis and biology of *Tilapia leucosticta* Trewavas in East Africa (Pisces, Cichlidae). Rev. Zool. Bot. Afr. 55: 353-373.

Lowe (McConnell), R.H. 1958. Observations on the biology of *Tilapia nilotica* L. in East African waters. Rev. Zool. Bot. Afr. 57: 129-170.

Lowe-McConnell, R.H. 1959. Breeding behaviour patterns and ecological differences between *Tilapia* species and their significance for evolution within the genus *Tilapia* (Pisces: Cichlidae). Proc. Zool. Soc. London 132: 1-30.

Lowe-McConnell, R.H. 1969. The cichlid fishes of Guyana, S. America, with notes on their ecology and breeding behaviour. Zool. J. Linn. Soc. London 48: 255-302.

Lowe-McConnell, R.H. 1975. Fish communities in tropical freshwaters: their distribution, ecology and evolution. Longman, London.

Lowe-McConnell, R.H. 1979. Ecological aspects of seasonality in fishes of tropical waters. Symp. Zool. Soc. London No. 44: 219-241.

Loya, L. and L. Fishelson. 1969. Ecology of fish breeding in brackish water ponds near the Dead Sea (Israel). J. Fish Biol. 1(3): 261-278.

Maar, A., M.A.E. Mortimer and I. van der Lingen. 1966. Fish culture in Central East Africa. FAO, Rome.

MacArthur, R.H. and E.O. Wilson. 1967. The theory of island biogeography. Princeton University Press, Princeton, New Jersey.

McBay, L.G. 1961. The biology of *Tilapia nilotica* (Linnaeus). Proc. Annu. Conf. Southeast Assoc. Game Fish Comm. 15: 208-218.

McCauley, R.W. 1980. The final preferendum as an index of the temperature optimum for growth in fish. Paper presented at the FAO/EIFAC Symposium on new development in the utilization of heated effluents and of recirculation systems for intensive aquaculture, May 28-30, 1980. Stavanger, Norway, Doc. E/76. 15 p.

McKaye, K.R. 1977. Competition for breeding sites between the cichlid fishes of Lake Jiloa, Nicaragua. Ecology 58: 291-302.

McLarney, B. 1978. Family-scale aquaculture in Colombia. Professor develops the Peasant Fish Culture Unit. Commer. Fish Farmer Aquaculture News 4(6): 28-29.

Magid, A. and M.M. Babiker. 1975. Oxygen consumption and respiratory behaviour of three Nile fishes. Hydrobiologia 46: 359-367.

Mahon, R. and E.K. Balon. 1977. Fish production in L. Kariba, reconsidered. Environ.

Biol. Fishes 1(2): 215-218.

Man, H.S.H. and I.J. Hodgkiss. 1977a. Studies on the ichthyo-fauna in Plover Cove Reservoir, Hong Kong. I. Sequence of population changes. J. Fish Biol. 10: 493-503.

Man, H.S.H. and I.J. Hodgkiss. 1977b. Studies on the ichthyo-fauna in Plover Cove Reservoir, Hong Kong: feeding and food relations. J. Fish Biol. 11: 1-13.

Marek, M. 1970. Culture of male tilapia at high densities with feeding on pellets. Dag Hadio 10: 5-9. (In Hebrew)

Marek, M. 1975. Revision of supplementary feeding tables for pond fish. Bamidgeh 27: 57-64.

Marlier, G. 1953. Etude biogéographique du bassin de la Ruzizi, basée sur la distribution des poissons. Ann. Soc. R. Zool. Belg. 84: 175-224.

Marshall, B.E. 1979a. Fish populations and fisheries potential of L. Kariba. S. Afr. J. Sci. 75: 485-488.

Marshall, B.E. 1979b. Observations on the breeding biology of *Sarotherodon macrochir* (Boulenger) in Lake McIlwaine, Rhodesia. J. Fish Biol. 14: 419-424.

Marshall, J.A. 1972. Influence of male sound production on oviposition in female *Tilapia mossambica* (Pisces: Cichlidae). Bull. Ecol. Soc. Amer. 53: 29.

Maruyama, T. 1958. An observation on *Tilapia mossambica* in ponds referring to the diurnal movement with temperature change. Bull. Freshwater Fish. Res. Lab., Tokyo 8(1): 25-32.

Maruyama, T. and R. Ishida. 1976. Effect of water depth in net cages on the growth and body shape of *Tilapia mossambica*. Bull. Freshwater Fish. Res. Lab., Tokyo 26(1): 11-19. (In Japanese with English summary)

Mawdesley-Thomas, L.E. 1972. Some tumours of fish. Symp. Zool. Soc. London 30: 191-283.

May, R.M., editor. 1976. Theoretical ecology: principles and applications. Blackwell Scientific Publ., Oxford.

Maynard Smith, J. 1977. Parental investment: a prospective analysis. Anim. Behav. 25: 1-9.

Meecham, K. 1975. Aquaculture in Malawi. Paper presented at the FAO/CIFA Symposium on Aquaculture in Africa, 30 Sept.-2 Oct. 1975. Accra, Ghana. CIFA/75/SC1. 6 p. FAO, Rome.

Mélard, Ch. and J.C. Philippart. 1980. Pisciculture intensive de *Sarotherodon niloticus* dans les effluents thermiques d'une centrale nucléaire en Belgique. Paper presented at the FAO/EIFAC Symposium on new development in the utilization of heated effluents and of recirculation systems for intensive aquaculture, May 28-30, 1980. Stavanger, Norway, EIFAC/80/Symp/Doc. E/11. 20 p.

Meschkat, A. 1967. The status of warm-water fish culture in Africa. FAO Fish. Rep. 44(2): 88-122.

Meyer, F.P. 1970. Seasonal fluctuations in the incidence of disease on fish farms. Spec. Publ. Amer. Fish. Soc. 5: 21-30.

Miller, J.W. and F.A. Pagán. 1973. Observations on sea cage culture of *Tilapia aurea* in waters off La Parguera, Puerto Rico. Paper presented at the 10th Annual Meeting of the Association of the Islands Marine Laboratories of the Caribbean, 4-7 October 1973. San Juan, Puerto Rico.

Miller, J.W. and D.I. Ballantine. 1974. Opercular algal growth on the cichlid fish *Tilapia aurea*, cultured in sea water. Aquaculture 4: 93-95.

Mires, D. 1973. A hatchery for forced spawning at Kibbutz Ein Hamifrats. Bamidgeh 25: 72-84.

Mires, D. 1974. On the high percent of tilapia males encountered in captive spawning and the effect of temperature on this phenomenon. Bamidgeh 26: 3-11.

Mires, D. 1977. Theoretical and practical aspects of the production of all-male *Tilapia* hybrids. Bamidgeh 29: 94-101.

Miranova, N.V. 1969. The biology of *Tilapia mossambica* Peters under natural and laboratory conditions. Probl. Ichthyol. 9: 506-514.

Miranova, N.V. 1977. Energy expenditure on egg production in young *T. mossambica* and their influence of maintenance conditions on their reproductive intensity. J. Ichthyol. 17: 627-633.

Moav, R., G. Wohlfarth, G.L. Schroeder, G. Hulata and H. Barash. 1977. Intensive polyculture of fish in freshwater ponds. 1. Substitution of expensive feeds by liquid cow manure. Aquaculture 10: 25-43.

Monfort, A. and J. Cl. Ruwet. 1968. Patrons moteurs et mouvements d'expressions chez les *Tilapia rendalli* (Boulenger 1896) et *tholloni congica* Poll et Thys 1960 (Poissons Cichlides). Rev. Zool. Bot. Afr. 77: 97-112.

Monfort-Braham, N. and J. Voss. 1969. Contribution à l'éthologie des poissons Cichlides: *Tilapia tholloni* (Sauvage 1884). Ann. Soc. R. Zool. Belg. 99: 59-68.

Moreau, J. 1971. Biologie comparée de *Tilapia rendalli* (Boulenger) (Pisces, Cichlidae) au lac Itasy et au lac de Montasoa. Cah. O.R.S.T.O.M. Sér. Hydrobiol. 5: 3-52.

Moreau, J. 1975. Interpretation trial of the annuli observed on scales of *Tilapia nilotica* and *Tilapia rendalli* of the lakes of Malagasy Highlands. I. Observations in Lake Itasy and in Lake Mantesoa. Ann. Hydrobiol. 5(2): 83-93.

Moreau, J. 1979a. Les introduction d'espèces étrangères dans les eaux continentales Africaines. Interêt et limites. *In* C. Levêque, M. Bruton and G.W. Ssontongo (eds.) The African freshwater fishes, their biology, recent trends. Nairobi SIL UNEP Symposium, December 1979.

Moreau, J. 1979b. Biologie et évolution des peuplements de Cichlides (Pisces) introduits dans les lacs malgaches d'altitude. Institut National Polytechnique de Toulouse, Toulouse Cedex, France. 301 + 41 p. Ph.D. dissertation.

Morgan, A. and M. Kalk. 1970. Seasonal changes in the waters of Lake Chilwa (Malawi). in a drying phase, 1966-68. Hydrobiologia 36: 81-103.

Morgan, P.R. 1972. Causes of mortality in the endemic Tilapia of Lake Chilwa (Malawi). Hydrobiologia 40(1): 101-119.

Moriarty, D.J.W. 1973. The physiology of digestion of blue-green algae in the cichlid fish *Tilapia nilotica*. J. Zool. 171: 25-40.

Moriarty, C.M. and D.J.W. Moriarty. 1973a. Quantitative estimation of the daily ingestion of phytoplankton by *Tilapia nilotica* and *Haplochromis nigripinnis* in Lake George, Uganda. J. Zool. 171: 15-24.

Moriarty, D.J.W. and C.M. Moriarty. 1973b. The assimilation of carbon from phytoplankton by the two herbivorous fishes: *Tilapia nilotica* and *Haplochromis nigripinnis*. J. Zool. 171: 41-56.

Moriarty, D.J.W., J.P.E.C. Darlington, I.G. Dunn, C.M. Moriarty and M.P. Tevlin. 1973. Feeding and grazing in Lake George, Uganda. Proc. R. Soc. London, Ser. B 184: 299-319.

Morissens, P. 1977. Male hybrid tilapia production in Israel. Cent. Tech. For. Trop. Nogent-sur-Marne, France. Notes Doc. 14 : 7-20. (In Portuguese with English summary)

Mortimer, M.A.E. 1960. Observations on the biology of *Tilapia andersonii* (Castelnau), (Pisces, Cichlidae), in Northern Rhodesia. Rep. Jt. Fish. Res. Org. (Lusaka) 9: 41-67.

Mugiya, Y. 1978. Effects of estradiol-Beta-17 on bone and otolith calcification in goldfish, *Carassius auratus*. Bull. Jap. Soc. Sci. Fish. 44: 1217-1221.

Munro, J.L. 1967. The food of a community of East African freshwater fishes. J. Zool. 151(3): 389-415.

Myers, G.S. 1938. Freshwater fishes and West Indian zoogeography. Smithson. Rep. for 1927 (Publ. 3465): 339-364.

Myrberg, A.A., E. Kramer and P. Heinecke. 1965. Sound production by Cichlid fishes. Science 149: 555-558.

Nagase, G. 1964. Contribution of the physiology of digestion in *Tilapia mossambica*. Digestive enzymes and the effect of diet on their activity. Z. Vergl. Physiol. 49: 270-284.

Naik, I.U. 1973. Studies on *Tilapia mossambica* Peters in Pakistan. Agric. Pak. 24(1): 47-76.

Nakamura, M. and H. Takahashi. 1973. Gonadal sex differentiation in *Tilapia mossambica* with special regard to the time of estrogen treatment effective in inducing complete feminization of genetic males. Bull. Fac. Fish. Hokkaido Univ. 24: 1-13.

Nanne, H. 1979. Un proyecto piscicola dirigido hacia la solución de un problema socio economico en Costa Rica. Rev. Lat. Amer. Acuicultura 1: 17-18.

Nei, M. 1972. Genetic distance between populations. Amer. Nat. 106: 282-292.

Neil, E.H. 1966. Observations on the behaviour of *Tilapia mossambica* (Pisces: Cichlidae) in Hawaiian ponds. Copeia 1: 50-56.

Nelson, S.G., A.C. Anderson, M.H. Momeni and R.R. Neo. 1976. Attempted sterilization of sexually undifferentiated fry of *Tilapia zillii* by cobalt gamma ray irradiation. Prog. Fish-Cult. 38: 131-134.

Nicholls, T.J. and G.P. Graham. 1972. The ultrastructure of lobule boundary cells and Leydig cell homologs in the testis of cichlid fish, *Cichlasoma nigrofasciatum*. Gen. Comp. Endocrinol. 19: 133-146.

Nicholls, T.J. and G. Maple. 1972. Ultrastructural observations on possible sites of steroid biosynthesis in the ovarian follicular epithelium of two species of Cichlid fishes, *Cichlasoma nigrofasciatum* and *Haplochromis multicolor*. Z. Zellforsch. Mikrosk. Anat. 128: 317-335.

Nigrelli, R.F. and H. Vogel. 1963. Spontaneous tuberculosis in fishes and in other cold-blooded vertebrates with special reference to *Mycobacterium fortuitum* (Cruz) from fish and human lesions. Zoologica (N.Y.) 48: 130-143.

Noakes, D.L.G. 1978a. Social behaviour as it influences fish production, p. 360-382. *In* S.D. Gerking (ed.) Ecology of freshwater fish production. Blackwell Sci. Publ., Oxford.

Noakes, D.L.G. 1978b. Ontogeny of behavior in fishes: a survey and suggestions, p. 103-125. *In* G.M. Burghardt and M. Bekoff (eds.) Development of behavior: comparative and evolutionary aspects. Garland, New York.

Noakes, D.L.G. 1979. Parent-touching behavior by young fishes: incidence, function and causation. Environ. Biol. Fishes 4: 389-400.

Noakes, D.L.G. 1981. Comparative aspects of behavioral ontogeny: a philosophy from fishes, p. 491-508. *In* K. Immelmann, G.W. Barlow, L. Petrinovitch and M. Man (eds.) The interdisciplinary study of early development. Cambridge University Press, Cambridge.

Noakes, D.L.G. and G.W. Barlow. 1973a. Ontogeny of parent-contacting behavior in young *Cichlasoma citrinellum* (Pisces: Cichlidae). Behaviour 45: 221-225.

Noakes, D.L.G. and G.W. Barlow. 1973b. Cross-fostering and parent-offspring responses in *Cichlasoma citrinellum* (Pisces: Cichlidae). Z. Tierpsychol. 33: 147-152.

Noble, R.L., R.D. Germany and C.R. Hall. 1976. Interaction of blue tilapia and large-mouth bass in a power plant cooling reservoir. Proc. Annu. Conf. Southeast. Assoc. Game Fish Comm. 29: 247-251.

Nomura, H. 1976. Present development and future perspectives in the intensive and extensive culture in São Paulo State. Ciênc. Cult. (Sao Paulo) 28(10): 1097-1107.

Nomura, H. 1977. Principales especies de peces cultivadas en el Brazil. FAO Fish. Rep. 159 (Vol. 1): 211-219.

Noreña, S.J. 1977. Analisis ictiofaunistico para el desarrollo de la piscicultura en el occidente Colombiano. FAO Fish. Rep. 159 (Vol. 1): 266-278.

N'Zimasse, D. 1979. Première approche de pisciculture intensive en cages en Centrafrique. Mém. Fin Etud., Inst. Univ. Technol. Agron., Mbaiki (Bangui), Rép. Centafricaine. 34 p.

Odum, W.E. 1968. The ecological significance of fine particle selection by striped mullet *Mugil cephalus*. Limnol. Oceanogr. 13: 92-98.

Okorie, O.O. 1975. Environmental constraints to aquaculture development in Africa. Paper presented at the FAO/CIFA Symposium on Aquaculture in Africa, 30 Sept.-2 Oct. 1975. Accra, Ghana. CIFA/75/SE 7. 7 p. FAO, Rome.

Olivereau, M. 1977. Données récentes sur le contrôle endocrinien de la reproduction chez le Téléostéens. Invest. Pesq. 41: 69-94.

Olivereau, M. and J. Olivereau. 1979. Effect of estradiol-17β on the cytology of the liver, gonads and pituitary, and on plasma electrolytes in the female freshwater eel. Cell Tissue Res. 199: 431-454.

Oppenheimer, J.R. 1970. Mouthbreeding in fishes. Anim. Behav. 18: 493-503.

Oppenheimer, J.R. and G.W. Barlow. 1968. Dynamics of parental behaviour in the black-chinned mouthbreeder *Tilapia melanotheron* (Pisces: Cichlidae). Z. Tierpsychol. 25: 889-914.

Ottobo, F.O. 1976. The evaluation of the beach seine as a fishing gear for Tilapia harvest in Kainji Lake, Nigeria. Arch. Hydrobiol. 78(3): 375-382.

Pagán, F.A. 1969. Cage culture of tilapia. FAO Fish Cult. Bull. 2: 6.

Pagán, F.A. 1973. Potential for cage culture of the cichlid fish, *Tilapia aurea*. Paper presented at the 10th Annual Meeting of the Association of the Islands Marine Laboratories of the Caribbean, 4-7 October 1973. San Juan, Puerto Rico.

Pagán, F.A. 1975. Cage culture as a mechanical method for controlling reproduction of *Tilapia aurea* (Steindachner). Aquaculture 6: 243-247.

Pagán-Font, F.A. 1977. Desarrollo y estado de la acuicultura en Puerto Rico. FAO Fish. Rep. 159 (Vol. 3): 14-18.

Pantastico, J.B. and J.P. Baldia. 1979. Supplemental feeding of *Tilapia mossambica*, p. 587-593. *In* J.E. Halver and K. Tiews (eds.) Finfish nutrition and fishfeed technology. Vol. 1. Heenemann Verlagsgessellschaft mbH, Berlin.

Paperna, I. 1960. Studies on monogenetic trematodes in Israel. 2. Monogenetic trematodes of cichlids. Bamidgeh 12: 20-33.

Paperna, I. 1964. Metazoan parasite fauna of Israel inland water fishes. Bamidgeh 16: 3-66.

Paperna, I. 1969. Parasitic crustacea from fishes of the Volta basin and South Ghana. Rev. Zool. Bot. Afr. 80: 208-216.

Paperna, I. 1970. Infection by *Ichthyophthirius multifiliis* of fish in Uganda. Prog. Fish-Cult. 34: 162-164.

Paperna, I. 1974. Lymphocystis in fish from East African lakes. J. Wildl. Dis. 9(4): 331-335.

Paperna, I. and J.P. Thurston. 1968. Monogenean parasites of cichlid fishes in Uganda. Bull. Off. Int. Epiz. 69: 1197.

Pauly, D. 1975. On the ecology of a small West-African Lagoon. Ber. Dtsch. Wiss. Komm. Meeresforsch. 24(1): 46-52.

Pauly, D. 1976. The biology, fishery and potential for aquaculture of *Tilapia melanotheron* in a small West African lagoon. Aquaculture 7(1): 33-49.

Pauly, D. 1981. The relationships between gill surface area and growth performance in fish: a generalization of von Bertalanffy's theory of growth. Meeresforschung 28(4): 251-282.

Payne, A.I. 1970. An experiment on the culture of *Tilapia esculenta* (Graham) and *T. zillii* (Gervais) (Cichlidae) in fish ponds. J. Fish Biol. 3: 325-340.

Payne, A.I. 1974. Some characteristics of the fish fauna of a dam in the Lake Victoria region of Tanzania including the effect of multispecific stocking with *Tilapia* species. Afr. J. Trop. Hydrobiol. Fish. 3(2): 111-122.

Payne, A.I. 1978. Gut pH and digestive strategies in estuarine grey mullet (Mugilidae) and tilapia (Cichlidae). J. Fish Biol. 13(5): 627-629.

Payne, A.I. and R.I. Collinson. A comparison of the biological characteristics of *Sarotherodon aureus* (Stdr) with those of *S. niloticus* (L.) and other tilapia of the delta and Lower Nile. (In press)

Pellegrin, J. 1921. Les poissons des eaux douces de l'Afrique du Nord Francaise. Mém. Soc. Sci. Nat. Phys. Maroc 1(2): 1-216.

Perrone, M., Jr. 1978. Mate size and breeding success in a monogamous cichlid fish. Environ. Biol. Fishes 3(2): 193-201.

Perrone, M., Jr. and T.M. Zaret. 1979. Parental care patterns of fishes. Amer. Nat. 113(3): 351-361.

Peter, R.E. 1978. Regulating mechanisms of the anterior pituitary, p. 341-352. *In* P.J. Gaillard and H.H. Boer (eds.) Comparative endocrinology. Elsevier North-Holland Biomedical Press, Amsterdam.

Peters, H.M. 1959. Über die Beziehungen zwischen Eizahl und Eigewicht bei Fischen. Z. Naturforsch. 14.b: 584-592.

Peters, H.M. 1963. Eizahl, Eigewicht und Gelegeentwicklung in der Gattung *Tilapia* (Cichlidae, Teleostei). Int. Rev. Gesamt. Hydrobiol. 48: 547-576.

Peters, H.M. 1965. Über larvale Haftorgane bei *Tilapia* (Cichlidae, Teleostei) und ihre Rückbildung in der Evolution. Zool. Jahrb. Abt. Allg. Zool. Physiol. Tiere 71: 287-300.

Peters, H.M. 1971. Testis weight in *Tilapia* (Pisces: Cichlidae). Copeia 1: 13-17.

Peters, H.M. 1973. Anatomie und Entwicklungsgeschichte des Lateralissystems von *Tilapia* (Pisces, Cichlidae). Z. Morphol. Tiere 74: 89-161.

Peters, H.M. 1979. Warum ich nicht fur *Sarotherodon* bin. Dtsch. Cichl. Ges. Info. 10(8): 141-146.

Peters, H.M. and S. Berns. 1978. Über die Vorgeschichte der maulbrütenden Cichliden. Aquarien Mag. 1978(5): 211-217 and (7): 324-331.

Petr, T. 1968. The establishment of lacustrine fish populations in the Volta Lake in Ghana during 1964-1966. Bull. Inst. Fondam. Afr. Noire, Ser. A. 30(1): 257-269.

Petr, T. 1969. Fish populations changes in the Volta Lake over the period January 1965-September 1966, p. 220-234. *In* L. Obeng (ed.) Man-made lakes: the Accra symposium. Ghana University Press, Accra.

Petr, T. 1975. Limnology and fisheries of the Nyumba ya Mungu, a man-made lake in Tanzania. Afr. J. Trop. Hydrobiol. Fish. 4(1): 39-50.

Philippart, J-C., Ch. Mélard and J-C. Ruwet. 1979. La pisciculture dans les effluents thermiques industriels. Bilan et perspectives d'une année de recherche à la centrale nucléaire de Tihange sur la Meuse, p. 779-791. *In* L. Calembert (ed.) Problématique et gestion des eaux intérieures. Ed. Derouaux, Liege.

Pianka, E.R. 1978. Evolutionary ecology. Harper and Row, New York.

Pickford, G.E. and J.W. Atz. 1957. The physiology of the pituitary gland of fishes. Zool. Soc., N.Y.

Pienaar, U.V. 1968. The freshwater fishes of the Krüger national park. Koedoe 11: 1-82.

Pillai, T.G. 1972. Pests and predators in coastal aquaculture: systems of the Indo-Pacific region, p. 456-470. *In* T.V.R. Pillay (ed.) Coastal aquaculture in the Indo-Pacific region. Fishing News Books Ltd., West Byfleet, Surrey, U.K.

Pillay, T.V.R. 1979. The state of aquaculture 1976, p. 1-10. *In* T.V.R. Pillay and W.A. Dill (eds.) Advances in aquaculture. Fishing News Books Ltd., Farnham, Surrey, England.

Piperno, Y. 1970a. Observations on the growth of tilapia on supplementary feed in storing pond conditions. Dag Hadio 10: 3-5. (In Hebrew)

Piperno, Y. 1970b. More on the growth of tilapia on supplementary feed in storing pond conditions. Dag Hadio 11: 1-3. (In Hebrew)

Planquette, P. 1974. Nile perch as a predator in tilapia ponds. FAO Aquaculture Bull. 6: 7.

Platt, S. and W.J. Hauser. 1978. Optimum temperature for feeding and growth of *Tilapia zillii*. Prog. Fish-Cult. 40: 105-107.

Polder, J. 1971. On gonads and reproductive behavior in the cichlid fish *Aequidens portalegrensis* (Hensel). Neth. J. Zool. 21: 265-365.

Poll, M. 1956. Poissons Cichlidae. Result. Sci. Explor. Hydrobiol. Lac Tanganika (1946-1947) 3(5B): 1-619.

Poll, M. 1957. Les genres des poissons d'eau douce de l'Afrique. Ann. Mus. R. Congo Belge. Ser. 8, Sci. Zool. 54: 1-191.

Poll, M. 1973. Nombre et distribution géographique des poissons d'eau douce africains. Bull. Mus. Nac. Hist. Nat. Paris (Ser. 3) Ecol. Gen. 6(150): 113-128.

Popma, T.J. 1978. Ensayo sobre el crecimiento de *Tilapia rendalli* Boulenger, enjaulada y alimentada con follaje de bore *(Alocasia macrorhiza)*. Cent. Piscic. Exp. Inf. Téc. 2: 43-50.

Popper, D. and T. Lichatowich. 1975. Preliminary success in predator contact of *Tilapia mossambica*. Aquaculture 5(2): 213-214.

Potts, W.T.W., M.A. Foster, P.P. Rudy and G.P. Howells. 1967. Sodium and water balance in the cichlid teleost, *Tilapia mossambica*. J. Exp. Biol. 47: 461-470.

Pretto-Malca, R.P. 1979. Methods of producing the female *Tilapia nilotica* x male *Tilapia hornorum* hybrid. Rev. Lat. Amer. Acuicultura 1: 29-32. (In Spanish with English abstract)

Prowse, G.A. 1972. Some observations on primary and fish production in experimental fish ponds in Malacca, Malaysia, p. 555-561. *In* Z. Kajak and A. Hillbricht-Ilkowska (eds.) Productivity problems of freshwaters. Polish Scientific Publishers, Warsaw and Krakow.

Pruginin, Y. 1965. Report to the Government of Uganda on the experimental fish culture project in Uganda, 1962-1964. FAO/UNDP (Technical Assistance). Reports on Fisheries. TA Reports 1960. 28 p. FAO, Rome.

Pruginin, Y. 1967. Report to the Government of Uganda on the experimental fish culture project in Uganda, 1965-66. FAO/UNDP (Technical Assistance). Reports on Fisheries. TA Reports 2446. 19 p. FAO, Rome.

Pruginin, Y. 1968. The culture of carp and Tilapia hybrids in Uganda. FAO Fish. Rep. 44(4): 223-229.

Pruginin, Y. and E.W. Shell. 1962. Separation of the sexes of *Tilapia nilotica* with a mechanical grader. Prog. Fish-Cult. 24: 37-40.

Pruginin, Y. and A. Arad. 1977. Fish farming in Malawi. FI:DP/MLW/71/516/10. 22 p. FAO, Rome.

Pruginin, Y., S. Rothbard, G. Wohlfarth, A. Halevy, R. Moav and G. Hulata. 1975. All-male broods of *Tilapia nilotica* x *T. aurea* hybrids. Aquaculture 6: 11-21.

Rabanal, H.R. and L.V. Hosillos. 1957. Control of less desirable fish competing with or harmful to desirable indigenous species in inland waters in the Philippines. Paper presented at the Indo-Pacific Fisheries Council, 7th Session, 13-27 May 1957. Bandung, Indonesia. IPFC/C 57/Tech./20. 23 p. FAO, Bangkok.

Rabelahatra, A. 1979. Rapport de Voyages d'Etudes sur L'aquaculture en Indonésie, Philippines et Indes. Rep. Dem. Malagasy, Proj. Développement Pêches Contin. et Aquaculture, Antananarivo. 68 p.

Radan, R.R. 1979. Tilapia: from nilotica and mossambica to a mutant called flamingo. Greenfields (Philippines) 9(10): 24-40.

Rappaport, A., S. Sarig and M. Marek. 1976. Results of tests of various aeration systems on the oxygen regime in Genosar experimental ponds and growth of fish there in 1975. Bamidgeh 28(3): 35-49.

Ray, L.E. 1978. Production of Tilapia in catfish raceways using geothermal water, p. 86-89. *In* R.O. Smitherman, W.L. Shelton and J.H. Grover (eds.) Culture of exotic fishes symposium proceedings. Fish Culture Section, American Fisheries Society, Auburn, Alabama.

Redner, B.D. and R.R. Stickney. 1979. Acclimation to ammonia by *Tilapia aurea*. Trans. Amer. Fish. Soc. 108: 383-388.

Reichle, D.F., R.V. O'Neill and W.F. Harris. 1975. Principles of energy and material exchange in ecosystems, p. 27-43. *In* W.H. van Dobben and R.H. Lowe-McConnel (eds.) Unifying concepts in ecology. Dr. W. Junk bv Publ., The Hague.

Reite, O.B., G.M.O. Maloiy and B. Aasehaug. 1974. pH, salinity and temperature tolerance of Lake Magadi *Tilapia*. Nature (London) 247: 315.

Ricardo, C.K. 1939. The fishes of Lake Rukwa. J. Linn. Soc. London, Zool. 40: 625-657.

Ricardo-Bertram, C.K. 1943. The fishes of the Bangweulu region. J. Linn. Soc. London, Zool. 41: 183-217.

Richards, W.J. 1976. Some comments on Balon's terminology of fish developmental intervals. J. Fish. Res. Board Can. 33: 1253-1254.

Ricklefs, R. 1973. Ecology. Chiron Press, Scranton, Pennsylvania.

Ridley, M. 1978. Paternal care. Anim. Behav. 26: 904-932.

Riedel, D. 1965. Some remarks on the fecundity of Tilapia (*T. mossambica* Peters) and its introduction into Middle Central America (Nicaragua) together with a first contribution towards the limnology of Nicaragua. Hydrobiologia 25: 357-388.

Rifai, S.A. 1979. The use of aquatic plants as feed for *Tilapia nilotica* in floating cages, p. 61-64. *In* Proc. Int. workshop on pen and cage culture of fish. International Development Research Centre, Canada and Aquaculture Dept. Southeast Asian Fisheries Development Center, Iloilo, Philippines.

Rifai, S.A. 1980. Control of reproduction of *Tilapia nilotica* using cage culture. Aquaculture 20: 177-185.

Roberts, R.J. 1975. Experimental pathogenesis of lymphocystis in the plaice *(Pleuronectes platessa)*, p. 431-442. *In* Wildlife diseases. Plenum Press, New York.

Roberts, R.J. and P. Matthiessen. 1979. Pathological findings in fishes of the Okavangu swamp, Botswana, following Endosulphan spraying. ODM Special report. UK Ministry of Overseas Development, London. 9 p.

Robertson, D.H. 1979. Host-parasite interactions between *Ichtyobodo necator* (Henneguy 1883) and farmed salmonids. J. Fish Dis. 2: 481-493.

Rodriguez, J.V. 1976. Primeros resultados del cultivo en jaulas flotantes de *Tilapia nilotica*. L. y *Tilapia mossambica* Peters, en Cuba. Contrib. Min. Ind. Pesq. Dir. Ramal de Aquicultura, Cuba. 13 p.

Rosenthal, H. 1976. Implication of transplantations to aquaculture and ecosystems. Paper presented at the FAO Technical Conference on Aquaculture, 26 May-2 June 1976. Kyoto, Japan. FIR: AQ/Conf/76/E.67. 19 p. FAO, Rome.

Rothbard, G., G. Hulata and J. Itzkovitch. 1980. Abnormalities in *Sarotherodon* larvae. J. Fish Dis. 3: 441-442.

Rothbard, S. 1979. Observations on the reproductive behavior of *Tilapia zillii* and several *Sarotherodon* spp. under aquarium conditions. Bamidgeh 31: 35-43.

Rothbard, S. and Y. Pruginin. 1975. Induced spawning and artificial incubation of *Tilapia*. Aquaculture 5: 315-321.

Rozin, P.M. and J. Mayer. 1961. Regulation of food intake in the goldfish. Amer. J. Physiol. 201: 968-974.

Rozin, P.M. and J. Mayer. 1964. Some factors influencing short-term food intake of the goldfish. Amer. J. Physiol. 206: 1430-1436.

Rüppell, E. 1852. Verz. Mus. Senckenberg. naturforsch. Ges. aufges. Sammlungen. IV. Fische. 40 p. Frankfurt-am-Main.

Russell-Hunter, W.D. 1970. Aquatic productivity: an introduction to some basic aspects of biological oceanography and limnology. Collier-Macmillan, London.

Russock, H.I. and M.W. Schein. 1977. Effect of age and experience on the filial behaviour of *Tilapia mossambica* fry (Pisces, Cichlidae). Behaviour 61(3-4): 276-303.

Ruwet, J.Cl. 1961. Contribution à l'études des problèmes piscicoles au lac de retenue de la Lufira (Katanga). Ed. FULREAC, Université de Liège.

Ruwet, J.Cl. 1961-62. Remarques sur le comportement de *Tilapia macrochir* Blgr et *T. melanopleura* Dum. (Pisces, Cichlidae) Ann. Soc. R. Zool. Belg. 92: 171-177.

Ruwet, J.Cl. 1962. La reproduction des *Tilapia macrochir* (Blgr) et *Tilapia melanopleura* (Dum.) au lac de barrage de la Lufira (Haut-Katanga). Rev. Zool Bot. Afr. 66: 244-271.

Ruwet, J.Cl. 1963a. *Tilapia melanopleura* (Dum.) poissons Cichlidès et la lutte contre la végétation semi-aquatique au lac-barrage de la Lufira (Haut-Katanga). Bull. Soc. R. Sci. Liège 32: 516-528.

Ruwet, J.Cl. 1963b. Observations sur le comportement sexuel de *Tilapia macrochir* Blgr (Poissons Cichlides) au lac retenue de la Lufira (Katanga). Behaviour 20: 242-250.

Ruwet, J.Cl. 1968. Familial behaviour of *Tilapia* and its implications. Nature (London) 217: 977.

Ruwet, J.Cl. and J. Voss. 1966. L'étude des mouvements d'expression chez les *Tilapia* (Poissons Cichlides). Bull. Soc. R. Sci. Liège 35: 778-800.

Ruwet, J.Cl. and J. Voss. 1974. Etude et culture des *Tilapia*, p. 197-227. *In* J.Cl. Ruwet (ed.) Zoologie et assistance technique. Ed. FULREAC, Université de Liège.

Ruwet, J.Cl., J. Voss and L. Hanon. 1975. Biologie et élevage des tilapia. FAO/CIFA Tech. Pap. 4 (Suppl. 1): 332-364.

Sanico, A.F. 1975. Effects of 17α-ethynyltestosterone and estrone on sex ratio and growth of *Tilapia aurea* (Steindachner). Kalikasan Philipp. J. Biol. 13: 1-17.

Sarig, S. 1969. Winter storage of *Tilapia*. FAO Fish Cult. Bull. 2(2): 8-9.

Sarig, S. 1971. Diseases of warmwater fishes. T.F.H. Publ., Neptune City, New Jersey.

Sarig, S. 1975. The status of information on fish diseases in Africa and possible means of their control. Paper presented at the Symposium on Aquaculture in Africa, 30 Sept.-2 Oct. 1975. Accra, Ghana. CIFA/75/SR5. 6 p. FAO, Rome.

Sarig, S. 1979. Fisheries and fish culture in Israel in 1978. Bamidgeh 31: 83-95.

Sarig, S. and M. Marek. 1974. Results of intensive and semi-intensive fish breeding techniques in Israel in 1971-1973. Bamidgeh 26: 28-48.

Sarig, S. and Y. Arieli. 1980. Growth capacity of tilapia in intensive culture. Bamidgeh 32(3): 57-65.

Schmein-Engberding, F. 1953. Die Vorzugstemperatur einiger knochenfische und ihre physiologische Bedeutung. Z. Fisch. 2: 125-155.

Schroeder, G.L. 1978. Autotrophic and heterotrophic production of microorganisms in intensely-manured fish ponds, and related fish yield. Aquaculture 14: 303-325.

Scott, P.W. 1977. Preliminary studies on disease in intensively farmed Tilapia in Kenya. University of Stirling, Stirling, Scotland. 159 p. M.S. thesis.

Shell, E.W. 1967. Relationship between rate of feeding, rate of growth and rate of conversion in feeding trials with two species of tilapia *Tilapia mossambica* Peters and *Tilapia nilotica* Linnaeus. FAO Fish. Rep. 44(3): 411-415.

Shell, E.W. 1968. Mono-sex culture of male *Tilapia nilotica* Linnaeus in ponds stocked at three rates. FAO Fish. Rep. 44(4): 353-356.

Shelton, W.L., K.D. Hopkins and G.L. Jensen. 1978. Use of hormones to produce monosex tilapia for aquaculture, p. 10-33. *In* R.O. Smitherman, W.L. Shelton and J.H. Grover (eds.) Culture of exotic fishes symposium proceedings. Fish Culture Section, American Fisheries Society, Auburn, Alabama.

Shine, R. 1978. Propagule size and parental care: the "safe harbor" hypothesis. J. Theor. Biol. 75: 417-424.

Siddiqui, A.Q. 1977a. Lake Naivasha (Kenya, East Kenya) fishery and its management together with a note on the food habits of fishes. Biol. Conserv. 12(3): 217-218.

Siddiqui, A.Q. 1977b. Reproductive biology, length-weight relationship and relative condition of *Tilapia leucosticta* (Trewavas) in Lake Naivasha, Kenya. J. Fish Biol. 10: 251-260.

Siddiqui, A.Q. 1979a. Reproductive biology of *Tilapia zillii* (Gervais) in Lake Naivasha, Kenya. Environ. Biol. Fishes 4(3): 257-262.

Siddiqui, A.Q. 1979b. Changes in fish species composition in Lake Naivasha, Kenya. Hydrobiologia 64(2): 131-138.

Silliman, R.P. 1975. Selective and unselective exploitation of experimental populations of *Tilapia mossambica*. Fish. Bull. 62: 21-52.

Silverman, H.I. 1978a. Effects of different levels of sensory contact upon reproductive activity of adult male and female *Sarotherodon (Tilapia) mossambicus* (Peters); Pisces: Cichlidae. Anim. Behav. 26: 1081-1090.

Silverman, H.I. 1978b. The effects of visual social stimulation upon age at first spawning in the mouth-brooding cichlid fish *Sarotherodon (Tilapia) mossambicus* (Peters). Anim. Behav. 26: 1120-1125.

Sklower, A. 1951. Fish farming and freshwater biology in Northern Rhodesia. Arch. Hydrobiol. 55: 284-303.

Smith, A. 1849. Illustrations of the zoology of South Africa. 40 (Lond.) vol. 4 Pisces. p. 5. Smith, Elder & Co., London.

Soin, S.G. 1968. Some developmental features of *Zoarces viviparus* (L.) in connection with livebearing. Vop. Ikthiol. 8: 283-293. (In Russian)

Solomon, D.J. 1977. A review of chemical communication in freshwater fish. J. Fish. Biol. 11: 363-376.

Sommerville, C. *Haplorchis pumilio* infections in tilapias. J. Fish Dis. (In press)

Soulé, M.E. 1980. Thresholds for survival: maintaining fitness and evolutionary potential, p. 151-170. *In* M.E. Soulé and B.A. Wilcox (eds.) Conservation: an evolutionary ecological perspective. Sinauer and Assoc., Sunderland, Mass.

Southwood, T.R.E. 1976. Bionomic strategies and population parameters, p. 26-48. *In* R.M. May (ed.) Theoretical ecology: principles and applications. Blackwell Sci. Publ., Oxford.

Spataru, P. 1978. Food and feeding habits of *Tilapia zillii* (Gervais) (Cichlidae) in Lake Kinneret (Israel). Aquaculture 14: 327-338.

Spataru, P. 1976. The feeding habits of *Tilapia galilaea* (Artedi) in Lake Kinneret (Israel). Aquaculture 9(1): 47-59.

Spataru, P. and M. Zorn. 1976. Some aspects of natural feed and feeding habits of *Tilapia galilaea* (Artedi) and *Tilapia aurea* (Steindachner) in Lake Kinneret. Bamidgeh 28: 12-17.

Spataru, P. and B. Hepher. 1977. Common carp predating on tilapia fry in high density polyculture fishpond system. Bamidgeh 29: 25-28.

Spataru, P. and M. Zorn. 1978. Food and feeding habits of *Tilapia aurea* (Steindachner) (Cichlidae) in Lake Kinneret (Israel). Aquaculture 13(1): 67-79.

Stearns, S.C. 1976. Life history tactics: a review of the ideas. Q. Rev. Biol. 51: 3-47.

Stearns, S.C. 1977. The evolution of life history traits: a critique of the theory and a review of the data. Annu. Rev. Ecol. Syst. 8: 145-171.

Steinitz, H. 1954. The distribution and evolution of the fresh water fishes of Palestine. Istanbul Univ. Fen Fak. Mecm. Hidrobiol. B. 1(4): 225.

Sterba, G. 1967. Freshwater fishes of the world. Studio Vista, E. Germany.

Stickney. R.R. and H.B. Simmons. 1977. Growth responses of *Tilapia aurea* to feed supplemented with dried poultry waste. Tex. J. Sci. 29(1/2): 93-99.

Stickney, R.R. and J.H. Hesby. 1978. Tilapia production in ponds receiving swine wastes, p. 90-101. *In* R.O. Smitherman, W.L. Shelton and J.H. Grover (eds.) Culture of exotic fishes symposium proceedings. Fish Culture Section, American Fisheries Society, Auburn, Alabama.

Stickney, R.R., L.O. Rowland and J.H. Hesby. 1977. Water quality—*Tilapia aurea* interactions in ponds receiving swine and poultry wastes. Proc. World Maricult. Soc. 8: 55-71.

Street, D.R. 1978. The socio-economic impact of fisheries programs in El Salvador. Res. Dev. Ser. No. 17. 14 p. Int. Cent. Aquaculture, Auburn Univ., Auburn, Alabama.

Sullivan, C.M. 1954. Temperature reception and responses in fish. J. Fish. Res. Board Can. 11: 153-170.

Sundararaj, B.I. and S. Goswami. 1977. Hormonal regulation of *in vivo* and *in vitro* oocyte maturation in the catfish, *Heteropneustes fossilis* (Bloch). Gen. Comp. Endocrinol. 32: 17-28.

Suwanasart, P. 1972. Effects of feeding, mesh size and stocking size on the growth of *Tilapia aurea* in cages. Annu. Rep. Int. Cent. Aquaculture, Auburn Univ., Auburn, Alabama 1971: 71-79.

Swingle, H.S. 1956. A repressive factor controlling reproduction fishes. Proc. Pac. Sci. Congr. 8: 865-871.

Swingle, H.S. 1960. Comparative evaluation of two tilapias as pondfishes in Alabama. Trans. Amer. Fish. Soc. 89: 142-148.

Swingle, H.S. 1961. Relationship of pH of pond waters to their suitability for fish culture. Proc. Pac. Sci. Congr. 9: 1-4.

Swingle, H.S. 1967. Fish kills caused by phytoplankton blooms and their prevention. FAO Fish. Rep. 44: 409-411.

T.V.A. (Tennessee Valley Authority). 1978. State-of-the-art waste heat utilization for agriculture and aquaculture. Tennessee Valley Authority, Chattanooga, Tennessee, U.S.A.

Tait, C.C. 1965. Mass fish mortalities. Zambia Fish. Res. Bull. 3: 28-30.

Tal, S. and I. Ziv. 1978. Culture of exotic fishes in Israel, p. 1-9. *In* R.O. Smitherman, W.L. Shelton and J.H. Grover (eds.) Culture of exotic fishes symposium proceedings. Fish Culture Section, American Fisheries Society, Auburn, Alabama.

Tandler, A. and F.W.H. Beamish. 1979. Mechanical and biochemical components of apparent specific dynamic action in large-mouth bass, *Micropterus salmoides* Lacépède. J. Fish Biol. 14: 343-350.

Tayamen, M.M. and W.L. Shelton. 1978. Inducement of sex reversal in *Sarotherodon niloticus* (Linnaeus). Aquaculture 14: 349-354.

Terkatin-Shimony, A. and Z. Yaron. 1978. Estrogens and estrogenic effects in *Tilapia aurea* (Cichlidae, Teleostei). Ann. Biol. Anim. Biochim. Biophys. 18: 1007-1012.

Thys van den Audenaerde, D.F.E. 1963. La distribution géographique des *Tilapia* au Congo. Acad. R. Sci. Outre-Mer (Brussels) Bull. Séances 9: 570-605.

Thys van den Audenaerde, D.F.E. 1964. Révision systématique des espèces congolaises du genre *Tilapia* (Pisces: Cichlidae). Ann. Mus. R. Afr. Cent. Ser. 8 Sci. Zool. 124: 1-155.

Thys van den Audenaerde, D.F.E. 1968. An annotated bibliography of *Tilapia* (Pisces: Cichlidae). Mus. R. Afr. Cent. Doc. Zool. No. 14. 406 p.

Thys van den Audenaerde, D.F.E. 1969. Description of *Tilapia louka* sp. nov. nom Sierra Leone, and some remarks on *Tilapia rheophila* Daget, 1962 (Pisces: Cichlidae). Rev. Zool. Bot. Afr. 79(1-2): 82-96.

Thys van den Audenaerde. D.F.E. 1971a. Some new data concerning the *Tilapia* species of the subgenus *Coptodon* (Pisces: Cichlidae). Rev. Zool. Bot. Afr. 84: 160-168.

Thys van den Audenaerde, D.F.E. 1971b. Some new data concerning the *Tilapia* species of the subgenus *Sarotherodon* (Pisces: Cichlidae). Rev. Zool. Bot. Afr. 84: 203-216.

Timms, A.M. and M.H.A. Keenleyside. 1975. The reproductive behaviour of *Aequidens paraguayensis* (Pisces: Cichlidae). Z. Tierpsychol. 39: 8-23.

Traore, K. and D. Campbell. 1976. La circulation des eaux du Lac de Kossou, ses causes, ses manifestations et ses conséquences. Rapp. Inf. Cent. Dév. Pêche Kossou. 13 p. (miméo)

Trewavas, E. 1933. Scientific results of the Cambridge Expedition to the East African lakes 1930-31. II. The cichlid fishes. J. Linn. Soc. London Zool. 38: 309-341.

Trewavas, E. 1965. *Tilapia aurea* (Steindachner) and the status of *Tilapia nilotica exul*, *T. monodi* and *T. lemassoni* (Pisces, Cichlidae). Isr. J. Zool. 14: 258-276.

Trewavas, E. 1966. A preliminary review of fishes of the genus Tilapia in the eastward-flowing rivers of Africa, with proposals of two new specific names. Rev. Zool. Bot. Afr. 74(3-4): 394-424.

Trewavas, E. 1973a. On the cichlid fishes of the genus *Pelmatochromis* with proposal of a new genus for *P. congicus*; on the relationship between *Pelmatochromis* and *Tilapia* and the recognition of *Sarotherodon* as a distinct genus. Bull. Br. Mus. (Nat. Hist.) Zool. 25: 1-26.

Trewavas, E. 1973b. A new species of cichlid fishes of rivers Quanza and Bengo, Angola, with a list of the known Cichlidae of these rivers and a note on *Pseudocrenilabris natalensis* Fowler. Bull. Br. Mus. (Nat. Hist.) Zool. 25(1): 27-37.

Trewavas, E. 1978. A discussion on *Tilapia* and *Sarotherodon*. Cichlidae 3: 127-131.

Trewavas, E. 1980. *Tilapia* and *Sarotherodon*. Buntbarsche Bull. 81: 1-6.

Trewavas, E. A review of the tilapiine fishes of the genera *Sarotherodon*, *Oreochromis* and *Danakilia*. Bull Br. Mus. (Nat. Hist.) Zool. (In press)

Trewavas, E. and G. Fryer. 1965. Species of *Tilapia* (Pisces, Cichlidae) from Lake Kitangiri, Tanzania., E. Africa. J. Zool. Soc. London 147: 108-118.

Trewavas, E., J. Green and S.A. Corbet. 1972. Ecological studies on crater lakes in West Cameroon. Fishes of Barombi Mbo. J. Zool. 167: 41-95.

Uchida, R.N. and J.E. King. 1962. Tank culture of tilapia. Fish. Bull. 62: 21-52.

Vaas, K.F. and A.E. Hofstede. 1952. Studies on *Tilapia mossambica* Peters (ikan mud-jair) in Indonesia. Contrib. Inland Fish. Res. Stn., Djakarta, Bogor, Indonesia. 1: 1-68.

Van der Lingen, M.I. 1959a. Some preliminary remarks on stocking rate and production of tilapia species at the Fisheries Research Centre, p. 54-62. *In* Proc. 1st Fisheries Day in Southern Rhodesia, August 1957. Government Printer, Salisbury.

Van der Lingen, M.I. 1959b. Some observations on the growth of *Tilapia mossambica* under different population densities and on different levels of feeding at the Fisheries Research Centre, p. 63-68. *In* Proc. 1st Fisheries Day in Southern Rhodesia, August 1957. Government Printer, Salisbury.

Van der Lingen, M.I. 1959c. Use of ducks in fish production at the Fisheries Research Centre, p. 69-73. *In* Proc. 1st Fisheries Day in Southern Rhodesia, August 1957. Government Printer, Salisbury.

Van der Lingen, M.I. 1973. Lake Kariba: early history and south shore, p. 132-143. *In* W.C. Ackermann, G.F. White and E.B. Worthington (eds.) Man-made lakes: their problems and environmental effects. Geophysical Monogr. No. 17. Amer. Geophys. Union, Washington, D.C.

Van Someren, V.D. 1962. The culture of *Tilapia nigra* (Günther) in ponds. VIII. The effect of cropping a breeding population by trapping. East Afr. Agric. For. J. 27(3): 176-185.

Van Someren, V.D. and P.J. Whitehead. 1959a. The culture of *Tilapia nigra* (Günther) in ponds. I. Growth after maturity in male *T. nigra*. East Afr. Agric. For. J. 25(1): 42-46.

Van Someren, V.D. and P.J. Whitehead. 1959b. The culture of *Tilapia nigra* (Günther) in ponds. II. The influence of water depth and turbidity on the growth of *T. nigra*. East Afr. Agric. For. J. 25(2): 66-72.

Van Someren, V.D. and P.J. Whitehead. 1960a. The culture of *Tilapia nigra* (Günther) in ponds. III. The early growth of males and females at comparable stocking rates, and the length/weight relationship. East Afr. Agric. For. J. 25(3): 169-173.

Van Someren, V.D. and P.J. Whitehead. 1960b. The culture of *Tilapia nigra* (Günther) in ponds. IV. The seasonal growth of male *T. nigra*. East Afr. Agric. For. J. 26(2): 79-86.

Van Someren, V.D. and P.J. Whitehead. 1961a. The culture of *Tilapia nigra* (Günther) in ponds. V. The effect of progressive alterations in stocking density on the growth of male *T. nigra*. East Afr. Agric. For. J. 26(3): 145-155.

Van Someren, V.D. and P.J. Whitehead. 1961b. The culture of *Tilapia nigra* (Gunther) in ponds. VI. The effect of selective cropping methods on breeding populations. East Afr. Agric. For. J. 26(4): 202-209.

Van Someren, V.D. and P.J. Whitehead. 1961c. The culture of *Tilapia nigra* (Günther) in ponds. VII. Survival and growth of tagged and untagged males of different sizes. East Afr. Agric. For. J. 27(1): 10-12.

Vanderpuye, C.J. 1972. Lake Volta. Afr. J. Trop. Hydrobiol. Fish. Spec. Issue 2: 115-134.

Vareschi, E. 1979. The ecology of Lake Nakuru (Kenya). II. Biomass and spatial distribution of fish (*Tilapia grahami* Boulenger = *Sarotherodon alcalicum grahami* Boulenger). Oecologia (Berl.) 37: 321-335.

Vincke, M.M.J. 1979. Rice paddy aquaculture: its situation and future role, p. 208-223. *In* T.V.R. Pillay and W.A. Dill (eds.) Advances in aquaculture. Fishing News Books Ltd., Farnham, Surrey, England.

Vollenweider, R.A. 1968. Scientific fundamentals of eutrophication of lakes and flowing waters, with particular reference to nitrogen and phosphorous as factors in eutrophication. DAS/CSI/68.27. OECD, Paris.

Von Kraft, A. and H.M. Peters. 1963. Vergleichende Studien über die Oogenese in der Gattung *Tilapia* (Cichlidae, Teleostei). Z. Zellforsch. Mikrosk. Anat. 61: 434-485.

Voss, J. 1969. Contribution à l'éthologie des Poissons Cichlides: *Tilapia zillii* (Gervais 1948). Rev. Zool. Bot. Afr. 79: 99-109.

Voss, J. 1972a. Contribution à l'éthologie des Poissons Cichlides: *Tilapia sparrmanii* Smith 1840. Rev. Zool. Bot. Afr. 85: 369-388.

Voss, J. 1972b. Caractères éthologiques et status systématique de *Tilapia sparrmanii* Smith 1840. Rev. Zool. Bot. Afr. 86: 31-49.

Voss, J. and J.Cl. Ruwet. 1966. Inventaire des mouvement d'expression chez *Tilapia guineensis* (Blkr. 1868) et *T. macrochir* (Blgr. 1912) (Poissons Cichlides). Ann. Soc. R. Zool. Belg. 96: 145-188.

Wales, J.H. 1970. Hepatoma in rainbow trout. Spec. Publ. Amer. Fish. Soc. 5: 351-365.

Weatherley, A.H. and B.M.G. Cogger. 1977. Fish culture: problems and prospects. Science 197(4302): 427-430.

Welcomme, R.L. 1964. The habitats and habitat preferences of the young of the Lake Victoria *Tilapia* (Pisces, Cichlidae). Rev. Zool. Bot. Afr. 70: 1-28.

Welcomme, R.L. 1966. Recent changes in the stocks of *Tilapia* in Lake Victoria. Nature (London) 212: 52-54.

Welcomme, R.L. 1967a. Observations on the biology of the introduced species of *Tilapia* in Lake Victoria. Rev. Zool. Bot. Afr. 76: 249-279.

Welcomme, R.L. 1967b. The relationship between fecundity and fertility in the mouthbrooding cichlid fish *Tilapia leucosticta*. J. Zool. 151: 453-468.

Welcomme, R.L. 1970. Studies on the effects of abnormally high water levels on the ecology of fish in certain shallow regions of Lake Victoria. J. Zool. 160: 405-436.

Welcomme, R.L. 1972. The inland waters of Africa. FAO/CIFA Tech. Pap. 1. 117 p.

Welcomme, R.L. 1979a. Fisheries ecology of the floodplain rivers. Longman, London.

Welcomme, R.L. 1979b. Preliminary record of international transfers of fish species. FAO Fish. Circ. No. 715. 31 p. (draft for comments)

Werren, J.H., M.R. Gross and R. Shine. 1980. Paternity and the evolution of male parental care. J. Theor. Biol. 82: 619-631.

West, G.J. and J. Glucksman. 1976. Introduction and distribution of exotic fish in Papua New Guinea. Papua New Guinea Agric. J. 27(1-2): 19-48.

Wheeler, A. and P.S. Maitland. 1973. The scarcer freshwater fishes of the British Isles. 1. Introduced species. J. Fish Biol. 5: 49-68.

Whitehead, P.J.P. 1962. The relationships between *Tilapia nigra* (Günther) and *Tilapia mossambica* Peters in the eastern rivers of Kenya. Proc. Zool. Soc. London 138: 605-637.

Whitehead, R.A. 1962. The life history and breeding habits of the west African cichlid fish *Tilapia mariae* and the status of *T. meeki*. Proc. Zool. Soc. London 139: 535-543.

Whitfield, A.K. and S.J.M. Blaber. 1976. The effects of temperature and salinity on *Tilapia rendalli* Boulenger 1896. J. Fish Biol. 9: 99-104.

Whitfield, A.K. and S.J.M. Blaber. 1978. Resource segregation among iliophagous fish in Lake St. Lucia, Zululand. Environ. Biol. Fishes 3(3): 293-296.

Whitfield, A.K. and S.J.M. Blaber. 1979. The distribution of the freshwater cichlid *Sarotherodon mossambicus* in estuarine systems. Environ. Biol. Fishes 4(1): 77-81.

Whyte, S.A. 1975. Distribution, trophic relationships and breeding habits of the fish populations in a tropical lake basin (Lake Bosumtwi, Ghana). J. Zool. 177: 25-56.

Wickler, W. 1965. Signal value of the genital tassel in the male *Tilapia macrochir* Blgr. (Pisces: Cichlidae). Nature (London) 208: 595-596.

Wickler, W. 1966a. Sexualdimorphismus, Paarbildung und Versteckbruten bei Cichliden (Pisces: Perciformes). Zool. Jahrb. Abt. Syst. Oekol. Geogr. Tiere 93: 127-138.

Wickler, W. 1966b. Über die biologische Bedeutung des Genital-Anhanges der Männlichen *Tilapia macrochir* (Pisces, Cichlidae). Senckenbergiana Biol. 47: 419-427.

Williams, R. 1971. Fish ecology of the Kafue River and floodplain environment. Fish. Res. Bull. (Zambia) 5: 305-330.

Wilson, E.O. and W.H. Bossert. 1971. A primer of population biology. Sinauer, Sunderland, Mass.

Winberg, G.G. 1956. Rate of metabolism and food requirements of fishes. Belorussian State University, Minsk. Fish. Res. Board Can. Transl. Ser. No. 194, 1960.

Worthington, E.B. and C.K. Ricardo. 1936. Scientific results of the Cambridge expedition to the east African lakes, 1930. 1. The fishes of Lake Rudolf and Lake Baringo. J. Linn. Soc. London, Zool. 39(15): 353-389.

Wu, S.Y. 1970. New bacterial disease of *Tilapia*. FAO Fish Cult. Bull. 2: 14.

Wynne-Edwards, V.C. 1962. Animal dispersion in relation to social behaviour. Oliver and Boyd, Edinburgh.

Yamamoto, T. 1969. Sex differentiation, p. 117-175. *In* W.S. Hoar and D.J. Randall (eds.) Fish physiology. Vol. III. Academic Press, New York.

Yaron, Z. 1966. Demonstration of 3β-hydroxysteroid dehydrogenase in the testis of *Tilapia mossambica* (Cichlidae, Teleostei). J. Endocrinol. 34: 127-128.

Yaron, Z. 1971. Observations on the granulosa cells of *Acanthobramaterrae-sanctae* and *Tilapia nilotica* (Teleostei). Gen. Comp. Endocrinol. 17: 247-252.

Yaron, Z., A. Terkatin-Shimony, Y. Shaham and H. Salzer. 1977. Occurrence and biological activity of estradiol-17β in the intact and ovariectomized *Tilapia aurea* (Cichlidae, Teleostei). Gen. Comp. Endocrinol. 33: 45-52.

Yashouv, A. 1958. Biological data on *Tilapia galilaea* and *Tilapia nilotica* in fish ponds. Bamidgeh 10: 46-52.

Yashouv, A. 1960. Effect of low temperatures on *Tilapia nilotica* and *Tilapia galilaea*. Bamidgeh 12(3): 62-66.

Yashouv, A. 1969. Mixed fish culture in ponds and the role of tilapia in it. Bamidgeh 21: 75-92.

Yashouv, A. and A. Halevy. 1967. Studies on growth and productivity of *Tilapia aurea* and its hybrid "Gan-Shmuel" in experimental ponds at Dor. Bamidgeh 24: 31-39.

Yoshikawa, H. and M. Oguri. 1978. Sex differentiation in a Cichlid, *Tilapia zillii*. Bull. Jap. Soc. Sci. Fish. 44: 313-318.

General Index

Entries for species of the genera *Sarotherodon* and *Tilapia* are included in the Species Index, and the more important entries for named water bodies are included in the Water Bodies Index.

Species Index

This index lists the more important entries to the species of the genera *Sarotherodon* and *Tilapia*. Entries for the two genera themselves, and for the heading Tilapia, are included in the General Index.

Water Bodies Index

This index includes the principal entries for named dams, floodplains, lagoons, lakes, reservoirs, rivers, river systems and swamps.

List of Participants

Chairmen

Dr. H.F. Henderson
Fishery Resources and Environment Division
FAO of the United Nations
Via delle Terme di Caracalla
00100 Rome
Italy

Dr. R.H. Lowe-McConnell
c/o Fish Section
British Museum (Natural History)
Cromwell Road
London, S.W. 7
England

Dr. D. Moriarty
Commonwealth Scientific and Industrial Research Organization
P.O. Box 120
Cleveland, Queensland 4163
Australia

Dr. R.S.V. Pullin
International Center for Living Aquatic Resources Management
MCC P.O. Box 1501
Makati, Metro Manila
Philippines

Participants

Dr. R.R. Avtalion
Department of Life Sciences
Bar-Ilan University
Ramat-Gan
Israel

*Dr. E.K. Balon**
Department of Zoology
College of Biological Science
University of Guelph
Guelph, Ontario
Canada N1G 2W1

Dr. S.H. Bowen
Department of Biological Sciences
Michigan Technological University
Houghton, Michigan 49931
U.S.A.

Dr. M.S. Caulton
King Prawns (Pvt.) Ltd.
P.O. Box 135
Kariba, Republic of Zimbabwe

Dr. J. Chervinski
Fish and Aquaculture Research Station
Dor, D.N. Hof-Hacarmel
Israel

Dr. A. Coche
Fishery Resources and Environment Division
FAO of the United Nations
Via delle Terme di Caracalla
00100 Rome
Italy

Dr. R.D. Guerrero III
Aquatic Biosystems
Bay, Laguna
Philippines

Dr. B. Hepher
Fish and Aquaculture Research Station
Dor, D.N. Hof-Hacarmel
Israel

Dr. B. Jalabert
Institut National De La Recherche Agronomique
Laboratoire de Physiologie des Poissons
Campus de Beaulieu
35042 Rennes-Cédex
France

Dr. L.L. Lovshin
International Center for Aquaculture
Auburn University
Auburn, Alabama 36844
U.S.A.

Mr. D. Mires
Ministry of Agriculture
Kibbutz Ein Hamifrats
D.N. Ashrat 25210
Israel

Dr. C.E. Nash
Kramer, Chin and Mayo, Inc.
1917 First Avenue
Seattle, Washington 98101
U.S.A.

Dr. D.L.G. Noakes
Department of Zoology
College of Biological Science
University of Guelph
Guelph, Ontario N1G 2W1
Canada

Dr. J-Cl. Philippart
Service d'ethologie animale
Institut Ed. van Beneden
Université de Lìege
Quai van Beneden, 22
B-4020 Lìege
Belgium

*Dr. Y. Pruginin**
Givat-Shapira
Israel

Prof. R.J. Roberts
Institute of Aquaculture
University of Stirling
Stirling, Scotland

*Dr. J-Cl. Ruwet**
Service d'ethologie animale
Institut Ed. van Beneden
Université de Lìege
Quai van Beneden, 22
B-4020 Liege
Belgium

*Dr. C. Sommerville**
Institute of Aquaculture
University of Stirling
Stirling, Scotland

Dr. E. Trewavas
Fish Section
British Museum (Natural History)
Cromwell Road, South Kensington
London, S.W. 7
England

*Dr. Y. Zohar**
Institut National De La Recherche Agronomique
Laboratoire de Physiologie des Poissons
Campus de Beaulieu
35042 Rennes-Cédex
France

***Co-author—did not attend the conference.**